PLANTS OF THE
SAN FRANCISCO BAY REGION

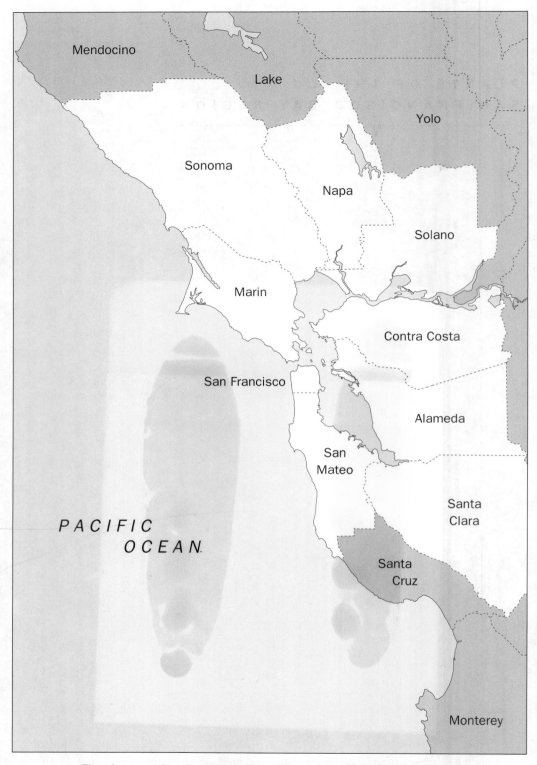

The nine counties shown here in white are those covered in this book.
Many of the plants we examine do exist in adjoining counties however.

PLANTS OF THE SAN FRANCISCO BAY REGION

Mendocino to Monterey

**Linda H. Beidleman and
Eugene N. Kozloff**

UNIVERSITY OF CALIFORNIA PRESS

Berkeley   Los Angeles   London

University of California Press
Berkeley and Los Angeles, California

University of California Press, Ltd.
London, England

Library of Congress Cataloging-in-Publication Data

Beidleman, Linda H.
    Plants of the San Francisco Bay region : Mendocino to Monterey / Linda H. Beidleman
and Eugene N. Kozloff.—2d ed.
        p.   cm.
    Rev. ed. of: Plants of the San Francisco Bay region / Eugene N. Kozloff and Linda H.
Beidleman. c1994.
    Includes bibliographical references (p.).
    ISBN 0-520-23172-4 (hardcover : alk. paper)—ISBN 0-520-23173-2 (pbk. : alk. paper)
    1. Botany—California—San Francisco Bay Area.   2. Plants—Identification.   I. Kozloff,
Eugene N.   II. Kozloff, Eugene N. Plants of the San Francisco Bay region.   III. Title.

QK149.B42   2003
581.9794'6—dc21

                                                                                2002029143

Printed in China
10 09 08 07 06 05 04 03
10 9 8 7 6 5 4 3 2 1

*Dedicated to all who appreciate
the beauty and diversity of the region's flora
and wish to know more about it.*

CONTENTS

PREFACE TO THE SECOND EDITION

We are pleased that the first edition of this guide, published in 1994, was received favorably, and we hope that this extensively revised edition will prove to be even more useful. The previously separate keys to trees and shrubs have been combined, and scientific names have been changed in accordance with corrections in the most recent printing of the *The Jepson Manual: Vascular Plants of California* (Hickman 1993).

Especially helpful suggestions that have led to improvements in some keys came from Richard Beidleman, the late Robert Ornduff, Alan Smith, and those who have used our guide in a field course: Joseph Balciunas, Lea Barker, Melissa Barrow, Diana Benner, David Bentley, Marie Bienkowski, Linda Brodman, Rachel Budelsky, Andy Butcher, Joy Cantley, Fran Carey, Terri Compost, Patrick Creehan, Jo Dankosky, Pat Eckhardt, Carolyn Flanagan, Dan Flanagan, Jon Goin, Melinda Groom, Kelley Haggerty, Katz Hasebe, Andrea Hitt, Amy Horne, John Hoskins, Laura Jerrard, Denise Kelly, Michelle Lee, Ed Leong, Corinna Lu, Terri McFarland, Patricia Monahan, Rebecca Peters, Helen Popper, Robert Prestegaard, Stanley Reichenberg, Jessie Schilling, Elizabeth Scott-Graham, Carol Souza, Andrea Taylor, Alynn Woodriff, Lani Yakabe, and Robert Young.

We would like to thank those who have helped us with the production of this edition, including Richard Beidleman, Natalie Bakopoulos, David Espinosa, Peggy Grier, and members of the staff of the University of California Press involved with our guide, especially Scott Norton, Phyllis Faber, Doris Kretschmer, and Nicole Stephenson.

Once again we ask users of this book to inform us of ambiguities, inaccuracies, or omissions that can be corrected in future printings.

Linda H. Beidleman
Eugene N. Kozloff

PREFACE TO THE FIRST EDITION

This book has been written for amateur naturalists, students, professional biologists, and others who need an easy-to-use guide to the flora of the San Francisco Bay Region. It covers approximately 2,027 species, subspecies, and varieties of higher plants. About 470 of these are aliens that have become firmly established since their accidental or willful introduction; the rest are native.

Nearly all of the plants dealt with are included in *A California Flora* and supplement, by Munz and Keck (1973). *The Jepson Manual: Vascular Plants of California* (1993), which provides an up-to-date treatment of the flora, lists additional native and nonnative species that have been reported from the region. We have included most of these.

In addition to a systematic coverage of higher plants, presented in the conventional order—ferns and fern allies, conifers and their relatives, and families of flowering plants—we provide keys for identification of shrubs and broad-leaved trees. Although all species in these last two categories are also included in keys to the families to which they belong, some users of this book will find it convenient to have shrubs and broad-leaved trees made more accessible by this treatment.

We expect that many persons will peruse the color photographs and line drawings of individual species at least as often as they refer to the text. Both sets of pictures are arranged in the same sequence as the sections that cover the major natural groups of plants and their families. Botanical manuals generally employ a rather extensive vocabulary. By using plain English whenever possible, we have eliminated much unnecessary terminology.

We are grateful for the encouragement we received while this book was developing. Peggy Grier and Myrtle Wolf were especially supportive. Together with Naomi Giddings and Shirley McPheeters, they also checked some keys with actual specimens, and Peggy proofread some of the manuscript. Richard Beidleman prepared the key to grasses, a difficult and time-consuming task, aided in the construction of several other keys, and assisted in proofreading. Josh Price provided computer expertise, which certainly speeded up this part of the project and greatly reduced frustration. Scott Ranney, at Alan Lithograph, expertly guided us through the printing process.

Most of our photography was accomplished in habitats where native or introduced species were growing wild. A substantial portion of this work was carried out, however, in the University of California Botanical Garden, Berkeley, where Roger Raiche was generous with information, and in the Botanic Garden of the East Bay Regional Park, where Steve Edwards gave valued help. A few pictures were taken in the collection of native plants at Merritt Community College and in other gardens. Our photographic coverage of the genus *Clarkia* would be far less complete than it is if it were not for the courtesy of Leslie Gottlieb, who provided seeds of several species. Ruth and Robert Athearn and Dianne and John Wilkinson generously allowed us to use portions of their sunny gardens for growing wildflowers native to our area.

In constructing and testing keys, extensive use was made of pressed herbarium specimens. Barbara Ertter, John Strother, and the late Lawrence Heckard at the University of California, Berkeley, and Sarah Gage, at the University of Washington, facilitated our use of the collections in their care. Early versions of some of our keys were improved by comments from Barbara and John, as well as from Lincoln Constance and Travis Columbus. Members of the team working on *The Jepson Manual*—especially the late James Hickman, Dieter Wilkin, and Susan D'Alcamo—generously allowed us to examine manuscript material and gave special advice when we asked for it.

To the University of Washington Press we are grateful for permission to use numerous illustrations in the excellent five-volume *Vascular Plants of the Pacific Northwest,* prepared by Hitchcock, Cronquist, Ownbey, and Thompson (1955–1969). Nearly all of the drawings taken from this publication were the work of Jeanne R. Janish. The University of California Press allowed us to use some illustrations from Jepson's *Manual of the Flowering Plants of California*. Other illustrations are from sources that are now in the public domain.

Users of this book will discover that there are many details packed into the space between its covers. While it is tempting to blame errors, omissions, and ambiguities on those who helped us, or on publications from which we drew information, we must say right now that we are personally responsible for all shortcomings that you may encounter. If you will be so kind as to tell us about problems you have discovered, you will have our thanks.

And now Linda and Gene thank each other. Our friendship and collegiality easily survived a few episodes of writer's block and some minor disagreements over construction of keys, general style, and organization. The ways in which we resolved our differences led to many improvements in the book. We hope that you enjoy using it, and that it will intensify your appreciation of the flora of this beautiful region of California.

Eugene N. Kozloff
Linda H. Beidleman

INTRODUCTION

Although much of the San Francisco Bay region is densely populated and industrialized, many thousands of acres within its confines have been set aside as parks and preserves. Most of these tracts were not rescued until after they had been altered. Construction of roads, modification of drainage patterns, grazing by livestock, and introduction of aggressive weeds are just a few of the factors that initiated irreversible changes in the plant and animal life. Yet on the slopes of Mount Diablo and Mount Tamalpais, in the redwood groves at Muir Woods, and in some of the regional parks, you can find habitats that probably resemble those that were present 200 years ago. Even tracts that are far from pristine have much that will bring pleasure to persons who enjoy the study of nature.

Visitors to our region soon discover that the area is diverse in topography, geology, climate, and vegetation. Hills, valleys, wetlands, and the seacoast are just some of the situations that have one or more well-defined assemblages of plants.

The counties that constitute the San Francisco Bay region include those that touch San Francisco Bay. Reading a map clockwise from Marin County, they are Marin, Sonoma, Napa, Solano, Contra Costa, Alameda, Santa Clara, San Mateo, and San Francisco. This book will also be useful in bordering counties, such as Mendocino, Lake, Santa Cruz, Monterey, and San Benito, because many of the plants dealt with also occur farther north, east, and south. For example, this book includes about three-quarters of the plants found in Monterey County and about half of those found in Mendocino County. Some species, in fact, are more common outside our region than within it. Nevertheless, it is important to keep in mind that the plants of the bordering counties are not comprehensively covered in this manual.

Scientific Names, Common Names, and Geographic Ranges of Plants

The Jepson Manual (Hickman 1993) will be, for many years to come, the definitive reference on the flora of California. We have therefore followed its scheme of classification and its names for species, subspecies, and varieties.

Latin names in brackets are those used by Munz and Keck in *A California Flora and Supplement* (1973). Unfortunately, common names for plants are not standardized as they are for birds and some other groups. We have attempted to provide one or two common names in use for each species.

Additional information that may be provided in connection with the Latin and common names of a plant consists of the region of origin (if the plant is not a native of California), the geographic range, and the extent to which it is rare or endangered. (See the abbreviations list for explanation of the symbols used.) Color plates are noted in parentheses following a species name. The black-and-white figures, except for those in the last section of the book, "Illustrations of Plant Structures," are grouped at the beginning of the individual family keys and are noted in the text by "(Fig.)".

A few words should be said about geographic range. If the range is not mentioned, it may be assumed that the plant has a wide distribution, at least in California, although it is not necessarily plentiful everywhere. When the range is given, it is primarily concerned with north-south distribution. For example, Ma-s means that Marin County is the northern limit of the species and that the range extends well into southern California, perhaps to the state border or beyond it. The species could also occur, however, in the Sierra Nevada or the Sacramento Valley, but this book is not concerned with this portion of its geographic distribution. Similarly, although SFBR indicates that the species is found through much of the San Francisco Bay region, the species' range may also include other areas of the state.

Persons already familiar with plants of the region will note that our scientific names reflect many changes that have been published since 1973. The changes are based on reexamination of each species in the light of names that have been applied to it, and there are some striking departures from names that have been in use for a long time. Now, with technologies available that can determine an organism's genetic makeup, or DNA, biologists are discovering that some plants once thought to be closely related based on anatomical characters are not very genetically similar. Undoubtedly, many changes in the way plants are grouped will occur in the future.

Some of the scientific names include the names of subspecies or varieties. The definitions for these entities have not been universally agreed upon by botanists. In general, the term *subspecies* (ssp.) should be applied to plants that differ rather clearly from those fitting the characteristics of the species but not to the extent of warranting a separate species name. The term *variety* (var.) should perhaps be restricted to plants that differ from a species or subspecies in relatively unimportant ways, such as the extent to which the leaves and stems are hairy.

Measurements

In the keys, the heights of plants and sizes of plant parts such as petals and leaves are given in metric units: meters (m), centimeters (cm), and millimeters (mm). These are routinely used in modern scientific work. Thus, one can directly relate measurements in this book to those in the *Jepson Manual* (Hickman 1993) and most other treatises. If you are not already familiar with the linear units of the metric system, a few comparisons with conventional units may be helpful. One meter, consisting of 100 cm, is slightly more than 39 in. One centimeter, consisting of 10 mm, is about two-fifths of an inch, so 2.5 cm equals 1 in, 5 cm equals 2 in, and 25 cm equals 10 in.

For temperature, altitude, long distances, and volumes, the conventional units—degrees Fahrenheit, feet, miles, and gallons, respectively—are used.

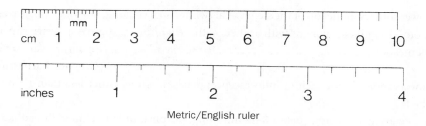

Metric/English ruler

How to Use a Key to Identify a Plant

A key for plant identification takes advantage of contrasting characteristics such as pink petals versus blue petals, leaf blades with toothed margins versus leaf blades with smooth margins, and so on. These characteristics are presented in a series of couplets. The two choices in the first couplet are 1a and 1b. If 1a is the better choice for the plant you are looking at, go to the next couplet—2a and 2b—under 1a. Don't wander into the territory under 1b. Similarly, if 1b happens to be the better choice, stick to the sequence of couplets under 1b.

The more you use keys of this type, the more quickly you will become familiar with the terms commonly used in plant classification. Experience, furthermore, will enable you to make judgments more quickly and to glide over choices that do not pertain to the specimen you are looking at.

If you have not used a key before, the little exercise provided here will help you. An "unknown" is shown in the illustration. This plant belongs to Saxifragaceae (Saxifrage Family). The pertinent portions of the key to this family are reproduced, and successive correct choices are underlined.

If you are not familiar with some of the terms you encounter, look these up in the glossary.

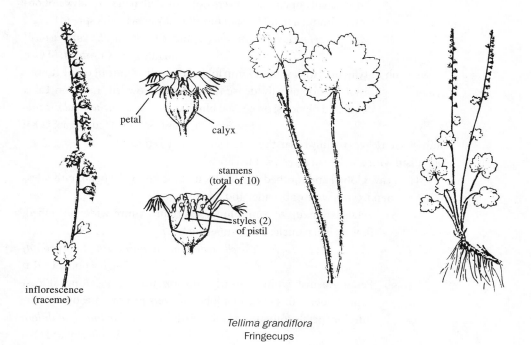

petal

calyx

stamens
(total of 10)

styles (2)
of pistil

inflorescence
(raceme)

Tellima grandiflora
Fringecups

1a Flower solitary at the top of the stem; petals white, 1–1.5 cm long, not lobed (all leaves basal, smooth margined; stamens with anthers 5, those without 5; stigmas 4; ovary superior; in wet meadows) . . . . *Parnassia californica [P. palustris var. californica]*
Grass-of-Parnassus; SB-n

1b Flowers more than 1 each in inflorescence; petals, if white, either less than 1 cm long or lobed

    2a Petals 4; stamens 3 (petals 8–12 mm long, purplish brown, almost threadlike; leaves mainly basal, lobed and toothed, some of them sprouting plants that may take root; ovary superior; on moist banks) . . . . *Tolmiea menziesii*
Piggy-back-plant; SFBR-n

    2b Petals 5; stamens 5 or 10

        3a Stamens 5 (leaves lobed and toothed; petals not more than 5 mm long; ovary half-inferior)

            4a Leaves scattered along the stems, as well as basal (leaf blades up to 8 cm wide, lobed and toothed; inflorescence usually more than 30 cm long; petals white; restricted to shady, wet banks and cliff sides) . . . . *Boykinia occidentalis [B. elata]*
Brookfoam

            4b Nearly all leaves basal, only 1 or 2 along each stem

                5a Petals yellow green, with 4–7 hairlike lobes, the corolla therefore resembling a snowflake (leaf blades up to 5 cm wide; in wet, shaded areas) . . . . *Mitella ovalis*
Mitrewort; Ma

                5b Petals white or pink, not lobed

                    6a Styles of pistil 2–4 mm long, protruding out of the flower; inflorescence usually open, the flowers not crowded; hairs on calyx not obvious without a hand lens; in moist, rocky areas (widespread) . . . . *Heuchera micrantha* [includes *H. micrantha* var. *pacifica*]
Smallflower Alumroot; SLO-n

                    6b Styles of pistil less than 2 mm long, not protruding out of the flower; inflorescence usually dense, the flowers crowded; hairs on calyx dense and obvious; on coastal bluffs . . . . *Heuchera pilosissima*
Seaside Alumroot, Seaside Heuchera; SLO-n

        3b Stamens 10 (leaves mainly basal, the stem leaves much reduced)

            7a Pistil with 2 styles (in moist, shaded areas)

                8a Leaves lobed and toothed, the blades up to 12 cm wide; petals either less than 1.5 mm wide or broader and lobed

                    9a Petals white, up to 4 mm long, less than 1.5 mm wide, not lobed; flowers in a panicle; ovary superior . . . . *Tiarella trifoliata* var. *unifoliata* [*T. unifoliata*]
Sugarscoop; SFBR-n

                    9b Petals greenish white to nearly crimson, up to 7 mm long, at least 2 mm wide, with 5–7 slender lobes; flowers in a raceme; ovary half-inferior (petals sometimes falling early) . . . . *Tellima grandiflora*
Fringecups; SLO-n

8b　Leaves toothed, the blades less than 10 cm wide; petals at least 2 mm wide, not lobed (petals white, sometimes with darker spots)

    10a　Leaf blades longer than wide, not more than 5 cm wide; filaments usually wider below than above; ovary half-inferior (petals 2–5 mm long; flowers often on 1 side of the inflorescence; widespread) .... *Saxifraga californica* California Saxifrage

    10b　Leaf blades not longer than wide, up to 10 cm wide; filaments usually wider above than below; ovary superior

        11a　Corolla irregular, 2 petals about 12 mm long, 3 petals 3–4 mm long; leaves up to 10 cm wide, with white veins on the upper surfaces and red veins on the undersides; new plants produced from prostrate stems that root at the nodes .... *Saxifraga stolonifera [S. sarmentosa]* Strawberry-geranium; as

        11b　Corolla regular, all petals 4–5 mm long; leaves up to 7 cm wide, with veins of the same color on both surfaces; new plants not produced from prostrate stems .... *Saxifraga mertensiana* Wood Saxifrage; Sn-n

7b　Pistil with 3 styles (petals white)

In certain choices, one or more characteristics, enclosed by parentheses, follow the basic features being contrasted. These characteristics give more information for that choice and all species under it. In choice 7b in the key example above, "(petals white)" applies to all the species under 7b. This information may be true for some of the species under 7a as well, but not for all of them.

The plants in a particular family, being related to one another, share many characteristics. Thus, there is not so much diversity in structure among the genera and species as there is among the numerous families. If you refer to the key to the families of herbaceous plants (p. 59), you can trace the path to get your unknown to Saxifragaceae (Saxifrage Family). First, you would determine that your plant belongs in group 4 (p. 64) (all other plants), since it is not parasitic, grasslike, or aquatic. The early choices (1b, 2b, 4b, 6b, 7b, 8b, 9b, 13b, 14b, and 15b) would lead you to group 4, subkey 3 (p. 73) and subsequently (choosing 1b, 4b, 18b, 37b, 46b, 55b, 58b, 59b, and 61a) to Saxifragaceae. That is when you would turn to the key for this family.

Conservation

The best way to save our native plants and animals is to set aside large, undisturbed tracts where the flora and fauna are typical. The need for protection is especially urgent in areas to which rare or endangered species are restricted. Among private nonprofit organizations, The Nature Conservancy, Trust for Public Land, Peninsula Open Space Trust, and other organizations have been particularly effective in acquiring lands with the goal of preserving natural communities of plants and animals. We also owe much to federal, state, and local agencies—including those that control the extensive acreages of watersheds—for protecting the fauna and flora in parks and other natural areas.

Botanical gardens, some of which are concerned entirely with the indigenous flora, are doing a good job of displaying native species and providing educational programs about them, thereby increasing public awareness and appreciation of the California flora.

California, like many other states, has laws to prevent the digging or cutting of native plants on public lands. Historically, characteristics of roots and bulbs have been used to identify some plants. The keys in this book concentrate on the habits of growth and features of stems, leaves, flowers, and fruit that can be seen with a hand lens or without magnification, causing little if any damage to the whole plant. In the field, it is important to note the plant's natural setting. What are the topography and soil like? What other plants grow nearby? Does the plant grow in shade or sun? These types of observations often help in identification, and they may assist you in relating the plant to a particular ecological situation.

Growing Native Plants

By 1850, numerous species of plants found in western North America, especially California, had been described and illustrated in botanical journals and grown in Europe. But the use of California plants as subjects for gardens and in revegetation of disturbed areas has developed slowly. In 1965, the California Native Plant Society was founded by some professional botanists and many enthusiastic amateurs. Although this organization was from its initiation concerned with protecting the flora, it has also raised public appreciation of native plants as garden subjects, especially of those species that are drought resistant. Seasonal sales conducted by regional chapters of the society and by botanical gardens are excellent sources of plants, bulbs, and seeds, as well as information and advice. Members of the local chapters receive notices of sales, field trips, classes, and programs about the California flora. Commercial growers, impressed by the turnout at native plant sales, have begun offering California native plants, and some nurseries now specialize in these plants. Furthermore, some landscape architects and garden installers now use native plants to a considerable extent.

References

References dealing with the plant communities, topography, climate, geology, and other physical attributes of our region are listed at the end of the next section.

Comprehensive Floras of California

Abrams, L. R., and R. S. Ferris. 1923–1960. *Illustrated flora of the Pacific states.* 4 vols. Stanford, Calif.: Stanford University Press.

Grillos, S. J. 1966. *Ferns and fern allies of California.* Berkeley: University of California Press.

Hickman, J. C., ed. 1993. *The Jepson manual: higher plants of California.* Berkeley: University of California Press.

Jepson, W. L. 1923, 1925. *A manual of the flowering plants of California.* Berkeley: Associated Students Store, University of California.

McMinn, H. E. 1939. *An illustrated manual of California shrubs.* Berkeley: University of California Press.

Metcalf, W. 1969. *Native trees of the San Francisco Bay region*. Berkeley: University of California Press.

Munz, P. H., and D. D. Keck. 1973. *A California flora and supplement*. Berkeley: University of California Press.

Floras of Specific Regions of Central and Northern California

Best, C., J. T. Howell, W. Knight, I. Knight, and M. Wells. 1996. *A flora of Sonoma County*. Sacramento: California Native Plant Society.

Bowerman, M. L., and B. J. Ertter. Forthcoming. *The flowering plants and ferns of Mount Diablo, California*. Sacramento: California Native Plant Society.

Ferris, R. S. 1968. *Native shrubs of the San Francisco Bay region*. Berkeley: University of California Press.

Ferris, R. S. 1970. *Flowers of the Point Reyes National Seashore*. Berkeley: University of California Press.

Howell, J. T. 1970. *Marin flora: manual of the flowering plants and ferns of Marin County, California*. 2d ed., with supplement. Berkeley: University of California Press.

Howell, J. T., P. H. Raven, and P. Rubtzoff. 1958. A flora of San Francisco, California. *Wasmann Journal of Biology* 16:1–157. Reprint, 1990. San Francisco: California Native Plant Society, Yerba Buena Chapter.

Keator, G. 1994. *Plants of the East Bay parks*. Niwot, Colo.: Roberts Rinehart Publishers, Inc.

Lyon, R., and J. Ruygt. 1996. *101 Napa County roadside wildflowers*. Napa, Calif.: Stonecrest Press.

Matthews, M. A. 1997. *An illustrated field key to the flowering plants of Monterey County*. Sacramento: California Native Plant Society.

McClintock, E., P. Reeberg, and W. Knight. 1990. *A flora of the San Bruno Mountains, San Mateo County, California*. Special publication no. 8. Sacramento: California Native Plant Society.

McHoul, L., and C. Elke. 1979. *Wild flowers of Marin*. Fairfax, Calif.: Tamal Land Press.

Peñalosa, J. 1963. A flora of the Tiburon Peninsula, Marin County, California. *Wasmann Journal of Biology* 21:1–74.

Sharsmith, H. K. 1965. *Spring wildflowers of the San Francisco Bay region*. Berkeley: University of California Press.

Sharsmith, H. K. 1982. *Flora of the Mount Hamilton Range of California*. Special publication no. 6. Sacramento: California Native Plant Society.

Smith, G. L., and C. R. Wheeler. 1990–1991. A flora of the vascular plants of Mendocino County, California. *Wasmann Journal of Biology* 48–49: i–iv, 1–387. Reprint, 1992. San Francisco: University of San Francisco.

Thomas, J. H. 1961. *Flora of the Santa Cruz Mountains of California*. Stanford, Calif.: Stanford University Press.

SAN FRANCISCO BAY REGION
PLANT COMMUNITIES
AND THEIR ENVIRONMENTS

Topography and Climate

Long before there was a San Francisco Bay, the westward flow from the large river system of the Central Valley, fed mostly by streams draining the Sierra Nevada, cut through two low points in the Coast Ranges, creating what are now Carquinez Strait and the Golden Gate. Smaller rivers from the Coast Ranges contributed to the scouring process. Later, at the close of the last ice age, when glaciers melted, the sea level began to rise. Over a period of several thousand years, water from the ocean gradually flooded much of the low-lying area behind the Golden Gate. Thus, San Francisco Bay was formed.

The Coast Ranges of the region consist of hills and substantial mountains whose orientation is from northwest to southeast. Close to the ocean, there are the Santa Cruz Mountains, Marin Hills (in which Mount Tamalpais is the highest peak), and Sonoma Mountains. Inland, there are the Mount Hamilton Range, Diablo Range, and Berkeley Hills, separated by Carquinez Strait and Suisun Bay from the Vaca Mountains and Mayacamas Range.

The Coast Ranges of the region, as mountains go, are not high—the top of Mount Hamilton is 4,261 ft above sea level, Mount Diablo has an altitude of 3,849 ft, and Mount Tamalpais reaches only 2,610 ft—but they are nevertheless barriers that impede the mixing of the inland air mass with the one that lies over the ocean. In general, therefore, the climate at the coast, influenced by the relatively stable temperature of the ocean, is more moist than the climate east of the mountains. Furthermore, it is cooler in summer and warmer in winter. But the topography of the Coast Ranges is complex. The valleys and ridges run in all directions, and there are numerous low points besides the huge gap at the Golden Gate. The large estuarine water mass of San Francisco Bay is another factor that conspires to create an extremely variable pattern of weather and perhaps more different microclimates within the nine counties of this area than are found in almost any other area of comparable size. For instance, on a particular summer day, residents of San Francisco may be shivering in temperatures of 50–60 degrees F while people in Walnut Creek are seeking refuge from the heat. Furthermore, the annual rainfall in the region may vary between 14 and 35 in.

One inescapable attribute of some parts of the region is fog. Most of our summer fogs originate when humid and relatively warm air from the Pacific approaches the coast. The coldness of the ocean water offshore causes the moisture in the air to condense into tiny airborne droplets.

Some fogs over San Francisco Bay are due partly to the daily tidal exchanges of water. After water has run out through the Golden Gate at low tide, the incoming rush of colder water at high tide may induce condensation of moisture in the overlying air. Shadows of clouds, which abruptly lower the temperature of the air, may bring about the same effect. Occasionally, fog develops when humid air from the ocean reaches the hills and is cooled as it is deflected upward. The result may be an extensive fog bank lying atop or against the hills or perhaps just small patches of fog that float not far overhead.

The higher ridges of the Coast Ranges may prevent low-lying fog from reaching the valleys farther east. At the Golden Gate, however, nothing stands in the way of a fog bank being pushed inland. The amount of water that moves through the gap in the form of fog is sometimes enormous. It could be as much as a million gallons per hour. The fog is essentially a cloud lying close to the surface of the water of the bay or of the ocean. If it rises, then it becomes a cloud of the conventional type.

In summer, when it is hot in the Central Valley, the rising of warm air causes cooler air and fog to be drawn in from the coast. Much of this eastward flow passes through the area, accounting for many of our summer fogs. After reaching the Central Valley, the fog is usually quickly evaporated by the daytime warmth and relative dryness. At night, the weather in the valley cools down a bit, and because there is less warm air to rise, the influx of air from the coast slows down or stops. The next day, however, the weather in the Central Valley heats up again, and the cycle is repeated. The result is more summer fog on the west side of the Coast Ranges.

In winter, air is cooler at higher elevations than in valleys. Cold air, being heavier than warm air, sinks down into the valleys. The Central Valley receives cold air from both the Sierra and the Coast Ranges. When this cold air comes in contact with moisture from warmer bodies of water in the valley, fog results. Unlike the summer fog on the west side of the Coast Ranges, which may soon dissipate, this Central Valley so-called tule fog cannot escape and in winter may remain for days at a time.

Fogs contribute to the moisture requirements of the vegetation. One reason for this is that high humidity and the presence of mist reduce the extent to which plants lose water to the atmosphere by evaporation. Furthermore, as mist condenses on foliage, some of the water drips down onto the soil, contributing to the moisture deposited by rain. In fact, a study of one area a few miles south of San Francisco showed that annual moisture added to the soil by drips from foliage was considerably more than the amount deposited by rain.

What about rain? Rain falls when the fine droplets of water in a cloud coalesce into larger drops heavy enough to fall. If this type of condensation takes place at a freezing temperature, there will be snow. If, on the other hand, an updraft causes raindrops to be carried upward until they are frozen, the result will be hail. Rain that freezes as it falls becomes sleet.

As previously mentioned, the complex topography of the region produces vastly different microclimates, which typically receive different amounts of rainfall. For example, San Rafael has an average annual rainfall of about 35 in, whereas San Mateo, about the same distance from the coast, has only 21 in. Martinez's yearly total is about 14 in, but Orinda, not far away, gets nearly twice that amount, and even more than Berkeley, which faces the Golden Gate. The disparity between the amount of rain falling on the western slopes of the mountains and hills near the coast and on some areas inland is even more striking. There are places in the Santa Cruz

Mountains where the annual rainfall reaches 60 in, yet in the Santa Clara Valley, just behind the mountains, the total rarely surpasses 15 in.

About 90% of the total annual rainfall occurs from October to April. Within this period, however, there may be dry spells that last several weeks, and in some years the amount of rainfall is much less than normal. The droughts may adversely affect the germination of seeds, the survival of seedlings, and even the growth, flowering, and fruiting of well-established plants. As a rule, heavy rainfall in the winter leads to a good display of spring flowers, especially annuals.

Soils

In the region under consideration, there are rocks of several distinctly different categories, including granite and basalt, which are of igneous origin, and shales, sandstones, and conglomerates, which are of sedimentary origin. Thus, because soil is primarily derived from rock, one may expect a wide variety of soil types to exist in the region.

As every gardener and farmer knows, not all plants grow well on all soils. Some of the attributes of a particular soil are shown by its general texture (coarseness or fineness), degree of acidity or alkalinity, the amount of organic matter it contains, its ability to retain moisture, the extent to which it allows air and water to penetrate, its supply of mineral nutrients, and the possible presence of substances that inhibit the growth of plants.

The majority of soils are mixtures of sand (particles from 0.02 to 2 mm), silt (particles from 0.002 to 0.02 mm), and clay (particles smaller than 0.002 mm). Fine particles have more surface area, in proportion to size, than do large particles. Thus, the more clay in the soil, the more water the soil retains. Furthermore, soil particles consisting largely of negatively charged compounds of aluminum, silicon, and oxygen attract positively charged mineral nutrients such as calcium and potassium. The nutrients are held so tightly to the particles that rainwater draining through the soil does not remove them, but they are available to plants.

One particular situation common in the area deserves special mention. This is the presence of a type of rock called serpentine, which consists, to a large extent, of magnesium silicate. It is typically somewhat shiny and slightly greenish, although after being weathered for a long time it becomes reddish brown, and the soil into which it eventually breaks down is usually reddish. Serpentine is abundant in California, and an act of the legislature named it the state rock. Outcrops of it can be found on Mount Tamalpais, Mount Diablo, and Mount Hamilton, in the hills directly east of Oakland and Berkeley, and in Edgewood Park in San Mateo County.

Soils in which serpentine is an important component are, as might be expected, rich in magnesium. Plants need this metal in order to synthesize chlorophyll, as well as for other processes. When magnesium is present in large quantities, however, it interferes with the uptake of calcium, which plants also require and which is scarce in serpentine soils. Inadequate supplies of other nutrients, including nitrogen, make life difficult for plants, too, and the toxic effect of nickel and chromium, sometimes present in unusually high concentrations, is still another problem. These features prevent many plants from growing on serpentine soils. Species that tolerate serpentine flourish in the absence of competition from other plants, and some are found only where it is present. Notable serpentine lovers are included in the following list.

TREES

Cupressus sargentii (Sargent Cypress)

SHRUBS

Quercus durata (Leather Oak)

HERBS

Allium fimbriatum (Fringed Onion)
Allium lacunosum (Wild Onion)
Aquilegia eximia (Serpentine Columbine)
Aspidotis densa (Indian's-dream)
Calystegia collina (Woolly Morning-glory)
Carex serratodens (Bifid Sedge)
Eriophyllum jepsonii (Jepson Woolly
   Sunflower)
Hesperolinon micranthum (Smallflower
   Dwarf Flax)
Monardella villosa ssp. *franciscana*
   (Coyotemint)

Muilla maritima (Common Muilla)
Parvisedum pentandrum (Mount
   Hamilton Parvisedum)
Polygonum douglasii ssp. *spergulariiforme*
   (Fall Knotweed)
Streptanthus breweri var. *breweri* (Brewer
   Streptanthus)
Streptanthus glandulosus ssp. *secundus*
   (One-sided Jewelflower)
Viola ocellata (Western Heart's-ease)
Zigadenus micranthus var. *fontanus*
   (Serpentine Star Lily)

Principal Plant Communities

A plant community, in the minds of most botanists, ecologists, and naturalists, is a relatively constant association of several to many species, certain of which predominate. A redwood forest is a good example, for *Lithocarpus densiflorus* (Tanbark Oak), *Polystichum munitum* (Western Sword Fern), *Oxalis oregana* (Redwood Sorrel), and some other plants are commonly associated with the redwood over much of its north-south range. Thus the entire assemblage may be called a community.

In classifying vegetational assemblages, even in a rather restricted geographic area, it is best to be flexible. One community may intergrade with another, and an association of plants that seems to fit a particular type of community in a general way may deviate from that type in certain respects. For instance, a Douglas-fir forest in Humboldt County, or farther north, has several species of plants that are not present in a Douglas-fir forest of Marin County. Nevertheless, it is convenient to be able to deal with the Douglas-fir forest as an entity, as long as we are prepared to accept some variation in its composition.

Valley and Foothill Woodland

In the San Francisco Bay Region, valley and foothill woodland occurs at elevations of about 300–3,500 ft and includes much of the area occupied by the Mount Diablo and Mount Hamilton ranges, and the hills in Napa and Solano counties. These inland habitats receive about 20 in of rain each year and are rarely touched by fog except in the winter. The summer weather is warm, with daytime temperatures often exceeding 90 degrees F.

The trees are generally somewhat scattered, so there is a well-lit understory, with a few shrubs and a wide variety of herbaceous plants. Some of the more common and conspicuous plants of this woodland community are listed below.

TREES

Aesculus californica (California Buckeye)
Pinus coulteri (Coulter Pine)
Pinus sabiniana (Foothill Pine)
Quercus agrifolia (Coast Live Oak)
Quercus chrysolepis (Canyon Live Oak)
Quercus douglasii (Blue Oak)
Quercus lobata (Valley Oak)
Quercus wislizeni var. *wislizeni* (Interior Live Oak)
Umbellularia californica (California Bay)

SHRUBS

Ceanothus cuneatus (Buckbrush)
Eriodictyon californicum (Yerba-santa)
Heteromeles arbutifolia (Toyon)
Rhamnus californica (California Coffeeberry)
Symphoricarpos albus var. *laevigatus* (Common Snowberry)
Toxicodendron diversilobum (Western Poison-oak)

HERBS

Cardamine californica var. *californica* (Common Milkmaids)
Collinsia heterophylla (Chinesehouses)
Dodecatheon hendersonii (Mosquito-bills)
Lithophragma heterophyllum (Hill Starflower)
Nemophila menziesii var. *menziesii* (Baby-blue-eyes)
Pentagramma triangularis (Goldback Fern)
Phoradendron villosum (Oak Mistletoe)
Ranunculus occidentalis (Western Buttercup)
Saxifraga californica (California Saxifrage)

Riparian Woodland

The complex of trees, shrubs, and herbaceous plants usually found along streams and rivers form the riparian woodland community. The vegetation consists mostly of species that require more soil moisture than do the oaks and other constituents of the valley and foothill woodland community. A few of the especially common plants of a riparian woodland in the region are listed below.

TREES

Acer macrophyllum (Bigleaf Maple)
Alnus rhombifolia (White Alder)
Platanus racemosa (Western Sycamore)
Populus fremontii (Fremont Cottonwood)
Salix lasiolepis (Arroyo Willow)
Salix lucida ssp. *lasiandra* (Shining Willow)
Umbellularia californica (California Bay)

SHRUBS

Cornus sericea ssp. *occidentalis* (Western Creek Dogwood)
Ribes sanguineum var. *glutinosum* (Pinkflower Currant)

Lonicera involucrata var. *ledebourii* (Black Twinberry)
Oemleria cerasiformis (Osoberry)
Rhododendron occidentale (Western Azalea)
Rhus trilobata (Skunkbrush)

Rosa californica (California Wild Rose)
Toxicodendron diversilobum (Western Poison-oak)
Vitis californica (California Wild Grape)

HERBS

Aralia californica (Elk-clover)
Claytonia perfoliata (Miner's-lettuce)
Cyperus eragrostis (Tall Cyperus)
Mimulus cardinalis (Scarlet Monkeyflower)

Thalictrum fendleri var. *polycarpum* (Manyfruit Meadowrue)
Woodwardia fimbriata (Giant Chain Fern)

Redwood Forest

Natural forests of *Sequoia sempervirens* (Coast Redwood) are found from Monterey County to southern Oregon. The best known and perhaps most nearly pristine grove in our area is the one in Muir Woods, Marin County, but there are a few other good examples. Furthermore, some of the groves that had been cut down for lumber have made a fairly good recovery and are found to have many of the plants that are characteristically associated with the Coast Redwood, including two trees, *Lithocarpus densiflorus* (Tanbark Oak) and *Umbellularia californica* (California Bay). As often happens when an area is disturbed, however, the balance of species changes. Some become rare or disappear entirely, others become proportionately more or less common, and opportunists not previously present may add themselves to the mix.

Natural groves are found in areas that receive considerable annual rainfall—at least 35 in—and that have frequent heavy fogs during the dry season. There are substantial plantations of redwoods in some places where they were not native, but these are not likely to have the usual associates. Many of the other plants typically found in redwood forests are listed below.

SHRUBS

Ceanothus thyrsiflorus (Blue-blossom)
Corylus cornuta var. *californica* (Hazelnut)
Gaultheria shallon (Salal)
Rhododendron macrophyllum (Rosebay)

Rosa gymnocarpa (Wood Rose)
Rubus parviflorus (Thimbleberry)
Vaccinium ovatum (Evergreen Huckleberry)

HERBS

Adenocaulon bicolor (Trailplant)
Anemone oregana (Western Wood Anemone)
Asarum caudatum (Wild-ginger)
Athyrium filix-femina var. *cyclosorum* (Western Lady Fern)
Clintonia andrewsiana (Red Bead Lily)
Disporum smithii (Largeflower Fairybells)
Oxalis oregana (Redwood Sorrel)

Polystichum munitum (Western Sword Fern)
Scoliopus bigelovii (Fetid Adder's-tongue)
Trillium ovatum (Western Trillium)
Vancouveria planipetala (Inside-out-flower)
Viola ocellata (Western Heart's-ease)
Viola sempervirens (Evergreen Violet)

Closed-cone Pine Forest

A closed-cone pine is one whose cones do not simply open up as soon as the seeds are mature. The scales of the cones are held tightly together by pitch, and they are not likely to separate for several years unless the pitch is melted by fire or exceptionally warm sunshine.

Within our area, the only species forming significant natural stands of a closed-cone pine is *Pinus muricata* (Bishop Pine). Its range, extending from Humboldt County to Baja California, overlaps that of *P. contorta* (Shore Pine) and *P. radiata* (Monterey Pine) (Santa Cruz, Monterey, and San Luis Obispo counties). Like these last two species, *P. muricata* is not found far from the coast. In Marin County, an extensive grove of this tree can be seen at Inverness Ridge, and there are specimens mixed with chaparral shrubs in the Carson Ridge area, northwest of Fairfax and San Anselmo. Other stands not too far away will be found in Sonoma, San Mateo, and Santa Cruz counties. Some common associates of the Bishop Pine are included in the following list.

TREES

Quercus agrifolia (Coast Live Oak)

SHRUBS

Arctostaphylos (manzanitas)
Baccharis pilularis (Coyotebrush)
Rhamnus californica (California Coffeeberry)

Toxicodendron diversilobum (Western Poison-oak)
Vaccinium ovatum (Evergreen Huckleberry)

Douglas-fir Forest

From southern Alaska to northern Sonoma County, there are dense forests in which *Pseudotsuga menziesii* (Douglas-fir), *Abies grandis* (Grand Fir), and *Tsuga heterophylla* (Western Hemlock) are the characteristic coniferous trees. Farther south—on Mount Tamalpais, on the Inverness Ridge, and in some other portions of our region—Douglas-fir forests include some other plants not typical of this community farther north. Characteristic members of our region's Douglas-fir forest are listed below.

TREES

Abies grandis (Grand Fir)
Arbutus menziesii (Pacific Madrone)
Lithocarpus densiflorus (Tanbark Oak)

Pseudotsuga menziesii (Douglas-fir)
Tsuga heterophylla (Western Hemlock)
Umbellularia californica (California Bay)

SHRUBS

Ceanothus thyrsiflorus (Blue-blossom)
Rhamnus californica (California Coffeeberry)

Toxicodendron diversilobum (Western Poison-oak)

HERBS

Polystichum munitum (Western Sword Fern)

Chaparral

Chaparral is a type of vegetation in which most of the obvious components are tough-leaved evergreen shrubs that are adapted for life in a relatively dry habitat. The word comes from Spain, where it has been used to refer to brushy places dominated by the chaparro, a kind of scrub oak. Although our chaparral does include a similar oak, it is not dominant, and a variety of other shrubs and herbs are common. Some of these plants are listed below.

SHRUBS

Adenostoma fasciculatum (Chamise)
Arctostaphylos glandulosa (Eastwood Manzanita)
Arctostaphylos glauca (Bigberry Manzanita)
Arctostaphylos manzanita ssp. *manzanita* (Parry Manzanita)
Arctostaphylos stanfordiana ssp. *stanfordiana* (Stanford Manzanita)
Arctostaphylos tomentosa ssp. *crustacea* (Brittleleaf Manzanita)
Ceanothus cuneatus (Buckbrush)
Ceanothus foliosus var. *foliosus* (Wavyleaf Ceanothus)

Ceanothus oliganthus var. *sorediatus* (Jimbrush)
Cercocarpus betuloides (Birchleaf Mountain-mahogany)
Eriodictyon californicum (Yerba-santa)
Heteromeles arbutifolia (Toyon)
Pickeringia montana (Chaparral Pea)
Quercus berberidifolia (California Scrub Oak)
Ribes malvaceum (Chaparral Currant)
Salvia mellifera (Black Sage)
Toxicodendron diversilobum (Western Poison-oak)

HERBS

Castilleja affinis ssp. *affinis* (Common Indian Paintbrush)

Pedicularis densiflora (Indian-warrior)
Salvia columbariae (Chia)

Chaparral is characteristic of hilly areas in which the soil, being gravelly or sandy, is well drained and does not hold water effectively. The vegetation is usually dense and often difficult to slog through. The shrubs are not often more than 2.5 m tall, but the foliage of most of them rubs us rather harshly. One of the few soft-leaved species in the list above is *Toxicodendron diversilobum* (Western Poison-oak), and contact with it is not recommended, either.

During the dry season, chaparral is extremely vulnerable to fire, partly because of the density of the vegetation and partly because the foliage is not juicy. A chaparral fire can be a very hot one; temperatures of over 1,000 degrees F have been recorded at ground level, and even 3 cm below the surface the temperature may exceed 300 degrees F. Nevertheless, chaparral is capable of rather quick recovery. One reason for this is that certain shrubs, after being burned to the ground, sprout new shoots from the crown. Others come back only if seeds have survived the fire or if seeds have been carried in from another area.

If there is ample rainfall during the growing season that follows a chaparral fire, there is likely to be a fine show of wildflowers, some of which may have been uncommon for several years preceding the fire. Some of the plants that grow luxuriantly in fire-ravaged chaparral are listed below.

SHRUBS

Adenostoma fasciculatum (Chamise)

Dicentra chrysantha (Golden Eardrops)
Emmenanthe penduliflora var. *penduliflora*
(Whispering-bells)

Papaver californicum
(Fire Poppy)

An interesting feature of chaparral in our region is that some of the shrubs begin to bloom in November or December. In the case of certain manzanitas, the opening of flowers is soon followed by production of new foliage; by late spring or early summer, the plants already have the buds of flowers that will not open until the arrival of the rainy season, several months later. Other shrubs of the chaparral, such as Chamise, flower in the spring, after they have produced new foliage.

Hill and Valley Grassland

Grassland is a treeless, shrubless vegetational assemblage that occupies large areas of California. Toward the end of the nineteenth century, three factors had begun to bring about an irreversible change in this kind of habitat. One of these was the conversion of grasslands into cultivated fields. Another was overgrazing by livestock first introduced by the Spanish in the 1700s. Still another was the introduction of plants, especially annual grasses, native to other parts of the world. Most of the native grasses were clump-forming perennials of the type called bunchgrasses. They were literally grazed to death, and their demise was hastened by lack of sufficient rain in certain years, which prevented them from making enough new growth to compensate for the grazing they suffered. The aggressive introduced species—their seeds arriving on the hair of domestic animals, on clothing, and by other means—joined a few native grasses in filling up the space that was opened up for them. The introduced grasses die by summer, giving California's grasslands a golden appearance. It is likely that before the flora was significantly altered, there would have been a substantial green or gray green component during the dry season.

It is now nearly impossible to find examples of pristine grassland. Nevertheless, there are many localities where bunchgrasses and their natural associates have persisted, even though they have been diluted by introduced species. And it is our grasslands that provide the most spectacular displays of native wildflowers, most of which have survived change better than the native bunchgrasses.

The following lists of herbaceous grassland plants is far from complete, but many of the common native and introduced species are included.

Achyrachaena mollis (Blow-wives)
Amsinckia menziesii var. *intermedia*
(Common Fiddleneck)
Brodiaea elegans (Elegant Brodiaea)
Calandrinia ciliata (Redmaids)
Castilleja densiflora (Common Owl's-clover)
Cirsium quercetorum (Brownie Thistle)

Lupinus bicolor (Minature
Lupine)
Micropus californicus (Slender
Cottonweed)
Microseris douglasii ssp. *douglasii*
(Douglas Microseris)
Nemophila menziesii var. *menziesii*
(Baby-blue-eyes)

Dichelostemma capitatum (Bluedicks)
Elymus multisetus (Big Squirreltail)
Eschscholzia californica (California Poppy)
Filago californica (California Fluffweed)
Hemizonia congesta ssp. *congesta* (Yellow Hayfield Tarweed)
Hordeum jubatum (Foxtail Barley)
Lasthenia californica (California Goldfields)
Lepidium nitidum (Common Peppergrass)
Linanthus parviflorus (Common Linanthus)

Plagiobothrys nothofulvus (Common Popcornflower)
Plantago erecta (California Plantain)
Platystemon californicus (Creamcups)
Ranunculus californicus (California Buttercup)
Sanicula bipinnatifida (Purple Sanicle)
Sidalcea malviflora ssp. *malviflora* (Common Checkerbloom)
Sisyrinchium bellum (Blue-eyed-grass)
Viola pedunculata (Johnny-jump-up)

NONNATIVE SPECIES

Avena fatua (Wild Oat)
Briza minor (Little Quaking Grass)
Bromus diandrus (Ripgut Grass)

Bromus hordeaceus (Soft Cheat Grass)
Cynosurus echinatus (Hedgehog Dogtail)
Erodium botrys (Broadleaf Filaree)

Coastal Prairie

Above cliffs along the coast, landward from the backshores of sandy beaches, and sometimes in other situations, there are areas in which perennial grasses, *Pteridium aquilinum* var. *pubescens* (Western Bracken), and several other types of herbaceous plants predominate. Typically there are neither trees nor large shrubs. Such coastal prairies nearly fit our definition of grassland, and the distinction between the two types of plant communities is not sharp. Common members of this community are listed below.

Calamagrostis nutkaensis (Pacific Reed Grass)
Calochortus luteus (Yellow Mariposa Lily)
Danthonia californica (California Oatgrass)
Deschampsia cespitosa ssp. *holciformis* (Pacific Hairgrass)
Dichelostemma capitatum (Bluedicks)
Dichelostemma congestum (Ookow)
Festuca californica (California Fescue)
Festuca idahoensis (Idaho Fescue)
Grindelia hirsutula var. *hirsutula* (Hairy Gumplant)

Heterotheca sessiliflora ssp. *bolanderi* (Bolander Golden Aster)
Iris douglasiana (Douglas Iris)
Lupinus formosus (Summer Lupine)
Lupinus variicolor (Varied Lupine)
Pteridium aquilinum var. *pubescens* (Western Bracken)
Ranunculus californicus (California Buttercup)
Sanicula arctopoides (Footsteps-of-spring)
Sisyrinchium bellum (Blue-eyed-grass)

As in the case of other plant communities, not all of the plants in the list are necessarily found at a particular site that qualifies as coastal prairie. Furthermore, some habitats that may seem to fit into this category are not genuine. They have developed in places that once were brushy or forested and that were cleared to promote the growth of native and introduced grasses that would provide favorable food for livestock.

Coastal Scrub and Coastal Forest

Close to the coast, from Monterey County to the Oregon border, there is a type of vegetation called coastal scrub. It consists of a mixture of grassland plants and low shrubs. Coastal scrub does not form a continuous north-south band; it is interrupted by closed-cone pine forest, coastal prairie, and other plant communities.

A few good examples of coastal scrub are found in San Mateo, Marin, and Sonoma counties. An introduced shrub, *Cytisus scoparius* (Scotch Broom), is widespread at these localities. Other plants common in this community are listed below.

SHRUBS

Artemisia californica (California Sagebrush)
Baccharis pilularis (Coyotebrush)
Heteromeles arbutifolia (Toyon)

Mimulus aurantiacus (Bush Monkeyflower)

HERBS

Anaphalis margaritacea (Pearly Everlasting)
Erigeron glaucus (Seaside Daisy)
Eriophyllum lanatum var. *achillaeoides* (Common Woolly Sunflower)

Scrophularia californica (Beeplant)
Sisyrinchium bellum (Blue-eyed-grass)
Triteleia laxa (Ithuriel's-spear)
Zigadenus fremontii (Common Star Lily)

Close to the shores of San Francisco Bay (in Oakland and Berkeley, for example), oak woodland mixes with the open coastal scrub described above. In both of these areas the climate is much more temperate than in the valley and foothill woodland. Some of the components of coastal forest are listed below.

TREES

Corylus cornuta var. *californica* (Hazelnut)
Quercus agrifolia (Coast Live Oak)

Sambucus mexicana (Blue Elderberry)

SHRUBS

Rhamnus californica (California Coffeeberry)
Rubus ursinus (California Blackberry)

Toxicodendron diversilobum (Western Poison-oak)

HERBS

Claytonia perfoliata (Miner's-lettuce)

Stachys ajugoides var. *rigida* (Rigid Hedgenettle)

Vernal Pools

A vernal pool is a low spot in a meadow that fills with water during the rainy season and then dries out by the beginning of summer. This type of habitat almost always has an interesting community of herbaceous plants, some of which do not occur anywhere else. A few of the more common species found in vernal pools are listed below.

Downingia concolor (Maroon-spotted Downingia)

Downingia pulchella (Flat-faced Downingia)

Lasthenia glabrata (Yellow-rayed Goldfields)

Limnanthes douglasii ssp. *douglasii* (Douglas Meadowfoam)

Freshwater Marsh

The vegetation of marshy areas around ponds, lakes, and slow-moving, shallow streams consists primarily of cattails, bulrushes, and sedges. There are, however, other types of aquatic or semi-aquatic herbaceous plants in this habitat, and the more commonly encountered species are listed here.

Carex obnupta (Slough Sedge)
Cyperus eragrostis (Tall Cyperus)
Juncus balticus (Baltic Rush)
Juncus lesueurii (Salt Rush)
Mimulus guttatus (Common Monkeyflower)
Oenanthe sarmentosa (Pacific Oenanthe)
Potentilla anserina ssp. *pacifica* (Pacific Cinquefoil)

Scirpus acutus var. *occidentalis* (Hardstem Bulrush)
Scirpus microcarpus (Smallfruit Bulrush)
Sparganium eurycarpum (Giant Bur-reed)
Typha latifolia (Broadleaf Cattail)
Veronica americana (American Brooklime)

The water and muck in a freshwater marsh are typically slightly acidic, and the availability of certain mineral nutrients that plants need—including phosphorus, nitrogen, and molybdenum—is often low. When this is the case, plant productivity is also low. The density of the vegetation is deceptive; it has been achieved slowly. Most of the common plants, incidentally, are perennials that are adept at vegetative propagation.

Our freshwater marshes are not so prevalent or so large as they once were. To a considerable extent, the decrease has been caused by the draining of marshes in order to make more land available for agricultural use and development, by diverting water needed for irrigation, and by building dams to create lakes.

Backshores of Sandy Beaches

Landward from the line reached by high tides on wave-swept sandy beaches, and also around certain bays, is an area on which loose sand is deposited by wind. This strip, ranging in width from a few meters to hundreds of meters, is called the backshore. The stability of the sand depends on plants that are rooted in it and on the amount of organic matter and clay that it contains. Much of the organic matter consists of the remains of previous generations of plants. Strong winds, especially when the tides are high, carry salt-laden droplets landward. Furthermore, the sand does not hold water very well and is generally poor in mineral nutrients that plants require. Its surface layer, on warm days, may absorb considerable heat from the sun. Being of a light color, the sand also reflects heat back at the plants. The grayish or whitish hairiness of the foliage of certain species is probably protective; it should help to reflect heat away from the plant. Some of the backshore plants are moderately to very succulent, storing up water that enables them to withstand extended periods during which there may be neither rain nor fog. As sand is blown about, it may accumulate in dunes, whose stability is constantly challenged. Even if a dune has been fairly thoroughly colonized by plants, a strong wind may uproot enough of the sand-binding plants to cause breakdown of the whole dune. The vegetation of backshores includes certain grasses, many other herbaceous species, and a few shrubs. There are no trees, for the substratum is not conducive to the success of tall plants that are frequently exposed to strong winds.

An introduced plant, *Ammophila arenaria* (European Beachgrass), is an effective sand binder that stabilizes dunes. It is more aggressive than the native *Leymus mollis* (American Dunegrass). Other efficient sand binders are *Juncus lesueurii* (Salt Rush), *Leymus pacificus* (Pacific Wild Rye), and *Poa douglasii* (Sand-dune Bluegrass).

Some plants found on the dunes often form tight low mats and thus stabilize the substratum. A plant that may be a California native, *Carpobrotus chilensis* (Iceplant), was widely planted for this purpose but is now being removed in some areas to encourage other native species. A few of these herbaceous species are listed below.

Abronia latifolia (Yellow Sand-verbena)
Abronia umbellata ssp. *umbellata* (Coast Sand-verbena)
Agoseris apargioides var. *eastwoodiae* (Coast Dandelion)
Ambrosia chamissonis (Silvery Beachweed)
Atriplex californica (California Saltbush)
Atriplex leucophylla (Beach Saltbush)
Calystegia soldanella (Beach Morning-glory)
Camissonia cheiranthifolia (Beach Primrose)
Erigeron glaucus (Seaside Daisy)
Eschscholzia californica (California Poppy)
Fragaria chiloensis (Beach Strawberry)
Heliotropium curassavicum (Seaside Heliotrope)
Lathyrus littoralis (Silky Beach Pea)
Lotus salsuginosus (Hooked-beak Lotus)
Plantago maritima (Sea Plantain)
Polygonum paronychia (Beach Knotweed)
Tanacetum camphoratum (Dune Tansy)

The most conspicuous shrubby plant of the backshore habitat is *Lupinus arboreus* (Yellow Bush Lupine), native in our area, but naturalized farther north. Its flowers are usually yellow, but there are forms with pale blue or pale lilac flowers. A little farther inland, where sand gives way to dirt, one may expect to see *Lupinus chamissonis* (Chamisso Bush Lupine), whose flowers are always bluish. Some other common shrubs are *Artemisia pycnocephala* (Coastal Sagewort), *Baccharis pilularis* (Coyotebrush), and *Ericameria ericoides* (Mock-heather).

Coastal Salt Marsh

Salt marshes are typically found around the edges of bays and river mouths, where there is little wave action but where fine sediment—silt and clay—can gradually accumulate. The deposition of sediment is followed by invasion of plants, arriving as seeds or fragments that can take root. Successful colonization leads to expansion of the plant community. The plants stabilize the substratum and trap more fine sediment.

The salt marsh is at a level that is submerged only during highest tides. The salinity varies according to the extent of tidal flooding, rainfall, runoff of freshwater, and evaporation. Unless the rate of evaporation cancels out rainfall, and unless there is very little freshwater draining toward the salt marsh, the salinity of the marsh will probably be greater on the seaward side than on the landward side. This accounts, in part, for the differences in vegetation.

The slope of an extensive salt marsh is usually so gradual that the surface of the marsh is nearly level. It is likely, however, to have shallow pools, as well as a system of tidal creeks formed by the erosive action of tidal flow into and out of the marsh.

The pressures of civilization have led to destruction of salt marshes in many parts of the world. In San Francisco Bay, large tracts of cheap acreage were needed for airports, industrial sites, housing developments, farmland, race tracks, and dumps, and today not more than 25% of the area once occupied by salt marshes remains in a nearly natural state. A cursory look at the shoreline in populated areas reveals the presence of industries whose wastes affect surviving marshes and ecological aspects of the rest of the bay.

Most of the flowering plants characteristic of salt marshes are found nowhere else. In the salt marshes of San Francisco Bay, Tomales Bay, Bodega Bay, and some other estuarine habitats within the region for which this book is written, the following species are especially common and widespread.

SHRUBS

Frankenia salina (Alkali-heath)

HERBS

Atriplex patula var. *obtusa* (Common Orache)

Cuscuta salina var. *major* (Salt Marsh Dodder)

Distichlis spicata (Saltgrass)

Jaumea carnosa (Fleshy Jaumea)

Limonium californicum (California Sea-lavender)

Plantago maritima (Sea Plantain)

Puccinellia nutkaensis (Alaska Alkali Grass)

Salicornia virginica (Virginia Pickleweed)

Triglochin maritima (Seaside Arrow-grass)

Castilleja ambigua (Johnny-nip), *Glaux maritima* (Sea-milkwort), and *Salicornia europaea* (Slender Glasswort) are less common or at least not widespread in salt marshes.

On the seaward side of the salt marsh, there will often be *Spartina foliosa* (California Cord Grass). It is submerged by tides lower than those required to inundate plants of the salt marsh proper. *Spartina alterniflora* (Salt-water Cord Grass), recently introduced from the Atlantic coast, is spreading aggressively in some places in San Francisco Bay.

On slightly drier and less saline soil landward of the salt marsh, members of the following characteristic assemblage of plants may be encountered, including the introduced *Tetragonia tetragonioides* (New Zealand Spinach).

SHRUBS

Baccharis pilularis (Coyotebrush)

HERBS

Atriplex patula var. *patula* (Spear Orache)

Grindelia stricta var. *platyphylla* (Pacific Gumplant)

Rumex maritimus (Golden Dock)

Spergularia macrotheca var. *macrotheca* (Perennial Sand-spurrey)

In mucky areas rarely, if ever, touched by high tides, the water is fresh or at most very slightly brackish. Here one will find *Scirpus acutus* var. *occidentalis* (Hardstem Bulrush), *Scirpus robustus* (Robust Bulrush), and *Typha latifolia* (Broadleaf Cattail).

References

All of the works listed below deal to a considerable extent with topography, climate, geology, and other physical attributes of our region, and some discuss plant communities. References concerned primarily with the flora of California, the San Francisco Bay region, and some other portions of the state are listed at the end of the introduction.

Baker, E. S. 1984. *An island called California.* 2d ed. Berkeley: University of California Press.

Barbour, M. G., and J. Major, eds. 1988. *Terrestrial vegetation of California.* 2d ed. Special publication no. 9. Sacramento: California Native Plant Society.

Barbour, M. G., R. B. Craig, F. R. Drysdale, and M. T. Ghiselin. 1973. *Coastal ecology: Bodega Head.* Berkeley: University of California Press.

Barbour, M., B. Pavlik, F. Drysdale, and S. Lindstrom. 1993. *California's changing landscapes.* Sacramento: California Native Plant Society.

Cooper, W. S. 1967. *Coastal dunes of California.* Memoir 104. Boulder, Colo.: Geological Society of America.

Dale, R. F. 1959. *Climates of the states: California.* Detroit: Gale Research Company.

Evens, J. G. 1988. *The natural history of the Point Reyes Peninsula.* Point Reyes, Calif.: Point Reyes National Seashore Association.

Gilliam, H. 1962. *Weather of the San Francisco Bay region.* Berkeley: University of California Press.

Gilliam, H., and M. Bry. 1966. *The natural world of San Francisco.* Garden City, N.Y.: Doubleday & Company.

Mount Diablo Interpretive Association. 2000. *The Mount Diablo Guide.* Berkeley: Berkeley Hills Books.

Munz, P. A., and D. D. Keck. 1949. California plant communities. *El Aliso* 2:87–105, 199–202.

Ornduff, R. 1974. *An introduction to California plant life.* Berkeley: University of California Press.

Sawyer, J. O., and T. Keeler-Wolf. 1995. *A manual of California vegetation.* Sacramento: California Native Plant Society.

Schoenherr, A. A. 1992. *A natural history of California.* Berkeley: University of California Press.

Smith, A. C. 1959. *Introduction to the natural history of the San Francisco Bay region.* Berkeley: University of California Press.

Sweeney, J. R. 1956. Responses of vegetation to fire. *University of California Publications in Botany* 28:143–250.

KEYS FOR IDENTIFICATION
OF SAN FRANCISCO
BAY REGION PLANTS

Major Plant Groups

Higher plants, or vascular plants, have tissues specialized for transport of water and the substances dissolved in it. Furthermore, except for ferns and fern relatives, higher plants have flowers, cones, or other structures that produce seeds.

The key below allows you to choose one of five major groups in which to begin your search for the identity of your plant. Proceed to the key for that group and decide on the family—or in some cases the genus or species—of your plant. Keep in mind that some families or genera will key out to several places because of the diversity of their members. Look in the index to find these multiple entries.

Color plates are noted in parentheses following a species name. The black-and-white figures, except for those in the section "Illustrations of Plant Structures," are grouped at the beginning of the individual family keys and noted in the text by "(fig.)". An (M) after a family name indicates that it is a monocotyledon. Further information about scientific and common names, plant ranges, measurements, and instruction on how to key a plant is given in the introduction.

1a Plant without flowers or cones and not producing seeds (certain fern allies have conelike structures, but these plants, unlike cone-bearing trees, are not woody)
    2a Leaves not compound, scalelike, grasslike, or rushlike . . . . FERN ALLIES, except *Marsilea*, Marsileaceae (p. 32)
    2b Leaves often compound, not scalelike, grasslike, or rushlike
        3a Plant aquatic or on edges of drying ponds; leaflets 4, palmately arranged . . . . FERN ALLIES, *Marsilea*, Marsileaceae (p. 32)
        3b Plant terrestrial; leaflets, if present, usually many more than 4 and pinnately arranged . . . . FERNS (p. 26)
1b Plant with flowers or cones, and producing seeds (in some species, the flower lacks petals, sepals, or both)
    4a Leaves needlelike or scalelike; plant without flowers, but bearing cones (pollen-producing cone relatively small, short lived; seed-producing cone, woody, leathery, or slightly fleshy) . . . . CONE-BEARING PLANTS (GYMNOSPERMS) (p. 35)

4b   Leaves usually not needlelike or scalelike; plant with flowers, but these sometimes lack petals, sepals, or both (in Betulaceae, the seed-producing catkins become woody and resemble cones)

  5a   Stem woody throughout (plant usually at least 50 cm tall) . . . .
. . . . FLOWERING PLANTS: TREES, SHRUBS, AND WOODY VINES (p. 41)

  5b   Stem not woody above the base . . . .
. . . . FLOWERING PLANTS: HERBACEOUS PLANTS (p. 59)

Nonflowering Plant Species and Families

Ferns

Ferns do not have flowers or seeds. Some or all of their leaves, or a specialized portion of a single leaf, bear structures called sporangia ("fig."). These are often clustered, forming sori, and the sori may be protected, at least for a time, by flaps or disks called indusia. The function of the sporangia is to produce microscopic spores, which are eventually released and scattered by wind. The spores that reach a suitable situation may develop into very small, short-lived plants called prothallia. These represent the sexual generation within the life cycle. Each prothallium produces a few eggs, or many sperm, or sometimes both. A film of water from rain or dew must be present to enable the sperm to swim to the eggs, which are located in little flask-shaped structures. An egg that has been fertilized develops into a new plant of the conspicuous spore-producing generation. At first there is only one tiny leaf, one stem, and one root, but soon more leaves and roots are formed, and the young fern becomes independent of the prothallium, which then dies.

The distinctions between families of ferns are based to some extent on the arrangements of sporangia and sori and on the form of the indusia, when these structures are present. Key 1 separates the species of ferns; key 2 enables you to appreciate the more obvious features that define the six families found in the area.

Key 1: Fern species

1a   Each mature plant annually producing a single leaf, this further divided into a vegetative portion and a branching structure that bears numerous globular sporangia (vegetative portion of the leaf persists for another year; generally in sphagnum bogs and wet meadows) . . . . *Botrychium multifidum* [includes *B. multifidum* ssp. *silaifolium*] (fig.)
    Leather Grape-fern; Mo-n; Ophioglossaceae, Adder's-tongue Family

1b   Plants not producing a single leaf that is divided into 2 very different portions (in some species, however, there are 2 types of leaves—vegetative and sporangia bearing)

  2a   Leaves lobed or once compound with the leaflets not deeply divided

    3a   Leaves compound (large ferns; leaves all similar, forming a crown)

      4a   Leaflets 3–8 cm long; indusia fringed with hairlike outgrowths (widespread) . . . . *Polystichum munitum* (fig.)
          Western Sword Fern; Dryopteridaceae, Wood Fern Family

      4b   Leaflets 2–3 cm long; indusia usually toothed but not fringed with hairlike outgrowths . . . . *Polystichum imbricans* [*P. munitum* var. *imbricans*]
          Imbricate Sword Fern; Mo-n; Dryopteridaceae, Wood Fern Family

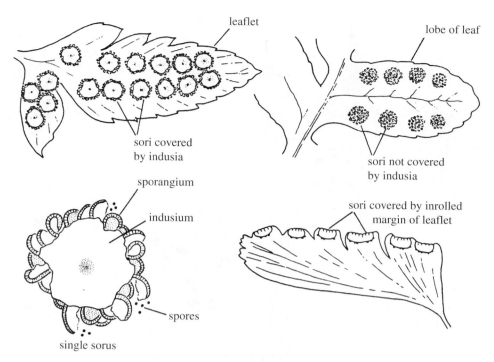

leaflet

lobe of leaf

sori covered
by indusia

sori not covered
by indusia

sporangium

indusium

sori covered by inrolled
margin of leaflet

spores

single sorus

Parts of a Fern

3b Leaves not compound but deeply lobed

    5a Leaves forming a crown and with dark, reddish brown or purplish brown peti-
        oles; leaves of 2 types: persistent vegetative leaves, generally more than 40 cm
        long with lobes 5–8 mm wide, and temporary (spring to summer) spore-
        producing leaves with lobes about 2 mm wide . . . . *Blechnum spicant* (fig.)

                                Deer Fern; SCr-n; Blechnaceae, Deer Fern Family

    5b Leaves arising from a creeping rhizome, not forming a crown, and not with
        dark petioles; leaves all of 1 type and all potentially spore producing, the sori
        being on the undersides

        6a Leaf blades leathery, the midribs not hairy on the upper surfaces (leaf
            lobes rounded at the tips; coastal, typically growing on cliffs and trunks
            of trees) . . . . *Polypodium scouleri*

                    Leather-leaf Fern; SCr-n; Polypodiaceae, Polypody Family

        6b Leaf blades not leathery, the midribs usually hairy on the upper surfaces

            7a Veins at tips of leaf lobes free, without cross connections; sori
                round; leaf lobes tapering to a point . . . . *Polypodium glycyrrhiza*

                      Licorice Fern; Mo-n; Polypodiaceae, Polypody Family

            7b At least some veins at tips of leaf lobes forming a network; sori usu-
                ally oval; leaf lobes usually rounded

                8a Leaf blade often widest at the base . . . .
                 . . . . *Polypodium californicum*

                  California Polypody; Polypodiaceae, Polypody Family

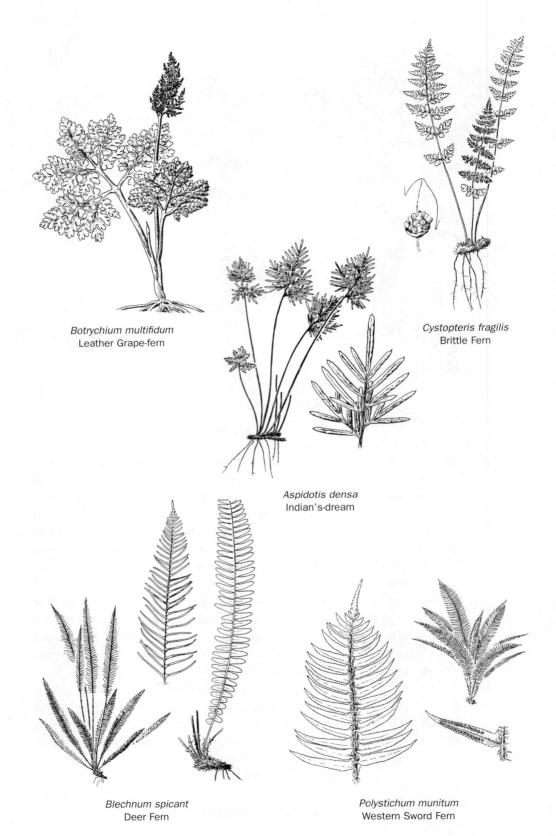

Botrychium multifidum
Leather Grape-fern

Cystopteris fragilis
Brittle Fern

Aspidotis densa
Indian's-dream

Blechnum spicant
Deer Fern

Polystichum munitum
Western Sword Fern

8b   Leaf blade often widest above the base (the most common polypody) .... *Polypodium calirhiza* (pl. 1)

Common Polypody; Polypodiaceae, Polypody Family

2b   Leaves compound 2–4 times, or if only once, the leaflets deeply lobed

9a   Leaflets somewhat fan shaped (leaves 2–4 times compound; petioles nearly black, polished; sori under inrolled leaflet margins; deciduous)

10a Petiole continued as a single rachis, which branches alternately; most leaflets less than twice as long as wide and not obviously lopsided, the stalk positioned in the middle (widespread)............... *Adiantum jordanii* (pl. 1)

California Maidenhair; Pteridaceae, Brake Family

10b Petiole branching into 2 rachises, each rachis branching equally again; most leaflets about twice as long as wide, appearing lopsided with respect to the position of the slender stalk ....

.... *Adiantum aleuticum [A. pedatum* var. *aleuticum]* (pl. 1)

Fivefinger Fern; Pteridaceae, Brake Family

9b   Leaflets not at all fan shaped

11a Petioles, at least their basal portions, brown or reddish brown, shiny (leaves 2–4 times compound)

12a Leaves with sori occupying much of the undersides of leaflets giving a granular or powdery appearance

13a Leaves either granular and green, or powdery and golden on the undersides (widespread) .... *Pentagramma triangularis [Pityrogramma triangularis* var. *triangularis]* (pl. 2)

Goldback Fern; Pteridaceae, Brake Family

13b Leaves granular and whitish on undersides ....

.... *Pentagramma pallida [Pityrogramma triangularis* var. *pallida]*

Silverback Fern; Pteridaceae, Brake Family

12b Leaves with sori at, or close to, the leaflet margins, undersides not granular or powdery

14a Leaflets with scales on the undersides and sometimes also on the upper surfaces (uncommon at elevations below 2,000 ft)

15a Scales on underside of leaflets divided into narrow lobes (scales not completely covering leaflets) ...... *Cheilanthes gracillima*

Lace Fern; SCl-n; Pteridaceae, Brake Family

15b Scales on undersides of leaflets usually not divided (scales widest at their bases)

16a Scales usually completely covering undersides of leaflets .... *Cheilanthes covillei* (pl. 1)

Coville Lace Fern; Me-Mo; Pteridaceae, Brake Family

16b Scales usually confined to the central portion of leaflets ....*Cheilanthes intertexta*

Coastal Lip Fern; Me-Mo; Pteridaceae, Brake Family

14b Leaflets without scales

17a Leaflets, or their divisions, rounded at the tips ....

.... *Pellaea andromedifolia [P. andromedaefolia]* (pl. 2)

Coffee Fern; Pteridaceae, Brake Family

17b Leaflets, or their divisions, pointed at the tips

    18a Most leaflets arranged in groups of 3 on rachis branches; outline of leaf blades elongated, not triangular . . . . *Pellaea mucronata* (pl. 2)

        Bird's-foot Fern; Pteridaceae, Brake Family

    18b Leaflets not arranged in groups of 3; outline of leaf blades not elongated, roughly triangular

        19a Leaves 3 times compound, most ultimate leaflets not deeply lobed; sori in longitudinal rows along length of ultimate leaflets (often on serpentine) . . . . *Aspidotis densa [Onychium densum]* (fig.)

          Indian's-dream; SLO-n; Pteridaceae, Brake Family

        19b Leaves nearly 4 times compound, the ultimate leaflets often deeply lobed; sori near leaflet margins (Coast Ranges) . . . . *Aspidotis californica*

          California Lace Fern; Pteridaceae, Brake Family

11b Petioles not brown or reddish brown and not polished

    20a Leaf blades triangular, usually 3 times compound; sori, if present, under inrolled leaflet margins (leaves often 1 m long, slightly hairy, deciduous, arising singly from a creeping underground rhizome; widespread) . . . . *Pteridium aquilinum* var. *pubescens* (pl. 2)

        Western Bracken; Dennstaedtiaceae, Bracken Family

    20b Leaf blades usually not triangular, only 1–2 times compound; sori not under inrolled leaflet margins

        21a Leaves usually less than 30 cm long, delicate, not forming a crown although they may be crowded; mostly in crevices between and beneath rocks (deciduous) . . . . *Cystopteris fragilis* (fig.)

          Brittle Fern, Bladder Fern; Dryopteridaceae, Wood Fern Family

        21b Leaves usually at least 30 cm long, not especially delicate, forming a crown; not limited to rocky habitats

          22a Leaf blade broadest near the middle, becoming narrower toward the base; indusia and sori more or less oblong or semicircular (deciduous)

            23a Leaves commonly more than 1 m long; sori arranged end to end in 2 rows; ultimate divisions of leaf blades narrowing gradually to pointed tips . . . . *Woodwardia fimbriata* (pl. 2)

              Giant Chain Fern; Blechnaceae, Deer Fern Family

            23b Leaves usually less than 1 m long; sori not arranged in 2 definite rows; ultimate divisions of leaf blades rounded (indusia oblong or J shaped; widespread) . . . . *Athyrium filix-femina* var. *cyclosorum* [*A. filix-femina* var. *sitchense*] (pl. 1)

              Western Lady Fern; SLO-n

              Dryopteridaceae, Wood Fern Family

22b Leaf blade widest at or near the base; indusia and sori more or less rounded

24a Ultimate divisions of leaf blades bristly at the tips; primary leaflet (those nearest the rachis) with a lobe at the base; indusia hairy

25a Leaves sometimes 3 times compound, with persistent scales on the main rachis; primary leaflets 5–12 cm long . . . . *Polystichum dudleyi*
Dudley Shield Fern; Ma-SLO
Dryopteridaceae, Wood Fern Family

25b Leaves not more than twice compound, with deciduous scales on the main rachis; primary leaflets 4–7 cm long . . . . *Polystichum californicum* (pl. 1)
California Shield Fern; Me-SCr
Dryopteridaceae, Wood Fern Family

24b Ultimate divisions of leaf blades not bristly at the tips; primary leaflet without a lobe at the base; indusia not hairy

26a Leaves at least partly 3 times compound . . . . *Dryopteris expansa*
Wood Fern; SM-n; Dryopteridaceae
Wood Fern Family

26b Leaves twice compound at most . . . . *Dryopteris arguta* (pl. 1)
Coastal Wood Fern; Dryopteridaceae
Wood Fern Family

Key 2: Fern families

1a Sporangia restricted to 1 of the 2 divisions of the single leaf produced during the growing season (the vegetative portion of the leaf of the preceding season may have persisted, however); sporangium without an annulus (a strip of thick-walled cells that straightens out in dry weather, tearing open the ripe sporangium); young leaves not coiled. . . . . . . *Botrychium*
Ophioglossaceae, Adder's-tongue Family

1b Sporangia either on vegetative leaves or on specialized leaves but not on a division of the only leaf produced during the growing season; sporangium with an annulus; young leaves coiled, unrolling as they enlarge

2a Sori located along the margins of ultimate leaflets or lobes (sometimes covered by inrolled margins, but not by indusia)

3a Leaves widely scattered, arising singly from a creeping underground rhizome; petioles green, not shiny, with a prominent groove; often more than 100 cm tall . . . . *Pteridium*
Dennstaedtiaceae, Bracken Family

3b Leaves produced in clusters; petioles brown to black, shiny, not grooved; rarely more than 50 cm tall . . . . *Adiantum, Aspidotis, Cheilanthes, Pellaea*
Pteridaceae, Brake Family

2b   Sori or scattered sporangia not located along the margins of ultimate leaflets or lobes

     4a   Leaves pinnately lobed but not fully compound

         5a   Leaves of 2 different forms, the ultimate lobes of those producing sporangia much narrower than those of strictly vegetative leaves; indusia present; sori oblong . . . . *Blechnum*

Blechnaceae, Deer Fern Family

         5b   Leaves all alike, the sporangia produced on vegetative leaves; indusia absent; sori round . . . . *Polypodium*

Polypodiaceae, Polypody Family

     4b   Leaves at least once compound

         6a   Sporangia scattered along the veins of leaflets, not concentrated in sori; indusia absent (leaves usually 2–3 times compound, less than 10 cm long) . . . . *Pentagramma*

Pteridaceae, Brake Family

         6b   Sporangia concentrated in sori; indusia present

             7a   Leaves often more than 100 cm long, once compound, with deeply lobed leaflets; sori oblong . . . . *Woodwardia*

Blechnaceae, Deer Fern Family

             7b   Leaves usually much less than 100 cm long, 1–4 times compound; sori round, but oblong in *Athyrium filix-femina* var. *cyclosorum* . . . . Dryopteridaceae, Wood Fern Family

Fern Allies

The plants called fern allies have life cycles comparable to those of ferns, and they are similar to ferns in that they have neither flowers, cones comparable to those of gymnosperms, nor seeds. They are, however, very diverse. The following key enables you to distinguish the few families represented in our region.

1a   Plants rarely more than 3 cm long, floating on freshwater, but sometimes stranded on mud above the water line (leaves scalelike, divided into 2 unequal lobes, the lower lobes serving as floats; stem obscure due to crowded, overlapping leaves but branching above every third leaf; roots short, inconspicuous, and unbranched) . . . . *Azolla filiculoides* (fig.)

Waterfern; Azollaceae, Mosquito-fern Family

1b   Plants not especially small and not floating

     2a   Plants rooted in mud or sand in shallow freshwater or slightly above the water line; leaves not scalelike and not as described in choice 2b

         3a   Leaves clustered on a short, slightly bulbous stem and nearly cylindrical, quill-like, with swollen bases; spores produced at leaf bases . . . . Isoëtaceae, Quillwort Family

         3b   Leaves arising singly or in groups along a creeping stem, either only 2–4 cm long and almost hairlike, or much longer, with hairy petioles and 4 palmately arranged, wedge-shaped leaflets; spores produced in short-stalked, egg-shaped, or spherical structures attached to stem . . . . Marsileaceae, Marsilea Family

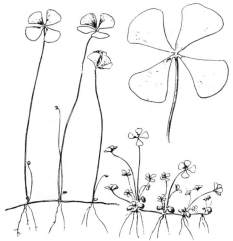

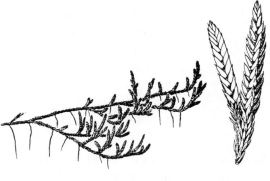

Marsilea vestita
Hairy Pepperwort

Selaginella wallacei
Little Clubmoss

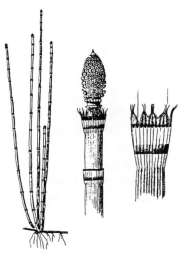

Equisetum hyemale ssp. *affine*
Common Scouring-rush

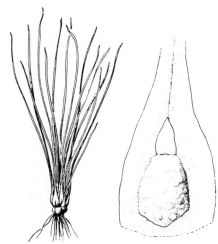

Isoetes nuttallii
Nuttall Quillwort

Azolla filiculoides
Waterfern

2b Plants generally not aquatic, although some are found in very wet places; leaves scale-like, either united to form sheaths that encircle the upright stem at the joints or crowded along branching stems that form mosslike growths

   4a Plants not mosslike; stem hollow, jointed, sometimes branched at the joints; leaves attached to one another, forming a sheath above each joint; spores produced in structures that resemble cones of coniferous trees . . . . Equisetaceae, Horsetail Family

   4b Plants mosslike; stem not hollow or jointed, branching extensively; leaves covering stem; spores produced in nearly quadrangular conelike structures (ripe sporangia usually orange and clearly evident amongst specialized leaves of each conelike structure) . . . . Selaginellaceae, Spikemoss Family

AZOLLACEAE (MOSQUITO-FERN FAMILY)   Our only representative of Azollaceae is *Azolla filiculoides* (fig.) (Waterfern). The family is sometimes included in Salviniaceae. The upper lobes of the leaves, which are the ones readily visible, are only about 1 mm long but nearly obscure the stem. The pale bluish green color of these lobes is due, in part, to the threads of *Anabaena azollae,* a symbiotic blue-green alga that grows within them. The plant usually becomes reddish in summer. This interesting relative of the ferns is often abundant in ponds and other freshwater situations where the water is quiet.

EQUISETACEAE (HORSETAIL FAMILY)   Plants called horsetails and scouring-rushes have underground stems from which the hollow, jointed, above-ground stems develop. Their scale-like leaves are attached to one another to form a sheath above each joint. Spores are produced in conelike structures. The cell walls of the epidermis contain considerable silica and may also have silicified projections that make the surface gritty. In times past, before commercial scouring powders had been developed for cleaning pots and pans, pulverized stems of equisetums were used for this purpose.

1a Green stems branched in whorls, without conelike reproductive structures at the tips (fertile stems brown, unbranched, short lived) . . . . *Equisetum telmateia* ssp. *braunii* (pl. 2)

Giant Horsetail

1b Green stems usually not branched, some usually with reproductive conelike structures at the tips

   2a Sheaths about as long as wide, with 2 black bands, one just below the teeth, the other near the base; above-ground stem perennial . . . . . . *Equisetum hyemale* ssp. *affine* (fig.)

Common Scouring-rush, Western Scouring-rush

   2b Sheaths decidedly longer than wide, with a single black band, this just below the teeth; above-ground stem usually annual . . . . *Equisetum laevigatum*

Smooth Scouring-rush, Braun Scouring-rush

ISOËTACEAE (QUILLWORT FAMILY)   Quillworts in our region grow in mud in and around ponds, including those that dry out in summer. They have a crown of slender, cylindrical, or three-sided leaves that arise from a short, bulbous stem. The swollen base of each fertile leaf contains a sporangium about 1 cm long within which spores are produced. The sporangium is covered to some extent by a translucent membrane that grows down over it. These species are difficult to tell apart without a microscope.

1a   Leaves flexible, 3–6 cm long (leaves 3–25, bright green with the bases white to brownish; sporangium covered at least 75% by a membrane) . . . . *Isoëtes orcuttii*

                                                               Orcutt Quillwort; CC-s

1b   Leaves rigid, usually much more than 6 cm long

    2a   Leaves 10–30, 5–28 cm long, bright green with the bases brownish to black; sporangium usually covered much less than 75% by a membrane . . . . *Isoëtes howellii*

                                                                   Howell Quillwort

    2b   Leaves 5–75, 3–20 cm long, light green or gray green with the bases white to brownish; sporangium covered at least 75% by a membrane . . . . *Isoëtes nuttallii* (fig.)

                                                                   Nuttall Quillwort

MARSILEACEAE (MARSILEA FAMILY)   Plants of Marsileaceae grow partly submerged in shallow freshwater or in mud that has been exposed by a drop in the water level. Their creeping stems produce single leaves or clusters of leaves at close intervals and also short-stalked globular or ovoid structures within which spores develop.

1a   Leaves similar to those of a four-leaf clover, up to 20 cm long, compound, with 4 palmately arranged, wedge-shaped leaflets . . . . *Marsilea vestita* (fig.)

                                                                   Hairy Pepperwort

1b   Leaves grasslike, rarely more than 4 cm long, not compound . . . . *Pilularia americana*

                                                              American Pillwort

SELAGINELLACEAE (SPIKEMOSS FAMILY)   The genus *Selaginella* is represented in the area by two species. These plants, which superficially resemble mosses, grow mostly in crevices between rocks on exposed hillsides. They are extremely drought resistant, appearing dead during dry weather, but become green after they have been well soaked by rains.

1a   Stem mostly prostrate, rooting at intervals, sometimes even near the tips . . . . *Selaginella wallacei* (fig.)

                                                                   Little Clubmoss; Ma-n

1b   Stem at least partly upright, rooting only near the base, where leaves do not persist . . . . *Selaginella bigelovii*

                                                         Spikemoss, Bigelow Moss-fern; Sn-s

Cone-bearing Plants (Gymnosperms)

Gymnosperms produce seeds, but they do not have flowers. The name of the group, meaning "naked seed," alludes to the fact that the seeds are not enclosed by a structure fully comparable to the fruit of a flowering plant. They are, in fact, generally exposed, at least by the time they are mature. Although there are exceptions with respect to this point, it is rarely difficult to decide whether a plant is a gymnosperm. Most gymnosperms—and all gymnosperms native to our area—have needlelike or scalelike leaves of types that are not often encountered among flowering plants.

In all of our native gymnosperms other than *Taxus brevifolia* (Pacific Yew), *Torreya californica* (California Nutmeg), and *Juniperus californica* (California Juniper), the seeds are produced in cones that become woody when they mature. Each cone consists of several to many scales, and it is on the surface of these scales that the seeds develop. In junipers, the scales are fleshy and so tightly fused together that the conelike structure is not immediately apparent. In yews, a single seed develops within a shallow cup; in the California Nutmeg, which also belongs to Taxaceae, the seed becomes completely enclosed, and the fleshy covering superficially resembles the fruit of an olive.

The cone-bearing plants in our region are separated in key 1. All of these species are either trees or very large shrubs and fall into one of four families. The distinctions between these four families are summarized in key 2.

Key 1: Gymnosperm species

1a Leaves scalelike, except on very small seedlings, on which they may be needlelike

2a Leaves in whorls of 3; seed-producing cone somewhat succulent (seed cone 1–1.5 cm wide, covered with a whitish deposit, and on trees separate from those that produce pollen) . . . . *Juniperus californica* (pl. 3)
California Juniper; Cupressaceae, Cypress Family

2b Leaves in pairs or whorls of 4; seed-producing cone woody when mature

3a Branchlets flattened; leaves in whorls of 4; unopened cone about twice as long as wide; cone scales without bumps, each with 2 seeds . . . .
. . . . *Calocedrus decurrens* (fig.)
Incense Cedar; Cupressaceae, Cypress Family

3b Branchlets cordlike, not flattened; leaves in pairs; unopened cone nearly globular; cone scales with prominent bumps, each scale with numerous seeds

4a Upper surfaces of older leaves without a resin pit or coating of resin

5a Top of tree conical; cone usually 2–2.5 cm long; new shoots not often more than 1.5 mm wide; up to 7 m tall; not commonly cultivated. . . . . . .
. . . . *Cupressus abramsiana*
Santa Cruz Cypress; SM-SCr; 1b; Cupressaceae, Cypress Family

5b Top of tree usually flattened or rounded; cone usually at least 2.5 cm long; new shoots more than 1.5 mm wide; often more than 20 m tall; native to Monterey County but often planted elsewhere . . . .
. . . . *Cupressus macrocarpa* (pl. 3)
Monterey Cypress; 1b; Cupressaceae, Cypress Family

4b Upper surfaces of older leaves with a resin pit and usually coated with resin

6a Bumps on cone scales low and conical, not elongated or curved; mountains near the coast (often on serpentine soils) . . . . . . *Cupressus sargentii*
Sargent Cypress; Me-s; Cupressaceae, Cypress Family

6b Bumps on cone scales, especially those of the end pair, elongated and curved; inland. . . . . *Cupressus macnabiana*
McNab Cypress; Sn-n; Cupressaceae, Cypress Family

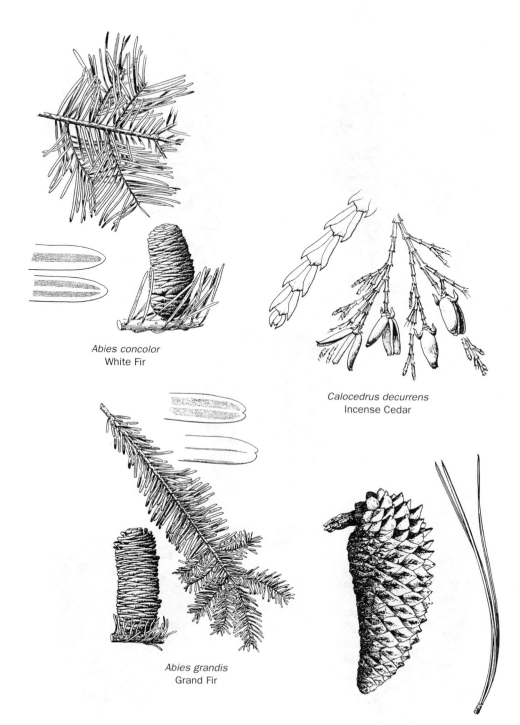

Abies concolor
White Fir

Calocedrus decurrens
Incense Cedar

Abies grandis
Grand Fir

Pinus attenuata
Knobcone Pine

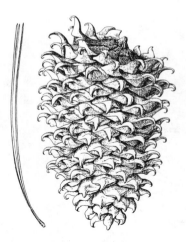

Pinus coulteri
Coulter Pine

Pinus lambertiana
Sugar Pine

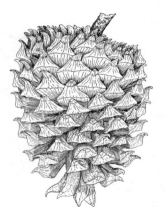

Pinus sabiniana
Foothill Pine

Pinus ponderosa
Pacific Ponderosa Pine

Pinus radiata
Monterey Pine

1b   Leaves needlelike

    7a   Needles in bundles of 2, 3, or 5 (base of bundle wrapped in a sheath)

        8a   Needles usually in bundles of 2 (cone lopsided; cone scales spinelike at tips, eroding with age; coastal)

            9a   Cone up to about 9 cm long, usually closed at maturity; needles 5–15 cm long .... *Pinus muricata* (pl. 3)

                      Bishop Pine; Pinaceae, Pine Family

            9b   Cone less than 5 cm long, open at maturity; needles 2–9 cm long .... *Pinus contorta*

                      Shore Pine; Pinaceae, Pine Family

        8b   Needles usually in bundles of 3 or 5

            10a   Needles in bundles of 5 (needles 5–11 cm long; cone up to 40 cm long, open at maturity) .... *Pinus lambertiana* (fig.)

                      Sugar Pine; Pinaceae, Pine Family

            10b   Needles in bundles of 3

                11a   Cone generally at least 15 cm long; needles often at least 25 cm long (cone open at maturity)

                    12a   Tip of cone scales pointed toward cone tip; needles 15–30 cm long, green; foliage dense; cone mostly 25–30 cm long, the stalk off center at the base .... *Pinus coulteri* (fig.)

                        Coulter Pine; CC-s; Pinaceae, Pine Family

                12b   Tip of cone scales pointed outward; needles 9–38 cm long; foliage not dense; cone mostly 15–20 cm long, the stalk centered at the base (widespread) .... *Pinus sabiniana* (fig.)

                        Foothill Pine, Gray Pine; Pinaceae, Pine Family

            11b   Cone rarely so much as 15 cm long; needles rarely so much as 25 cm long

                13a   Needles 12–26 cm long; cone open at maturity (cone scales spinelike at tips, eroding with age; Coast Ranges) .... *Pinus ponderosa* (pl. 3; fig.)

                        Pacific Ponderosa Pine; Pinaceae, Pine Family

                13b   Needles 6–16 cm long; cone opening if exposed to fire or prolonged dry, hot weather

                    14a   Cone about twice as long as wide; cone scales spinelike at tips, eroding with age (in mountains away from the coast) .... *Pinus attenuata* (fig.)

                        Knobcone Pine; SCL-n; Pinaceae, Pine Family

                    14b   Cone less than twice as long as wide; cone scales not spinelike at the tips (native to San Mateo County and south, but widely planted) .... *Pinus radiata* (fig.)

                        Monterey Pine; SM-s; 1b; Pinaceae, Pine Family

    7b   Needles arising individually, not in bundles

        15a   Needles usually notched at their tips (the only needle-leaved tree in our region that has this characteristic; seed-producing cone upright, 8–10 cm long, restricted to upper branches, and falling apart at maturity; needles less than 5 cm long) .... *Abies grandis* (fig.)

                      Grand Fir; Sn-n; Pinaceae, Pine Family

15b Needles not notched at the tips

    16a Needles with sharply pointed tips

        17a Needles about 1.5 cm long, dark green on both surfaces; seed-producing cone woody, about 2.5 cm long, on trees that also produce pollen trees; commonly taller than 15 m and sometimes up to 100 m (native to areas close to the coast, but widely planted elsewhere). . . . . *Sequoia sempervirens* (pl. 3)

            Coast Redwood; Taxodiaceae, Bald-cypress Family

        17b Needles usually more than 1.5 cm long, dark green on upper surfaces, either yellow green or with 3 green lines and 2 white lines on undersides; seed-producing cone fleshy and 1 seeded, on separate trees from those producing pollen; rarely 15 m tall

            18a Needles 2.5–7 cm long, with 3 green lines and 2 white lines on undersides; seed completely enclosed in a green or purple-streaked fruitlike covering about 2.5 cm long . . . . . *Torreya californica* (pl. 4)

                California Nutmeg; Me-Mo, Taxaceae, Yew Family

            18b Needles 1–2.5 cm long, yellow green on undersides; seed partly enclosed in a red cup about 1 cm long . . . . *Taxus brevifolia* (pl. 4)

                Pacific Yew; SCr-n; Taxaceae, Yew Family

    16b Needles rounded at the tips

        19a Needles mostly of 2 sizes, some about 1 cm long, others about 1.5 cm long; cone about 2 cm long . . . . *Tsuga heterophylla* (pl. 3)

            Western Hemlock; Sn-n; Pinaceae, Pine Family

        19b Needles not obviously of 2 sizes and usually at least 2 cm long; cone 6–7 cm long

            20a Cones hanging down, not limited to upper branches, not falling apart at maturity; seed-bearing scales alternating with 3-pronged sterile bracts; needles 2–4 cm long; from near sea level to 5,000 ft (widespread) . . . . *Pseudotsuga menziesii* (pl. 3)

                Douglas-fir; Mo-n; Pinaceae, Pine Family

            20b Cones upright, only on upper branches, falling apart at maturity; seed-bearing scales not alternating with sterile bracts; needles 3–9 cm long; at 3,000 ft and higher in elevation . . . . *Abies concolor* (fig.)

                White Fir; Pinaceae, Pine Family

Key 2: Gymnosperm families

1a Seeds produced singly, either incompletely surrounded by a fleshy cuplike structure, or completely enclosed within a fleshy, fruitlike covering without apparent coalesced scales; leaves needlelike . . . . *Taxus, Torreya*

            Taxaceae, Yew Family

1b Seeds produced on the surface of scales of a woody cone or within a berrylike structure that consists of a few coalesced scales; leaves needlelike or scalelike

    2a Leaves opposite or in whorls, scalelike (but needlelike on young seedlings); cone woody or fleshy . . . . *Calocedrus, Cupressus, Juniperus*

        Cupressaceae, Cypress Family

2b   Leaves alternate (or in bundles that are alternate), needlelike; cone woody

      3a   Bud of cone enclosed by obvious bracts; needles either in bundles or separate, normally falling separately after aging . . . . *Abies, Pinus, Pseudotsuga, Tsuga*

Pinaceae, Pine Family

      3b   Bud of cone not enclosed by obvious bracts; needles separate, falling attached to twigs after aging . . . . *Sequoia sempervirens* (pl. 3)

Coast Redwood; Taxodiaceae, Bald-cypress Family

Flowering Plant Families

Trees, Shrubs, and Woody Vines

1a   Trees or shrubs

    2a   Leaves opposite or in whorls . . . . TREES AND SHRUBS, GROUP 1

    2b   Leaves alternate

      3a   At least some leaves compound . . . . TREES AND SHRUBS, GROUP 2

      3b   Leaves not compound

         4a   Plant with thorns or prickles on the stem or branches . . . . TREES AND SHRUBS, GROUP 3

         4b   Plant without thorns or prickles on the stem or branches

            5a   Flowers both sessile and in composite heads like those of a daisy, thistle, or dandelion; fruit an achene, this often bearing a pappus consisting of bristles or scales (leaves without stipules; dried or drying flower heads, helpful in identification, often persist well into fall and winter) . . . . TREES AND SHRUBS, GROUP 4

            5b   Flowers not in composite heads as described in choice 5a; fruit not an achene

               6a   Usually at least some leaves palmately lobed . . . . TREES AND SHRUBS, GROUP 5

               6b   Leaves either pinnately lobed or not lobed

                  7a   Fruit a nut or acorn more than 1 cm long; staminate flowers in catkins, pistillate flowers solitary or in small clusters

                      8a   Nut either enclosed within a spiny bur, or an acorn set into a scaly cup (leaves usually tough; look on or under the tree for spiny burs, acorns, or scaly cups) . . . . Fagaceae, Oak Family

                      8b   Nut enclosed by a pair of united, somewhat papery bracts (stigmas bright red; leaf mostly 5–6 cm long, coarsely toothed, and with a prominent notch at the base) . . . . *Corylus cornuta* var. *californica* (fig.)

Hazelnut; Betulaceae, Birch Family

7b   Fruit not a nut or an acorn at least 1 cm long; flowers either not in catkins, or both pistillate and staminate flowers in catkins

     9a   Leaves deciduous; flowers small, in staminate and pistillate catkins on different plants; seeds covered with cottony hairs (look for the remains of pistillate catkins and seeds that may still be on the ground) .... Salicaceae, Willow Family

     9b   Plant not in every respect as in choice 9a .... TREES AND SHRUBS, GROUP 6

1b   Woody vines

   10a   Stem spiny or prickly except in *Rubus ulmifolius* var. *inermis* (but this species has the other characteristics of blackberries in *Rubus*)

     11a   Leaves not compound, some modified as tendrils; flower with 6 equal, greenish perianth segments; staminate and pistillate flowers on separate plants; fruit fleshy, but not an aggregate................................. *Smilax californica* Greenbrier; Na-n; Liliaceae, Lily Family (M)

     11b   Leaves compound, none modified as tendrils; flower with 5 white petals and 5 green sepals; all flowers with both stamens and pistils; fruit fleshy, an aggregate of separate fruits, as in a blackberry or raspberry........................... *Rubus* Rosaceae, Rose Family

   10b   Stem not spiny or prickly

     12a   At least some leaves compound (CAUTION: *Toxicodendron diversilobum*, Anacardiaceae, may cause a skin rash. Do not touch!)

       13a   Leaves alternate; fruit white, without a feathery style; sepals and petals present, yellow green (widespread; touching this plant—even the stem—may cause serious skin rash) .... *Toxicodendron diversilobum [Rhus diversiloba]* (pl. 4) Western Poison-oak; Anacardiaceae, Sumac Family

       13b   Leaves opposite; fruit brownish, with a persistent, feathery style more than 2 cm long; petals absent; sepals white .... *Clematis* Ranunculaceae, Buttercup Family

     12b   Leaves not compound; flower with a single pistil

       14a   Leaves opposite; blades not heart shaped (corolla tubular, cream colored, yellow, pink, or purple; fruit red)............................... *Lonicera* Caprifoliaceae, Honeysuckle Family

       14b   Leaves alternate; blades more or less heart shaped

         15a   Leaves with 3 or 5 lobes and coarsely toothed; flowers 1.5 mm long, many in each panicle, opposite the leaves; corolla and calyx regular; fruit fleshy .... *Vitis californica* (pl. 56) California Wild Grape; SLO-n; Vitaceae, Grape Family

         15b   Leaves smooth margined; flower 20–40 mm long, solitary in leaf axils; corolla absent, calyx irregular, resembling the bowl of a smoker's pipe; fruit a winged capsule .... *Aristolochia californica* (pl. 6) Dutchman's-pipe, Pipevine; Mo-n; Aristolochiaceae, Pipevine Family

Trees and Shrubs, Group 1:   Plant woody, with opposite or whorled leaves

1a   At least some leaves compound, with 3 or more completely separate leaflets; trees or large shrubs

    2a   Leaves palmately compound with 5–9 leaflets; inflorescence 10–20 cm long (leaflets 6–17 cm long; corolla white to pink, 1.5 cm long; fruit pear shaped, dry, 5–8 cm wide; widespread) . . . . *Aesculus californica* (pl. 30)
        California Buckeye; Hippocastanaceae, Buckeye Family

    2b   Leaves pinnately compound with 3–9 leaflets; inflorescence less than 10 cm long

        3a   Leaflets with 3–5 lobes, these toothed; fruit dry, consisting of 2 winged halves (end leaf lobe often wider than the other 2) . . . . . . *Acer negundo* var. *californicum* (fig.)
            Boxelder; Aceraceae, Maple Family

        3b   Leaflets not lobed, usually toothed; fruit not consisting of 2 winged halves

            4a   Pistillate and staminate flowers in leaf axils on separate plants; corolla consisting of 2 white petals or absent; leaflets 1–10 cm long; fruit dry, winged . . . . *Fraxinus*
                Oleaceae, Olive Family

            4b   Flowers at stem ends, all with a pistil and stamens; corolla 5 lobed, white or cream colored; leaflets 3–20 cm long; fruit a fleshy berry . . . . *Sambucus*
                Caprifoliaceae, Honeysuckle Family

1b   Leaves not compound (shrubs)

    5a   At least some stem leaves sessile (including those directly beneath flowers)

        6a   Corolla irregular (corolla more than 5 mm long; upper leaves often reduced)

            7a   Leaves with a strong minty odor; stem 4 sided; stamens 2 or 4, all with anthers; fruit consisting of 4 nutlets (widespread) . . . . *Lepechinia calycina, Salvia mellifera*
                Lamiaceae, Mint Family

            7b   Leaves without a strong minty odor; stem rounded; stamens 2, 4, or 5, sometimes 1 without an anther; fruit a capsule. . . . . Scrophulariaceae, Snapdragon Family

        6b   Corolla regular

            8a   Petals 5, 10–50 mm long; sepals separate, 5, 2 bractlike, smaller and slightly below other 3 (petals separate, white, pink, or purple, often with a dark or yellow blotch at the bases; leaves 2–8 cm long, not toothed) . . . . *Cistus*
                Cistaceae, Rockrose Family

            8b   Petals usually 4, not more than 13 mm long; sepals united, usually 4, all equal

                9a   Petals united, lilac to purple, often with an orange throat; leaves 5–30 cm long, sometimes toothed; stem not 4 sided . . . . *Buddleja davidii*
                  Summer Lilac, Butterfly-bush; as; Buddlejaceae, Buddleja Family

                9b   Petals separate, not lilac or purple; leaves 2.5–5 cm long, toothed; stem 4 sided . . . . *Haloragis erecta*
                  Seaberry; au; Haloragaceae, Water-milfoil Family

    5b   All stem leaves with at least a short petiole

        10a   Branches ending in stout thorns (leaves either not toothed, or with some irregular teeth; petioles 2–3 mm long; branches grayish; fruit 5–8 mm long, purplish black, with a grayish coating; large shrub) . . . . *Forestiera pubescens [F. neomexicana]* (fig.)
            Desert Olive; CC-s; Oleaceae, Olive Family

10b Branches not ending in thorns

    11a Leaves toothed, palmately lobed, or both (you may need a hand lens to see the teeth)

        12a Leaves palmately lobed, the lobes often with irregular teeth; tree (blades usually more than 10 cm wide; flowers greenish, in racemes; fruit dry, consisting of 2 winged halves; widespread) . . . . *Acer macrophyllum* (fig.)

                Bigleaf Maple; Aceraceae, Maple Family

        12b Leaves not lobed, but toothed; shrubs, sometimes large

            13a Leaf usually with 3, sometimes 5, prominent veins originating at the base of the blade (blades hairy, glandular, elliptical, often with a slight indentation at the base, toothed in upper half; deciduous; corolla white, 6–8 mm wide; flowers many in each dense cyme; fruit 10–12 mm long, red) . . . . *Viburnum ellipticum*

                Oval-leaf Viburnum; CC, Sn-n; 2
                Caprifoliaceae, Honeysuckle Family

            13b Leaf with only 1 prominent vein, the midrib

                14a Leaves with a strong odor of sage; corolla 2 lipped, more than 5 mm long; flower clusters at ends of branches; fruit consisting of 4 nutlets (evergreen; leaves 2–7 cm long, hairy on the undersides; corolla lavender or pale blue, sometimes white or pinkish; widespread) . . . . *Salvia mellifera* (pl. 33)

                Black Sage; CC-s; Lamiaceae, Mint Family

                14b Leaves without an odor of sage; corolla not 2 lipped, not more than 5 mm long; flower clusters in leaf axils; fruit a capsule with 2 or 3 lobes

                    15a Leaves not stiff, usually at least 7 cm long, normally deciduous, but sometimes persisting through winter (leaves tapering to a slender tip; branchlets angled; flower solitary or a few on peduncles 2–7 cm long; petals 5, brownish purple, with small light dots; seeds orange, adhering for a time to the fruit after it splits open . . . .
            . . . . *Euonymus occidentalis* (pl. 19)

                Western Burningbush; Mo-n
                Celastraceae, Staff-tree Family

                    15b Leaves rather stiff and usually less than 3 cm long, evergreen

                      16a Flowers 1–3 in each cluster; petals 4, brownish red; branchlets angled . . . . . . *Paxistima myrsinites* (pl. 19)

                Oregon Boxwood; Ma-n
                Celastraceae, Staff-tree Family

                    16b Flowers many in each cluster; petals 5, white or some shade of blue or lavender; branchlets not angled . . . . .
            . . . . *Ceanothus*

                Rhamnaceae, Buckthorn Family

11b Leaves not toothed, and if lobed, not palmately so

    17a Staminate and pistillate flowers on separate plants, and in drooping catkins 5–20 cm long (leaves leathery, the blades oval or elliptical; evergreen; pistillate catkins persist for many months)....................
.... Garryaceae, Silk-tassel Family

    17b Stamens and pistils present in all flowers, these not in catkins

        18a Largest leaf blades more than 5 cm long and up to 12 cm long (petioles 4–12 mm long; deciduous; in moist places)

            19a Leaves hairy, especially on the undersides; fruit white, cream colored, or red (leaves with 4–7 pairs of prominent secondary veins; flowers sometimes with showy bracts) .... *Cornus*
Cornaceae, Dogwood Family

            19b Leaves not hairy; fruit not white, cream colored, or red

                20a Flowers in dense, spherical heads 1–3 cm wide; fruit 3–4 mm long, widest near the top (corolla 7–8 mm long, white or pale yellow, with 4 lobes) ....
.... *Cephalanthus occidentalis* var. *californicus* (fig.)
California Button-willow; Na-n
Rubiaceae, Madder Family

                20b Flowers not in dense, spherical heads; fruit more than 5 mm long, not widest near the top

                    21a Leaves with a spicy odor when crushed; flower solitary, without bracts; corolla with many perianth segments, 2–5 cm long, reddish brown; fruit at least 20 mm long, leathery, becoming hard, brownish ....
.... *Calycanthus occidentalis* (pl. 17)
Spicebush, Sweetshrub; Na-n, e
Calycanthaceae, Spicebush Family

                    21b Leaves without a spicy odor when crushed; flowers in pairs, with 2 bracts beneath; corolla about 1.5 cm long, yellow, with 5 lobes; fruit 6–10 mm long, fleshy, nearly black when ripe ....
.... *Lonicera involucrata* var. *ledebourii* (pl. 18)
Black Twinberry; Caprifoliaceae
Honeysuckle Family

        18b Largest leaf blades usually not more than 5 cm long, but sometimes up to 7 cm in Cistaceae (blades usually at least slightly hairy)

            22a Leaves deciduous; fruit white (corolla pink or white)

                23a Leaf blades gray green, 2–5 cm long, with prominent secondary veins; inflorescence 2–5 cm wide, with numerous flowers (in moist areas) .... *Cornus glabrata*
Brown Dogwood; Cornaceae, Dogwood Family

                23b Leaf blades not gray green, 1–3 cm long, without prominent secondary veins; inflorescence less than 1 cm wide, usually with fewer than 15 flowers .... *Symphoricarpos*
Caprifoliaceae, Honeysuckle Family

22b Leaves evergreen; fruit, if fleshy, black when mature (petals separate)

24a Flowers or fruit many in each inflorescence; stamens 5 (sepals 5; petals 5, white or pale blue; fruit a capsule) . . . . .
. . . . . *Ceanothus cuneatus* [includes *C. ramulosus*] (pl. 47)
Buckbrush; Rhamnaceae, Buckthorn Family

24b Flowers or fruit 1–7 in each inflorescence; stamens many

25a Flowers 1–7 at stem tips; sepals 5; petals 5, 2–3 cm long, rose or purple; fruit a capsule, dry when mature . . . . .
. . . . *Cistus creticus* [*C. villosus* var. *corsicus*]
Crete Rockrose; me; Cistaceae, Rockrose Family

25b Flower solitary in leaf axils; sepals 4; petals 4, cream with pink tinge; fruit fleshy . . . .
. . . . *Luma apiculata* [*Eugenia apiculata*]
Temu; sa; Myrtaceae, Myrtle Family

Trees and Shrubs, Group 2: Trees and shrubs with alternate, compound leaves

1a Plant with thorns at the branch tips or prickles along the stem (shrubs)

2a Plant with thorns at the branch tips; corolla red purple, irregular, resembling that of a pea; leaves nearly sessile, evergreen, with leaflets 1–2 cm long; fruit greenish, a pod constricted between seeds (corolla 15–18 mm long; in chaparral) . . . .
. . . . *Pickeringia montana* (pl. 26)
Chaparral Pea; Me-s; Fabaceae, Pea Family

2b Plant with prickles along the stem; corolla pink to red purple, regular; leaves not sessile, deciduous, with leaflets often more than 2 cm long; fruit yellow to red, fleshy, not a pod

3a Leaflets usually 3; corolla white or deep pink to red purple; fruit an aggregate of fleshy units (in moist places). . . . . *Rubus leucodermis, Rubus spectabilis*
Rosaceae, Rose Family

3b Leaflets 5–9; corolla pink; fruit fleshy, but not an aggregate . . . . *Rosa*
Rosaceae, Rose Family

1b Plant without thorns or prickles at branch tips or along the stem, but in *Robinia* (Fabaceae) the stipules are spines

4a Most leaves on plant with 3 leaflets (sometimes with an additional pair of leaflets or lobes)

5a Leaflets shallowly lobed or with prominent teeth (corolla regular; leaves deciduous; shrubs) (CAUTION: touching *Toxicodendron diversilobum* may cause serious skin rash.)

6a Leaves without stipules; fruit pulpy, white or orange red, 2–8 mm long; corolla greenish or yellowish, 2–5 mm long . . . . *Toxicodendron, Rhus*
Anacardiaceae, Sumac Family

6b Leaves with stipules; fruit raspberry-like, yellow or red, 15–20 mm long; corolla deep pink to purplish red, 10–15 mm long . . . .
. . . . *Rubus spectabilis* [includes *R. spectabilis* var. *franciscanus*] (pl. 49)
Salmonberry; SCl-Sn; Rosaceae, Rose Family

5b   Leaflets not lobed, sometimes with very fine teeth

    7a   Leaves with a prominent citrus odor when crushed; corolla regular, with 4 or 5 lobes, 4–5 mm long, greenish white; fruit dry, winged, flattened; tree or shrub (central leaflet larger than other 2) . . . . *Ptelea crenulata* (pl. 51)

                                                                   Hoptree; SCl-n; Rutaceae, Rue Family

    7b   Leaves without a prominent citrus odor when crushed; corolla irregular, like that of a pea, mostly entirely yellow; fruit not as described in choice 7a; shrubs

        8a   Flowers 4–10 in clusters at the ends of main stems or short side branches; calyx hairy; fruit densely hairy (corolla 10–20 mm long, on a pedicel 1–3 mm long; invasive). . . . . *Genista monspessulana [Cytisus monspessulanus]* (pl. 24)

                                          French Broom; me; Fabaceae, Pea Family

        8b   Flowers 2–7 in clusters in leaf axils; calyx not hairy; fruit not hairy or only on the margins

            9a   Corolla 7–12 mm long, nearly sessile; stem not obviously ridged, usually sparingly leafy; rarely more than 1 m tall . . . . *Lotus scoparius* (pl. 25)

                        Deerweed, California Broom; Fabaceae, Pea Family

            9b   Corolla 15–22 mm long, not sessile; stem obviously ridged, usually densely leafy; often more than 2 m tall (invasive) . . . . *Cytisus scoparius* (pl. 24)

                               Scotch Broom; eu; Fabaceae, Pea Family

4b   Most leaves on plant with more than 3 leaflets

  10a   Leaves palmately compound, with 5–12 leaflets (corolla irregular, like those of a pea, the lower 2 petals united to form a keel; leaves with stipules; fruit hairy; shrubs) . . . . *Lupinus*

                                              Fabaceae, Pea Family

  10b   Leaves pinnately compound

    11a   Leaves twice compound, the leaflets less than 1 cm long (petals yellow or yellow and orange)

        12a   Leaves with 3–6 pairs of primary leaflets (primary leaflets with 12–20 pairs of secondary leaflets, these 5–7 mm long; corolla regular, less than 5 mm long; stamens many; tree) . . . . *Acacia baileyana*

                          Cootamundra Wattle; au; Fabaceae, Pea Family

        12b   Leaves usually with more than 7 pairs of primary leaflets

            13a   Primary leaflets with 15–35 pairs of secondary leaflets, these 5–15 mm long; corolla regular, less than 5 mm long, in several rounded clusters; stamens many; fruit with conspicuous constrictions; tree . . . . *Acacia decurrens*

                               Green Wattle; au; Fabaceae, Pea Family

            13b   Primary leaflets with 7–10 pairs of secondary leaflets, these less than 8 mm long; corolla irregular, over 35 mm long, not in clusters; stamens 10; fruit without obvious constrictions; shrub (corolla yellow, orange, and red; stamens red, 8–9 cm long) . . . . *Caesalpinia gilliesii*

                          Bird-of-paradise; sa; Fabaceae, Pea Family

11b Leaves once compound, the leaflets more than 1 cm long

    14a Leaflets with at least some teeth

        15a Teeth spine tipped; shrubs (leaves 25–45 cm long, with 5–21 leaflets; corolla yellow; fruit bluish, covered with a powdery bloom) . . . . *Berberis*

            Berberidaceae, Barberry Family

        15b Teeth not spine tipped; trees (leaflets 5–13 cm long; fruit 4–5 cm long, with 1 seed)

            16a Leaflets with a few coarse teeth near the base; fruit flattened, the seed forming a bulge near the middle (leaflets 13–25, 8–13 cm long) . . . . *Ailanthus altissima*

                Tree-of-heaven; as; Simaroubaceae, Quassia Family

            16b Leaflets finely toothed around most of the margins; fruit rounded, leathery

                17a Leaflets 11–19, 5–10 cm long, not smelling peppery; fruit 30–35 mm wide, brownish . . . . *Juglans californica* var. *hindsii* [*J. hindsii*]

                    Northern California Black Walnut

                    Na, CC; 1b; Juglandaceae, Walnut Family

                17b Leaflets 15–30, 1–6 cm long, smelling peppery; fruit about 5 mm wide, reddish (corolla 1–2 mm long, white; pistillate and staminate flowers on separate plants) . . . . *Schinus molle*

                  Peruvian Peppertree; sa; Anacardiaceae, Sumac Family

    14b Leaflets not toothed

        18a Leaflets 4–6 (width of leaflets less than half their length; flowers 2–7 in clusters in leaf axils; corolla 7–12 mm long, resembling those of a pea, mostly yellow) . . . . *Lotus scoparius* (pl. 25)

            Deerweed, California Broom; Fabaceae, Pea Family

        18b Leaflets at least 8

            19a Leaves with an even number of leaflets, without an unpaired end leaflet (evergreen; leaflets 12–16, up to 4 cm long; corolla 8–12 mm long, irregular, yellow; fruit 8–12 cm long) . . . . *Senna multiglandulosa* [*Cassia tomentosa*]

                  Senna; mx; Fabaceae, Pea Family

            19b Leaves with an odd number of leaflets, one of these at the end

                20a Leaves and fruit smelling peppery; stipules absent; evergreen; fruit pinkish, round, less than 1 cm long; corolla regular; pistillate and staminate flowers on separate plants (corolla 1–2 mm long, white; leaflets 3–5 cm long; tree) . . . . *Schinus molle*

                  Peruvian Peppertree; sa; Anacardiaceae, Sumac Family

20b Leaves and fruit not smelling peppery; stipules present, spiny or bristly; deciduous; fruit neither pink nor round, 5–10 cm long; corolla irregular; all flowers with both pistils and stamens

    21a Petals 5, 15–20 mm long, white, the 2 lower petals fused to form a keel; stipules forming spines; leaflets and fruit not glandular; tree ... *Robinia pseudoacacia* Black Locust, Desert Locust; na; Fabaceae, Pea Family

    21b Petal 1, about 5 mm long, red purple; stipules bristly; leaflets and fruit glandular; shrub . . . . *Amorpha californica* var. *napensis* (fig.)

False Indigo; Ma; 1b; Fabaceae, Pea Family

Trees and Shrubs, Group 3: Trees and shrubs with alternate leaves that are not compound; plant with thorns or prickles on the stem or branches

1a Leaves of 2 types: young leaves at base of plant narrow, smooth margined; mature leaves on branches whose tips function as thorns (corolla yellow, irregular, resembling that of a pea, 15–18 mm long; fruit a pod, 15–18 mm long) . . . . *Ulex europaea* [*U. europaeus*]

Gorse, Furze; eu; Fabaceae, Pea Family

1b Leaves not as described in choice 1a

  2a Leaf blade with 3 or more prominent veins originating from the base

    3a Leaves lobed or coarsely toothed; usually deciduous; flowers in clusters of 1–5, in leaf axils; fruit a juicy berry. . . . . *Ribes*

Grossulariaceae, Gooseberry Family

    3b Leaves not lobed or coarsely toothed, sometimes with fine teeth; evergreen; flowers many in each cluster, at end of the stem or branchlets; fruit a capsule, dry at maturity. . . . . *Ceanothus*

Rhamnaceae, Buckthorn Family

  2b Leaf blade with 1 prominent vein, the midrib

    4a Leaves often lobed, usually not toothed

      5a Leaves up to 2 cm long; corolla irregular, lavender red; stamens 10; fruit a green pod . . . . *Alhagi pseudalhagi* [*A. camelorum*]

Camelthorn; as; Fabaceae, Pea Family

      5b Leaves often more than 10 cm long; corolla regular, white or blue purple; stamens 5; fruit a fleshy berry, orange or yellow . . . . *Solanum lanceolatum, Solanum marginatum*

Solanaceae, Nightshade Family

    4b Leaves usually toothed, usually not lobed

      6a Some leaves with 1 or 2 teeth larger than other teeth (leaves 3–12 cm long; deciduous; flowers few in each cluster at stem ends; petals 7–11 mm long, white; stamens many; fruit up to 1.5 cm long, yellow or red; pedicel at least 2 cm long . . . . *Malus fusca* (fig.)

Oregon Crab Apple; Sn, Na-n; Rosaceae, Rose Family

6b   Leaves usually with rather evenly developed teeth, sometimes smooth margined

     7a   Leaves with teeth only in upper two-thirds of the blade (flowers several in each raceme; petals 4–5 mm long, white; stamens usually 20; leaves 2–9 cm long, deciduous; fruit fleshy, 7–8 mm long, blackish) . . . . *Crataegus suksdorfii*

            Black Hawthorn; Ma-n; Rosaceae, Rose Family

     7b   Leaves either with teeth around most of the margins or with teeth absent

         8a   Leaf blades at least twice as long as wide, definitely hairy on the undersides (flowers several in each cluster at branch ends; petals white; stamens 20; leaves 2.5–5 cm long, evergreen, sometimes toothed; fruit 6–8 mm long, yellow orange or red) . . . . *Pyracantha angustifolia*

                Firethorn; as; Rosaceae, Rose Family

         8b   Leaf blades usually less than twice as long as wide, either not hairy or scarcely so on the undersides

            9a   Leaves not leathery; flowers 5–10 mm long; stamens about 15 (deciduous; flowers 1–7 in each cluster on branch ends; petals white; leaves 3–6 cm long; fruit 15–25 mm long, yellow to dark red) . . . . *Prunus subcordata* (fig.)

                   Sierra Plum; Mo-n; Rosaceae, Rose Family

            9b   Leaves leathery; flower less than 5 mm long; stamens 4 or 5

               10a   Deciduous (leaves 1.5–4 cm long, with minute teeth, if any; staminate and pistillate flowers on separate plants; corolla absent; calyx 4 lobed; fruit dry, purple black) . . . . *Forestiera pubescens [F. neomexicana]* (fig.)

                      Desert Olive; CC-s; Oleaceae, Olive Family

              10b   Evergreen

                 11a   Leaves 1–2 cm long; flower clusters about 1 cm long, with 1–6 flowers; petals absent, sepals 4, greenish; fruit fleshy, not lobed, red . . . . *Rhamnus crocea* (pl. 47)

                         Spiny Redberry; La-s

                         Rhamnaceae, Buckthorn Family

                 11b   Leaves up to 5 cm long; flower clusters up to 15 cm long, with numerous flowers; petals and sepals 5, blue or white; fruit dry, 3 lobed, not red . . . . *Ceanothus spinosus*

                       Greenbark Ceanothus; SLO-n

                       Rhamnaceae, Buckthorn Family

Trees and Shrubs, Group 4:   Trees and shrubs with alternate leaves that are not compound; plant not spiny or prickly; flowers both sessile and in composite heads like those of a daisy, thistle, or dandelion; fruit an achene, often bearing a pappus consisting of bristles or scales (dried or drying flower heads, helpful in identification, often persist well into fall and winter) . . . . Asteraceae, Sunflower Family

1a Leaves not lobed, usually less than 5 times as long as wide (head with disk flowers only; pappus consisting of bristles)

2a Flower head not glandular; leaves not toothed and without glands, of 2 types: young leaves densely hairy and about 10 mm long and mature leaves not hairy and 2–3 mm long (involucre 4–6 mm high; corolla yellow; in washes) . . . . *Lepidospartum squamatum*
Scalebroom, Nevada Broomshrub; SCl-s

2b Flower head glandular; leaves usually toothed and with glands, all similar

3a Involucre 3–5 mm high; staminate and pistillate flowers on separate plants (corolla white; leaves 5–15 mm long, not hairy; common in chaparral) . . . . *Baccharis pilularis* [includes *B. pilularis* var. *consanguinea*] (pl. 8)
Coyotebrush, Chaparral-broom

3b Involucre 8–12 mm high; most flowers with stamens and a pistil

4a Involucre 8–10 mm high; leaves 10–40 mm long, not hairy, often clustered; corolla yellow (plant woody only at the base; in sandy coastal areas) . . . . *Isocoma menziesii* ssp. *vernonioides* [*Haplopappus venetus* ssp. *vernonioides*]
Coast Goldenbush; SF-s

4b Involucre 10–12 mm high; leaves 10–60 mm long, slightly hairy, often clustered; corolla white . . . . *Brickellia californica*
California Brickellia, California Brickelbush

1b Leaves either lobed or much more than 5 times as long as wide

5a Involucre more than 7 mm high (corolla yellow)

6a Phyllaries in 2 series, those of upper series much larger than those of lower; leaves woolly (leaves 3–10 cm long, usually with lobes 1–3 mm wide; heads many in each cluster, with ray and disk flowers; pappus consisting of bristles) . . . . *Senecio flaccidus* var. *douglasii* [*S. douglasii*]
Bush Groundsel, Shrubby Butterweed; Me-s

6b Phyllaries in 2–4 overlapping series, all about the same size; leaves sometimes hairy, but not woolly

7a Flower head with ray and disk flowers (leaves 1–5.5 cm long; head 1 on each peduncle; leaves slightly or not at all hairy, about 3 mm wide, not lobed; pappus consisting of bristles) . . . . *Ericameria linearifolia* [*Haplopappus linearifolius*] (pl. 10)
Interior Goldenbush, Stenotopsis; La-s

7b Flower head with disk flowers only

8a Leaves lobed, 2.5–4 cm long; pappus lacking (head solitary; phyllaries distinctly ridged on back; leaves densely hairy) . . . . *Santolina chamaecyparisus* [*S. chamaecyparissus*]
Lavender-cotton; me

8b Leaves not lobed, 1.5–3 cm long; pappus consisting of bristles or scales

9a Flower head about 3 times as high as wide, many in each cluster; phyllaries distinctly ridged on back; leaves hairy; pappus consisting of bristles . . . . *Chrysothamnus nauseosus* ssp. *mohavensis*
Rubber Rabbitbrush; SCl(MH)-s

9b   Flower head not appreciably higher than wide, 1 or few in each cluster; phyllaries not distinctly ridged on back; leaves scarcely if at all hairy; pappus consisting of 5–8 scales (bark of older stem grayish or whitish, shredding) . . . . *Eastwoodia elegans*
Yellow Mock Aster; Al-s, e

5b   Involucre not more than 7 mm high

10a   Leaves less than 1.5 cm long, not lobed (leaves often clustered; heads many in each cluster, with yellow disk and ray flowers; pappus consisting of bristles; on backshores of beaches) . . . . *Ericameria ericoides [Haplopappus ericoides]* (fig.)
Mock-heather

10b   Leaves usually more than 1.5 cm long, sometimes lobed

11a   Leaves and stem grayish, with a sagebrush odor; leaves often clustered; heads about 10 on each racemelike branch (head with pale yellow disk flowers only; phyllaries in 3 or 4 series; pappus consisting of minute scales; leaves lobed; common) . . . . *Artemisia californica* (pl. 8)
California Sagebrush; Ma, Na-s

11b   Leaves and stem neither grayish nor with a sagebrush odor; leaves not clustered; heads usually more than 10 at panicle-like branch ends

12a   Young leaves woolly, becoming hairless, sometimes lobed; phyllaries in 1 series; head usuallly with ray and disk flowers; pappus consisting of minute scales (leaves 3–7 cm long; corolla yellow) . . . .
. . . . *Eriophyllum staechadifolium*
[includes *E. staechadifolium* var. *artemisiaefolium*]
Seaside Woolly Sunflower, Lizardtail; SCr-Me

12b   All leaves scarcely or not at all hairy, not lobed; phyllaries in 3 or 4 series; head with disk flowers only; pappus consisting of bristles

13a   Leaves up to 15 cm long, sometimes toothed; petioles winged; corolla white; staminate and pistillate flowers on different plants; near stream banks and ditches . . . .
. . . . *Baccharis salicifolia [B. viminea]* (pl. 8)
Mulefat, Seep-willow

13b   Leaves 3–6 cm long, not toothed; petioles not winged; corolla yellow; most flowers with stamens and pistils; widespread . . . .
. . . . *Ericameria arborescens [Haplopappus arborescens]*
Goldenfleece

Trees and Shrubs, Group 5:   Trees and shrubs with alternate leaves that are not compound, but usually at least some palmately lobed; plant not spiny or prickly, except sometimes the fruit; flowers not in composite heads like those of a sunflower, thistle, or dandelion

1a   Largest leaves usually more than 10 cm wide (stipules present)

2a   Leaf lobes not toothed (leaves deciduous, with star-shaped hairs; flowers in hanging globular heads; bark smooth, pale, with a pattern; usually near water; tree) . . . .
. . . . *Platanus racemosa* (fig.)
Western Sycamore; Platanaceae, Sycamore Family

2b Leaf lobes toothed

    3a Leaf lobes much longer than wide; inflorescence with pistillate flowers above the staminate; petals absent; fruit usually with spinelike outgrowths (sepals less than 1 cm long; leaves not hairy) . . . . *Ricinus communis* (pl. 24)

                      Castor-bean; eu; Euphorbiaceae, Spurge Family

    3b Leaf lobes about as long as wide; inflorescence with flowers having both pistils and stamens; petals present; fruit without spinelike outgrowths

        4a Petals pink or purple, with darker veins; filaments fused into a column; leaf lobes often with star-shaped hairs . . . . *Lavatera*

                      Malvaceae, Mallow Family

        4b Petals white; filaments not fused into a colum; leaf lobes either not hairy or without star-shaped hairs . . . . *Physocarpus capitatus, Rubus parviflorus*

                      Rosaceae, Rose Family

1b Largest leaves rarely more than 6 cm wide but sometimes wider in species of *Malacothamnus*

    5a Flowers irregular (leaves evergreen, with dense branched hairs; sepals mostly red; petals green; fruit about 1 cm long) . . . . *Castilleja foliolosa* (pl. 52)

              Woolly Indian Paintbrush; Scrophulariaceae, Snapdragon Family

    5b Flowers regular

        6a Leaves without stipules; ovary inferior (leaves deciduous; fruit fleshy and sometimes also prickly) . . . . *Ribes*

                      Grossulariaceae, Gooseberry Family

        6b Leaves with stipules, these sometimes deciduous; ovary superior

            7a Leaf lobes not toothed; petals absent, sepals orange yellow, often red tinged; stamens 5; fruit 2–4 cm long, bristly (evergreen; flower mostly 3–5 cm wide; leaves with star-shaped hairs) . . . . *Fremontodendron californicum* [includes *F. californicum* sspp. *crassifolium* and *napense*] (pl. 55)

                Flannelbush, Fremontia; Sterculiaceae, Cacao Family

            7b Leaf lobes toothed; petals present; sepals green; stamens at least 20; fruit usually less than 2 cm long, not bristly

                8a Leaves evergreen, with star-shaped hairs; filaments fused into a column; petals pink or rose; fruit about 3 mm long (leaf lobes shallow, sometimes absent) . . . . *Malacothamnus*

                      Malvaceae, Mallow Family

                8b Leaves deciduous, if hairy, without star-shaped hairs; filaments not fused; petals white; fruit 5–10 mm long . . . . *Physocarpus capitatus, Rubus parviflorus*

                      Rosaceae, Rose Family

Trees and Shrubs, Group 6: Trees and shrubs with alternate leaves that are not compound, either pinnately lobed or not lobed; plant not spiny; flowers not both sessile and in composite heads

1a Blades of most leaves at least 3 times as long as wide, sometimes 4–5 times, or leaves often absent in *Spartium junceum* (Fabaceae)

2a   Most leaves distinctly and regularly toothed, sometimes smooth margined in *Eriodic-tyon* (Hydrophyllaceae)

       3a   Young branches and usually upper surfaces of leaves sticky; corolla white to pur-plish, 1–1.5 cm long, tubular with 5 lobes (evergreen; leaves 4–15 cm long, often blackened by a fungus; widespread) . . . . *Eriodictyon californicum* (pl. 31)

                                  Yerba-santa; SB-n; Hydrophyllaceae, Waterleaf Family

       3b   Young branches and upper surfaces of leaves not sticky; corolla not as described in choice 3a

          4a   Marginal teeth unequal, the larger ones, in general, alternating with 1 or 2 smaller ones (evergreen; leaves bright green, not hairy, often aromatic when bruised; staminate and pistillate flowers on the same plant but in separate catkinlike inflorescences; fruit purplish, covered with whitish wax) . . . .
. . . . *Myrica californica* (fig.)

                             Wax-myrtle; Myricaceae, Wax-myrtle Family

          4b   Marginal teeth almost equal

              5a   Leaves hairy, deciduous; each marginal tooth tipped by a tapering hair; flower funnel shaped, with 5 lobes; mostly in moist habitats (corolla at least 3 cm long, whitish or pink, sometimes with yellow on the upper-most lobe) . . . . *Rhododendron occidentale* (pl. 23)

                     Western Azalea; SCr-n; Ericaceae, Heath Family

              5b   Leaves not hairy, evergreen; marginal teeth without hairs; flower not funnel shaped, the petals separate; in dry habitats

                  6a   Leaves gray green, the teeth small; petals 4, yellow, 20–30 mm long; fruit narrow, up to 10 cm long, not red . . . .
. . . . *Dendromecon rigida* (pl. 39)

                     Bush Poppy; Sn-s; Papaveraceae, Poppy Family

                  6b   Leaves dark green, the teeth prominent; petals 5, white, about 3 mm long; fruit rounded, up to 1 cm long, bright red when ripe (wide-spread) . . . . *Heteromeles arbutifolia* (pl. 48)

                     Toyon, Christmas-berry; Rosaceae, Rose Family

2b   Leaves either not toothed, or with a few irregular teeth (except perhaps on young seedlings, or on atypical leaves that sometimes appear on a few branches of a mature plant)

       7a   Leaves up to 10 mm long, not more than 1 mm wide (leaves clustered; flowers in panicles 4–12 cm long; corolla white, less than 5 mm long; widespread) . . . .
. . . . *Adenostoma fasciculatum* (pl. 48)

                     Chamise; Me-s; Rosaceae, Rose Family

       7b   Leaves mostly more than 10 mm long, considerably more than 1 mm wide

          8a   Leaf blade or apparent leaf blade (modified petiole), with 2–5 prominent veins originating at the base

              9a   Leaves spine tipped; flowers white to pink; shrub . . . .
. . . . *Polygonum bolanderi*

             Bolander Knotweed; Na-n; Polygonaceae, Buckwheat Family

9b Leaves not spine tipped; flowers cream colored or yellow; shrubs or trees

    10a Flowers cream colored, in clusters along branches; fruit curved or twisted; tree . . . . *Acacia melanoxylon*

        Blackwood Acacia; au; Fabaceae, Pea Family

    10b Flowers bright yellow, solitary along branches; fruit straight; shrub or small tree . . . . *Acacia longifolia*

        Sydney Golden Wattle; au; Fabaceae, Pea Family

8b Leaf blade with only 1 prominent vein, the midrib, originating at the base

    11a Leaves either decidedly aromatic or with a strong odor when crushed (evergreen; trees)

        12a Leaves olive green, 3–10 cm long, abruptly pointed; flowers yellow green, less than 1 cm wide, in clusters at ends of branches; ovary superior; stamens 9; fruit fleshy, with a single large seed (widespread) . . . . *Umbellularia californica* (pl. 34)

            California Bay; Lauraceae, Laurel Family

        12b Leaves pale green, 10–20 cm long, gradually narrowed to a point; flowers cream colored, 2 cm wide, mostly solitary in leaf axils; ovary inferior; stamens more than 25; fruit a woody capsule . . . . *Eucalyptus globulus*

            Blue Gum; au; Myrtaceae, Myrtle Family

    11b Leaves neither aromatic nor with a strong odor when crushed

        13a Perianth irregular; leaves sometimes absent, 2–3 mm wide (corolla 20–25 mm long, yellow; leaves 10–30 mm long, not succulent; widespread and invasive) . . . . *Spartium junceum* (pl. 27)

            Spanish Broom; me; Fabaceae, Pea Family

        13b Perianth regular; leaves present, usually much more than 3 mm wide except in *Eriogonum* (Polygonaceae)

            14a Leaves succulent (leaves 5–35 mm long; flower without petals, usually with both pistils and stamens) . . . . *Suaeda*

                Chenopodiaceae, Goosefoot Family

            14b Leaves not succulent

                15a Flower usually more than 1 cm long (petals united; flowers with pistils and stamens, not in racemes)

                    16a Leaves stiff, often more than 12 cm long, evergreen; corolla mostly pink or rose; fruit becoming a dry capsule (corolla 3–4 cm long) . . . . *Rhododendron macrophyllum* (pl. 23)

                      Rosebay; Mo-n; Ericaceae, Heath Family

                  16b Leaves not stiff, most less than 12 cm long, sometimes summer deciduous; corolla white, blue, or purple; fruit a fleshy berry . . . . *Solanum*

                    Solanaceae, Nightshade Family

15b Flower not more than 1 cm long (flower white, yellow, pink, blue, or violet)

    17a Leaves silvery, covered with small scales; shrub or tree (corolla absent; calyx 4 lobed, yellow) . . . . *Elaeagnus angustifolius [E. angustifolia]* Oleaster, Russian-olive; as Elaeagnaceae, Oleaster Family

    17b Leaves not silvery, without small scales; shrubs

        18a Flowers in racemes, the staminate and pistillate ones on separate plants; in shade (leaves 5–13 cm long; fruit becoming blue black) . . . . *Oemleria cerasiformis [Osmaronia cerasiformis]* (fig.) Osoberry; Rosaceae, Rose Family

        18b Flowers in heads, umbels, or panicles, all with stamens and a pistil; in sunny areas

            19a Flowers up to 3 mm long, in heads or umbels, white to pink; leaves up to 1.5 cm long; fruit an achene . . . . *Eriogonum fasciculatum* (pl. 43) California Buckwheat; SLO-n Polygonaceae, Buckwheat Family

            19b Flowers 5–9 mm long, in panicles, blue to violet; leaves 6–20 cm long; fruit consists of 4 nutlets . . . . *Echium candicans* Viper's-bugloss; af Boraginaceae, Borage Family

1b Blades of leaves rarely so much as 3 times as long as wide

    20a Leaves and young branches covered with grayish scales; restricted to saline and alkaline soils (leaf blades almost triangular; flowers staminate or pistillate, and in separate clusters; fruit 3–4 mm long; shrub). . . . . *Atriplex lentiformis* [includes *A. lentiformis* ssp. *breweri*] Big Saltbush, Quailbush; SF-s, e; Chenopodiaceae, Goosefoot Family

    20b Leaves and young branches not covered with scales; not primarily in saline and alkaline soils

        21a Flowers either staminate or pistillate, and in separate catkins (leaves deciduous, 3–15 cm long, coarsely and unevenly toothed; trees). . . . . *Alnus* Betulaceae, Birch Family

        21b All flowers with stamens and a pistil, and these not in catkins

            22a Evergreen (except *Ceanothus integerrimus*); leaf often stiff, either with a prominent raised midrib, or with 3 prominent veins diverging from the base; flowers small, the inflorescences either in the leaf axils or at branch ends . . . . Rhamnaceae, Buckthorn Family

22b Often evergreen, but rarely conforming to all characteristics described in choice 22a

  23a Leaves toothed, usually not lobed

    24a Corolla saucer shaped, the petals separate; stamens usually more than 10; leaves not tough or leathery, except in *Heteromeles arbutifolia* and *Prunus ilicifolia* .... Rosaceae, Rose Family

    24b Corolla urn shaped, the petals united; stamens 8–10; leaves tough and leathery (evergreen; corolla white; fruit fleshy, blue black)

      25a Leaf 2–3 cm long, without an indentation at the base; ovary inferior; often more than 1 m tall ..... *Vaccinium ovatum* (pl. 23)
      Evergreen Huckleberry, California Huckleberry
      Ericaceae, Heath Family

      25b Leaf 5–10 cm long, often with a slight indentation at the base; ovary superior; usually less than 1 m tall ....
      .... *Gaultheria shallon* (pl. 23)
      Salal; Ericaceae, Heath Family

  23b Leaves not toothed, but rarely so in *Arbutus, Arctostaphylos,* and *Vaccinium parviflorum* (Ericaceae)

    26a Bark of much of trunk and main stems smooth, reddish brown, and peeling away (leaves often stiff, evergreen; corolla usually less than 1 cm long, urn shaped, with 5 lobes, and white or pink) ....
    .... *Arbutus, Arctostaphylos*
    Ericaceae, Heath Family

    26b Bark of trunk and main stems not smooth, reddish brown, or peeling

      27a Leaves less than 5 mm long, scalelike (flower regular, 1–2 mm long, green or pink)

        28a Stem and branchlets succulent, green; leaves evergreen, not excreting salt; corolla absent, calyx green; in alkali areas ...
        .... *Allenrolfea occidentalis*
        Iodinebush; Al, CC; Chenopodiaceae, Goosefoot Family

        28b Stem and branchlets not succulent, not green; leaves deciduous, often excreting salt; corolla pink or white, calyx green; usually near streams and rivers ....
        .... *Tamarix ramosissima [T. pentandra]*
        Salt-cedar, Tamarisk; as; Tamaricaceae, Tamarisk Family

      27b Leaves usually more than 5 mm long, not scalelike

        29a Leaves with 3–5 prominent veins; corolla irregular, resembling that of a pea, reddish purple; fruit 5–8 cm long, a dry pod (corolla 8–12 mm long; leaf blade with a notch at the base; deciduous) .... *Cercis occidentalis* (pl. 24)
        Western Redbud; Sl-n; Fabaceae, Pea Family

29b Leaf blades with only 1 prominent vein, the midrib; corolla, if present, regular, not reddish purple; fruit less than 2 cm long, not a pod

30a Each leaf dark green on the upper surface and woolly on the underside; stamens about 20 (flowers in clusters; petals separate, white, pink, or rose; fruit about 8 mm long, red; evergreen; ovary inferior) . . . .
. . . . *Cotoneaster*
Rosaceae, Rose Family

30b Each leaf not both dark green on the upper surface and woolly on the underside; stamens not more than 10

31a New branches extremely flexible; corolla absent; calyx yellow (calyx about 1 cm long, tubular, 4 lobed; flowers often appearing in winter; leaf blades 3–7 cm long, deciduous; fruit green or brown, 8–10 mm long; ovary superior) . . . .
. . . . *Dirca occidentalis* (pl. 55)
Western Leatherwood; SFBR; 4
Thymelaeaceae, Daphne Family

31b New branches not extremely flexible; corolla present; calyx green

32a Most branches angled lengthwise and green; deciduous; corolla urn shaped, pink or green; ovary inferior (leaves not more than 2.5 cm long) . . . .
. . . . *Vaccinium parvifolium* (pl. 23)
Red Huckleberry
SCl-n; Ericaceae, Heath Family

32b Older branches not angled lengthwise and not green; evergreen or sometimes summer deciduous; corolla not urn shaped, not pink or green; ovary superior

33a Leaves stiff, 1–3.5 cm long; petals separate or nearly so, white or yellow; stamens 8–10; in swamps and bogs . . . .
. . . . *Ledum glandulosum* [includes *L. glandulosum* ssp. *columbianum*] (fig.)
Western Labrador-tea; SCr-n
Ericaceae, Heath Family

33b Leaves not stiff, often at least 10 cm long; petals united, white, yellow, blue, or purple; stamens 5; usually in dry areas . . . . *Nicotiana glauca, Solanum*
Solanaceae, Nightshade Family

Herbaceous Plants

To speed up finding the family to which a particular herbaceous plant belongs, the keys are arranged in four groups. These are organized partly on the basis of structural features and partly on superficial features and on places where the plants grow. If you become familiar with the characteristics of these groups, you can go directly to the right one and start keying to find the plant's family. Then, in the "Flowering Plant Species" section, turn to that family.

A family, genus, or species may appear in the key more than once. This is because a family or genus is so diverse or because a species is so variable that it cannot always be identified by using a single sequence of choices. Look in the index to find these multiple entries. If the couplet leads to only one or two genera in a family, these are given with the family name. An (M) after a family name indicates that it is a monocotyledon.

1a Plant either attached to above-ground portions of other plants or lacking perceptible chlorophyll and therefore not green (terrestrial) . . . . HERBACEOUS PLANTS, GROUP 1
1b Plant neither attached to above-ground portions of other plants nor lacking chlorophyll
   2a Plant grasslike (terrestrial or aquatic, often tufted; leaves usually narrow with parallel veins; flower usually lacking showy perianth segments and often without any) . . . .
      . . . . HERBACEOUS PLANTS, GROUP 2
   2b Plant not grasslike
      3a Plant aquatic (either floating or partly or wholly submerged; floating leaves some- times different from those submerged) . . . . HERBACEOUS PLANTS, GROUP 3
      3b Plant not aquatic . . . . HERBACEOUS PLANTS, GROUP 4

Herbaceous Plants, Group 1:   Plants terrestrial, either attached to above-ground parts of other plants or lacking perceptible chlorophyll and therefore are not green
1a Plant attached to and parasitic on above-ground portions of other plants
   2a Plant slender, usually with orange or yellow stems attached to herbs or shrubs; leaves reduced to scales; flower with a pistil and stamens; corolla 2–6 mm long, white, 5 lobed; ovary superior; fruit dry when mature . . . . Cuscutaceae, Dodder Family
   2b Plant usually stout, with green, yellow green, orange yellow, or brownish stems at- tached to branches of trees or shrubs; leaves either scalelike or well developed; flower either pistillate or staminate; corolla absent; ovary inferior; fruit fleshy when mature, white, bluish, or purplish. . . . . Viscaceae, Mistletoe Family
1b Plant not attached to above-ground portions of other plants (plant not green; leaves re- duced to scales)
   3a Corolla regular; stamens 8–10 (petals 4 or 5; sepals usually 4 or 5, sometimes none; ovary superior; saprophytes or parasites in forest soils) . . . . . . Ericaceae, Heath Family
   3b Corolla irregular; stamens 4 or fewer
      4a Petals 3, separate, the lower one decidedly different from the upper ones; sepals 3, separate or united at the bases; anther-bearing stamen 1; ovary inferior; sapro- phytes in forest soils. . . . . *Cephalanthera, Corallorhiza*
                                                                Orchidaceae, Orchid Family (M)

4b   Petals 5, united, 3 forming a lower lip and 2 forming an upper lip; sepals 5, united; anther-bearing stamens 4; ovary superior; parasites attached to underground parts of other plants. . . . . Orobanchaceae, Broomrape Family

Herbaceous Plants, Group 2:   Plant terrestrial or aquatic, often tufted; leaves usually narrow with parallel veins; flowers usually lacking showy perianth segments and often without any; all monocotyledons

1a   Plant strictly marine, growing in salt water (plant submerged except at low tide; inflorescence 1 sided, located within a boat-shaped bract; leaves narrow, the margins nearly parallel) . . . . Zosteraceae, Eelgrass Family (M)

1b   Plant not growing in salt water, either terrestrial or in fresh or brackish water

    2a   Flowers in a dense, almost smooth cylindrical inflorescence, 1–2 cm wide, the staminate flowers above the pistillate flowers (plant rooted in mud or muck, but only the lower portion under water) . . . . *Typha*
                              Typhaceae, Cattail Family (M)

    2b   Inflorescence not as described in choice 2a

        3a   Flowers in terminal racemes, with 3–6 concave perianth segments; fruit 2 or 3 times as long as wide, the 3–6 divisions splitting apart at maturity; usually growing in salt marshes; leaves basal, succulent (stamens 1–6, with very short filaments) . . . . *Triglochin*
                         Juncaginaceae, Arrow-grass Family (M)

        3b   Flowers and fruit not as described in choice 3a; plant, if growing in salt marshes, without succulent leaves

            4a   Inflorescence consisting of several globular heads of either all-staminate or all-pistillate flowers; pistillate heads soon resembling burs (plant sometimes mostly submerged, but the inflorescence held above the water surface) . . . . .
               . . . . *Sparganium*
                               Typhaceae, Cattail Family (M)

            4b   Inflorescence not as described in choice 4a

                5a   In freshwater and brackish habitats: ditches, streams, and vernal pools; inflorescence raised above the water, most flowers with one purplish perianth segment; submerged pistillate flowers near base of plant enclosed by leaf sheaths and with styles at least 5 cm long; leaves basal, mostly cylindrical, but the lower portions broadened into open sheaths . . . . *Lilaea scilloides*
                    Flowering Quillwort; Juncaginaceae, Arrow-grass Family (M)

                5b   Plant, if growing in freshwater and brackish habitats, not as described in choice 5a

                    6a   Stem typically 3 sided, but sometimes cylindrical (stem usually solid; bracts of inflorescence often leaflike, sometimes absent; flower without definite perianth segments, at least partly enclosed by a bract; leaf sheaths closed; mostly growing in wet areas) . . . .
                      . . . . Cyperaceae, Sedge Family (M)

6b Stem cylindrical or flattened

    7a Flower with 6 perianth segments; bracts of inflorescence leaflike or stemlike, often much larger than the flower cluster; leaves without ligules; leaf sheaths open or closed; stem usually solid (mostly growing in wet areas) . . . .
. . . . Juncaceae, Rush Family (M)

    7b Flower without definite perianth segments; bracts of inflorescence not larger than the flower cluster; leaves with ligules; leaf sheaths open; stem usually hollow . . . .
. . . . Poaceae, Grass Family (M)

Herbaceous Plants, Group 3:   Plants aquatic, either floating or partly or wholly submerged; floating leaves sometimes different from those submerged (this group does not include aquatic plants that are fern allies)

1a Leaves either opposite, in whorls, or absent

    2a Leaves in whorls or absent

        3a Leaves absent (stem floating, usually flattened and rarely more than 5 mm long or wide, sometimes with microscopic flowers located along the edges; roots, if present, unbranched, and sometimes only 1; plant reproducing vegetatively; sometimes found on wet soil after a drop in the water level) . . . .
. . . . Lemnaceae, Duckweed Family (M)

        3b Leaves present, most in whorls of at least 3

            4a Leaves slender, not divided into lobes

                5a Leaves usually 8–10 in each whorl, extending stiffly outward almost at a right angle to the stem, smooth margined; plant usually only partly submerged . . . . *Hippuris vulgaris* (fig.)
Mare's-tail; Hippuridaceae, Mare's-tail Family

                5b Leaves rarely more than 8 at each node, usually 3–6, not stiff or extending outward, usually toothed; plant submerged . . . .
. . . . Hydrocharitaceae, Waterweed Family (M)

            4b Leaves, at least the submerged ones, divided into slender lobes (stem long, weak, mostly or wholly submerged)

                6a Leaves firm, the lobes with small teeth, thus somewhat rough to the touch . . . . *Ceratophyllum demersum* (fig.)
Hornwort; Ceratophyllaceae, Hornwort Family

                6b Submerged leaves delicate, the lobes nearly hairlike and not toothed (corolla white or purplish) . . . . Haloragaceae, Water-milfoil Family

    2b Leaves opposite

        7a Leaves not more than 1 mm wide

            8a Leaf, at least near the base, with finely toothed margins; in freshwater; ovary inferior. . . . . Hydrocharitaceae, Waterweed Family (M)

            8b Leaf smooth margined; in fresh or brackish water; ovary superior . . . .
. . . . *Zannichellia palustris* (fig.)
Horned-pondweed; Zannichelliaceae, Horned-pondweed Family (M)

7b Most leaves more than 2 mm wide, sometimes less in Callitrichaceae

9a Tip portion of flowering stem extending above the water surface; leaves with petioles, the blades up to 2.5 cm long; petals absent; sepals 4; stamens 4; ovary inferior; fruit plump, slightly 4 sided, about 4 mm long, containing many small seeds.................... *Ludwigia palustris [L. palustris* var. *pacifica]*
Common Water-primrose; Onagraceae, Evening-primrose Family

9b Tip portions of all stems submerged or floating; leaves sessile, rarely so much as 1.5 cm long; petals and sepals absent; stamen 1; ovary superior; fruit flattened, about 2 mm long, breaking apart into 4 1-seeded units (flowers staminate or pistillate) .... Callitrichaceae, Water-starwort Family

1b Leaves, or structures appearing to be leaves, either alternate, basal, or arising from horizontal stems

10a Leaves without blades and consisting entirely of hollow, cross-barred petioles (leaves arising in clusters from a creeping stem; flowers in small umbels; ovary inferior).....
....*Lilaeopsis*
Apiaceae, Carrot Family

10b Leaves, or structures appearing to be leaves, with blades, but these may be divided into slender lobes

11a Submerged leaves, or structures appearing to be leaves, divided into slender, nearly hairlike lobes (there may also be floating leaves with broad blades)

12a Some so-called leaves (these are in fact modified stems) with small bladders that trap microscopic aquatic organisms; flowering stem raised above the water, other stems and leaves completely submerged; corolla yellow, irregular; pistil 1........................ Lentibulariaceae, Bladderwort Family

12b Leaves without bladders; all stems and leaves completely submerged or floating, not out of the water; corolla white, regular; pistils many .... *Ranunculus*
Ranunculaceae, Buttercup Family

11b None of the leaves with slender, nearly hairlike lobes

13a Leaves compound (leaflets 3, sometimes more than 7 cm long; petioles stout, normally holding blades well above the water level; corolla white to pale purple, with scalelike hairs on the inner surface) .... *Menyanthes trifoliata* (fig.)
Buckbean; SFBR-n; Menyanthaceae, Buckbean Family

13b Leaves not compound

14a Leaf blades arrowhead shaped, circular or broadly heart shaped (stamens many)

15a Leaf blades arrowhead shaped, raised well above the water level; flowers rarely more than 2.5 cm wide, in whorls on stems rising out of the water; petals and sepals 3 .... *Sagittaria*
Alismataceae, Water-plantain Family (M)

15b Leaf blades circular or broadly heart shaped, generally floating; flower at least 5 cm wide, solitary and either floating or raised slightly above the water; petals and sepals many (sepals sometimes petal-like) .... Nymphaeaceae, Waterlily Family

14b Leaf blades not arrowhead shaped or heart shaped, but sometimes circular in *Hydrocotyle* (Apiaceae)

16a Leaf blades either almost circular or approximately kidney shaped and lobed (leaf blades about 4 cm wide, the circular ones with petioles attached near the center, as in a nasturtium; stamens 5; ovary inferior) . . . . *Hydrocotyle*
Apiaceae, Carrot Family

16b Leaf blades neither almost circular nor kidney shaped and lobed

17a Leaves basal (blades oval)

18a All leaf blades floating; perianth segment 1; stamens many (perianth white; anthers purple; leaves with many lengthwise veins and cross veins; inflorescence branched in twos) . . . . *Aponogeton distachyon [A. distachyus]*
Cape-pondweed; af
Aponogetonaceae, Cape-pondweed Family (M)

18b Leaf blades usually raised well above the water level; perianth segments 6; stamens 6 (petioles usually at least as long as the blades; pistils 6 to many; fruit an achene)

19a Petioles inflated, bladelike; perianth segments often more than 2 cm long, white, lilac, or blue; fruit a capsule . . . . *Eichhornia crassipes*
Water Hyacinth; sa
Pontederiaceae, Pickerel-weed Family (M)

19b Petioles not inflated; perianth segments less than 1 cm long, white, yellow, pink, or rose; fruit an achene . . . .
. . . . Alismataceae, Water-plantain Family (M)

17b Leaves not basal

20a Leaves usually either floating or out of the water, not sessile (leaves more than 5 mm wide)

21a Stipules united around the stem; sepals 5, united, pink or brownish; petals absent; stamens 3–8; ovary superior (flowering stem raised well above the water level)
. . . . *Polygonum*
Polygonaceae, Buckwheat Family

21b Stipules, if present, not united around the stem; sepals 5 or 6, separate, green; petals 5 or 6, white or yellow; stamens 10 or 12; ovary inferior . . . . *Ludwigia*
Onagraceae, Evening-primrose Family

20b At least some leaves submerged and sessile (leaf usually with a sheath at the base)

22a Stem not much branched; perianth segments 6, united, pale yellow, the tubular portion at least 15 mm long (leaves usually at least 7 cm long but not more than 0.5 cm wide; flowers opening at the surface, each within a rolled-up bract) . . . . *Heteranthera dubia* (fig.)
Water Stargrass; Sl, Me-n
Pontederiaceae, Pickerel-weed Family (M)

22b Stem often much branched; perianth segments none or 4, separate, green, less than 5 mm long (floating leaves, if present, with broad blades; submerged leaves nearly threadlike; inflorescence generally raised slightly above the water level) . . . .
. . . . Potamogetonaceae, Pondweed Family (M)

Herbaceous Plants, Group 4: Plant terrestrial, although sometimes growing in wet places

1a Sunflowers, thistles, dandelions, and so forth: flowers both sessile and concentrated in composite heads, the base of these surrounded by somewhat leaflike or scalelike bracts (phyllaries); corolla sometimes tubular, sometimes drawn out into a petal-like ray; ovary inferior; fruit an achene, often bearing a pappus consisting of bristles or scales; style branches 2 (dry flower heads, usually helpful in identification, often persist well into fall and winter). . . . . Asteraceae, Sunflower Family

1b Flowers, if in dense heads, not as described in choice 1a

   2a Flowers densely crowded on a solid, elongated inflorescence at least 1 cm wide, this either partly enveloped by a single yellow or white bract or its base encircled by several white, petal-like bracts (leaves mostly basal)

      3a Inflorescence usually at least 5 cm long, partly enveloped by a yellow or white bract up to 20 cm long; each flower without a bract beneath it; leaf blades 15–150 cm long. . . . . Araceae, Arum Family (M)

      3b Inflorescence not more than 4 cm long, its base encircled by several white, petal-like bracts, 1–3 cm long, often tinged with red on the undersides; each flower, except for the lowermost ones, with a small white bract beneath it; leaf blades about 15 cm long (usually in low areas where the soil is alkaline or slightly salty) . . . .
. . . . *Anemopsis californica* (pl. 51)
Yerba-mansa; SCl-s; Saururaceae, Lizard's-tail Family

   2b Flowers, if densely crowded on a solid inflorescence, not as described in choice 2a

      4a Plant growing in salt marshes, inland alkaline habitats, or sphagnum bogs

         5a Plant growing in salt marshes and some inland alkaline habitats; plant succulent, with opposite branches that appear to be jointed; leaves reduced to scales, without glands; flowers protruding above scalelike leaves at stem tips; calyx and corolla absent . . . . *Salicornia*
Chenopodiaceae, Goosefoot Family

         5b Plant growing in sphagnum bogs; plant not succulent, branched, or jointed; leaves basal, not scalelike, with red, gland-tipped hairs that trap insects; flowers at the top of a leafless stalk; calyx and corolla present. . . . .
. . . . *Drosera rotundifolia*
Roundleaf Sundew; Sn-n; Droseraceae, Sundew Family

      4b Plant, if growing in the habitats listed in choice 4a, not like *Drosera* or *Salicornia*

         6a Leaf blades usually at least 75 cm wide, with fleshy petioles often 1 m long arising from creeping underground stems (inflorescence more than 1 m long; petals usually 2, hoodlike, less than 5 mm long; ovary inferior; fruit red) . . . .
. . . . *Gunnera tinctoria [G. chilensis]*
sa; Gunneraceae, Gunnera Family

6b Leaf blades not so much as 75 cm wide, but even if close to this size, not on petioles arising from creeping stems

    7a Corolla, calyx, or both decidedly irregular ....

        .... HERBACEOUS PLANTS, GROUP 4, SUBKEY 1

    7b Corolla, calyx, or both essentially regular (both sometimes absent in Euphorbiaceae)

        8a Leaves up to 20 cm long, alternate, but lower ones may seem to be opposite; lower leaves lobed and toothed; flowers generally pistillate or staminate, these on separate plants; petals absent; calyx lobes 3–9; stamens 8–12; styles 3, each 2 lobed; ovary inferior; up to 1 m tall; usually growing along dry stream beds *Datisca glomerata* (fig.)

                Durango-root; Datiscaceae, Datisca Family

        8b Plant not as described in choice 8a

           9a Plant with milky sap (break a leaf tip to see)

              10a Petals and sepals absent; flowers pistillate or staminate, the staminate flowers, consisting of 1 stamen, located around each pistillate flower (leaves opposite or alternate) ....

                  .... *Chamaesyce, Euphorbia*

                      Euphorbiaceae, Spurge Family

              10b Petals, sepals, or both present; flowers with both stamens and a pistil and not as in choice 10a

                  11a Flowers in umbels; leaves usually at least 75 cm long, alternate, 1–3 times compound; stem surrounded by a sheath consisting of fused stipules; ovary inferior (petals white to green; fruit a black berry; growing along streams in shaded canyons) ....

                    .... *Aralia californica* (pl. 6)

                  Elk-clover, Spikenard; Araliaceae, Ginseng Family

                  11b Flowers not in umbels; leaves usually less than 10 cm long, opposite or in whorls, smooth margined; stipules absent or very small; ovary superior (seeds usually with tufts of hair)

                    12a Flower with 5 concave hoodlike lobes between the corolla lobes and stamens; stamens united to form a tube around the pistils; leaves in whorls or opposite ..... Asclepiadaceae, Milkweed Family

                    12b Flower not as described in choice 12a; stamens not united to form a tube around the pistils; leaves opposite ....

                    .... Apocynaceae, Dogbane Family

           9b Plant without milky sap

              13a At least some leaves opposite or in whorls ....

              .... HERBACEOUS PLANTS, GROUP 4, SUBKEY 2

              13b Leaves alternate or basal

14a Leaves usually smooth margined; flower less than 10 mm long; either stipules present, membranous and fused around stem nodes, or flowers in a cuplike involucre; petals absent; sepals 5 or 6, united, petal-like, and often in 2 whorls; stamens 3–9; styles 3; ovary superior; fruit often winged . . . . Polygonaceae, Buckwheat Family

14b Plant not in every respect as in choice 14a

15a Inflorescence a head or an umbel, usually compound; leaves usually compound, sometimes lobed, often aromatic when bruised; base of petiole expanded, sheathing the stem; sepals sometimes absent; petals 5, separate; stamens 5; styles 2; ovary inferior . . . . . . .Apiaceae, Carrot Family

15b Plant not as described in choice 15a . . . . HERBACEOUS PLANTS, GROUP 4, SUBKEY 3

Herbaceous Plants, Group 4, Subkey 1: Corolla, calyx, or both decidedly irregular; plant not grasslike, primarily green and terrestrial, even if occasionally in very wet places or inundated by flooding or high tides

1a Leaves opposite or in whorls (petals united)

2a Bracts present beneath inflorescence and also beneath each flower; bracts sometimes spiny; stem not 4 sided (stamens 4; corolla 7–10 mm long, white, pink, blue, or purple; ovary inferior; fruit an achene) . . . . Dipsacaceae, Teasel Family

2b Bracts sometimes present beneath either inflorescence or each flower, but not both, and not spiny; stem often 4 sided

3a Corolla with a spur at the base; calyx usually appearing to be absent because there are no obvious lobes; stamens 1 or 3; fruit an achene (stigma not lobed; ovary inferior) . . . . Valerianaceae, Valerian Family

3b Corolla often without a spur; calyx distinct; stamens 2, 4, or 5; fruit a berry, capsule, or developing into separate nutlets

4a Fruit a berry or a capsule; stamens 2, 4, or 5, 1 often antherless; plant usually without a strong aroma; stem sometimes 4 sided

5a Upper lip of corolla 4 lobed, the lower lip with 1 lobe; stamens 5, all with anthers; ovary inferior; fruit a red berry (corolla pale yellow; plant vinelike, woody at the base) . . . . *Lonicera subspicata* var. *denudata* [*L. subspicata* var. *johnstonii*] Southern Honeysuckle; CC(MD); SCl(MH); Caprifoliaceae, Honeysuckle Family

5b Upper lip of corolla 2 lobed, lower lip 3 lobed; stamens 2, 4, or 5, 1 often antherless; ovary superior; fruit a capsule . . . . Scrophulariaceae, Snapdragon Family

4b Fruit developing into 2 or 4 nutlets; stamens 4, all with anthers; plant usually with a strong aroma; stem usually 4 sided (corolla white, red, lavender, purple, or blue, rarely light yellow)

6a   A single bract present beneath each flower, but absent beneath inflorescence; aroma of plant not minty; leaves not glandular; stamens not protruding from the corolla (corolla 2–6 mm long; leaves toothed or lobed) . . . . Verbenaceae, Verbena Family

6b   Bracts absent beneath each flower but often present beneath inflorescence; aroma of plant often minty; leaves often glandular; stamens often protruding from the corolla. . . . . Lamiaceae, Mint Family

1b   Leaves alternate or mainly basal

    7a   Leaves compound, with at least 3 leaflets (sepals 5, united; petals 5, the 2 lower ones united by their edges to form a keel and the uppermost one enlarged to form a banner; stamens usually 10) . . . . Fabaceae, Pea Family

    7b   Leaves not compound

        8a   Sepals 3, usually petal-like; petals 3; veins of leaves nearly parallel (1 petal usually with a pouch or spur; ovary inferior; fruit a capsule) . . . . Orchidaceae, Orchid Family (M)

        8b   Sepals 2 or 4–6, usually not petal-like; petals 1 or 4–6, except 3 in *Polygala;* veins of leaves usually not parallel

           9a   Petals united, at least at their bases or tips

              10a   Ovary inferior or half-inferior (stamens 5; corolla 5 lobed, forming 2 lips; plant found either on dry slopes or around drying vernal pools or ditches) . . . . *Downingia, Nemacladus*
                    Campanulaceae, Bluebell Family

              10b   Ovary superior

                  11a   Sepals 2, separate, soon deciduous; petals 4, the 2 outer petals separate and different from the 2 inner petals, these united at the tips; stamens 6, all with anthers (1 or 2 petals with pouches or spurs at their bases) . . . . *Dicentra, Fumaria*
                        Papaveraceae, Poppy Family

                  11b   Sepals usually 5, united; petals 5, 2 united to form an upper lip and 3 united to form a lower lip; stamens 2, 4, or 5, one often antherless . . . . Scrophulariaceae, Snapdragon Family

           9b   Petals separate

              12a   Stamens more than 10, usually many

                  13a   Sepals 4–6, not petal-like, without spurs; petals 4–6, white, not much smaller than the sepals, each deeply lobed but without spurs; stamens all on one side of the corolla . . . . *Reseda luteola, Reseda odorata*
                    Resedaceae, Mignonette Family

                  13b   Sepals 5, petal-like, the uppermost one with a spur; petals 1 or 4, usually not white, much smaller than the sepals, usually not lobed, those of the upper pair with spurs; stamens not all on one side of the corolla . . . . *Consolida, Delphinium*
                    Ranunculaceae, Buttercup Family

              12b   Stamens not more than 10

14a Flower superficially resembling that of a pea; sepals 5, the 2 at the sides of the flower winglike; petals usually 3, the lowermost one keel-like (stamens 6–8; petals pink or rarely white) . . . .
. . . . *Polygala californica* (pl. 43)
California Milkwort; SLO-n; Polygalaceae, Milkwort Family

14b Flower not resembling that of a pea; sepals 5, not winglike; petals 4 or 5, not keel-like

15a Petals white, cream colored, yellow, blue, or violet, the lower-most one with a pouch or spur; sepals without spurs; stamens 5; leaf blades usually not more than 5 cm long, usually not rounded
. . . . .Violaceae, Violet Family

15b Petals white, orange, or brownish purple, without pouches or spurs; sepals sometimes with spurs; stamens 3, 8, or 10; leaf blades often more than 5 cm long, often rounded

16a Petals orange; uppermost sepal with a spur; stamens 8; petiole attached in the center of the circular blade . . . .
. . . .*Tropaeolum majus*
Garden Nasturtium; sa
Tropaeolaceae, Nasturtium Family

16b Petals white or brownish purple; sepals without spurs; stamens 3 or 10; petiole not attached in the center of the blade
. . . . *Saxifraga stolonifera, Tolmiea menziesii*
Saxifragaceae, Saxifrage Family

Herbaceous Plants, Group 4, Subkey 2:   At least some leaves opposite or in whorls; corolla, calyx, or both essentially regular; plant not grasslike, primarily green and terrestrial, but occasionally in very wet places or inundated by flooding or high tides

1a At least some leaves in whorls

2a Flower lacking either petals or sepals (sepals so inconspicuous in Rubiaceae that they appear to be absent)

3a Stem leaves 3, in a single whorl; flower solitary on each stem; pistils many; style and stigma 1; stamens more than 20 (leaves with 3 leaflets or lobes; sepals 5, petal-like, white, blue, purple, or red; ovary superior; in shaded areas) . . . .
. . . . *Anemone oregana [A. quinquefolia* var. *oregana]* (fig.)
Western Wood Anemone, Western Wind-flower
Mo-n; Ranunculaceae, Buttercup Family

3b Stem leaves usually in more than 1 whorl; flowers several on each stem; pistil 1; styles or stigmas 2–5; stamens not more than 20

4a Petals 3 or 4, united; sepals appearing to be absent because there are no obvious calyx lobes; stamens 3 or 4; styles 2; ovary inferior; stem 4 sided (leaves usually not succulent). . . . . *Galium*
Rubiaceae, Madder Family

4b Petals absent; sepals 5 or 6, petal-like; stamens 3–20; stigmas or styles 3–5; ovary superior; stem not 4 sided

5a Leaves succulent; sepals 5, separate; stamens 3–20; stigmas or styles 3–5
. . . . Molluginaceae, Carpetweed Family

5b Leaves not succulent; sepals 6, united; stamens 3 or 9; styles 3 . . . .
. . . . Polygonaceae, Buckwheat Family

2b Flower with sepals and petals, but sometimes the sepals are similar to the petals

6a Petals 3, sepals 3, often similar to each other; style or stigma 3 lobed (leaves smooth margined, in 1–13 whorls, the upper ones sometimes alternate; flowers 1 to several on each stem; stamens 3 or 6; ovary superior) . . . . Liliaceae, Lily Family (M)

6b Petals 4–7, sepals 2 or 4–7, and not similar to the petals; style or stigma not 3 lobed

7a Leaves toothed (leaves in more than 1 whorl; sepals 5; petals 5, white to dark pink; stamens 10; ovary superior; stigma 5 lobed; in coniferous forests) . . . . .
. . . . *Chimaphila*
Ericaceae, Heath Family

7b Leaves not toothed

8a Sepals 2 (whorl of leaves single, with the lower leaves opposite or alternate; petals 4–6, yellow; stamens 5–20; ovary inferior; style with 4–6 lobes; prostrate) . . . . *Portulaca oleracea* (fig.)
Common Purslane; eu; Portulacaceae, Purslane Family

8b Sepals 4–7

9a Stem leaves in 1 whorl just below the flowers (petals 5–7, white or pink; stamens 5–7; ovary superior) . . . . *Androsace, Trientalis*
Primulaceae, Primrose Family

9b Leaves either in more than 1 whorl or some leaves alternate

10a Petals 4–6, separate, red purple; stamens usually 12; ovary superior; in wet areas . . . . *Lythrum salicaria*
Purple Loosestrife; eu; Lythraceae, Loosestrife Family

10b Petals usually 4, united, pink or lavender; stamens 4 or 5; ovary inferior; not restricted to wet areas . . . .
. . . . *Sherardia arvensis* (pl. 50)
Field Madder; me; Rubiaceae, Madder Family

1b Leaves not in whorls, at least some opposite (CAUTION: species of Urticaceae have stinging hairs.)

11a Flower without petals and sometimes without sepals

12a Leaves palmately lobed (leaf lobes 2; stamens 6; calyx with 5 or 6 lobes, about 1 mm long, pale yellow, pink, or rose; fruit an achene; stems prostrate, forming mats; grows in shade) . . . . *Pterostegia drymarioides*
Polygonaceae, Buckwheat Family

12b Leaves not lobed

13a Either most leaves basal or some alternate, usually the upper ones

14a Upper leaves alternate; leaves usually either scaly or powdery, sometimes finely toothed; flowers often pistillate or staminate, the pistillate ones without sepals but with bracts below them . . . . *Atriplex, Kochia*
Chenopodiaceae, Goosefoot Family

14b Most leaves basal; leaves not scaly or powdery, usually not toothed; flowers with a pistil and stamens, each flower with 6 united sepals (stamens 3 or 9; styles 3) . . . . *Chorizanthe*
Polygonaceae, Buckwheat Family

13b   All leaves opposite

15a   Leaves toothed (most flowers pistillate or staminate)

16a   Leaves coarsely toothed and with stinging hairs; sepals 2–4; pistillate and staminate flowers in leaf axils of the same plant; often more than 100 cm tall . . . . *Hesperocnide, Urtica*
Urticaceae, Nettle Family

16b   Leaves finely toothed and without stinging hairs; sepals 3; staminate flowers at stem tips, pistillate flowers in the leaf axils of separate plants; up to 30 cm tall . . . . *Mercurialis annua*
Mercury; eu; Euphorbiaceae, Spurge Family

15b   Leaves not toothed

17a   Calyx tubular, at least 13 mm long; each flower or cluster of flowers above a cup of united bracts (calyx white, yellow, magenta, or red; stamens usually 3–5; stem nodes often swollen) . . . .
. . . . Nyctaginaceae, Four-o'clock Family

17b   Calyx, if tubular, not more than 7 mm long; flowers and clusters of flowers either without bracts or the bracts not forming a cup

18a   Sepals 5, separate; stem nodes often swollen (stipules often present but not surrounding the stem; stamens 3–5; leaves with 1 prominent vein) . . . . Caryophyllaceae, Pink Family

18b   Sepals 4 or 5, united; stem nodes not swollen

19a   Plant erect or prostrate, forming mats; stipules either absent or surrounding the stem; stamens 3 or many (leaves with 1 prominent vein; growing on dry margins of wetlands or at the edges of saline areas) . . . . *Cypselea, Sesuvium*
Aizoaceae, Seafig Family

19b   Plant erect, not forming mats; stipules absent; stamens 5

20a   Leaf not succulent, usually hairy, with 3–5 prominent veins; annual . . . . *Kochia scoparia*
. . . . [includes K. scoparia var. subvillosa]
Summer-cypress; eua; Chenopodiaceae
Goosefoot Family

20b   Leaf succulent, not hairy, with 1 prominent vein from the base; perennial (restricted to coastal salt marshes)
. . . . *Glaux maritima*
Sea-milkwort; SLO-n; Primulaceae, Primrose Family

11b   Flower with petals and sepals

21a   Petals united, at least at their bases

22a   Stamens 2 or many

23a   Stamens 2; ovary superior; petals appearing to be 4 (the two upper ones are united), white, lavender, blue, or violet; corolla slightly irregular; leaves not succulent, often toothed. . . . . *Veronica*
Scrophulariaceae, Snapdragon Family

23b Stamens many; ovary inferior; petals many, pink or purple; corolla not irregular; leaves succulent, not toothed (in coastal habitats) . . . . *Aptenia cordifolia [Mesembryanthemum cordifolium]*

Baby Sunrose; af; Aizoaceae, Seafig Family

22b Stamens 3–5 (ovary superior)

24a Stem often 4 sided; bract present beneath each flower (corolla with 4 or 5 lobes, white, red, lavender, purple, or blue; stamens 4; leaves toothed or lobed; fruit consisting of 4 nutlets) . . . . . . Verbenaceae, Verbena Family

24b Stem not 4 sided; bract absent beneath flowers

25a Leaves lobed (corolla 5 lobed; fruit a capsule)

26a Leaves palmately lobed; style 1, stigmas 3 (corolla white, yellow, or pink) . . . . *Linanthus*

Polemoniaceae, Phlox Family

26b Leaves pinnately lobed; style 1, 2 lobed . . . . Hydrophyllaceae, Waterleaf Family

25b Leaves not lobed

27a Pistil 1, stigma not lobed (corolla 5 lobed; stamens 5)

28a Leaves hairy, the upper ones alternate; corolla white, sometimes yellow inside; fruit consisting of 4 nutlets . . . . Boraginaceae, Borage Family

28b Leaves not hairy, usually all stem leaves opposite; corolla pinkish orange or blue; fruit a capsule . . . . *Anagallis arvensis* (pl. 45)

Scarlet Pimpernel, Poor-man's Weatherglass; eu
Primulaceae, Primrose Family

27b Pistils 3–5 or stigma with 2 or 3 lobes

29a Pistils 3–5; leaves succulent (corolla with 3–5 lobes) . . . . *Crassula*

Crassulaceae, Stonecrop Family

29b Pistil 1, stigma with 2 or 3 lobes; leaves not succulent

30a Corolla not twisted in bud, white, yellow, or pink, 5 lobed; stigma 3 lobed . . . . *Phlox gracilis, Phlox speciosa* ssp. *occidentalis*

Polemoniaceae, Phlox Family

30b Corolla usually twisted in bud, yellow, pink, or blue, with 4 or 5 lobes; stigma usually 2 lobed . . . . Gentianaceae, Gentian Family

21b Petals separate

31a Pistils 3–5, or deeply 5 lobed (ovary superior)

32a Pistil deeply 5 lobed; stamens 10, 5 sometimes antherless; leaves not succulent, but toothed, lobed, or palmately compound; corolla lavender or red, sometimes pale . . . . Geraniaceae, Geranium Family

32b Pistils 3–5; stamens 4 or 5, all with anthers; leaves succulent, smooth margined; corolla usually yellow . . . . *Crassula, Parvisedum pentandrum*

Crassulaceae, Stonecrop Family

31b Pistil 1, not deeply lobed

33a Leaves lobed or compound (corolla white, yellow, or green)

34a Leaves compound, with 6–12 leaflets; ovary superior (stamens 10; style 1; fruit spiny, separating into 5 units; plant prostrate, forming mats up to 1 m wide) .... *Tribulus terrestris* (pl. 56)
Puncture-vine, Caltrop; me; Zygophyllaceae, Caltrop Family

34b Leaves lobed; ovary inferior or half-inferior

35a Leaves mostly basal but with 2 opposite stem leaves; stamens 10; styles 3; fruit not prickly or bumpy, not separating into units (seeds spiny) .... *Lithophragma cymbalaria*
Mission Starflower, Missionstar; SCl-s
Saxifragaceae, Saxifrage Family

35b Leaf pairs mostly scattered along the stem; stamens 5; styles 2; fruit sometimes prickly or bumpy, separating into 2 units (calyx without lobes, apparently absent) ....
.... *Apiastrum, Bowlesia*
Apiaceae, Carrot Family

33b Leaves not lobed or compound

36a Plant growing at the edges of salt marshes, on sand dunes, or alkaline habitats (leaves not toothed; plant to at least some extent prostrate, forming mats)

37a Petals 4–6, pink, each with a small, tonguelike outgrowth near the middle; leaves not succulent; stamens usually 3; style 3 lobed; ovary superior; plant woody at the base ....
.... *Frankenia salina [F. grandifolia]* (pl. 28)
Alkali-heath; Ma, Sl-s; Frankeniaceae, Frankenia Family

37b Petals many, magenta, pink, or yellow, lacking outgrowths; leaves succulent; stamens many; stigmas 4–20; ovary inferior; plant not woody .... *Carpobrotus, Drosanthemum*
Aizoaceae, Seafig Family

36b Plant usually not growing in the habitats described in choice 36a, or if so, not prostrate

38a Sepals with small teeth alternating with them (stem sometimes 4 sided; leaves not toothed; sepals 4–6; petals 4–6, purple or lavender; stamens 4–6 or 12; ovary superior) ....
.... Lythraceae, Loosestrife Family

38b Sepals without small teeth alternating with them

39a Sepals 2–4

40a Sepals usually 3, usually falling away as the flower opens (leaves not toothed; petals 4 or 6, white, cream colored, or yellow; stamens 8 or more; stigmas either many or 1 and 3 lobed; ovary superior) ....
.... *Meconella californica, Platystemon californicus*
Papaveraceae, Poppy Family

40b Sepals 2 or 4, not falling away as the flower opens

41a Stamens 2 or 8; sepals 2 or 4; petals 2 or 4, white, lavender, or red; stigma either not lobed or 2 or 4 lobed; leaves not succulent, sometimes alternate on upper part of stem; ovary inferior . . . .
. . . . *Circaea, Epilobium*
Onagraceae, Evening-primrose Family

41b Stamens 1–3 or 5–20; sepals 2; petals 4–6, white, pink, or sometimes yellow; stigmas 3; leaves succulent, all opposite on stem, not toothed; ovary superior or half-inferior (annuals) . . . *Claytonia, Montia fontana*
Portulacaceae, Purslane Family

39b Sepals usually 5 (petals usually 5)

42a Leaves with black glands (petals yellow or salmon colored; stamens more than 14, usually in a few clusters; style single, 3 lobed; leaves not toothed; ovary superior)
. . . . Hypericaceae, St. John's-wort Family

42b Leaves without black glands

43a Leaves with 3 prominent veins; ovary half-inferior (leaves toothed; petals white, soon becoming green or falling; plant woody at the base, forming a mat, stems prostrate with tips rising; in woods) . . . . *Whipplea modesta* (fig.)
Yerba-de-selva, Modesty; Mo-n
Philadelphaceae, Mock-orange Family

43b Leaves with 1 prominent vein; ovary superior

44a Petals white, pink, lavender, rose, or red; leaves not toothed; stamens usually 10; styles 2–5 or single and 2 lobed; stem nodes often swollen . . . . . Caryophyllaceae, Pink Family

44b Petals yellow; some leaves toothed; stamens 5; styles 2; stem nodes not swollen (annual; in moist meadows) . . . . *Sclerolinon digynum*
Yellow Flax; Linaceae, Flax Family

Herbaceous Plants, Group 4, Subkey 3: Leaves alternate or basal; corolla, or calyx if corolla absent, essentially regular; plant not grasslike, primarily green and terrestrial, even if occasionally in very wet places or inundated by flooding or high tides

1a Leaf usually with several prominent nearly parallel veins; flower parts in cycles of 3 or 6, except for *Maianthemum* in Liliaceae

    2a Ovary inferior (flowers with 3 sepals similar to the 3 petals; bases of leaves forming sheaths around the stem; plant usually upright) . . . . Iridaceae, Iris Family (M)

    2b Ovary superior

        3a Flower with 3 green sepals and 3 white petals; bases of leaves forming sheaths around the stem; plant creeping, rooting at the nodes . . . . *Tradescantia fluminensis*
Spiderwort; sa; Commelinaceae, Spiderwort Family (M)

3b   Flower either with 3 sepals different from the 3 petals or with 6 petal-like perianth segments that are all alike; bases of leaves not forming sheaths; plant usually upright . . . . Liliaceae, Lily Family (M)

1b Leaf usually with a single prominent vein that originates at the base of the blade, then branches, or with 3 or more prominent veins that diverge from the base of the blade; flower parts generally not in cycles of 3 or 6, but there are some exceptions

4a   Flower with petals or sepals absent, sometimes both

5a   At least some leaves compound, with at least 3 leaflets (ovary superior)

6a   Leaves basal, with petioles about 40 cm long; leaflets 3, fan shaped, pale green, about 8 cm long, with a vanilla-like odor; petals and sepals absent (in moist woodland habitats) . . . . *Achlys triphylla* (pl. 15)

Vanilla-leaf; Sn-n; Berberidaceae, Barberry Family

6b   Leaves not as described in choice 6a; sepals present

7a   Sepals 3–7, separate, often not green; floral tube absent; stamens 10 to many, usually not dark . . . . Ranunculaceae, Buttercup Family

7b   Sepals 4, greenish, united at their bases; floral tube present; stamens 4, purple or black (leaves mainly basal, with 11–17 leaflets, each with 3–7 lobes; floral tube bristly; pistil 1; in open, sandy habitats) . . . .
. . . . *Acaena pinnatifida* var. *californica* [*A. californica*]

California Acaena; Sn-s; Rosaceae, Rose Family

5b   Leaves not compound, except sometimes in *Lepidium* (Brassicaceae)

8a   Leaves and stem usually either scaly or powdery and often at least slightly succulent, or many of the leaves slender, rigid, and prickle tipped (staminate and pistillate flowers sometimes on separate plants) . . . .
. . . . Chenopodiaceae, Goosefoot Family

8b   Leaves and stem generally not scaly or powdery, and if there are slender, rigid leaves, these not prickle tipped

9a   Leaves palmately lobed

10a   Plant small, upright; leaves fan shaped, usually with 3 main lobes, these toothed or lobed; stamen 1; fruit not spiny; calyx 4 lobed; corolla absent; ovary superior . . . .
. . . . *Aphanes occidentalis* [*Alchemilla occidentalis*] (pl. 48)

Western Dewcup, Lady's-mantle; Rosaceae, Rose Family

10b   Plant large, climbing over other plants; leaves not fan shaped; stamens 3 or 4; fruit with soft spines; calyx apparently absent because it lacks conspicuous lobes; corolla 5 lobed; ovary inferior . . .
. . . . *Marah*

Cucurbitaceae, Gourd Family

9b   Leaves either not lobed or not palmately lobed

11a   Some leaves usually toothed or lobed, the basal ones sometimes compound (sepals 4, separate; stamens 2, 4, or 6; ovary superior)
. . . . *Lepidium*

Brassicaceae, Mustard Family

11b   Leaves smooth margined

12a Leaves heart shaped or more or less triangular; ovary inferior or half-inferior (ground cover, the stem prostrate or the tip rising)

13a Leaf blades heart shaped, up to 10 cm wide, with the aroma of ginger when bruised; sepals 3, united, brownish purple and drawn out into slender tails more than 2 cm long; stamens 12; fruit a fleshy capsule; in moist woods . . . .
. . . . *Asarum caudatum* (pl. 6)
Wild-ginger; SCl-n; Aristolochiaceae, Pipevine Family

13b Leaf blades more or less triangular, about 1 cm wide, without the aroma of ginger; sepals 4 or 5, separate, greenish yellow, without tails; stamens 1 to many; fruit dry, with 2–5 horns; usually near the coast, especially around salt marshes . . . . *Tetragonia tetragonioides* (pl. 4)
New Zealand Spinach; au; Aizoaceae, Seafig Family

12b Leaves neither heart shaped nor more or less triangular; ovary superior

14a Each flower with a sharp-tipped bract beneath it (flower pistillate or staminate; sepals 1–5, separate or united at their bases; stamens 3–5) . . . .
. . . . Amaranthaceae, Amaranth Family

14b Each flower usually without a bract, but if bracts present, these not sharp tipped (bracts sometimes present beneath flower clusters)

15a Leaves up to 30 cm long; sepals 5; often more than 200 cm tall (racemes up to 20 cm long; sepals united, white or pink; stamens 10; fruit about 1 cm wide, fleshy, purplish black) . . . . *Phytolacca americana*
Pokeweed, Pigeonberry; na
Phytolaccaceae, Pokeweed Family

15b Leaves not more than 15 cm long; sepals 3–6; usually not more than 50 cm tall

16a Inflorescence not entirely in leaf axils (staminate and pistillate flowers in separate clusters either on the same plant or on separate plants; sepals none or 5 or 6, separate; fruit a capsule) . . . .
. . . . *Croton, Eremocarpus*
Euphorbiaceae, Spurge Family

16b Inflorescence in leaf axils

17a Flowers staminate or pistillate, in separate clusters on same plant; sepals 4, united; fruit an achene (plant without stinging hairs; some flowers with a pistil and stamens) . . . .
. . . . *Parietaria judaica*
Western Pellitory; eua
Urticaceae, Nettle Family

17b Flowers with a pistil and stamens; sepals 5, separate; fruit a capsule (mostly in dried soil bordering ponds) .... *Glinus lotoides*

                sa; Molluginaceae, Carpetweed Family

4b Flower with petals and sepals

18a Petals united, at least at their bases (peel back sepals to examine this)

19a Either each pistil with more than 1 style, or the stigma or style lobed (sepals 5; petals 5)

20a Ovary inferior

21a Plant a vine, with tendrils; leaves lobed; corolla yellow green; stigmas 3, each 2 lobed; stamens 1–5; fruit a red or orange berry .... *Bryonia dioica*

             White Bryony; eu; Cucurbitaceae, Gourd Family

21b Plant not a vine, without tendrils; leaves sometimes toothed but not lobed; corolla white, blue, or violet; style with 2–5 lobes; stamens 5; fruit a capsule (stem 4 sided) .... Campanulaceae, Bluebell Family

20b Ovary superior

22a Styles 2, or if single, the stigma or style 2 lobed (stamens usually 5)

23a Sepals separate or nearly so; plant usually sprawling, but sometimes erect; corolla often more than 2 cm long, the lobes twisted together in bud and with distinct pleats after opening; leaf blade often triangular or with 2 lobes at the base .... Convolvulaceae, Morning-glory Family

23b Sepals usually united; plant usually erect; corolla usually less than 2 cm long, the lobes neither twisted nor pleated; leaf blade rarely triangular or with only 2 lobes .... Hydrophyllaceae, Waterleaf Family

22b Styles 3 or 5, or if single, the stigma or style with 3–5 lobes

24a Styles 5, or if single, the style 5 lobed (sepals 5; petals 5)

25a Leaves compound with 3 leaflets, these lobed; sepals separate or united at their bases; stamens 10; sometimes growing near the coast, but not in salt marshes .... Oxalidaceae, Oxalis Family

25b Leaves not compound; sepals united; stamens 5; mostly growing in salt marshes or on bluffs near the shore (leaves mainly basal) .... Plumbaginaceae, Leadwort Family

24b Styles 3 or if single, the stigma 3 lobed (stamens 5)

26a Leaves coarsely toothed, lobed, or compound; sepals united, less than 1 cm long; corolla less than 3 cm long .... Polemoniaceae, Phlox Family

26b Leaves not toothed, lobed, or compound; sepals separate, more than 1 cm long; corolla 5–6 cm long (plant climbing over other plants) .... *Ipomoea purpurea*

          Garden Morning-glory; sa
          Convolvulaceae, Morning-glory Family

19b Each pistil or pistil lobe with 1 style, neither the stigma nor the style lobed (ovary superior)

    27a Pistils 4 or 5, or 4 or 5 lobed (leaves succulent, not toothed or lobed; corolla 4 or 5 lobed, white, yellow, or red; stamens 8 or 10). . . . .
. . . . *Dudleya, Sedum*
Crassulaceae, Stonecrop Family

    27b Pistil 1, not lobed

        28a Leaves entirely basal (ovary superior)

            29a Corolla 5 lobed

                30a Plant growing on muddy shorelines; leaves not toothed; corolla 1–3 mm long, white, pink, lavender, or bluish, the lobes not turned back; stamens 4; up to 7 cm tall . . . *Limosella*
Scrophulariaceae, Snapdragon Family

                30b Plant not growing on muddy shorelines; leaves sometimes toothed; corolla at least 6 mm long, magenta or white, the lobes turned back; stamens 5; usually more than 10 cm tall
. . . . *Dodecatheon*
Primulaceae, Primrose Family

            29b Corolla 4 lobed

                31a Corolla magenta or white, 6–23 mm long, the lobes turned back; stamens 4 (widespread) . . . .
. . . . *Dodecatheon hendersonii* [includes
D. hendersonii ssp. *cruciatum*] (pl. 45)
Mosquito-bills, Sailor-caps; SB-n
Primulaceae, Primrose Family

                31b Corolla not magenta or white, 1–9 mm long, the lobes not turned back; stamens 2 or 4

                    32a Leaf blades heart shaped; sepals separate; corolla blue to lavender, not dry and membranous; stamens 2 . . . . *Synthyris reniformis*
[includes *S. reniformis* var. *cordata*]
Snowqueen; Ma-n; Scrophulariaceae
Snapdragon Family

                    32b Leaf blades not heart shaped; sepals united; corolla colorless, dry and membranous; stamens usually 4 . . . . . . Plantaginaceae, Plantain Family

        28b Leaves not entirely basal (pistil 1)

            33a Fruit developing into 4 nutlets (look at the base of the pistil to see them); inflorescence often slightly coiled and 1 sided (leaves not toothed or lobed; plant usually bristly) . . . .
. . . . Boraginaceae, Borage Family

            33b Fruit a capsule or berry; inflorescence not coiled or 1 sided

                34a Corolla yellow, without purple spots (corolla 1.5–3 cm wide; inflorescence a raceme; longest leaves 25–50 cm; 1–2 m tall) . . . . *Verbascum*
Scrophulariaceae, Snapdragon Family

34b Corolla either white or purplish, or if yellow, with purple
spots inside

35a Corolla less than 3 mm long, white or pink, with 4 or 5
lobes (inflorescence a raceme; leaves 2–5 cm long; up
to 40 cm tall; growing in moist areas) . . . .
. . . . Primulaceae, Primrose Family

35b Corolla at least 4 mm long, often much longer, white,
purplish, or yellow with purple spots, 5 lobed

36a Leaves 1–5 cm long, the blades rounded and the
petioles longer than the blades; corolla lobes not
twisted in bud; up to 40 cm tall (leaves toothed or
lobed; corolla 5–12 mm long, white with a yellow
center; in moist, rocky habitats near the coast) . .
. . . . *Romanzoffia californica [R. suksdorfii]*
Mistmaiden; SCr-n; Hydrophyllaceae
Waterleaf Family

36b Leaves often more than 5 cm long, the blades not
rounded and the petioles shorter than the blades;
corolla lobes often twisted together in bud; often
up to 150 cm tall . . . .
. . . . Solanaceae, Nightshade Family

18b Petals separate

37a At least some leaves compound (ovary superior)

38a Pistil 1, not lobed

39a Stamens many more than 10 (sepals white or purple green, sometimes pink at the tips; petals white; leaves 3–4 times compound, the
leaflets toothed; fruit a red or white berry) . . . .
. . . . *Actaea rubra* [includes *A. rubra* ssp. *arguta*] (fig.)
Baneberry; SLO-n; Ranunculaceae, Buttercup Family

39b Stamens not more than 10 (fruit a capsule)

40a Petals 6, each with 3 lobes, white or lavender tinged; sepals
12–15, in 2 whorls, the outer 6–9 bractlike; stamens 6 (in coniferous forests) . . . . *Vancouveria planipetala* (pl. 15)
Inside-out-flower, Redwood-ivy; Mo-n
Berberidaceae, Barberry Family

40b Petals 4 or 5, without 3 lobes, white, yellow, pink, or rose; sepals
4 or 5, not in 2 whorls; stamens 6 or 10

41a Sepals 4; petals 4; stamens 6; style 1, the stigma with 1 or 2
lobes . . . . Brassicaceae, Mustard Family

41b Sepals 5; petals 5; stamens 10; styles 5, not lobed (leaflets 3,
each lobed) . . . . Oxalidaceae, Oxalis Family

38b Pistil either more than 1, or nearly separate with 5 lobes

42a Pistil 1, 5 lobed (petals pink to red purple; leaves with stipules; sta-
mens 10; styles 5, elongating in fruit and coiling) . . . .
. . . . *Geranium robertianum* (pl. 29)
Herb-Robert; eu; Geraniaceae, Geranium Family

42b Pistils more than 1, not lobed

    43a Leaves with stipules; sepals alternating with bracts; sepals and petals attached to a floral tube (petals white, cream colored, or yellow; pistils 10 to many; stamens 10–35) . . . . Rosaceae, Rose Family

    43b Leaves without stipules; bracts absent; sepals and petals not attached to a floral tube

        44a Petals thick, maroon or brownish red (petals 8–13 mm long; pistils 2–5; stamens many) . . . *Paeonia brownii* (fig.) Western Peony; SCl-n; Paeoniaceae, Peony Family

        44b Petals thin, white, yellow, or red

            45a Pistils 4 or 5 and petals without spurs; stamens 8 or 10 (in moist places) . . . . Limnanthaceae, Meadowfoam Family

            45b Pistils 5 or many, if 5, then petals with spurs; stamens 5–15 or many . . . . *Aquilegia, Ranunculus* Ranunculaceae, Buttercup Family

37b Leaves not compound

    46a Stamens more than 10

        47a Leaves not toothed or lobed

            48a Sepals 2 or 6–8, all similar; petals white, pink, or red (petals 5 or 10–19; styles with 2–8 lobes; ovary superior or half-inferior) . . . . *Calandrinia ciliata, Lewisia rediviva* Portulacaceae, Purslane Family

            48b Sepals 5, 2 sometimes narrower; petals yellow

                49a Petals many, without orange spots; stigma with 10–20 lobes (sepals unequal; leaves 10–12 mm wide; ovary at least half-inferior) . . . . *Conicosia pugioniformis* af; Aizoaceae, Seafig Family

                49b Petals 5, each with an orange spot at the base; stigma either 3 lobed or not lobed

                    50a Sepals all similar; stigma 3 lobed; ovary inferior . . . . *Mentzelia dispersa* Nevada Stickleaf, Nada Stickleaf Loasaceae, Blazing-star Family

                    50b Outer 2 sepals narrower than the others; stigma not lobed; ovary superior (leaves up to 3 mm wide; in sandy or gravelly areas) . . . . *Helianthemum scoparium* [includes *H. scoparium* var. *vulgare*] (fig.) Peak Rushrose; Me-s; Cistaceae, Rockrose Family

        47b Leaves toothed or lobed

            51a Pistils many (petals 1–22, yellow, sometimes becoming white; ovary superior) . . . . *Ranunculus* Ranunculaceae, Buttercup Family

51b　Pistil 1

　　　52a　Pistil with 2–7 prominent lobes (leaves lobed; petals white, 4–6 mm long, each 3 lobed; stamens 12–15, these on 1 side of the flower; ovary superior) .... *Reseda alba*
　　　　　White Mignonette; me; Resedaceae, Mignonette Family

　　　52b　Pistil without prominent lobes

　　　　　53a　Stamens united to form a cylinder around the pistil (stamens many; sepals 5, united; petals 5; ovary superior) .... Malvaceae, Mallow Family

　　　　　53b　Stamens not united to form a cylinder around the pistil

　　　　　　　54a　Sepals 2 or 3, sometimes forming a cap over the other flower parts until the flower opens; petals 4 or 6, white, yellow, cream colored, orange, or red; ovary superior; stigma with 4–20 lobes; stamens 12 to many ...Papaveraceae, Poppy Family

　　　　　　　54b　Sepals 5, not forming a cap; petals 5, yellow; ovary inferior; stigma 3 lobed; stamens at least 20 .... Loasaceae, Blazing-star Family

46b　Stamens 10 or fewer

　　55a　Pistils either many, 4 or 5, or 1 with 5 nearly separate lobes (ovary superior)

　　　　56a　Pistils many, on a long, slender receptacle; each sepal with a spur near the base (petals 1–3 mm long, white, green, or yellow; stamens usually 10; leaves smooth margined; in wet areas; up to 12 cm tall) .... *Myosurus minimus* (fig.)
　　　　　Common Mousetail; Ranunculaceae, Buttercup Family

　　　　56b　Pistils 4 or 5, or 1 with 5 nearly separate lobes, the receptacle not long and slender; sepals without spurs

　　　　　　57a　Leaves succulent, not toothed or lobed; petals white or yellow; pistils 4 or 5, sometimes fused at their bases; stamens 8 or 10 .... *Sedum*
　　　　　　　Crassulaceae, Stonecrop Family

　　　　　　57b　Leaves not succulent, but lobed; petals white, pink, purple, or red; pistil 1 with 5 nearly separate lobes; stamens 10 .... *Geranium*
　　　　　　　Geraniaceae, Geranium Family

　　55b　Pistils 1 or 2

　　　　58a　Sepals 2 (leaves not toothed or lobed; petals white, yellow, pink, red; stamens 3–10; style or stigma 2 or 3 lobed; ovary superior) .... Portulacaceae, Purslane Family

　　　　58b　Sepals 4–7

　　　　　　59a　Petals 4

　　　　　　　60a　Stamens 4 or 8; ovary inferior; stigma 4 lobed or not lobed; petals white, yellow, pink, red, or lavender .... Onagraceae, Evening-primrose Family

60b Stamens 3 or 6, sometimes 2 or 4; ovary superior; stigma 2 lobed or not lobed; petals white, cream colored, yellow, or purple ....
.... Brassicaceae, Mustard Family
59b Petals 5
61a Leaves not sessile, mostly basal, the stem leaves very reduced or absent; leaf blades often rounded and toothed or lobed; petals white or pink; antherless stamens, if present, lobed; ovary superior or half-inferior .... Saxifragaceae, Saxifrage Family
61b Leaves sessile, not basal; leaf blades narrow, not toothed or lobed; petals white, yellow, pink, or blue; antherless stamens, if present, not lobed; ovary superior .... Linaceae, Flax Family

Flowering Plant Species

This section includes all the higher plant families found in the region except for the ferns, fern allies, and cone-bearing plants, which are included in earlier parts of the book.

The families are arranged alphabetically in one of the following two sections, "Monocotyledonous Families" or "Dicotyledonous Families," based on the criteria in the key below. When the seed of a monocotyledonous plant germinates, a single leaflike structure (cotyledon) emerges, whereas in a dicotyledonous plant two cotyledons emerge. Other features that separate the two categories are found in the microscopic anatomy of the stems.

1a Flower parts (petals, sepals, stamens) commonly in cycles of 4, 5, or numerous; leaf usually with a single prominent vein (midrib) that originates at the base of the blade, then branches, or with 3 or more prominent veins that diverge from the base of the blade......
.... FLOWERING PLANTS, DICOTYLEDONOUS FAMILIES
1b Flower parts commonly in cycles of 3 or 6; leaf usually with several prominent nearly parallel veins......................... FLOWERING PLANTS, MONOCOTYLEDONOUS FAMILIES

Dicotyledonous Families

ACERACEAE (MAPLE FAMILY) Aceraceae consists of trees and shrubs with deciduous, opposite leaves and opposite branches. In the species in our area, the flower has five separate sepals and either five separate petals or none. The flowers, generally in drooping racemes, may be strictly staminate or strictly pistillate or may have stamens and a pistil. The pistil, with two styles, develops into a dry fruit that consists of two broad-winged structures called samaras. Each of these usually contains a single seed. The samaras, while still together or after separating, are scattered with the help of wind.

Besides *Acer macrophyllum* and *A. negundo* var. *californicum*, California has *A. glabrum* and *A. circinatum*. All are excellent subjects for gardens in places that are not especially dry. *A. circinatum* (Vine Maple) is particularly valuable because its height usually does not exceed 6 m and because its leaves, before falling in autumn, often turn orange or red.

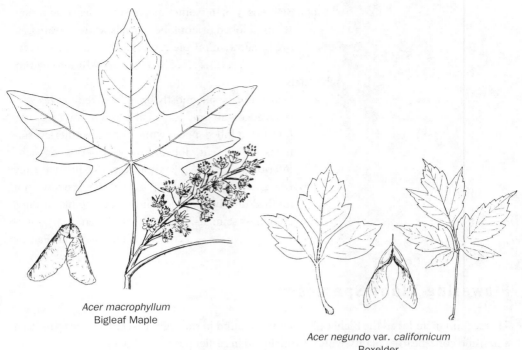

Acer macrophyllum
Bigleaf Maple

Acer negundo var. californicum
Boxelder

1a   Leaves palmately lobed; flower with petals; seed-containing portion of each samara about one-fifth the length of the samara (leaves up to about 25 cm wide; widespread) . . . . *Acer macrophyllum* (fig.) Bigleaf Maple

1b   Leaves pinnately compound, with 3–5 leaflets; flower without petals; seed-containing portion of each samara about one-third the length of the samara (leaflets 3–10 cm long) . . . . *Acer negundo* var. *californicum* (fig.) Boxelder

AIZOACEAE (SEAFIG FAMILY)   The Seafig or Fig-marigold Family has many species, mostly natives of southern Africa, but these are apportioned to relatively few genera. Nevertheless, the group is difficult to define, partly because some of the genera show affinities with some genera in the family Molluginaceae. The following summary of characters applies to our representatives.

In most species, the leaves are succulent and the ovary is at least half-inferior. Some species have many narrow petals, but others have none. Stamens tend to be numerous, and there are several stigmas or styles.

Tetragonia tetragonioides (New Zealand Spinach), cultivated to some extent as a vegetable, has become established in many places. Two species of *Carpobrotus*, commonly known as iceplant, have distinctive three-sided, succulent leaves. These plants should be known to just about everyone familiar with the region, for they have been extensively planted for control of erosion on banks of highways. One species, *C. chilensis*, grows on coastal bluffs and the backshores of sandy beaches; it may have been on the Pacific coast before European explorers arrived, though the rest of the species of *Carpobrotus* and those closely related genera are native to southern Africa.

1a   Flower without petals, but the 3–5 sepals may be petal-like

    2a   Leaves alternate, 2–5 cm long, with nearly arrowhead-shaped blades; leaves and other parts of the plant with glassy bumps (leaves succulent; stamens many; sepals greenish yellow; ovary inferior; usually sprawling to some extent; near the shore) . . . . *Tetragonia tetragonioides* (pl. 4)
        New Zealand Spinach; au

    2b   Leaves opposite, 0.5–4 cm long, with oval or spatula-shaped blades; leaves and other parts of the plant without glassy bumps

        3a   Leaves spatula shaped, succulent, up to about 2.5 cm long; sepals rose; stamens many; ovary half-inferior; in saline habitats, inland as well as at the coast; up to about 50 cm tall. . . . . *Sesuvium verrucosum*
            Western Sea-purslane

        3b   Leaves narrowly oval, not obviously succulent, about 1 cm long; sepals greenish; stamens usually 3; ovary superior; on mud banks, inland; compact, prostrate, forming small mats . . . . *Cypselea humifusa;* sa

1b   Flower with numerous narrow petals (stamens many; ovary inferior or half-inferior)

    4a   Leaves alternate (leaves 15–20 cm long; petals yellow, separate) . . . . *Conicosia pugioniformis;* af

    4b   Leaves opposite

        5a   Leaves differentiated into blade and petiole, the blade oval; petals united at their bases; corolla about 1.5 cm wide (leaves 1–3 cm long; petals purple; along the coast) . . . . *Aptenia cordifolia [Mesembryanthemum cordifolium]*
            Baby Sunrose; af

        5b   Leaves not obviously differentiated into blade and petiole, nearly cylindrical or 3 sided; petals separate; corolla generally at least 2 cm wide (on sand dunes and coastal bluffs)

            6a   Leaves nearly cylindrical, less than 1.5 cm long (petals bright pink or rose) . . . . *Drosanthemum floribundum [Mesembryanthemum floribundum]*
                Rosea Iceplant, Magic-carpet; af

            6b   Leaves 3 sided, 4–10 cm long

                7a   Leaves 6–10 cm long, slightly curved, the lower edges with fine teeth; corolla 7–10 cm wide, pink or yellow when fresh, but pink or purple after wilting (introduced for binding sand and now well established) . . . . *Carpobrotus edulis [Mesembryanthemum edule]* (pl. 4)
                    Hottentot-fig, Iceplant; af

                7b   Leaves 4–7 cm long, straight, the lower edges smooth margined; corolla 3–5 cm wide, magenta (may be native). . . . . *Carpobrotus chilensis [Mesembryanthemum chilense]* (pl. 4)
                    Iceplant, Seafig; af?

AMARANTHACEAE (AMARANTH FAMILY)   All the amaranths in our area belong to a single genus, and all of them are annual with alternate, smooth-margined leaves. Many or these plants are weedy immigrants. The flowers, concentrated in dense inflorescences, lack petals, but they have one to five sepals, which are sometimes united at their bases. Amaranths also have some sharp-tipped bracts just below the sepals, and they have three to five stamens. The fruit is dry when mature, and it contains a single, shiny, dark seed. The ovary is superior.

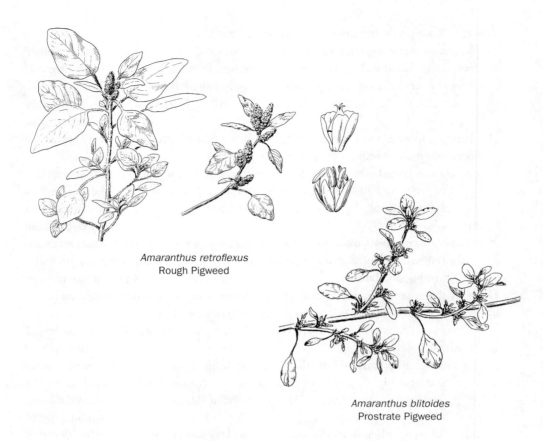

Amaranthus retroflexus
Rough Pigweed

Amaranthus blitoides
Prostrate Pigweed

Distinctions between species are based to a considerable extent on the form and proportion of the sepals and bracts of the staminate and pistillate flowers. In fact, the name of the family comes from a word meaning "unfading," a reference to the persistent sepals and bracts.

Celosia argentea (cockscomb) is grown for its lush, usually pink or red inflorescences, and species of *Amaranthus* have been harvested by Native Americans for food. But this family is generally not popular with gardeners.

1a   Plant low, the stem mostly lying on the ground
    2a   Inflorescence up to 10 cm long, at the ends of stems; blades of larger leaves generally widest in the lower half (widespread, sometimes growing in cracks in pavement) . . . .
. . . . *Amaranthus deflexus* (pl. 4)
Low Amaranth; eu
    2b   Inflorescence not more than 1.5 cm long, in the leaf axils; blades of larger leaves generally widest near or above the middle
       3a   Staminate flower usually with 3 sepals, pistillate flower with only 1 well-developed sepal; staminate flower with 1–3 stamens . . . . *Amaranthus californicus*
California Amaranth
       3b   Staminate and pistillate flowers usually with 4 or 5 sepals; staminate flower with 3 or 4 stamens. . . . . *Amaranthus blitoides* (fig.)
Prostrate Pigweed

1b   Plant upright, even if sometimes branching mostly from the base

    4a   Blades of larger leaves not often so much as 3 cm long and usually widest above the middle; plant bushy, with many branches arising at the base; inflorescences not more than 1.5 cm long, in the leaf axils; flower with 3 sepals (stem pale) . . . . *Amaranthus albus*
                                                        Tumbleweed; sa

    4b   Blades of larger leaves commonly more than 4 cm long (sometimes 10 cm long), and usually widest near or below the middle; plant not bushy, usually branching above rather than at the base; inflorescences usually much more than 1.5 cm long, sometimes at the tip of the stem as well as in the leaf axils; flower with 5 sepals (these of unequal length)

        5a   Sepals blunt or notched at the tips and often with a sharp extension

            6a   Styles upright or curving inward; inflorescence green, not drooping (widespread). . . . . *Amaranthus retroflexus* (fig.)
                                                    Rough Pigweed; sa

            6b   Styles curving downward; inflorescence usually red or reddish, drooping (garden flower, sometimes escaping) . . . . *Amaranthus caudatus*
                                    Quilete, Love-lies-bleeding; sa

        5b   Sepals narrowing gradually to a point at the tips

            7a   Styles usually curving downward. . . . . *Amaranthus powellii*
                                        Powell Amaranth

            7b   Styles upright

                8a   Bracts of pistillate flower about 1–1.5 times the length of the sepals; sepals 1.5 mm long . . . . *Amaranthus cruentus*
                                  South American Amaranth; sa

            8b   Bracts of pistillate flower nearly or fully twice the length of the sepals; sepals up to 2 mm long. . . . . *Amaranthus hybridus*
                        Green Amaranth, Slender Pigweed; eua

ANACARDIACEAE (SUMAC FAMILY)   Shrubs, vines, and trees of the Sumac or Cashew Family have a milky or resinous, often pungent sap. The leaves are alternate, and the flowers are small. In some species, they have stamens and a pistil; in others, staminate and pistillate flowers are on separate plants. This family includes *Mangifera indica* (mango; southeast Asia), *Pistacia vera* (pistachio; probably Iran), and *Anacardium occidentale* (cashew; Central and South America).

    Our representatives have compound leaves and flowers with a five-lobed calyx and a five-lobed corolla; the sepals and petals are usually united at their bases. There are 5 or 10 stamens. *Toxicodendron diversilobum* (Western Poison-oak), a plant that grows throughout our region, should be avoided because contact with it brings on mild to severe dermatitis in most people.

1a   Trees up to 25 m tall, evergreen; leaflets usually more than 15, peppery smelling (leaves 10–30 cm long; fruit leathery, reddish; extensively cultivated and sometimes escaping) . . . . *Schinus molle*
                                    Peruvian Peppertree; sa

1b   Shrubs or vines not more than 3 m tall, deciduous; leaflets 3 or occasionally 5, not peppery smelling, but may be aromatic

2a   Leaflets rounded, more or less equal; corolla and fruit whitish; shrub or vine (widespread; CAUTION: contact with any part of this plant, even its bare stem, may result in a severe rash.) . . . . *Toxicodendron diversilobum [Rhus diversiloba]* (pl. 4)

Western Poison-oak

2b   Leaflets triangular, the end 1 larger than the 2 side ones; corolla greenish, fruit red; shrub. . . . . *Rhus trilobata* [includes *R. trilobata* vars. *malacophylla* and *quinata*] (pl. 4)

Skunkbrush, Squawbush

APIACEAE (CARROT FAMILY)   Apiaceae, or Umbelliferae, includes many familiar vegetables and herbs such as carrot, parsley, coriander, parsnip, celery, fennel, dill, and anise. Some species in this family are deadly poisonous, and others may irritate the skin. It is therefore prudent to handle unfamiliar plants carefully.

Members of the Carrot Family usually have alternate leaves, with petioles that are dilated near their bases into sheaths that clasp the stems. The stems are often ribbed and hollow. At least parts of these plants usually have an aroma due primarily to various oils.

The flowers are nearly always concentrated in flat-topped inflorescences called umbels, which are usually compound. The rays of the primary umbels give rise to secondary umbels that consist of pedicels bearing the flowers. The primary umbels may have bracts beneath them, and the secondary umbels may have bractlets. Each flower has five stamens, two styles, and five petals, which may be slightly unequal, especially in flowers that are located along the outer edges of the umbel. Sepals are often absent. The ovary is inferior. The fruit consists of a pair of one-seeded divisions that are for a time joined tightly to a central partition. They are often conspicuously ribbed, and in certain species, the ribs, or some of them, are drawn out into membranous wings.

1a   Plant somewhat resembling thistles, the bracts and often the leaves stiff and prickly; inflorescence a dense head (corolla greenish; usually in wet places, such as vernal pools)

    2a   Bracts of primary umbels with smooth margins, prickly only at the tips; coastal . . . . *Eryngium armatum* (fig.)

Coyote-thistle, Coast Eryngo

    2b   Bracts of primary umbels with spiny margins; Coast Ranges

        3a   Styles on fruit much longer than the sepals . . . . *Eryngium aristulatum* var. *aristulatum*

Button Celery

        3b   Styles on fruit about the length of the sepals . . . . . *Eryngium aristulatum* var. *hooveri*

Hoover Button Celery; SFBR; 1b

1b   Plant not resembling thistles, neither leaves nor bracts stiff and prickly; inflorescence usually not a dense head

    4a   Leaves reduced to hollow petioles, which are cross barred and appear jointed (corolla white or maroon; small, creeping perennial; in marshlands)

        5a   Leaves cylindrical or flattened, usually more than 1 mm thick, with 4–13 well defined cross walls. . . . . *Lilaeopsis occidentalis* (fig.)

Western Lilaeopsis; Sl, Ma-n

        5b   Leaves cylindrical, usually less than 1 mm thick, with 3–8 poorly defined cross walls. . . . . *Lilaeopsis masonii*

Mason Lilaeopsis; SFBR; 1b

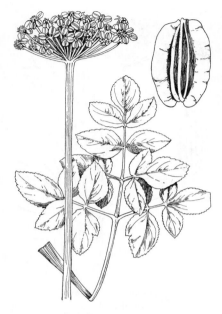

Angelica hendersonii
Henderson Angelica

Apiastrum angustifolium
Wild Celery

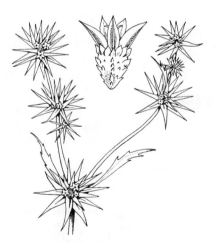

Eryngium armatum
Coyote-thistle

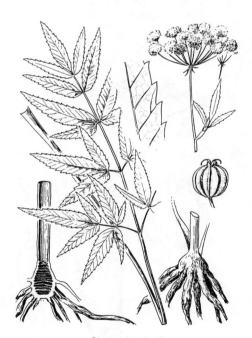

Cicuta douglasii
Douglas Water Hemlock

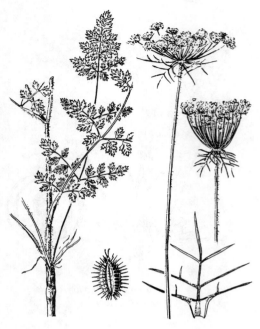

Daucus carota
Queen-Anne's-lace

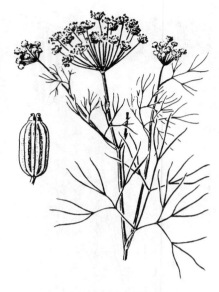

Foeniculum vulgare
Sweet Fennel

Perideridia gairdneri
Gairdner Yampah

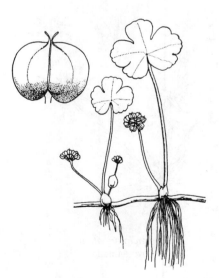

Hydrocotyle ranunculoides
Water Pennywort

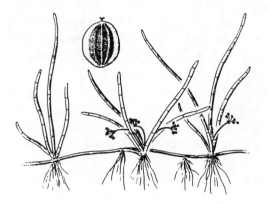

Lilaeopsis occidentalis
Western Lilaeopsis

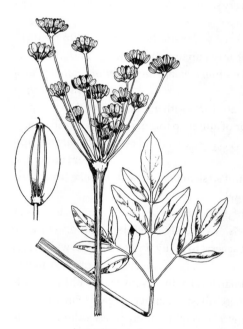

Lomatium nudicaule
Pestle Parsnip

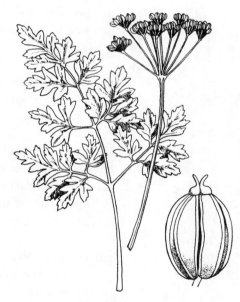

Ligusticum apiifolium
Lovage

4b Leaves with well-developed blades

    6a Leaves usually opposite (blades mostly 5 lobed; fruit with star-shaped hairs; umbels in leaf axils, with 1 or few flowers; corolla yellowish green; annual; usually growing in shade) . . . . *Bowlesia incana;* Sn

    6b Leaves alternate, except sometimes in *Apiastrum* (subkey 1)

        7a Plant small, creeping, and aquatic, with nearly circular or kidney-shaped leaf blades (leaves either lobed or with shallow, rounded teeth)

            8a Leaf blades kidney shaped, lobed; petiole attached at the point of separation of 2 leaf lobes . . . . *Hydrocotyle ranunculoides* (fig.)

                        Water Pennywort, Floating Marsh Pennywort

            8b Leaf blades nearly circular, with shallow rounded teeth; petiole attached near the center of the blade

                9a Inflorescence a simple umbel . . . . *Hydrocotyle umbellata*

                                  Marsh Pennywort

                9b Inflorescence consisting of several whorls of flowers . . . .
. . . . *Hydrocotyle verticillata* [includes *H. verticillata* var. *triradiata*]

                              Whorled Marsh Pennywort; SFBR

        7b Plant, if aquatic, not especially small and not with nearly circular leaf blades

            10a Ultimate divisions of leaves very slender, threadlike (bracts and bractlets usually absent; leaves not hairy)

                11a Corolla white; rays 1–3, each with 6–20 flowers; leaves up to 10 cm long, without an odor of anise; up to 0.5 m tall . . . .
. . . . *Ciclospermum leptophyllum* [*Apium leptophyllum*]

                                  Marsh Parsley; sa

                11b Corolla yellow; rays 15–40, each with 18–25 flowers; leaves up to 30 cm long, with an odor of anise; up to 2 m tall (widespread) . . . . .
. . . . *Foeniculum vulgare* (fig.)

                                  Sweet Fennel; me

            10b Ultimate divisions of leaves not so slender as to be threadlike (they may, however, be extremely narrow)

                12a Leaves similar to those of parsley in having small ultimate leaflets or lobes, most of these divisions usually much less than 0.5 cm wide and 1.5 cm long and not several to many times as long as wide . . . . .
. . . . APIACEAE, SUBKEY 1

                12b Leaves not like those of parsley, having relatively few ultimate leaflets or lobes, most of these divisions either at least 1 cm wide and 1.5 cm long, or, if narrower, several to many times as long as wide . . . . APIACEAE, SUBKEY 2

Apiaceae, Subkey 1: Leaves similar to those of parsley in having small ultimate leaflets or lobes, most of these divisions usually much less than 0.5 cm wide and 1.5 cm long and not several to many times as long as wide

1a Corolla usually white, sometimes reddish or purple

2a   Bracts present beneath primary umbels

    3a   Stem usually purple spotted; stem and leaves not hairy; fruit without bristles or spines; often more than 2 m tall (bracts not lobed or compound; leaves 15–30 cm long; widespread; CAUTION: extremely poisonous, the deadly hemlock of ancient literature; may also cause a skin rash if handled) . . . . *Conium maculatum*
Poison-hemlock; eu

    3b   Stem not purple spotted; stem and leaves hairy; fruit with bristles or spines; usually less than 1 m tall

        4a   Bracts 1 or 2, not lobed or compound; leaves up to 6 cm long; pedicel up to 4 mm long (bractlets several; rays more or less equal; fruit with hooked bristles; up to 1 m tall) . . . . *Torilis arvensis*
Hedge Parsley; eu

        4b   Bracts 2 or several, lobed or compound; leaves usually more than 5 cm long; pedicel up to 14 mm long

            5a   Rays 1–9, unequal; leaves 4–10 cm long; fruit with hooked bristles; bracts and bractlets not more than 5 (up to 40 cm tall) . . . .
. . . . *Yabea microcarpa [Caucalis microcarpa]*
California Hedge Parsley

            5b   Rays many, more or less equal; leaves 7–35 cm long; fruit with spines; bracts and bractlets usually more than 5

                6a   Bracts shorter than rays, and with undivided lobes; secondary umbels usually with more than 15 flowers; 1 corolla in each umbel purple; biennial, often more than 75 cm tall (widespread) . . . .
. . . . *Daucus carota* (fig.)
Queen-Anne's-lace, Wild Carrot; eu

                6b   Bracts as long as rays and usually with divided lobes; secondary umbels with fewer than 13 flowers; all corollas white; annual, rarely more than 50 cm tall (coastal plants more compact and fleshier than inland plants) . . . . *Daucus pusillus*
Rattlesnake-weed

2b   Bracts absent beneath primary umbels

    7a   Bractlets beneath secondary umbels 0–5

        8a   Bractlets absent (leaves sometimes opposite; inflorescence branched; primary umbels usually sessile in upper leaf axils; secondary umbels with only a few flowers; fruit about 1.5 mm long, heart shaped, without a beak; annual) . . . . *Apiastrum angustifolium* (fig.)
Wild Celery; La-s

    8b   Bractlets 2–5

        9a   Bractlets conspicuous; fruit 30–70 mm long, usually bristly, the beak 20–60 mm long; up to 30 cm tall. . . . . *Scandix pecten-veneris* (pl. 5)
Shepherd's-needle, Venus'-needle; me

        9b   Bractlets inconspicuous; fruit 2–5 mm long, with both hooked bristles and rounded bumps, without a beak; often somewhat sprawling. . . . . . .
. . . . *Torilis nodosa*
Knotted Hedge Parsley; eua

7b   Bractlets beneath secondary umbels more than 5

    10a   Bractlets turned down (fruit with slender, hooked bristles and with a beak about one-third the length of fruit body) .... *Anthriscus caucalis* [*A. scandicina*]

Bur-chervil; eua

    10b   Bractlets not turned down

        11a   Leaves scattered along the stem; fruit 3–5 mm long, with hooked bristles and rounded bumps; annual .... *Torilis arvensis*

Hedge Parsley; eu

        11b   Leaves mostly basal; fruit more than 5 mm long, without hooked bristles or rounded bumps; perennial

            12a   Pedicel 2–8 mm long; corolla purple (mature fruit twice as long as wide; plant grayish; on serpentine) .... *Lomatium ciliolatum* var. *hooveri*

Hoover Lomatium; Na; 4

            12b   Pedicel up to 20 mm long; corolla white, greenish, or purple

                13a   Corolla and fruit not hairy; mature fruit sometimes more than 3 times as long as wide .... *Lomatium macrocarpum*

Sheep Parsnip, Largefruit Lomatium; SLO-n

                13b   Corolla and fruit hairy; mature fruit rarely more than twice as long as wide .... *Lomatium dasycarpum* (pl. 5)

Hog Fennel

1b   Corolla yellow or straw colored (leaves mostly basal; biennial or perennial)

    14a   Inflorescence branched, the umbels scattered along the branches; bracts 2; fruit not ribbed, but with scattered tall bumps; secondary umbels with fewer than 12 flowers

        15a   Corolla straw or salmon colored; bumps on upper part of fruit ending in bristles or hooks; limited to elevations above 3,000 ft .... *Sanicula saxatilis*

Rock Sanicle; CC(MD), SCl(MH); 1b

        15b   Corolla yellow; bumps on fruit not ending in bristles or hooks; not limited to higher elevations. .... *Sanicula tuberosa*

Tuberous Sanicle, Turkey-pea

    14b   Inflorescence not branched, the umbels at the end of a peduncle; bracts absent; fruit ribbed, but without tall bumps; secondary umbels often with more than 12 flowers

        16a   Plant not hairy; bractlets not more than 1 mm wide; all ribs of fruit at least slightly extended as wings (leaves gray green, the ultimate divisions 1 mm wide)

Cymopterus terebinthinus var. *californicus*
[*Pteryxia terebinthina* var. *californica*]; Sn-n

        16b   Plant usually at least somewhat hairy; bractlets usually at least 2–3 mm wide; only 2 ribs of fruit extended as wings

            17a   Bractlets without papery margins; corolla pale yellow; pedicel up to 14 mm long; ultimate divisions of leaves not more than 7 mm long................ .... *Lomatium macrocarpum*

Sheep Parsnip, Largefruit Lomatium; SLO-n

            17b   Bractlets with papery margins; corolla bright yellow; pedicel up to 9 mm long; ultimate divisions of leaves often more than 10 mm long

18a   Plant with an obvious stem, usually with 2 or 3 leaves along the stem (widespread)..........................*Lomatium utriculatum* (pl. 5)

Spring-gold, Bladder Parsnip

18b   Plant nearly stemless, with only 1 leaf along the stem.................

....*Lomatium caruifolium*

Carawayleaf Lomatium; Me-SLO

Apiaceae, Subkey 2:   Leaves not like those of parsley, with relatively few ultimate leaflets or lobes, most of these divisions either at least 1 cm wide and 1.5 cm long or, if narrower, several to many times as long as wide

1a   Leaves deeply lobed, not truly compound (inflorescence branched, the umbels scattered along the branches; bracts and bractlets present)

    2a   Lower leaves pinnately lobed; corolla usually purple, sometimes yellow (fruit covered with hooked prickles; plant sometimes almost stemless; widespread)..............

....*Sanicula bipinnatifida* (pl. 5)

Purple Sanicle

    2b   Lower leaves palmately lobed; corolla yellow

        3a   Leaf lobes not toothed; maturing fruit prickly, but only in the lower third.......

....*Sanicula maritima*

Adobe Sanicle, Salt Marsh Sanicle; SF-SLO; 1b

        3b   Leaf lobes toothed; maturing fruit prickly to the base

            4a   Leaves yellow green throughout the normal growing season (prostrate; leaf teeth pointed; umbels compact, 1–1.5 cm wide; coastal)...................

....*Sanicula arctopoides* (pl. 5)

Footsteps-of-spring, Yellow-mats; Mo-n

            4b   Leaves green during the normal growing season

                5a   Plant typically branching well above the base; leaf blades 3–12 cm long, the margins evenly toothed; widespread.....*Sanicula crassicaulis* (pl. 5)

Snakeroot, Gambelweed

                5b   Plant typically branching only at the base; leaf blades 1–4 cm long, margins irregularly toothed; coastal....................*Sanicula laciniata*

Coast Sanicle; SLO-n

1b   Leaves compound, at least some divisions clearly separate to the rachis

    6a   Corolla yellow or purple

        7a   Bracts present beneath the primary umbels

            8a   Inflorescence branched, the umbels scattered along the branches (bractlets 1–2 mm wide; corolla yellow; pedicel less than 2 mm long; fruit with hooked bristles; widespread)................................*Sanicula bipinnata*

Poison Sanicle; Me-s

            8b   Inflorescence not branched, the umbels at peduncle tips

                9a   Bractlets beneath secondary umbels present, not threadlike; mostly around drying pools and in grassland (bractlets often toothed; corolla usually yellow, sometimes purplish)...........*Lomatium caruifolium*

Carawayleaf Lomatium; Me-SLO

9b Bractlets beneath secondary umbels sometimes present, threadlike;
mostly on dry slopes

    10a Corolla yellow; pedicel 2–6 mm long; inland . . . .
. . . . *Lomatium marginatum* var. *marginatum*
Yellow Hartweg Lomatium; Na-n

    10b Corolla purple; pedicel 5–12 mm long; Coast Ranges . . . .
. . . . *Lomatium marginatum* var. *purpureum*
Purple Hartweg Lomatium; Na-n

7b Bracts absent beneath the primary umbels, but sometimes present in *Pastinaca*

    11a Bractlets beneath secondary umbels absent, but sometimes present in *Pastinaca*

        12a Leaves not more than once compound, 10–30 cm long; leaflets sometimes deeply lobed; corolla yellow or orange. . . . . *Pastinaca sativa*
Parsnip; eu

        12b Leaves at least partly twice compound, 9–20 cm long; leaflets, if lobed, usually only at the tips; corolla yellow or purple . . . .
. . . . *Lomatium nudicaule* (fig.)
Pestle Parsnip

    11b Bractlets beneath secondary umbels present, but sometimes absent in *Lomatium californicum*

        13a Rays of primary umbels more than 10 (fruit 8–15 mm long; leaflets toothed, lobed, or both)

            14a Some leaves present on the stem above the base; leaflets usually not more than 2 cm wide; fruit not obviously notched at both ends; bractlets, if present, 1–2 mm wide . . . . *Lomatium californicum*
Chu-chu-pate, California Lomatium

            14b Leaves usually entirely basal; leaflets 1–6 cm wide; fruit obviously notched at both ends; bractlets 2–8 mm wide . . . *Lomatium repostum*
Napa Lomatium; Na; 4

        13b Rays of primary umbels 2–10

            15a Some leaves present on the stem above the base; rays 2–5 (fruit 12–20 mm long; mostly in woods) . . . . *Osmorhiza brachypoda*
California Sweet-cicely; CC-s

            15b Leaves basal; rays 5–10

                16a Bractlets usually longer than pedicels; mature fruit 4–7 mm long . . . . *Tauschia hartwegii*
Hartweg Tauschia; CC-s

                16b Bractlets about the same length as, or shorter than, pedicels; mature fruit 3–5 mm long . . . . *Tauschia kelloggii*
Kellogg Tauschia; SCr-n

6b Corolla usually white, sometimes reddish

  17a Leaves only once compound

    18a Leaflets 3, usually more than 10 cm long and deeply lobed; umbels 10–25 cm wide; often more than 1 m tall; in moist habitats (widespread; CAUTION: contact with leaves or stem may lead to a skin irritation) . . . *Heracleum lanatum* (pl. 5)
Cow Parsnip; Mo-n

18b Leaflets more than 3, generally less than 5 cm long, toothed but not lobed; umbels 2–3 cm wide; usually less than 1 m tall; in shallow water or very wet places .... *Berula erecta*
Cutleaf Water Parsnip

17b Leaves at least twice compound (leaflets more than 3)

  19a Bracts beneath primary umbels present (bractlets present)

    20a Nearly all leaflets at least 10 times as long as wide and not lobed or toothed (rays 9–20, 2–6.5 cm long; styles diverging widely, but not curving downward along sides of the conical elevations from which they originate)..................................... *Perideridia kelloggii*
Kellogg Yampah; Mo-n

    20b Most leaflets only a few times as long as wide and usually lobed or toothed (end leaflet sometimes long and narrow and not toothed or lobed)

      21a Rays of primary umbels 5–10, 3–8 cm long; end leaflet long, narrow, not toothed or lobed, other leaflets broader and shallowly lobed .... *Perideridia californica*
California Yampah; CC-SLO

      21b Rays of primary umbels more than 10, up to 3 cm long; all leaflets essentially alike, toothed and sometimes lobed .... *Oenanthe sarmentosa*
Pacific Oenanthe, Water Parsley

  19b Bracts beneath primary umbels usually absent, but sometimes present in all choices under 19a

    22a Bractlets beneath secondary umbels present (leaves 20–60 cm long; rays of primary umbels 1.5–8 cm long; fruit 2–5 mm long)

      23a Nearly all leaflets at least 10 times as long as wide and not lobed or toothed; rays 7–16 (bractlets 1–4 mm long; leaflets 2–12 cm long; styles curving downward along sides of the conical elevations from which they originate) .... *Perideridia gairdneri* (fig.)
Gairdner Yampah; 4

      23b Most leaflets only a few times as long as wide and usually lobed or toothed; rays 12–30

        24a Petioles about as long as leaf blades; leaflets lobed, 1–5 cm long (bractlets 3–7 mm long) .... *Ligusticum apiifolium* (fig.)
Lovage; SM-n

        24b Petioles much shorter than leaf blades; leaflets usually finely toothed but not lobed, 1–10 cm long (secondary veins of leaflets mostly directed toward clefts between the marginal teeth; CAUTION: both species extremely poisonous)

          25a Rays 2–4.5 cm long; bractlets 2–5 mm long; ribs of fruit narrower than darker intervals between them; in coastal salt marshes .... *Cicuta maculata* var. *bolanderi* [*C. bolanderi*]
Bolander Water Hemlock; Ma, Sl, CC

25b Rays 2–8 cm long; bractlets 2–15 cm long; ribs of fruit broader than darker intervals between them; in freshwater habitats . . . . *Cicuta douglasii* (fig.)
Douglas Water Hemlock, Western Water Hemlock

22b Bractlets beneath secondary umbels absent, but sometimes present in *Osmorhiza chilensis*

26a Rays of primary umbels 3–16

27a Rays 3–8, 2–12 cm long; flowers or fruit in each secondary umbel not more than 10; fruit more than 4 times as long as wide (mostly in woods; widespread) . . . . *Osmorhiza chilensis* (pl. 5)
Wood Sweet-cicely, Mountain Sweet-cicely

27b Rays 7–16, not more than 2.5 cm long; flowers or fruit in each secondary umbel usually more than 20; fruit not more than 3 times as long as wide . . . . *Apium graveolens*
Celery, Smallage; eua

26b Rays of primary umbels 20–65 (rays 2–11 cm long)

28a Undersides of leaves usually woolly; coastal . . . . *Angelica hendersonii* (fig.)
Henderson Angelica; Mo-n

28b Undersides of leaves sometimes hairy, but not woolly; inland . . . . *Angelica tomentosa*
California Angelica, Woolly Angelica; SCr-n

APOCYNACEAE (DOGBANE FAMILY) Commonly cultivated plants that belong to the Dogbane Family are *Nerium oleander* (oleander), a shrub or small tree of Mediterranean Europe, and two species of *Vinca* (periwinkle), which are ground covers that sometimes become so well established along roadsides that they look wild. The stems of *Apocynum cannabinum* (Indian Hemp) have long fibers that Native Americans have used for making ropes and cords. Although the genus name refers to its use as a dog poison, some of these plants have also been used for medical purposes.

Most members of the family are perennial with opposite, smooth-margined leaves and milky sap. Both the calyx and corolla have five lobes, but the lobes of the calyx are sometimes separated nearly to their bases, whereas the corolla has a substantial tube. The ovary is superior. The five stamens, alternating with the corolla lobes, are attached to the tube. As the fruit ripens, the two halves of it separate, each half forming a pod that splits open after it has dried.

1a Flower solitary; corolla about 4 cm wide, blue or violet, the 5 lobes all skewed slightly in 1 direction; rooting at the nodes or at the tips of the stem (widely cultivated as a ground cover and often escaping, forming large masses, especially in shaded areas) . . . . *Vinca major* (fig.)
Greater Periwinkle; eu

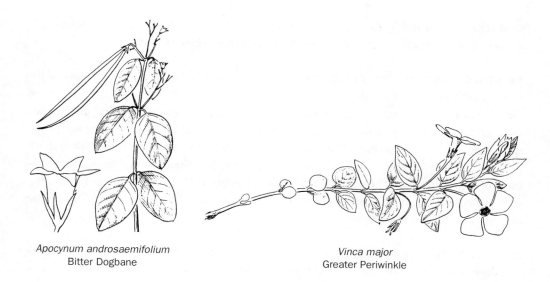

Apocynum androsaemifolium
Bitter Dogbane

Vinca major
Greater Periwinkle

1b Flowers in panicles; corolla less than 1 cm wide, greenish white, white, or pink, the lobes not skewed; bushy, the stem not spreading and taking root (pods up to 10 cm long, releasing seeds with tufts of white hair; sap somewhat poisonous)

    2a Corolla white or greenish white, less than 5 mm long; leaves often more than 3 times as long as wide and not drooping. . . . . *Apocynum cannabinum* . . . . [includes *A. cannabinum* var. *glaberrimum* and *A. sibiricum* var. *salignum*]

                                                        Indian Hemp

    2b Corolla pink, sometimes very pale, mostly 5–10 mm long; leaves rarely more than twice as long as wide, usually drooping . . . . *Apocynum androsaemifolium* [includes *A. medium* var. *floribundum* and *A. pumilum* vars. *pumilum* and *rhomboideum*] (fig.)

                        Bitter Dogbane, Mountain Hemp; Na, Sn-n

ARALIACEAE (GINSENG FAMILY)   Araliaceae, which has about 50 genera, contains many species that are cultivated primarily for their leaves. *Hedera helix* (English ivy), a native of Eurasia, is known to almost everyone, but the family also includes the numerous house plants (species of *Aralia*), as well as *Fatsia japonica,* commonly grown outdoors in milder parts of California. The several species of ginseng, dug for their roots, which have medicinal properties, belong to the genus *Panax.* They are natives of Asia and eastern North America.

The only member of the family growing wild in our area is *Aralia californica* (pl. 6) (Elk-clover or Spikenard). Its large, compound leaves are typically separated into three divisions, each of these with three to five leaflets, these often more than 10 cm long and slightly notched at the base. The sap is milky, and the flowers are usually small and borne in umbels. There are generally five separate petals, five sepals united at their bases, and five stamens. The ovary is inferior. The black fruit is slightly fleshy and about 5 mm wide. Elk-clover is found in moist, tree-shaded canyons, where it reaches a height of almost 3 m. In spite of its large size, however, it qualifies as an herb rather than a shrub, for its stems do not become woody.

ARISTOLOCHIACEAE (PIPEVINE FAMILY) Two very different-looking plants, *Aristolochia californica* (Dutchman's-pipe) and *Asarum caudatum* (Wild-ginger), represent the Pipevine Family in our region. Wild-ginger grows as an evergreen ground cover in shady woodland. *Aristolochia californica,* a deciduous woody vine, climbs over other shrubs. The two plants share several features, however. The alternate leaves have heart-shaped blades with smooth margins. The flower lacks petals, but the sepals, which may be united or not, resemble petals because of their coloration. The ovary is inferior or half-inferior. There are 6 stamens fused to the style in *Aristolochia* and 12 stamens free of the style in *Asarum*. In *Aristolochia californica* the calyx is a curved, two-lipped tube.

1a   Small herbs, forming a ground cover in moist woods; leaf blade wider than long, with a deep cleft at the base; calyx tube cup shaped, the lobes equal and several times as long as wide, purplish brown (plant with aroma of ginger) . . . . *Asarum caudatum* (pl. 6)
Wild-ginger; SCl-n

1b   Woody vines, climbing over shrubs; leaf blade mostly longer than wide, with a shallow cleft at the base; calyx tube deep, curved, thus somewhat resembling the bowl of a smoker's pipe, the lobes not quite equal and not much longer than wide, greenish with purple veins . . . . .
. . . . *Aristolochia californica* (pl. 6)
Dutchman's-pipe, Pipevine; Mo-n

ASCLEPIADACEAE (MILKWEED FAMILY) Milkweeds, so called because of their white sap, are perennial plants mainly with opposite or whorled leaves. The flower is rather intricate. After it opens, the five united sepals and five united petals, turn sharply downward. The five stamens, attached near the base of the corolla, are joined by their filaments to form a tube around the pistil. Just outside this tube is a crownlike structure whose five lobes are called hoods. In some of the species of our only genus, *Asclepias,* there is a further complication: the inner face of each hood has an outgrowth called a horn. Because of the unique flower, the pollination process for milkweeds is highly specialized. The ovary is superior. As the fruit dries it cracks open and releases seeds that have tufts of long, silky hairs.

Some milkweeds are food for caterpillars of certain butterflies, including the Monarch Butterfly *(Danaus plexippus).* The caterpillars and the soon-to-be adult butterflies become distasteful to predators because of toxins accumulated from milkweed sap.

1a   Stem reclining and usually undulating or zigzagging, distinctly flattened; corolla less than 4 mm long (corolla reddish purple; leaves opposite; on serpentine outcrops) . . . .
. . . . *Asclepias solanoana*
Prostrate Milkweed, Solanoa; Na, Sn-n; 4

1b   Stem upright, not flattened; corolla at least 4 mm long
  2a   Hoods without horns (corolla dark red purple or purplish, at least 8 mm long)
    3a   Leaves whitish, woolly, not sessile; hoods with a slitlike opening extending from the top down to about the middle of the outer side (widespread) . . . .
. . . . *Asclepias californica* [includes *A. californica* ssp. *greenei*] (pl. 6)
California Milkweed, Round-hooded Milkweed

3b Leaves green with short hairs, sessile and clasping the stem; hoods open at the top and on the inner side . . . . *Asclepias cordifolia*
Purple Milkweed; Sl-n

2b Hoods with horns

4a Hoods slender, extending far above the stamens and stigma; tube formed by the united filaments scarcely evident; corolla rose purple, at least 8 mm long; leaves opposite . . . . *Asclepias speciosa*
Showy Milkweed, Greek Milkweed; Sl-n

4b Hoods not extending appreciably above the stamens and stigma; tube formed by the filaments well developed; corolla cream or greenish white, less than 6 mm long; leaves opposite or whorled

5a Leaves dark green, at most only slightly hairy, usually at least 6 times as long as wide and rarely more than 1 cm wide, whorled; corolla greenish white . . . .
. . . . *Asclepias fascicularis* (pl. 6)
Narrowleaf Milkweed, California Milkweed

5b Leaves whitish, very hairy, less than 3 times as long as wide and generally at least 3 cm wide, sometimes whorled; corolla cream colored. . . . *Asclepias eriocarpa*
Kotolo, Indian Milkweed; Me-s

ASTERACEAE (SUNFLOWER FAMILY)   In California, as well as in many other parts of the world, Asteraceae, or Compositae, has more species than any other family of flowering plants. Keying an unfamiliar member of this group to genus and species may therefore be a longer process and require more attention to detail than would be necessary to identify a mint or a saxifrage, for example. But recognition of the family itself is easy because of the distinctive structure of the flower head. Each head consists of several to many small flowers attached to a disk-shaped, conical, or concave receptacle. Below and around this receptacle are leaflike bracts called phyllaries. When you are served a cooked Artichoke, you nibble off the more nutritious portions of the phyllaries, then scrape away the flowers and eat the receptacle.

There are two main types of flowers: one in which the corolla is tubular and usually five lobed and one in which one side of the corolla is drawn out into a petal-like structure. In a daisy, flowers with the tubular type of corolla occupy the central part of the head and are called disk flowers; those with the petal-like type of corolla are distributed around the margins and are called ray flowers. In many members of the family there are only disk flowers or only flowers of the ray-flower type. Thus it is convenient to break up the large key for identifying plants of Asteraceae into three main groups: those that have only ray flowers, those that have only disk flowers, and those that have both (fig.).

Sepals, as we know them in other families, are absent or are replaced by scales or hairs that form a structure called a pappus. If present, the pappus usually facilitates dispersal of the small, dry fruit by wind, animals, or other means. The fruit, called an achene, is below the pappus and corolla and is fused tightly to the single seed inside it. When stamens are present, they are attached to the corolla, and their anthers are united into a tube around the style. Certain flowers of the head may lack stamens, others may lack a pistil, and some may lack both.

Of the numerous species of Asteraceae growing in our area, many have been introduced. Some of the introductions are among the most aggressive of weeds. Yet the family has given us hundreds of kinds of garden flowers, food plants, and sources of valuable oils, dyes, and medicines.

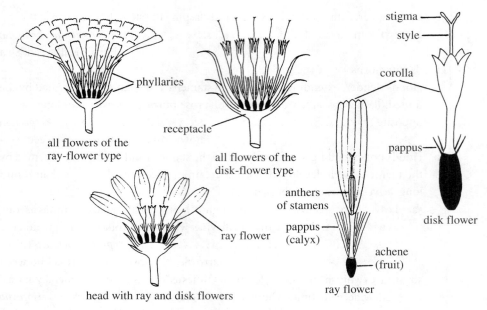

phyllaries

all flowers of the
ray-flower type

receptacle

all flowers of the
disk-flower type

stigma

style

corolla

pappus

anthers
of stamens

pappus
(calyx)

disk flower

ray flower

achene
(fruit)

ray flower

head with ray and disk flowers

ray flower

Asteraceae (Sunflower Family)

Some perennial species that are good subjects for gardens devoted to indigenous plants are listed below.

Baccharis pilularis (Coyotebrush; especially valuable for erosion control)
Erigeron glaucus (Seaside Daisy)
Eriophyllum confertiflorum (Golden Yarrow)
Heterotheca sessiliflora ssp. *bolanderi* (Golden Aster)
Tanacetum camphoratum (Dune Tansy)

Among the many annuals that effectively cover bare ground and make a colorful showing, *Coreopsis calliopsidea* (Leafystem Coreopsis), *Lasthenia californica* (California Goldfields), and *Layia platyglossa* (Tidytips) are of particular interest.

The first step in identifying a member of Asteraceae is to use the preliminary key, which will lead you to one of the three main groups of the family. The shrubs of the Sunflower Family in our area, although included in the following key, are also together in "Trees, Shrubs, and Woody Vines, Group 4."

1a Flowers in each head either all of the ray-flower type or primarily of the disk-flower type (ray flowers, if present, inconspicuous; see choice 2b)

    2a Flowers all of the ray-flower type; plant usually with milky sap (break a leaf tip to see) . . . . ASTERACEAE, GROUP 1

    2b Flowers primarily of the disk-flower type, the rays, if present, either less than 2 mm long or not extending more than 2 mm beyond the phyllaries; plant without milky sap . . . . ASTERACEAE, GROUP 2

1b At least some of the flower heads with disk flowers and conspicuous ray flowers . . . . ASTERACEAE, GROUP 3

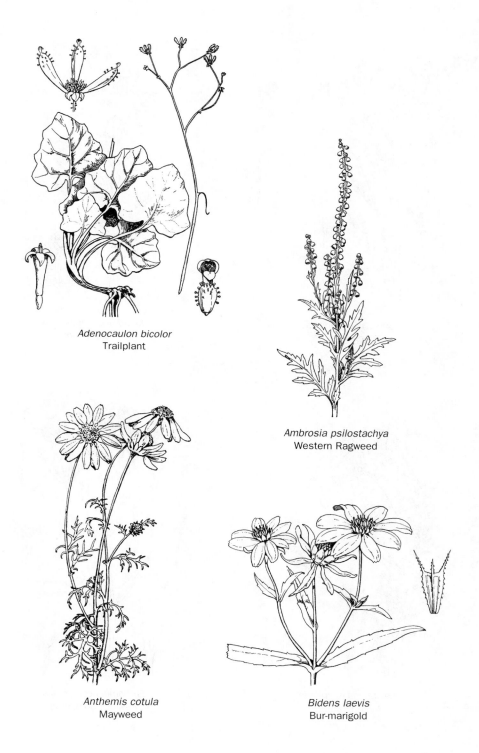

Adenocaulon bicolor
Trailplant

Ambrosia psilostachya
Western Ragweed

Anthemis cotula
Mayweed

Bidens laevis
Bur-marigold

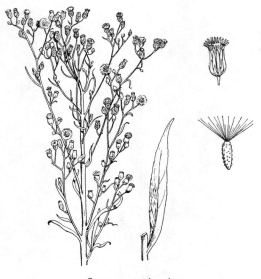

Conyza canadensis
Horseweed

Chamomilla suaveolens
Pineappleweed

Ericameria ericoides
Mock-heather

Euthamia occidentalis
Western Goldenrod

Hieracium albiflorum
Whiteflower Hawkweed

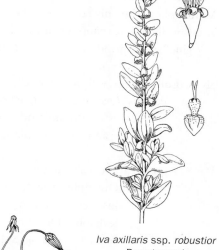

Iva axillaris ssp. *robustior*
Povertyweed

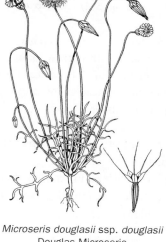

Microseris douglasii ssp. *douglasii*
Douglas Microseris

Petasites frigidus var. *palmatus*
Coltsfoot

Xanthium spinosum
Spanish Thistle

Asteraceae, Group 1: Flowers all of the ray-flower type; plant usually with milky sap (break a leaf tip to see)

1a   Mature plant with only 1 or 2 flower heads at the end of an unbranched stem

    2a   Corolla purple (pappus consisting of straight, brownish bristles; leaves smooth margined; widespread). . . . . *Tragopogon porrifolius* (pl. 15)

                                                         Salsify, Oysterplant; eu

    2b   Corolla orange, yellow, or white

        3a   Pappus consisting of straight bristles at the end of a slender beak

            4a   Leaves not toothed or lobed, scattered along the flowering stem (pappus white; annual or biennial). . . . . *Tragopogon dubius*

                                                         Yellow Salsify; eu

            4b   Leaves usually toothed or lobed, and mostly basal

                5a   Outer phyllaries turned back; achene with short spines at the upper end, just below the slender beak (leaf lobes or teeth pointed toward base of leaf; lobe at leaf tip not much longer than other lobes; usually less than 20 cm tall; widespread) . . *Taraxacum officinale* [includes *T. laevigatum*]

                                                          Common Dandelion; eu

                5b   Phyllaries generally not turned back; achene without spines

                    6a   Leaf blades with narrow lobes or teeth, some pointed toward the base of the leaf (lobe at leaf tip usually twice the length of other lobes; flower head 20–60 mm high; perennial; up to 50 cm tall; at elevations above 2,500 ft) . . . . *Agoseris retrorsa*

                                                           Spearleaf Dandelion

                    6b   Lobes or teeth, if present on the leaf blades, usually not pointed toward the base of the leaf

                        7a   Peduncle of flower head often more than 2–3 times the length of the leaves; annual (leaves 5–15 cm long; flower head 10–25 mm high) . . . . *Agoseris heterophylla*

                                                           Annual Dandelion

                    7b   Peduncle of flower head usually less than twice the length of the leaves; perennial

                        8a   Phyllaries unequal in size and shape, those of the uppermost series taller and narrower than the others; flower head 25–40 mm high; up to 60 cm tall (leaves 10–25 cm long) . . . . *Agoseris grandiflora* (pl. 7)

                                            California Dandelion, Largeflower Agoseris

                        8b   Phyllaries mostly equal in size and shape; flower head 15–20 mm high; up to 45 cm tall (coastal)

                            9a   Beak of achene slightly longer than the rest of the achene; leaves 6–30 cm long, with deeply separated, narrow lobes . . . *Agoseris apargioides* var. *apargioides*

                                                           Seaside Dandelion

9b Beak of achene half the length of the rest of the achene; leaves 2–15 cm long, toothed or with irregular, wide lobes . . . . *Agoseris apargioides* var.*eastwoodiae [A. apargioides var. eastwoodae]* (pl. 6)

Coast Dandelion, Eastwood Agoseris

3b Pappus consisting of flattened scales that end in bristles, the scales sometimes absent in *Microseris douglasii* ssp. *tenella*

10a Some phyllaries 2–3 cm long; flower head upright; achene and pappus together often more than 2.5 cm long (annual; up to 60 cm tall; widespread) . . . . *Uropappus lindleyi [Microseris lindleyi]* (pl. 15)

Silverpuffs

10b Phyllaries not more than 1.5 cm long; flower head nodding in the bud stage, upright when fully open; achene and pappus together usually less than 2.5 cm long

11a Flower head often more than 2 cm wide; perennial; up to 80 cm tall

12a Pappus white, 12–20 mm long; Coast Ranges . . . . *Microseris laciniata*

Cutleaf Microseris; Na, Sn-n

12b Pappus brown or yellowish, 8–13 mm long; coastal . . . . *Microseris paludosa*

Coast Microseris, Marsh Scorzonella; Sn-Mo

11b Flower head less than 1.5 cm wide, except sometimes in *Microseris douglasii;* annual; mostly less than 60 cm tall

13a Pappus scales with 2 small lobes (pappus white to brownish, 14–19 mm long, the scales 2–5 mm long; achene 6–8 mm long; up to 40 cm tall; coastal) . . . . *Stebbinsoseris decipiens [Microseris decipiens]*

Santa Cruz Microseris, Thomas Microseris; Ma; 1b

13b Pappus scales not lobed

14a Pappus 8–18 mm long, white to brownish, the scales 4–11 mm long; flower head with up to 50 flowers; up to 30 cm tall (inland) . . . . *Microseris acuminata*

Sierra Foothills Microseris; Al-n

14b Pappus 4–14 mm long, silvery to blackish, the scales 1–6 mm long; flower head with more than 100 flowers; up to 60 cm tall

15a Mainly coastal, in sandy soil; corolla yellow or orange (pappus scales 1–4 mm long; achene widest near the middle) . . . . *Microseris bigelovii*

Bigelow Microseris, Coast Microseris

15b Mainly inland, in serpentine and clay soils; corolla yellow or white

16a Achene widest near the tip; pappus scales 1–6 mm long . . . . *Microseris douglasii* ssp. *douglasii* (fig.)

Douglas Microseris

16b Achene widest near the middle; pappus scales, if present, less than 1 mm long . . . .
. . . . *Microseris douglasii* ssp. *tenella*
Delicate Douglas Microseris; SFBR-s

1b Mature plant with a few to many flower heads on or near the end of a branched or unbranched stem

17a Corolla purple, blue, pink, cream colored, or white

18a Corolla bright blue, occasionally pink or white; flower head at least 3 cm wide when open (they close in the afternoon); pappus consisting of minute scales (flower heads scattered along stem; up to 1 m tall; widespread) . . . .
. . . . *Cichorium intybus* (pl. 9)
Chicory, Blue-sailors; eu

18b Corolla purple, pale blue, pink, or white; flower head less than 3 cm wide; pappus consisting of bristles

19a Flower head 2–3 cm wide, solitary at the ends of branches (corolla white; upper leaves reduced, sessile; often more than 1 m tall) . . . . . . *Rafinesquia californica*
California Chicory

19b Flower heads 1–1.5 cm wide, not solitary at the ends of branches

20a Flower heads almost sessile along the leafless branches (corolla pink or white; sometimes more than 1 m tall; widespread) . . . .
. . . . *Stephanomeria virgata* (pl. 14)
Tall Stephanomeria

20b Flower heads on prominent peduncles at the ends of branches

21a Lower leaves not distinctly toothed or lobed (corolla white; leaves mostly basal, the petioles winged; less than 1 m tall; widespread) . . .
. . . . *Hieracium albiflorum* (fig.)
Whiteflower Hawkweed

21b Lower leaves toothed or lobed

22a Leaves not mainly basal, not hairy; corolla pale blue or cream colored; often more than 1.5 m tall . . . . *Lactuca biennis*
Tall Blue Lettuce; na

22b Leaves mainly basal with tufts of woolly hairs on the undersides near the margins; corolla white or pink; not more than 0.5 m tall
. . . . *Malacothrix floccifera*
Woolly Malacothrix

17b Corolla yellow or greenish yellow

23a Leaves mostly basal, the stem leaves, if present, generally reduced

24a Leaves bristly, rough to the touch (pappus consisting of bristles)

25a Flower head 2–3 cm high; leaves up to 20 cm long; flowering stem sometimes branched at least once; up to 80 cm tall (widespread) . . . .
. . . . *Hypochaeris radicata* (pl. 12)
Rough Cat's-ear, Hairy Cat's-ear; eu

25b Flower head up to 1.5 cm high; leaves up to 12 cm long; flowering stem not branched; up to 30 cm tall . . . . . . *Leontodon taraxacoides* [*L. leysseri*]
Hairy Hawkbit; eu

24b Leaves not bristly, although they may be hairy
    26a Pappus either absent or of 1–8 bristles (less than 50 cm tall)
        27a Stem branching from the base; leaves with tufts of woolly hairs on the undersides near the margins . . . . *Malacothrix floccifera*
                Woolly Malacothrix
        27b Stem not branching from the base, but there may be branches above; leaves not hairy . . . . *Malacothrix clevelandii*
                Cleveland Malacothrix
    26b Pappus consisting of more than 25 bristles
        28a Involucre 10–16 mm high
            29a Longest leaves 15–40 cm, lobed; up to 70 cm tall (at elevations above 1,500 ft) . . . . *Crepis intermedia*
                Intermediate Hawk's-beard
            29b Longest leaves 2–6 cm, usually not lobed; up to 40 cm tall (opening of flower head limited to periods of bright sunshine) . . . . *Hypochaeris glabra*
                Smooth Cat's-ear; eu
        28b Involucre usually 5–9 mm high, sometimes 11 mm
            30a Leaves toothed or lobed, the lobe at the end of the blade usually at least 2–3 times as long as wide; up to 90 cm tall (widespread) . . . . *Crepis capillaris* (pl. 10)
                Smooth Hawk's-beard; eu
            30b Leaves with definite lobes, the one at the end of the blade not much longer than wide; up to 35 cm tall . . . . . . *Crepis bursifolia*
                Italian Hawk's-beard; me
23b With at least some substantial leaves above the base of the plant
    31a Margins or midveins of leaves, and sometimes other parts of the plant, prickly or bristly (upper leaves clasping the stem; pappus consisting of bristles)
        32a Leaves prickly on midveins and often elsewhere (leaves along the stem and also basal; phyllaries not prickly, in 2 or several series; at least 50 cm tall)
            33a Leaves not prickly on margins, more than 10 times as long as wide . . . . *Lactuca saligna*
                Willow Lettuce; eu
            33b Leaves prickly on margins, less than 5 times as long as wide
                34a Stem not prickly, at least in upper part; leaves scarcely or not at all lobed . . . . *Lactuca virosa*
                    Wild Lettuce; eu
                34b Stem prickly; leaves usually deeply lobed (widespread) . . . . *Lactuca serriola* (pl. 13)
                    Prickly Lettuce; eu
        32b Leaves not prickly on midveins
             35a Phyllaries not prickly; stem not prickly (leaves on the stem and basal; phyllaries in 3 series; often more than 1 m tall; widespread) . . . . *Sonchus asper*
                Prickly Sow Thistle; eu

35b Phyllaries prickly; stem usually prickly

   36a Leaves along stem but not basal and without whitish spots; phyllaries in 1 series, united at their bases; less than 50 cm tall

                    Urospermum picroides; eu

   36b Leaves basal and also along stem and with whitish spots; phyllaries in 2 or 3 series, the outer separate; up to 80 cm tall (widespread) . . . . *Picris echioides* (pl. 13)

                    Bristly Ox-tongue; eu

31b Margins or midveins of leaves not prickly or bristly

  37a Upper leaves clasping the stem; often more than 1 m tall (pappus consisting of bristles)

    38a Leaves either smooth margined or with fewer than 10 lobes or teeth; upper portion of flower stem not branched; flowers 8–12 in each head (involucre 10–17 mm high; basal leaves few) . . . *Lactuca saligna*

                    Willow Lettuce; eu

    38b Leaves with more than 10 teeth or lobes; upper portion of flower stem branched; flower head with more than 15 flowers (widespread)

      39a Stem leaves much reduced compared with basal leaves; leaves lobed, the lobe at the tip not arrowhead shaped; involucre 5–8 mm high . . . . *Crepis capillaris* (pl. 10)

                    Smooth Hawk's-beard; eu

      39b Stem leaves not much reduced compared with basal leaves; leaves toothed and usually lobed, the lobe at the tip often arrowhead shaped; involucre 10–13 mm high . . . . *Sonchus oleraceus*

                    Common Sow Thistle; eu

  37b Upper leaves not clasping the stem; less than 1 m tall

    40a Leaves not more than 1 cm wide (pappus consisting of 1–6 bristles; leaves sometimes toothed or lobed, not hairy) . . . . *Malacothrix clevelandii*

                    Cleveland Malacothrix

    40b Some leaves over 2 cm wide

      41a Pappus consisting of many more than 10 bristles (leaves lobed or toothed; plant with long, glandular hairs) . . . . *Crepis monticola*

                    Mountain Hawk's-beard; SCl-n

      41b Pappus absent

        42a Stem with glandular hairs at the base; flower head 5–10 mm high; upper leaves with obvious teeth; lower leaves lobed as well as toothed . . . . *Lapsana communis*

                    Nipplewort; eu

        42b Stem without glandular hairs at the base (there may be other hairs, however); flower head 10–13 mm high; upper leaves scarcely toothed; lower leaves lobed but not toothed . . . . *Rhagadiolus stellatus;* eu

Asteraceae, Group 2: Flowers primarily of the disk-flower type, the rays if present, either less than 2 mm long or not extending more than 2 mm beyond the phyllaries; plant without milky sap

1a   Involucre with sharp spines (see exceptions in 1b regarding *Ambrosia* and *Soliva*)

    2a   Plant as a whole not spiny except for the involucre

       3a   Involucre with hooked spines

          4a   Flower head decidedly higher than wide, sometimes nearly twice as high as wide; corolla not pink or purple, sometimes absent; pappus absent

             5a   Leaves up to 20 cm long, usually toothed and lobed; stem often spotted red or black . . . . *Xanthium strumarium*

[includes *X. strumarium* vars. *canadense* and *glabratum*]

Cocklebur

             5b   Leaves less than 3 cm long, not toothed or lobed; stem not spotted (staminate and pistillate flowers in the same head; pistillate flowers enclosed by a woolly, boat-shaped bract) . . . . *Ancistrocarphus filagineus [Stylocline filaginea]*

Woolly Fish-hooks, Hooked Stylocline; Me-s

          4b   Flower head about as high as wide; corolla pink or purple; pappus consisting of bristles (leaves up to 30 cm long)

             6a   Blades of larger leaves usually somewhat pointed at the tip; petioles of larger leaves hollow, without angular ridges . . . . . . *Arctium minus* (pl. 7)

Common Burdock; eu

             6b   Blades of larger leaves usually rounded at the tip; petioles of larger leaves solid, usually with angular ridges . . . . *Arctium lappa*

Great Burdock; eu

       3b   Involucre with straight spines

          7a   Spines of involucre 1–2 mm long; corolla blue purple; pappus bristles feathery (leaves lobed; uncommon escape from cultivation) . . . . *Cynara scolymus*

Artichoke; me

          7b   Spines of involucre often more than 5 mm long; corolla white, yellow, pink, or purple; pappus bristles, if present, not feathery (basal leaves sometimes absent at flowering time)

             8a   Corolla usually purple or pink, sometimes white; leaves not rough to the touch (spines of involucre 10–25 mm long)

                9a   Involucre 8–14 mm high; basal leaves 15–25 cm long, with a few wide lobes, these usually not toothed; corolla usually pink, sometimes white; pappus bristles 1 mm long . . . . *Centaurea iberica*

Iberian Knapweed, Iberian Star-thistle; me

                9b   Involucre 6–8 mm high; basal leaves less than 10 cm long, with many narrow lobes, these usually toothed; corolla purple; pappus absent . . . . *Centaurea calcitrapa*

Purple Star-thistle; eu

8b   Corolla yellow; leaves usually rough to the touch (leaves less than 15 cm long, usually lobed, but not toothed; pappus bristles 2–4 mm long; common)

    10a   Spines of involucre purple, 5–10 mm long, slender .... *Centaurea melitensis*

Tocalote; me

    10b   Spines of involucre yellow, 10–25 mm long, stout .... *Centaurea solstitialis* (pl. 9)

Yellow Star-thistle; me

2b   Plant spiny on the stem, leaves, or both, as well as on the involucre

    11a   Stem with trios of spines; leaves not spiny; involucre with hooked spines (leaves 3–8 cm long, very hairy on the undersides, 3–5 lobed; spines on stem straw colored, 1.5–2 cm long; pappus absent; widespread) ...... *Xanthium spinosum* (fig.)

Spanish Thistle, Spiny Cocklebur; eu

    11b   Stem without trios of spines; leaves with spines; involucre with straight spines

        12a   Leaves with white mottling (flower head up to 5 cm wide, with spines about as long; corolla pink to purple; pappus bristles 15–20 mm long; up to 2 m tall; widespread)................................. *Silybum marianum* (pl. 14)

Milk Thistle; me

        12b   Leaves without white mottling, but the leaf veins may be whitish

            13a   Stem with spiny "wings" that extend from one leaf base nearly to the next (corolla pink or purplish)

                14a   Upper surfaces of leaves rough to the touch; involucre 3–4 cm high; pappus bristles feathery, 20–30 mm long (widespread) .... *Cirsium vulgare*

Bull Thistle; eu

                14b   Leaves not rough to the touch; involucre 1.5–2 cm high; pappus bristles not feathery, 10–15 mm long

                    15a   Branches usually ending in only 1–5 flower heads; outer surface of phyllaries roughened (widespread) .... *Carduus pycnocephalus* (pl. 9)

Italian Thistle; me

                    15b   Branches usually ending in more than 5 flower heads (sometimes at least 20); outer surface of phyllaries smooth .... *Carduus tenuiflorus*

Slenderflower Thistle; me

        13b   Stem without spiny wings or with only short ones

            16a   Upper half of most phyllaries with a lengthwise resinous ridge (involucre 1.5–3 cm high; margins of phyllaries slightly toothed or fringed with hairs; pappus bristles feathery; in wet areas)

                17a   Leaves densely white woolly, especially on the undersides; involucre densely hairy; corolla white to purplish red (phyllaries often purplish at the tips; sometimes on serpentine) .... *Cirsium douglasii*

Swamp Thistle; Me-Mo

17b Leaves hairy, but not densely so; involucre scarcely or not at all hairy; corolla pale pink or rose

 18a Upper leaves with many narrow lobes, those just below upper flower heads 1–2 cm long; in brackish marshes . . . . .
  . . . . *Cirsium hydrophilum* var. *hydrophilum*
  Suisun Thistle; Sl; 1b

 18b Upper leaves either smooth margined or with a few wide lobes, those just below upper flower heads 2–4 cm long; wet areas on serpentine . . . .
  . . . . *Cirsium hydrophilum* var. *vaseyi*
  Mount Tamalpais Thistle; Ma (MT); 1b

16b Phyllaries without a lengthwise resinous ridge

 19a Margins of phyllaries with fine or jagged teeth

  20a Leaves mostly basal; not more than 30 cm tall (phyllaries with lighter margins; corolla purple or white; involucre 2.5–5 cm high; pappus bristles feathery) . . . .
   . . . . *Cirsium quercetorum* [*C. walkerianum*] (pl. 9)
   Brownie Thistle; SFBR-n

  20b Stem leaves well developed; usually much more than 30 cm tall (involucre often very hairy)

   21a Margins of phyllaries with fine teeth; involucre 1.5–5 cm high (pappus bristles feathery; corolla white, rose, purple, or red) . . . . *Cirsium occidentale*
    var. *californicum* [*C. californicum*]
    California Thistle; CC-s

   21b Margins of phyllaries with jagged teeth; involucre 2–3.5 cm high

    22a Involucre often very hairy; phyllaries more or less equal, none or them leaflike; corolla cream colored to pink; pappus bristles feathery . . . .
     . . . . . *Cirsium remotifolium* [includes *C. callilepis* vars. *callilepis* and *pseudocarlinoides*]
     Fringebract Thistle; SFBR-n

    22b Involucre very hairy; outer phyllaries leaflike, definitely longer than inner; corolla yellow; pappus consisting of scales 10–13 mm long . . . .
     . . . . *Carthamus lanatus*
     Woolly Distaff Thistle; me

 19b Margins of phyllaries not toothed

  23a Phyllaries not hairy

   24a Leaves mostly basal; not more than 30 cm tall (phyllaries with lighter margins, not bent back, or bent back only at the tips; corolla purple or white; pappus bristles feathery) . . . *Cirsium quercetorum* [*C. walkerianum*] (pl. 9)
    Brownie Thistle; SF-n

24b Stem leaves well developed; often more than 50 cm tall

25a Phyllaries bent back for more than half their length; corolla white; in wet areas on serpentine (pappus bristles feathery) ....
.... *Cirsium fontinale* var. *campylon*
[*C. campylon*]
Mount Hamilton Thistle; SCl (MH); 1b

25b Phyllaries not bent back; corolla yellow or blue purple; in dry, disturbed areas

26a Involucre 3–6 cm high, not hidden by leaves, with unbranched spines; corolla blue purple; pappus bristles feathery (widespread) .... *Cynara cardunculus* (pl. 10)
Cardoon, Artichoke Thistle; me

26b Involucre 2–4 cm high, almost hidden by the leaves below it, with branched spines; corolla yellow; pappus bristles not feathery ....
.... *Cnicus benedictus*
Blessed Thistle; eu

23b Phyllaries hairy, but sometimes only on the margins (pappus bristles feathery)

27a Spines on phyllaries at least 10 mm long (corolla purple or red to white; involucre densely hairy; in moist areas)

28a Leaf below each flower head at least as long as the head; involucre 25–35 mm high; coastal and in Coast Ranges .... *Cirsium brevistylum*
Indian Thistle

28b Leaf below each flower head shorter than the head; involucre 15–30 mm high; coastal bluffs and slopes, sometimes on serpentine (not common) .... *Cirsium andrewsii*
Franciscan Thistle; Sn-SM; 1b

27b Spines on phyllaries less than 10 mm long, usually not more than 5 mm long

29a Corolla white (involucre 15–30 mm high)

30a Phyllaries reddish, usually bent back, with short hairs mainly on the inner surfaces; on wet serpentine slopes ....
.... *Cirsium fontinale* var. *fontinale*
Fountain Thistle; SM; 1b

30b Phyllaries greenish, not regularly bent back, the outer surfaces hairy; usually in dry areas, sometimes on serpentine ....
.... *Cirsium cymosum*
Peregrine Thistle; Mo-n

29b Corolla crimson, purplish, or lavender, sometimes white in *Cirsium arvense*

31a Involucre 1–2 cm wide and high; heads of pistillate and staminate flowers on separate plants (phyllaries sometimes hairy; corolla purplish pink or lavender, sometimes white; perennial; spreading by underground rhizomes; at low elevations; widespread) . . . . *Cirsium arvense*

Canada Thistle; eu

31b Involucre 1.5–5 cm wide and high, sometimes more; most flowers with a pistil and stamens

32a Corolla crimson; involucre sometimes densely hairy (on slopes at elevations of up to 8,500 ft; widespread) . . . . *Cirsium occidentale* var. *venustum* (pl. 9)

Venus Thistle, Red Thistle

32b Corolla usually purple; involucre densely hairy (in sandy soil along the coast and on adjacent hills) . . . . *Cirsium occidentale* var. *occidentale* [*C. coulteri*]

Cobweb Thistle

1b Involucre without sharp spines (exceptions: in species of *Ambrosia*, there are spines on the involucre of the pistillate flower head, located in the leaf axils below the racemes of staminate flower heads; in *Soliva*, each flattened achene, which could be confused with a phyllary, ends in spines)

33a All leaves alternate

34a Upper half of each phyllary with 12–15 pointed marginal teeth; peripheral flowers much larger than those near the center of the head and with unequal corolla lobes; corolla usually deep blue, but sometimes white, pink, or purple (leaves sometimes lobed; pappus bristles 2–3 mm long). . . . . *Centaurea cyanus*

Bachelor's-button, Cornflower; me

34b Flower head and corolla not as described in choice 34a . . . . ASTERACEAE, GROUP 2, SUBKEY

33b Some or all leaves opposite

35a Leaves opposite below, alternate above

36a Leaves usually smooth margined, sometimes with a few teeth

37a Involucre 10–20 mm high; corolla yellow, orange, or red; rays not extending more than 2 mm beyond the phyllaries; pappus scales 10, white, 3–11 mm long; plant not woody; widespread . . . . *Achyrachaena mollis* (pl. 6)

Blow-wives

37b Involucre 5–6 mm high; corolla white; rays absent; pappus absent; plant sometimes woody; in salt marshes or alkaline places . . . .
. . . . *Iva axillaris* ssp. *robustior* (fig.)

Povertyweed; Sl, Al, CC

36b Leaves lobed or with many teeth (flower head less than 8 mm high, the staminate flower heads in racemes at the ends of branches, the pistillate flower heads in the leaf axils below the racemes; pappus absent)

38a Plant forming mats; stem prostrate, the tip rising (involucre of pistillate flower head spiny; in sandy seacoast habitats). . . . .
. . . *Ambrosia chamissonis* [includes *A. chamissonis* ssp. *bipinnatisecta*] (pl. 7)

Silvery Beachweed, Beachbur; Mo-n

38b Plant not forming mats; stem erect

39a Leaves pinnately lobed, the lobes not separated nearly to the midrib; involucre of pistillate flowers sometimes with pronounced bumps, but without distinct spines . . . . *Ambrosia psilostachya* (fig.)

Western Ragweed, Common Ragweed

39b Leaves twice pinnately lobed, the primary lobes separated nearly to the midrib; involucre of pistillate flowers with hooked spines . . . . .
. . . . *Ambrosia confertiflora*

Ragweed, Weakleaf Burweed; SF

35b All leaves opposite

40a Lower leaves with petioles

41a Leaves compound, with 3–5 leaflets; plant not glandular, annual; pappus bristles usually 2 (up to 120 cm tall, branched; in damp places). . . . .
. . . . *Bidens frondosa* (pl. 8)

Sticktight; Sn

41b Leaves not compound; plant glandular; perennial; pappus bristles many

42a Corolla yellow; involucre 10–17 mm high; stem not purplish; up to 60 cm tall . . . . *Arnica discoidea* [includes *A. discoidea* vars. *alata* and *eradiata*] (pl. 7)

Coast Arnica, Rayless Arnica; Mo-n

42b Corolla whitish; involucre not more than 5 mm high; stem purplish; up to 150 cm tall . . . .
. . . . *Ageratina adenophora* [*Eupatorium adenophorum*]

Sticky Eupatorium; mx

40b Leaves sessile

43a Flower head not woolly; leaves 3–10 cm long; corolla with rays 1 mm long; stem weak but usually upright, up to 35 cm tall; pappus consisting of scales, the flowers without scales beneath them; in wet clay soils (corolla yellow) . . . . *Lasthenia glaberrima*

Smooth Goldfields; Al-n

43b Flower head woolly; leaves not more than 2.5 cm long; corolla without rays; stem prostrate, up to 15 cm long; pappus absent, but some flowers with a scale beneath them; mostly in hard-packed soil, sometimes in dry mud of vernal pools

44a Leaves often extending 5–10 mm beyond the flower heads; leaves 6–10 times as long as wide (leaves up to 20 mm long) . . . . *Psilocarphus oregonus*
Oregon Woolly-heads

44b Leaves not extending more than 2–3 mm beyond the flower heads; leaves 1–6 times as long as wide

45a Flower head 6–10 mm wide; leaves up to 25 mm long . . . . *Psilocarphus brevissimus* var. *multiflorus*
Delta Woolly-marbles; SFBR; 4

45b Flower head 2–5 mm wide; leaves up to 15 mm long . . . . *Psilocarphus tenellus*
Slender Woolly-heads

Asteraceae, Group 2, Subkey: Flowers primarily of the disk-flower type, the rays, if present, inconspicuous; involucre without spines; all leaves alternate

1a Corolla mostly not yellow (species in choices 1a and 1b cannot be separated rigorously by color, so try both choices to determine best fit)

2a Shrubs (corolla white; flower heads in clusters of several to many; pappus consisting of bristles)

3a Involucre 10–12 mm high; most flowers with a pistil and stamens . . . . *Brickellia californica*
California Brickellia, California Brickelbush

3b Involucre 3–5 mm high; pistillate and staminate heads on separate plants

4a Larger leaves up to 15 cm long, not distinctly toothed, with 3 prominent veins; in moist habitats . . . . *Baccharis salicifolia [B. viminea]* (pl. 8)
Mulefat, Seep-willow

4b Leaves not more than 4 cm long, usually toothed, with a single prominent vein; in dry habitats (widespread) . . . . *Baccharis pilularis* [includes *B. pilularis* var. *consanguinea*] (pl. 8)
Coyotebrush, Chaparral-broom

2b Herbs

5a Leaves usually lobed, sometimes twice, or with a few broad teeth

6a Leaf blades whitish on the undersides, nearly or fully as wide as long, either palmately lobed or with a few prominent teeth (corolla white or pale pink)

7a Leaf blades nearly circular, more than 10 cm long, palmately lobed; pappus consisting of bristles . . . . *Petasites frigidus* var. *palmatus [P. palmatus]* (fig.)
Coltsfoot; SM-n

7b Leaf blades nearly triangular, 3–25 cm long, with broad teeth; pappus absent . . . . *Adenocaulon bicolor* (fig.)

Trailplant; SCr-n

6b Leaf blades not especially whitish on the undersides, if nearly as wide as long, 1–2 times pinnately lobed

8a Plant prostrate; some leaves compound, the leaflets lobed; involucre less than 5 mm high; pappus absent; corolla green (achene flattened, with broad, membranous wings; style persisting as a rigid spine) . . . . *Soliva sessilis* [includes *S. daucifolia*]

Common Soliva; sa

8b Plant erect; leaves not compound; involucre at least 10 mm high; pappus consisting of bristles; corolla white to blue

9a Involucre 6–15 cm high; pappus bristles 30–55 mm long; corolla blue (not common outside of cultivation) . . . . *Cynara scolymus*

Artichoke; me

9b Involucre 1–1.5 cm high; pappus bristles 6–8 mm long; corolla white to blue . . . . *Acroptilon repens* [*Centaurea repens*]

Russian Knapweed; eua

5b Leaves not lobed and either without teeth or only with small teeth

10a Stem and leaves without long, woolly hairs, but there may be short hairs, glandular hairs, or both (pappus consisting of bristles)

11a Flower heads in dense clusters of more than 10; peduncle less than 1 cm long (plant often with a strong odor)

12a Flower clusters in the leaf axils; corolla purple (lower leaves with petioles; in moist areas, including saline habitats) . . . . *Pluchea odorata*

Salt Marsh Fleabane; SFBR

12b Flower clusters at the ends of the main stems or branches; corolla white, green, or pink; all leaves sessile; in dry habitats

13a Leaves not all sessile; staminate and pistillate heads on separate plants; in moist areas (widest leaves about 20 mm; may be woody at the base) . . . . *Baccharis douglasii*

Marsh Baccharis, Douglas Baccharis

13b Leaves all sessile; staminate and pistillate flowers, if present, not in separate heads or on separate plants; in dry areas

14a Phyllaries white; widest leaves 10–15 mm; involucre about as long as wide (widespread) . . . . *Gnaphalium californicum* (pl. 11)

California Everlasting

14b Phyllaries white, green, or pink; widest leaves 3–7 mm; involucre longer than wide (widespread) . . . . *Gnaphalium ramosissimum*

Pink Everlasting; Ma-s

11b Flower head solitary; peduncle 1–5 cm long (corolla pink, lavender, or violet)

15a Flower heads both in the leaf axils and at the ends of the main stems or branches; leaves not in a basal cluster; upper leaves not glandular or hairy, reduced upward but not scalelike (in wet habitats, including salt marshes) .... *Aster subulatus* var. *ligulatus* [*A. exilis*]
Slim Aster; SFBR-s

15b Flower heads at the ends of the main stems or branches; larger leaves in a basal cluster, but these may wither early; upper leaves glandular or hairy, sometimes scalelike

16a Phyllaries glandular, but not especially hairy or woolly; upper leaves glandular .... *Lessingia ramulosa*
Sonoma Lessingia; SN, Sl

16b Phyllaries, at least the lower ones, hairy or woolly, but not glandular; upper leaves hairy or woolly (at least when young)

17a Involucre 5–13 mm high; flowers at the edge of the head considerably larger than those in the center; basal leaves usually persisting through the period of flowering .... *Lessingia hololeuca*
Woolly-headed Lessingia; Na, Ma-SCl

17b Involucre 5–8 mm high; flowers at the edge of the head not obviously larger than those in the center; basal leaves usually withering and falling away before the period of flowering .. .... *Lessingia arachnoidea* [*L. hololeuca* var. *arachnoidea*]
Crystal Springs Lessingia; SM-SCr; 1b

10b Stem and 1 or both surfaces of leaves with long woolly hairs, these sometimes not very dense

18a One leaf surface much more hairy than the other (pappus consisting of bristles)

19a Leaves up to 12 cm long, the upper surfaces green and scarcely hairy, the undersides whitish and densely woolly; flower heads in dense clusters; corolla white, green, or pink .... *Gnaphalium purpureum* [includes *G. peregrinum*] (pl. 11)
Purple Cudweed

19b Leaves less than 6 cm long, the upper surfaces densely woolly compared to the undersides; flower head solitary; corolla white to lavender (basal leaves absent after flowering)

20a Phyllaries and upper leaves glandular .... *Lessingia micradenia* var. *micradenia*
Tamalpais Lessingia; Ma(MT)-n; 1b

20b Phyllaries and upper leaves not glandular .... *Lessingia micradenia* var. *glabrata*
Smooth Lessingia; SCl; 1b

18b Leaves uniformly hairy on both surfaces, but sometimes not densely so (flower head solitary or in clusters)

21a Leaves just below each cluster of flower heads either absent or much shorter than the cluster (pappus consisting of bristles)

    22a Flower heads solitary at the ends of branches; stem leaves markedly reduced compared with basal leaves (corolla pink to lavender)

        23a Phyllaries and upper leaves glandular ... *Lessingia ramulosa* Sonoma Lessingia; SN, Sl

        23b Phyllaries and upper leaves not glandular .... *Lessingia hololeuca* Woolly-headed Lessingia; Na, Ma-SCl

    22b Flower heads grouped in slightly rounded or nearly flat-topped clusters; stem leaves not markedly reduced

        24a Stem and leaves yellow green; phyllaries white, yellow green, or brown; in moist areas (leaves 2–6 mm wide, 20–50 mm long; corolla yellow or tan) .... *Gnaphalium stramineum [G. chilense]* Cotton-batting-plant

        24b Stem and leaves whitish because of woolliness; phyllaries white or brown; mostly in dry areas

            25a Leaves up to 4 mm wide and 40 mm long; phyllaries brown; corolla tipped with red .... *Gnaphalium luteo-album* Weedy Cudweed; eua

            25b Leaves 4–10 mm wide and up to 80 mm long; phyllaries white, sometimes tan; corolla not tipped with red (stem usually extensively branched in the middle part of the plant) .... *Gnaphalium canescens* ssp. *microcephalum [G. microcephalum]* White Everlasting; Ma-s

21b Leaves below each cluster of flower heads as long as, or longer than, the cluster (sometimes hidden by woolliness)

    26a Flower heads solitary in the leaf axils; petioles 1–1.5 cm long; pappus absent (corollas enclosed by rigid scales; phyllaries absent; about 15 cm tall, forming mounds; on serpentine) .... *Hesperevax sparsiflora [Evax sparsiflora]* Erect Hesperevax

    26b Flower heads in clusters at the ends of branches; leaves sessile; pappus of each fertile flower consisting of bristles

        27a Leaves 10–50 mm long and 3–5 mm wide, sometimes larger (in moist habitats)

            28a Leaves 20–50 mm long; involucre 5–6 mm high, yellow green; corolla yellow or tan .... *Gnaphalium stramineum [G. chilense]* Cotton-batting-plant

28b Leaves 10–30 mm long; involucre 3 mm high, brown; corolla white, green, or pink ... *Gnaphalium palustre*
Lowland Cudweed

27b Leaves usually not more than 15 mm long or 3 mm wide (clusters of flower heads usually less than 1 cm wide)

29a Leaves just below each cluster of flower heads extending more than 3 mm beyond the cluster; pappus bristles 19–30 (up to 30 cm tall; widespread) ....
.... *Filago gallica*
Fluffweed; me

29b Leaves just below each cluster of flower heads not extending beyond the cluster; pappus bristles 1–12

30a Flower heads densely woolly

31a Flower head 3.5–5 mm high ...... *Micropus amphibolus [Stylocline amphibola]*
Mount Diablo Cottonweed; Ma-Al

31b Flower head 2–4 mm high ....
.... *Micropus californicus*
Slender Cottonweed

30b Flower heads hairy but not woolly (phyllaries absent; a scale present at the base of each pistillate flower)

32a Tips of corollas purplish, visible above the scales (widespread) ....
.... *Filago californica* (pl. 11)
California Fluffweed

32b Corollas hidden by the scales ....
.. *Stylocline gnaphaloides [S. gnaphalioides]*
Everlasting Neststraw; Ma

1b Corolla mostly yellow (species in choices 1a and 1b cannot be separated rigorously by color, so try both choices to determine best fit)

33a Largest leaves at or near the base of the plant, the stem leaves few and reduced or numerous and scalelike (phyllaries often dark tipped; pappus consisting of bristles)

34a Leaves sessile (lower leaves elliptic, densely hairy; upper leaves scalelike, scarcely hairy)..................................... *Lepidospartum squamatum*
Scalebroom, Nevada Broomshrub; SCl-s

34b Some leaves with petioles

35a Petioles of basal leaves winged; plant whitish (in wet places) ....
.... *Senecio hydrophilus*
Alkali-marsh Butterweed; SFBR-n

35b Petioles of basal leaves not winged; plant not whitish .... *Senecio aronicoides*
California Butterweed, Groundsel; SM-n

33b Stem with well-developed leaves

36a Involucre very glandular (leaves narrow)

    37a Phyllaries widest at the middle, tapering abruptly at the tips; pappus either absent or scarcely evident; leaves sometimes toothed, bristly, or both

        38a Involucre 4–6 mm high; glands on phyllaries obvious (corolla sometimes purple tinged; at elevations above 3,500 ft) . . . . . . *Madia glomerata*
Mountain Tarweed

        38b Involucre 2–3 mm high; glands on phyllaries hidden among long hairs, not obvious without a hand lens (widespread) . . . . *Madia exigua*
Threadstem Madia, Small Tarweed

    37b Phyllaries widest at their bases and tapering gradually to the tips; pappus consisting of many bristles; leaves usually smooth margined and not bristly

        39a Leaves and stem hairy . . . . . . *Erigeron petrophilus* var. *petrophilus* (pl. 10)
Rock Daisy; Mo-n

        39b Upper leaves and stem not hairy

            40a Leaves 2–6 mm wide; lower stem and leaves hairy or glandular
. . . *Erigeron petrophilus* var. *viscidulus* [*E. inornatus* var. *viscidulus*]
Klamath Rock Daisy; Ma(PR); 4

            40b Leaves less than 2 mm wide; lower stem and leaves not hairy or glandular . . . .
. . . . *Erigeron reductus* var. *angustatus* [*E. inornatus* var. *angustatus*]
Northern Rayless Daisy; SM-n

36b Involucre not glandular

    41a Involucre less than 5 mm high

        42a Flower heads solitary at the ends of branches that are usually at least 3 cm long (leaves sessile, clasping)

            43a Leaves twice pinnately lobed

                44a Flower head 2–5 mm wide; leaves hairy; pappus absent . . . .
. . . . *Cotula australis*
Southern Brassbuttons, Australian Cotula; au

                44b Flower head 6–10 mm wide; leaves not hairy; pappus a minute crown

                    45a Flower head 6–9 mm high (plant with an odor of pineapple; widespread) . . . . *Chamomilla suaveolens*
[*Matricaria matricarioides*] (fig.)
Pineappleweed, Rayless Chamomile; na

                    45b Flower head 10–11 mm high . . . .
. . . . *Chamomilla occidentalis* [*Matricaria occidentalis*]
Valley Pineappleweed

            43b Leaves either not lobed or only irregularly lobed, definitely not twice pinnately lobed (flower head 8–10 mm wide)

                46a Leaves not hairy, irregularly lobed; flower head nearly smooth, resembling a button, about 1 cm wide; rays absent; pappus absent; in wet places, including margins of salt marshes . . . .
. . . . *Cotula coronopifolia* (pl. 10)
Brassbuttons; af

46b Leaves hairy, not toothed or lobed; flower head not resembling a button; rays if present, not more than 2 mm long; pappus bristles 3; in dry grassland and woods (corolla often reddish)

    47a Involucre 3–5 mm high; stem not branching or branching only from the base … *Pentachaeta exilis* [*Chaetopappa exilis*] Meager Pentachaeta

    47b Involucre up to 3 mm high; stem branching above the base .... *Pentachaeta alsinoides* [*Chaetopappa alsinoides*] Tiny Pentachaeta

42b Flower heads many, in elongated racemes or extensive panicles whose branches are mostly less than 2 cm long

  48a Flower heads in elongated racemes, these about 2 cm wide; rays absent; pappus absent or a minute crown

    49a Larger leaves either not lobed or with a few lobes or coarse teeth, but most of these divisions not separated nearly to the midrib (leaves densely hairy beneath)

      50a Involucre hairy (widespread) .... *Artemisia douglasiana* (pl. 7) Mugwort

      50b Involucre not hairy (coastal) .... *Artemisia suksdorfii* Coastal Mugwort; Sn-n

    49b Larger leaves lobed, most of the primary lobes separated nearly to the midrib

      51a Leaves not hairy; lobes of larger leaves regularly toothed; annual or biennial, not woody; up to 30 cm tall .... *Artemisia biennis* Biennial Sagewort; eu

      51b Leaves hairy; lobes of larger leaves not toothed; perennial; woody, at least at the base; often more than 70 cm tall

        52a Stem and leaves usually whitish, the hairs not matted; involucre 4–5 mm high, hairy; woody only at the base (in coastal habitats) .... *Artemisia pycnocephala* (pl. 8) Coastal Sagewort, Beach Sagewort; Mo-n

        52b Stem and leaves grayish, the hairs matted; involucre 2–3 mm high, not hairy; woody throughout (widespread) .... *Artemisia californica* (pl. 8) California Sagebrush; Ma, Na-s

  48b Flower heads in spreading panicles, these often 10–15 cm wide; very small rays sometimes present; pappus consisting of bristles

    53a Longest leaves 30 mm; upper leaves mostly not more than 5 mm long; pappus bristles reddish (outer disk flowers sometimes much larger than those near the center; corolla often with a dark band inside; leaves sometimes toothed or lobed)

54a Margins of outer phyllaries densely hairy but not glandular .... *Lessingia tenuis [L. germanorum* var. *parvula]*
Spring Lessingia; SCl-SLO; 4

54b Margins of outer phyllaries not densely hairy but glan-
dular .... *Lessingia glandulifera*
[*L. germanorum* var. *glandulifera*]
Valley Lessingia, Common Lessingia; SFBR-s

53b Longest leaves often more than 60 mm; upper leaves at least
10 mm long; pappus bristles whitish or tan

55a Vine or shrub

56a Vine, the stem climbing over other plants; leaf blades
about as long as wide, palmately 5–9 lobed (very invasive) .... *Senecio mikanioides*
German-ivy; af

56b Shrub, the stem upright; leaf blades much more than 5
times as long as wide, smooth margined (inland;
common) ....
... *Ericameria arborescens [Haplopappus arborescens]*
Goldenfleece

55b Herb (plant upright; leaf blades at least twice as long as
wide)

57a Leaves with lobes or teeth of rather uniform size;
lower leaves up to 8 cm long (involucre hairy, 2–3 mm
high; pappus bristles white; up to 1 m tall) ....
.... *Conyza coulteri*
Coulter Horseweed; SCl-s

57b Leaves either smooth margined or with a few irregular
lobes or teeth; lower leaves up to 10 cm long (involucre 3–4 mm high)

58a Phyllaries without brown or black midveins, but
they may be purple; involucre hairy; pappus bristles tan, turning red (extremely small rays present) .... *Conyza bonariensis*
South American Horseweed; sa

58b Phyllaries with brown or black midveins; involucre not hairy or only scarcely so; pappus bristles
dirty white

59a Hairs usually present on both leaf surfaces
and also on the margins; very small rays present (widespread) .... *Conyza canadensis* (fig.)
Horseweed

59b Hairs present only on the leaf margins; rays
absent .... *Conyza bilbaoana*
Rayless Horseweed; sa

41b Involucre at least 5 mm high

  60a Some leaves, often only the lower ones, toothed or lobed

    61a Pappus lacking or of scales

      62a Stem not woody, sprawling with the tip rising (pappus consisting of scales; on coastal dunes) ....
.... *Tanacetum camphoratum* (pl. 14)
Dune Tansy; SF

      62b Stem woody, at least at the base, upright

        63a Leaves evenly toothed, up to 3.5 cm long; pappus lacking; established in coastal areas ....
.... *Santolina chamaecyparisus* [*S. chamaecyparissus*]
Lavender-cotton; me

        63b Leaves irregularly lobed, 2–7 cm long; pappus scales very small; on coastal bluffs ...... *Eriophyllum staechadifolium*
[includes *E. staechadifolium* var. *artemisiaefolium*]
Seaside Woolly Sunflower, Lizardtail; SCr-Me

    61b Pappus consisting of bristles

      64a Plant woody at the base; leaves often in bunches ....
.... *Isocoma menziesii* ssp. *vernonioides*
[*Haplopappus venetus* ssp. *vernonioides*]
Coast Goldenbush; SF-s

      64b Plant not woody; leaves not in bunches

        65a All leaves sessile

          66a Pappus bristles white; plant with 5 or fewer branches, sometimes not branched

            67a Leaves not hairy; up to 25 cm tall ....
.... *Senecio aphanactis*
Rayless Ragwort; Sl-s; 2

            67b Leaves sparsely to densely hairy on the undersides; up to 80 cm tall .... *Senecio sylvaticus*
Wood Groundsel; eu

          66b Pappus bristles reddish; plant with multiple branches that are again branched (outer disk flowers sometimes raylike)

            68a Stem and leaves not glandular; on sand dunes and in other coastal habitats ....
.... *Lessingia germanorum*
San Francisco Lessingia; SFBR-s; 1b

            68b Stem and leaves glandular; inland

              69a Outer phyllaries not glandular, but hairy, inner phyllaries glandular ... *Lessingia tenuis*
[*L. germanorum* var. *parvula*]
Spring Lessingia; SCl-SLO; 4

69b Phyllaries glandular, sometimes hairy . . . . . .
. . . . *Lessingia glandulifera*
[*L. germanorum* var. *glandulifera*]
Valley Lessingia; SFBR-s

65b At least lower leaves with petioles

70a Leaves toothed

71a Leaves 7–20 cm long; stem and leaves scarcely or not at all hairy; pappus white . . . .
. . . . *Erechtites minima* [*E. prenanthoides*]
Australian Burnweed, Coast Fireweed; au

71b Leaves up to 10 cm long; stem and leaves definitely hairy; pappus tan, turning red (phyllaries may have purple midveins; extremely small rays present) . . . . *Conyza bonariensis*
South American Horseweed; sa

70b Leaves lobed

72a Phyllaries black tipped; base of involucre with small, black-tipped bracts (widespread) . . . .
. . . . *Senecio vulgaris*
Common Groundsel; eua

72b Phyllaries not black tipped; base of involucre without small bracts

73a Involucre and stem only slightly if at all hairy; small rays present . . . . *Senecio sylvaticus*
Wood Groundsel; eu

73b Lower part of involucre and upper stem woolly; rays absent (coastal) . . . .
. . . . *Erechtites glomerata* [*E. arguta*]
Bushman's Burnweed; au

60b Leaves smooth margined

74a Phyllaries ridged on back (shrub, with a strong odor; pappus consisting of bristles) . . . . . . *Chrysothamnus nauseosus* ssp. *mohavensis*
Rubber Rabbitbrush; SCl (MH)-s

74b Phyllaries not ridged on back

75a Pappus consisting of scales

76a Plant not woody; flower head globose (in wetlands) . . . . . .
. . . . *Helenium puberulum* (pl. 11)
Rosilla, Sneezeweed

76b Plant woody at the base; flower head conical

77a Bark not whitish; on coastal bluffs . . . .
. . . . *Eriophyllum staechadifolium*
[includes *E. staechadifolium* var. *artemisiaefolium*]
Seaside Woolly Sunflower, Lizardtail; SCr-Me

77b Bark whitish; on inland hillsides . . . .
. . . . *Eastwoodia elegans*
Yellow Mock Aster; Al-s, e

75b Pappus consisting of bristles
    78a Flower head 10–15 mm high; up to 60 cm tall
        79a Upper leaves sparse, some of them reduced to a length of 1 cm or less .... *Heterotheca oregona* var. *scaberrima* [*Chrysopsis oregona* var. *scaberrima*] Sticky Oregon Golden Aster; Al, SCl-Mo
        79b Upper leaves crowded, 2–5 cm long
            80a Stem and leaves with stout, long hairs; not confined to stream beds; inland .... *Heterotheca oregona* var. *rudis* [*Chrysopsis oregona* var. *rudis*] Inland Oregon Golden Aster; SF-n
            80b Stem and leaves without hairs; in dry stream beds; coastal .... *Heterotheca oregona* var. *oregona* [*Chrysopsis oregona* var. *oregona*] Oregon Golden Aster; Ma-n
    78b Flower head 4–8 mm high; often over 100 cm tall
        81a Plant woody at the base; longest leaves up to 4 cm (involucre 5–7 mm high; coastal) .... *Isocoma menziesii* ssp. *vernonioides* [*Haplopappus venetus* ssp. *vernonioides*] Coast Goldenbush; SF-s
        81b Plant not woody; longest leaves up to 10 cm
            82a Leaves green on the upper surfaces, with white hairs on the undersides; phyllaries pearly white; perennial (widespread) .... *Anaphalis margaritacea* (pl. 7) Pearly Everlasting
            82b Leaves similar on both surfaces; phyllaries not pearly white; annual .... *Conyza bonariensis* South American Horseweed; sa

Asteraceae, Group 3:   At least some of the flower heads with disk flowers and significant ray flowers (species in this group often vary with respect to the color of their ray flowers, so try both choices to determine best fit)

1a Ray flowers white, pink, red, blue, violet, purple, or lavender
    2a Ray flowers usually pink, red, blue, violet, purple, or lavender, but sometimes white (leaves alternate or basal; perennial, or sometimes biennial in *Erigeron philadelphicus*)
        3a Tips of phyllaries all at the same level
            4a Flower head with more than 100 rays (pappus consisting of bristles; leaves usually toothed)
                5a Flower head 6–15 mm wide; ray flowers pink or white, the rays less than 1 mm wide; not confined to coastal locations. . . . *Erigeron philadelphicus* Philadelphia Daisy

5b Flower head 15–35 mm wide; ray flowers purple, blue, or white, the rays 2–3 mm wide; on coastal bluffs . . . . *Erigeron glaucus* (pl. 10)

Seaside Daisy; SLO-n

4b Flower head with not more than 60 rays

6a Leaves toothed or lobed; involucre more than 8 mm high; ray flowers purple or violet

7a Leaves 10–20 cm long, toothed; involucre 12–15 mm high; rays 30–60; pappus scales 4; on open areas, including burns . . . . *Hulsea heterochroma*

Red-rayed Hulsea; SCl-s

7b Leaves 3–8 cm long, the lower ones deeply lobed; involucre about 10 mm high; rays about 13; pappus consisting of bristles; established along the coast . . . . *Senecio elegans* (pl. 13)

Purple Ragwort; af

6b Leaves not toothed or lobed; involucre 4–7 mm high; ray flowers usually blue (rays 15–60; pappus consisting of bristles)

8a Leaves 3–6 cm long, 1–2 mm wide, usually bunched on 1 side of stem; along riverbanks or in oak woodland . . . . *Erigeron foliosus* var. *hartwegii*

Hartweg Daisy; Mo-n

8b Leaves 2–5 cm long, often wider than 2 mm, not bunched on 1 side of stem; open areas or pine forests . . . . *Erigeron foliosus* var. *foliosus*

Leafy Daisy; SFBR

3b Tips of phyllaries at 2 or more different levels

9a Leaves mostly grayish green, due to woolliness, but this sometimes absent; ray flowers not producing achenes (ray flowers pink, purple, or white; mainly coastal)

10a Flower head 2–3.5 cm wide; leaves toothed . . . . *Lessingia filaginifolia* var. *californica*

[includes *Corethrogyne californica* vars. *californica* and *obovata*]

California Aster, California Beachaster; Mo-n

10b Flower head rarely more than 2.5 cm wide; leaves sometimes toothed . . . . *Lessingia filaginifolia* var. *filaginifolia* [includes *Corethrogyne filaginifolia* vars. *hamiltonensis* and *virgata* and *C. leucophylla*]

Common Lessingia, Common Beachaster; Al-s

9b Leaves not grayish green and not woolly, but there may be sparse hairs; ray flowers producing achenes, these with a bristly pappus

11a Lower leaves 2–3 times as long as wide, toothed, 4–10 cm long; involucre 6–9 mm high; rays 10–15, white to pale violet; in dry forests . . . . *Aster radulinus*

Broadleaf Aster, Roughleaf Aster; SLO-n

11b Most leaves 6–8 times as long as wide, usually not toothed, 5–15 cm long; involucre 5–7 mm high; rays 20–35, violet or purple; usually in open areas

12a Plant not hairy; in salt marshes and other saline habitats around San Francisco Bay . . . .
. . . . . *Aster lentus* [includes *A. chilensis* vars. *lentus* and *sonomensis*]
Suisun Marsh Aster; SF, Sn, Na, SCl; 1b

12b Tip of peduncle, and sometimes also leaves and stems, hairy; mostly outer Coast Ranges, but sometimes in moist areas of the inner Coast Ranges (widespread) . . . .
. . . . *Aster chilensis* [includes *A. chilensis* var. *invenustus*] (pl. 8)
Common California Aster

2b Ray flowers usually white but sometimes pink, red, or purple

13a Plant with glandular hairs, or the phyllaries sticky because of glandular secretions (pappus of ray achene absent or a crown of scales)

14a Glands on phyllaries saucer shaped; flower heads at the ends of branches or scattered along stems (leaves alternate or basal; pappus of disk achene a crown of scales; annual)

15a Glands on phyllaries yellow; rays 5–11; leaves sometimes toothed; up to 180 cm tall . . . . *Blepharizonia plumosa*
Big Tarweed; Al-e; 1b

15b Glands on phyllaries black; rays 1–5; leaves not toothed; not more than 70 cm tall (ray flowers white, pink, or red)

16a Phyllaries with only a few hairs; rays 1 or 2, each with 3 lobes separate nearly to the base of the ray; flower head less than 0.5 cm wide; leaves up to 5 cm long . . . . *Calycadenia pauciflora*
Smallflower Rosinweed; Sn-n

16b Phyllaries densely hairy on the margins; rays 2–5, each with 3 lobes separate for about half the length of the ray; flower head up to 2 cm wide; leaves up to 8 cm long . . . . *Calycadenia multiglandulosa*
[includes *C. hispida* ssp. *reducta* and *C. multiglandulosa* sspp. *cephalotes* and *robusta*] (pl. 9)
Sticky Rosinweed, Hispid Calycadenia; Me-SCl

14b Glands on phyllaries globular; flower heads mostly at the ends of branches

17a Achene of each ray flower only partly enclosed by the adjacent phyllary (leaves sometimes toothed, alternate, the lower ones about as wide as the upper; pappus absent; annual)

18a Flower heads in both the leaf axils and at the ends of branches
. . . . *Hemizonia congesta* ssp. *clevelandii*
Cleveland Spikeweed; Sn, Na-n

18b All flower heads at the ends of branches . . . .
. . . . *Hemizonia congesta* ssp. *luzulifolia*
[includes *H. luzulaefolia* sspp. *luzulaefolia* and *rudis*] (pl. 12)
Hayfield Tarweed; La-s

17b Achene of each ray flower completely enclosed by the adjacent phyllary

19a Leaves opposite below, alternate above, sometimes toothed, but not lobed; pappus, if present on disk achene, consisting of 1–5 bristles; perennial (involucre up to 5 mm high; rays up to 4.5 mm long, each 3 lobed) . . . . *Holozonia filipes*

Hareleaf; Na, Ma-SCl

19b All leaves alternate, the lower usually lobed; pappus of disk achene consisting of 10–35 bristles; annual

    20a Rays 6–15 mm long; involucre up to 10 mm high; pappus bristles 10–15; on sandy soil, inland . . . . *Layia glandulosa*

White Layia; CC-s

    20b Rays up to 2.5 mm long; involucre up to 7.5 mm high; pappus bristles 24–35; only on coastal sand dunes . . . . . . *Layia carnosa*

Beach Layia; SF; 1b

13b Plant without glandular hairs or sticky secretions

  21a Flower head more than 1.5 cm wide

    22a Leaves mostly basal (leaves sometimes toothed; pappus absent; ray flowers white to purple; widespread) . . . . *Bellis perennis*

English Daisy; eu

    22b Stem leaves well developed, alternate

      23a Leaves lobed, the lobes less than 1 mm wide

        24a Flower head 2.5–3.5 cm wide; stem reddish; pappus absent . . . .
. . . . *Chamaemelum fuscatum [Anthemis fuscata]*

Dogfennel; me

        24b Flower head 1.5–2 cm wide; stem green; pappus a crown of scales (widespread) . . . . *Anthemis cotula* (fig.)

Mayweed, Dogfennel; eu

      23b Leaves, if lobed or compound, with lobes or leaflets at least 3 mm wide

        25a Flower heads 2.5–5 cm wide, solitary on each stem; leaves toothed or lobed; pappus absent or a crown of scales . . . .
. . . *Leucan themum vulgare [Chrysanthemum leucanthemum]*

Ox-eye Daisy; eu

        25b Flower heads usually less than 2 cm wide, several on each stem; leaves compound; pappus a crown of scales . . . .
. . . . *Tanacetum parthenium [Chrysanthemum parthenium]*

Feverfew; eu

  21b Flower head less than 1.5 cm wide

    26a Leaves all basal, the blades about 10 cm wide, palmately lobed (pappus bristles many; some flower heads without rays) . . . .
. . . . *Petasites frigidus* var. *palmatus [P. palmatus]* (fig.)

Coltsfoot; Mo-n

    26b Leaves not all basal, the blades not more than 3 cm wide, if lobed or compound, pinnately so

27a Leaves lobed or compound, the lobes or leaflets less than 1 mm wide (flower head 4–6 mm high; rays 2–5 mm long, white to pink; leaves alternate; pappus absent; widespread) . . . . *Achillea millefolium* [includes *A. borealis* sspp. *arenicola* and *californica*] (pl. 6) Common Yarrow, Milfoil

27b Leaves not lobed or compound, but sometimes toothed

    28a Leaves alternate; pappus bristles 0–20 (lower leaves 1–3.5 cm long; up to 20 cm tall) . . . . *Pentachaeta bellidiflora* [*Chaetopappa bellidiflora*] White-rayed Pentachaeta; Ma, SM, SCr; 1b

    28b Leaves opposite; pappus a crown of scales

        29a Flower head 5–10 mm wide; leaves and stem covered with short hairs; leaves sometimes toothed (in freshwater marshes) . . . . *Eclipta prostrata* [*E. alba*] False Daisy; SFBR

        29b Flower head 3–4 mm wide; leaves and stem usually not hairy, but the leaves sometimes with a few hairs on the veins; leaves toothed . . . . *Galinsoga parviflora* Littleflower Quickweed; sa

1b Ray flowers completely or partly yellow, yellow green, orange, or red (*Blennosperma nanum,* in group 3, subkey, sometimes has white ray flowers)

  30a Leaves mostly basal, with stem leaves absent or reduced (in species of *Lagophylla,* in group 3 subkey, the leaves are sometimes lacking)

    31a Flower heads at least 4 on each main stem; pappus consisting of bristles (rays fewer than 10)

      32a Phyllaries apparently in 1 series, the tips of all of them at the same level

        33a Petioles of lower leaves winged; plant whitish; in damp places (leaf blades up to 20 cm long, sometimes toothed; phyllaries 4–9 mm long, the tips black) . . . . *Senecio hydrophilus* Alkali-marsh Butterweed; SFBR-n

        33b Petioles not winged; plant not whitish; in dry areas

          34a Lower leaves up to 10 cm long, usually deeply lobed, and the lobes toothed; phyllaries 8–12 mm long, the tips green (sometimes on serpentine) . . . . *Senecio eurycephalus* Cutleaf Butterweed; Sn-n

          34b Lower leaves up to 20 cm long, sometimes toothed; phyllaries 4–8 mm long, the tips black . . . . *Senecio aronicoides* California Butterweed, Groundsel; SM-n

      32b Phyllaries in 2 or more series, the tips at different levels

        35a Flower head 1.5–3 cm wide; rays 5–12 mm long (leaves sometimes toothed) . . . . *Pyrrocoma racemosa* [*Haplopappus racemosus*] Racemose Pyrrocoma; Na

        35b Flower head less than 1 cm wide; rays 2–3 mm long

36a Leaves toothed, usually sticky; phyllaries oblong, blunt (mostly coastal) .... *Solidago spathulata* (pl. 14)

Coast Goldenrod, Dune Goldenrod; Mo-n

36b Leaves not toothed, not sticky; phyllaries slender, pointed .... *Solidago guiradonis*

Guirado Goldenrod; 4

31b Flower heads usually solitary at the ends of main stems; pappus absent or a crown of scales

37a Leaves much more than 2 cm wide

38a Leaves 1–2 times lobed (pappus absent) .... *Balsamorhiza macrolepis*

Bigscale Balsamroot; SF-e; 1b

38b Leaves sometimes toothed, but not lobed

39a Outer phyllaries enlarged, extending beyond the disk and the rays (pappus a crown of scales)

40a Most leaves with a woolly or cottony coating, usually not toothed; on sunny slopes .... *Wyethia helenioides*

Gray Mule-ears

40b Leaves without a woolly or cottony coating, but sometimes hairy or rough to the touch, sometimes toothed; on shady slopes (widespread) .... *Wyethia glabra* (pl. 15)

Smooth Mule-ears

39b Outer phyllaries not enlarged, not extending beyond the disk and rays

41a Leaves sometimes toothed, the stem leaves alternate; phyllaries with hairs about 2 mm long, these mainly concentrated at the margins; pappus scales 1–4 (widespread) .... *Wyethia angustifolia* (pl. 15)

Narrowleaf Mule-ears

41b Leaves not toothed, the lower stem leaves opposite; phyllaries with hairs less than 1 mm long, these concentrated at the tips; pappus absent .... *Helianthella californica*

California Helianthella; SCl-n

37b Leaves not more than 1.5 cm wide, sometimes much less

42a Phyllaries without membranous margins, those of the upper and lower series similar (pappus scales 4–8)

43a Length of each flower peduncle above last stem leaves at least twice the length of these leaves; pappus scales 4, in 1 series; largest leaves smooth margined or lobed ..... *Chaenactis glabriuscula* var. *lanosa*

Yellow Pincushion; SF-s

43b Length of each flower peduncle above last stem leaves less than twice the length of these leaves; pappus scales 7 or 8, in 2 series; largest leaves usually twice lobed (sometimes on serpentine) .... *Chaenactis glabriuscula* var. *heterocarpha*

[*C. glabriuscula* var. *gracilenta* and *C. tanacetifolia*]

Inland Yellow Pincushion; Na-s

42b Phyllaries of the upper series with membranous margins and often yellowish, marked with darker lines, or both, those of the lower series green and leaflike (leaves deeply lobed)

44a Leaves rarely with more than 2 lobes (pappus absent) . . . . *Coreopsis douglasii*
Douglas Coreopsis; SCl-s

44b Leaves usually with at least 4 lobes, these sometimes divided again

45a Stem leafy for about one-third or one-half its length; pappus scales 2–5 mm long . . . . *Coreopsis calliopsidea* (pl. 9)
Leafystem Coreopsis; Al-s

45b Leaves all basal; pappus scales lacking or only 1 mm long

46a Leaves up to 10 cm long, with lobes 4–8 mm long, these not crowded; at elevations of 100–3,000 ft . . . . *Coreopsis stillmanii*
Stillman Coreopsis; CC-s

46b Leaves up to 5 cm long, with lobes 2–3 mm long, these crowded; at elevations of 1,800–4,500 ft . . . . *Coreopsis hamiltonii*
Mount Hamilton Coreopsis; CC (MD), SCl (MH); 1b

30b Leaves present along the stem, not insignificant compared to the basal leaves (*Coreopsis calliopsidea*, in choice 30a, may have well-developed leaves in the lower half of the stem)

47a At least some leaves alternate . . . . ASTERACEAE, GROUP 3, SUBKEY

47b All leaves opposite

48a Phyllaries in 2 or 3 series, the tips not all at the same level

49a Leaves fleshy, not toothed; pappus absent or consisting of 1–5 bristles less than 1 mm long; not more than 15 cm tall; in salt marshes (phyllaries not hairy; rays 3–5 mm long) . . . . *Jaumea carnosa* (pl. 12)
Fleshy Jaumea; SFBR-n

49b Leaves not fleshy, often toothed or with shallow lobes; pappus of scales or bristles 2–5 mm long; often 100 cm tall; not usually in salt marshes

50a Phyllaries with long hairs, the outer series not different from inner; disk flowers reddish; rays about 10 mm long; pappus scales 2, 2 mm long (leaves rough to the touch because of bumps or short hairs) . . . . *Helianthus ciliaris*
Blueweed; na

50b Phyllaries without long hairs, the outer series leaflike, the inner membranous; disk flowers yellow; rays 6–8 mm long; pappus bristles 2–4, 2–5 mm long (in freshwater wetlands)

51a Rays 8–15 mm long, pale yellow; flower head nodding as it ages . . . . *Bidens cernua*
Nodding Bur-marigold; SF-n

51b Rays 15–30 mm long, bright yellow; flower head not nodding as it ages . . . . *Bidens laevis* (fig.)
Bur-marigold; Ma-s

48b Phyllaries in 1 or 2 series, the tips all at the same level

52a Phyllaries united at their bases to form a cup (leaves not toothed; rays 6–13)

53a Rays scarcely visible, shorter than the height of the involucre; phyllaries hairy; pappus consisting of 1–9 scales ... *Lasthenia glaberrima*
Smooth Goldfields; Al-n

53b Rays obvious, longer than the height of the involucre; phyllaries not hairy; pappus absent (mostly in vernal pools and alkaline flats)

54a Achene not hairy, but the surface may be roughened; rays 4–14 mm long . . . . *Lasthenia glabrata*
Yellow-rayed Goldfields

54b Achene with short hairs; rays 6–10 mm long . . . .
. . . . *Lasthenia ferrisiae*
Ferris Goldfields; CC-SLO; 4

52b Phyllaries usually free from one another, sometimes slightly fused at their bases

55a Leaves triangular, the blades less than 2 times as long as wide; petioles of lower leaves 4–7 cm long (phyllaries with long hairs; rays 10–15, up to 30 mm long; pappus consisting of many bristles; leaves toothed) . . . . *Arnica cordifolia*
Heartleaf Arnica

55b Leaves elliptic, the blades at least 2–4 times as long as wide; petioles none

56a Phyllaries 3–6, not hairy; flower head higher than wide; rays usually 4, about 1 mm long; pappus scales 1–3 (leaves usually not toothed; on shaded slopes) . . . . *Lasthenia microglossa*
Small-rayed Goldfields; CC-s

56b Phyllaries at least 8, hairy; flower head usually about as high as wide; rays 6–16, more than 1 mm long; pappus absent or consisting of 1–8 bristles or scales

57a Leaves 4–8 mm wide, either with long hairs on the margins or with several lobes, these mainly at the tips

58a Leaves with several lobes concentrated at the tips, but without hairs; basal leaves absent; rays not more than 3 mm long (somewhat succulent; coastal) . . . .
. . . . *Lasthenia maritima* [*L. minor* ssp. *maritima*]
Maritime Goldfields; Mo-n

58b Leaves not lobed, but with long hairs on the margins; basal leaves prominent; rays 5–16 mm long . . . .
. . . . *Lasthenia macrantha*
Seacoast Goldfields; Ma-n; 1b

57b Leaves rarely more than 3 mm wide, without long hairs and either not lobed or with a few lobes, these not mainly at the tips

59a Leaves lobed, the lobes less than 1 mm wide (rays 5–7 mm long; often near drying pools) .... *Lasthenia fremontii*
Fremont Goldfields; Sl, CC

59b Leaves, if lobed, with lobes at least 2 mm wide

 60a Leaves not lobed; rays 5–10 mm long; on grassy slopes (widespread) .... *Lasthenia californica*
[*L. chrysostoma* sspp. *chrysostoma* and *hirsutula*] (pl. 13)
California Goldfields, Coast Goldfields

 60b Lower leaves lobed; rays 4–8 mm long; in moist areas .... *Lasthenia minor*
Woolly Goldfields; Sn-s

Asteraceae, Group 3, Subkey: Flower head with both ray flowers and disk flowers; ray flowers completely or partly yellow, yellow green, orange, or red; leaves present along the stem, at least some of them alternate

1a Plant woody, at least at the base

 2a Leaves neither toothed nor lobed

  3a Undersides of young leaves densely hairy (rays 6–9, 3–5 mm long; pappus scales 1 mm long) ....
... *Eriophyllum staechadifolium* [includes *E. staechadifolium* var. *artemisiaefolium*]
Seaside Woolly Sunflower, Lizardtail; SCr-Me

  3b Undersides of leaves not densely hairy

   4a Rays 13–18, 8–15 mm long (leaves 10–55 mm long and up to 2.5 mm wide; pappus bristles 5–7 mm long, white) ....
.... *Ericameria linearifolia* [*Haplopappus linearifolius*] (pl. 10)
Interior Goldenbush, Stenotopsis; La-s

   4b Rays 11 or fewer, not more than 5 mm long

    5a Rays 2–6; leaves 4–12 mm long, almost cylindrical, about 1 mm wide, and in clusters; pappus consisting of many bristles (on backshores of beaches) .... *Ericameria ericoides* [*Haplopappus ericoides*] (fig.)
Mock-heather

    5b Rays 8–11; leaves up to 50 mm long, flat, mostly wider than 1 mm, not in clusters; pappus consisting of 10–12 scales .... *Gutierrezia californica*
California Matchweed, San Joaquin Matchweed; SFBR-n + e

 2b At least some leaves toothed or lobed

  6a Leaves not hairy, toothed, rarely lobed; phyllaries glandular, the flower head very sticky (rays 20–60, 12–20 mm long; pappus bristles 1–6, more than 1.5 mm long; at borders of salt marshes and on seaside bluffs)

   7a Lower leaves with petioles; involucre 10–15 mm wide; not often more than 50 cm tall, woody only near the base (widespread) ....
.... *Grindelia stricta* var. *platyphylla* [includes *G. stricta* ssp. *venulosa*]
Pacific Gumplant; Ma-n

7b   All leaves sessile; involucre 12–23 mm wide; up to 150 cm tall, woody throughout .... *Grindelia hirsutula* var. *hirsutula* [includes *G. humilis*]

Hairy Gumplant; SFBR

6b   Leaves, at least young ones, woolly on the undersides, at least some lobed; phyllaries not glandular, the flower head not sticky

8a   Rays 10–20 mm long; pappus consisting of bristles (involucre more than 5 mm wide; flower heads in clusters of 3–20) ....

.... *Senecio flaccidus* var. *douglasii* [*S. douglasii*]

Bush Groundsel, Shrubby Butterweed; Me-s

8b   Rays 2–10 mm long; pappus absent or consisting of 1–15 scales less than 1.5 mm long

9a   Involucre not more than 5 mm wide; flower heads in clusters of 5 or more, each head on a peduncle not more than 1 cm long

10a   Rays (6–8) 6–10 mm long .... *Eriophyllum latilobum*

San Mateo Woolly Sunflower; SM; 1b

10b   Rays 2–5 mm long

11a   Phyllaries 5 or 6; rays 4–6; on dry slopes in the Coast Ranges (widespread) .... *Eriophyllum confertiflorum*

[includes *E. confertiflorum* var. *laxiflorum*] (pl. 10)

Golden Yarrow

11b   Phyllaries 8–11; rays 6–9; on coastal bluffs ....

.... *Eriophyllum staechadifolium*

[includes *E. staechadifolium* var. *artemisiaefolium*]

Seaside Woolly Sunflower, Lizardtail; SCr-Me

9b   Involucre at least 10 mm wide; flower head mostly solitary, each on a peduncle at least 2 cm long (rays 6–10 mm long)

12a   Rays 5–9, orange yellow (involucre 5–7 mm high; sometimes on serpentine) .... *Eriophyllum jepsonii*

Jepson Woolly Sunflower; CC-SB; 4

12b   Rays 8–15, yellow

13a   Involucre 5–8 mm high; leaves up to 1.5 cm wide, the unlobed portion nearest the petiole 2–3 mm wide; Coast Ranges (widespread) ....

.... *Eriophyllum lanatum* var. *achillaeoides* (pl. 11)

Common Woolly Sunflower; SCl, Mo-n

13b   Involucre 8–10 mm high; leaves up to 2 cm wide, the unlobed portion nearest the petiole about 5 mm wide; coastal bluffs .... *Eriophyllum lanatum* var. *arachnoideum*

Coastal Woolly Sunflower

1b   Plant not woody

14a   Plant with glandular hairs or with phyllaries that are sticky because of glandular secretions

15a Leaves opposite below, alternate above (leaves with long, nonglandular hairs, and usually with a strong odor, the lower ones sometimes toothed; achene of each ray flower completely enclosed within a phyllary)

16a Pappus of each disk achene consisting of scales, these sometimes only 1 mm long, pappus of each ray achene a crown of scales or absent (involucre 4–7 mm high; disk flowers 8–30; anthers yellow or brown)

17a Leaves opposite well up the stem; glandular hairs gold colored or brown; rays 8–15, 4–10 mm long; up to 65 cm tall; perennial (pappus scales 1 mm long; at the edges of wooded areas) . . . . *Madia madioides* Woodland Madia

17b Leaves alternate except for the lowest pair; glandular hairs black or dark brown; rays 3–8, 2–7 mm long; up to 25 cm tall; annual (rays often with a reddish blotch at their bases)

18a Leaves evenly spaced on stem; pappus scales 2–4 mm long; plant generally branched; flower head nodding in bud stage; on volcanic ash . . . . *Madia nutans* Nodding Madia; Na, Sn

18b Leaves mainly in crowded clusters; pappus scales 1 mm long; plant with few if any branches; flower head not nodding in bud stage; on serpentine . . . . *Madia hallii* Hall Madia; Na

16b Neither ray nor disk achenes with a pappus

19a Rays less than 5 mm long (rays usually yellow, sometimes red tinged, but without a red blotch)

20a Disk flowers usually not more than 6; rays 1–7

21a Involucre 2–4 mm high; leaves 1–3 cm long; anthers yellow; up to 15 cm tall (rays 3–5, 1 mm long) . . . . *Madia minima* Small Madia; SCl(MH)

21b Involucre 5–9 mm high; leaves up to 9 cm long; anthers black; up to 70 cm tall

22a Rays 1–3, 1–3 mm long; leaves up to 9 cm long; at elevations of 3,500–8,800 ft . . . . *Madia glomerata* Mountain Tarweed

22b Rays 3–7, 3–4 mm long; leaves up to 7 cm long; at elevations up to 500 ft . . . . *Madia anomala* Plumpseed Madia; CC-n

20b Disk flowers at least 11, except sometimes in *Madia gracilis;* rays 3–12 (anthers black or at least dark)

23a Leaves lemon scented, the longest up to 9 cm (plant glandular only in the upper half but hairy throughout; rays 4–10 mm long; involucre 6–8 mm high) . . . . *Madia citriodora* Lemon-scented Tarweed; Na-n

23b Leaves not lemon scented but with a resinous odor, the longest more than 9 cm (widespread)

24a Plant glandular and hairy throughout; leaf more than 10 mm wide at the base; rays 6–9, 1–4 mm long; involucre 7–15 mm high .... *Madia sativa* [includes *M. capitata*]
Coast Tarweed

24b Plant glandular only in the upper half, but hairy throughout; leaf 5 mm wide at the base; rays 3–9, 1–8 mm long; involucre 6–10 mm high (disk flowers 2–12) ... *Madia gracilis* (pl. 13)
Slender Madia, Slender Tarweed

19b Rays 6–20 mm long

25a Rays 3–12, 6–10 mm long (longest leaves 10 cm; plant glandular only in the upper half, but hairy throughout; anthers black or at least dark)

26a Leaves not lemon scented but with a resinous odor; disk flowers 2–12 (widespread) .... *Madia gracilis* (pl. 13)
Slender Madia, Slender Tarweed

26b Leaves lemon scented; disk flowers 15–50 .... *Madia citriodora*
Lemon-scented Tarweed; Na-n

25b Rays 8–20, 6–20 mm long

27a Plant without conspicuous nonglandular hairs; rays without a red blotch; anthers yellow or black; longest leaves 10 cm (involucre 4–7 mm high; disk flowers 20–65; rays 8–16; annual; at elevations below 3,000 ft) .... *Madia radiata*
Showy Madia, Golden Madia; CC-s; 1b

27b Plant with conspicuous nonglandular hairs; rays often with a red blotch; anthers black or purple black; longest leaves 20 cm (stem often reddish or dark purple)

28a Basal cluster of leaves well developed; glands on upper stem and leaves usually obvious without a hand lens; flowers produced August to November (involucre 6–12 mm high; disk flowers 15 to many; rays 12–20, 10–20 mm long; at elevations below 3,000 ft; widespread) ....
.... *Madia elegans* ssp. *densifolia*
Common Madia

28b Basal cluster of leaves either lacking or not well developed; glands on upper stem and leaves not obvious without a lens; flowers produced March to August

29a Rays 12–14, 10–20 mm long; disk flowers 25–50; involucre 8–11 mm high; at elevations below 3,000 ft ...
.... *Madia elegans* ssp. *vernalis*
Springtime Madia

29b Rays 8–16, 6–15 mm long; disk flowers sometimes fewer than 25; involucre 4–10 mm high; at elevations of 3,000–8,000 ft ..... *Madia elegans* ssp. *elegans* (pl. 13)
Elegant Madia

15b All leaves alternate

    30a Upper leaves often clustered, ending abruptly in an open gland (rays 4–6 mm long; lower leaves toothed, upper leaves reduced and usually smooth margined; pappus absent)

        31a Anthers yellow (rays 3–10)

            32a Stem and leaves in upper part of plant densely hairy; each phyllary with 25–50 slender, gland-tipped structures . . . . *Holocarpha heermannii* (pl. 12)

                Heermann Tarplant; CC-SCl (MH)

            32b Stem and leaves in upper part of plant not densely hairy; each phyllary with 5–20 stout, gland-tipped structures . . . . *Holocarpha obconica* [includes *H. obconica* ssp. *autumnalis*]

                San Joaquin Tarplant; Al, CC-s

        31b Anthers black

            33a Ray flowers 3–7; flower heads scattered in a raceme . . . . *Holocarpha virgata*

                Virgate Tarweed; Na, SCl

            33b Ray flowers 8–16; flower heads densely clustered . . . . *Holocarpha macradenia*

                Santa Cruz Tarweed; Ma, Al, SCr; 1b

    30b Upper leaves not clustered and not ending in an open gland; however, there may be glands elsewhere on the plant

        34a Leaves smooth margined (leaves not more than 5 mm wide)

            35a Achene of each ray flower partly enclosed by the adjacent phyllary (leaves 6–15 cm long and 0.5 cm wide; rays 5–13, often 3 lobed; plant sometimes glandular on the lower portion, but usually hairy throughout; pappus absent) . . . . *Hemizonia congesta* ssp. *congesta* [includes *H. lutescens* and *H. multicaulis* sspp. *multicaulis* and *vernalis*] (pl. 12)

                Yellow Hayfield Tarweed; Me-Ma, Al, Sn

            35b Achene of each ray flower not enclosed by a phyllary (pappus of each ray achene absent)

                36a Leaves up to 5 cm long; very small, globular glands present on the stem and leaves; involucre 7–10 mm high; rays 3–30, yellow, not 3 lobed; pappus of each disk achene consisting of scales and many bristles; perennial (plant hairy and camphor scented) . . . . *Heterotheca sessiliflora* ssp. *echioides* [*Chrysopsis villosa* vars. *camphorata* and *echioides*] (pl. 12)

                Bristly Golden Aster; Sn, Al, SCl (MH)-s

36b Leaves often more than 8 cm long; prominent, saucer-shaped glands present at the tips of the upper leaves and sometimes on the phyllaries; involucre 4–7 mm high; rays 2–6, yellow or rose, sometimes with a red dot at their bases, often 3 lobed; pappus of each disk achene absent or consisting of 7–12 scales; annual

37a Leaves and stem not hairy, except the lower leaves; saucer-shaped glands usually absent on the phyllaries; rays yellow (anthers sometimes purple) .... *Calycadenia truncata*
Rosinweed; Me-SCl

37b Leaves and stem usually hairy; saucer-shaped glands usually present on the phyllaries; rays yellow or rose ....
.... *Calycadenia multiglandulosa* [includes *C. hispida* ssp. *reducta* and *C. multiglandulosa* sspp. *cephalotes* and *robusta*] (pl. 9)
Sticky Rosinweed, Hispid Calycadenia; Me-SCl

34b Leaves usually toothed, lobed, or compound, except some species of *Grindelia* may have smooth-margined leaves

38a Achene of each ray flower completely enclosed by the adjacent phyllary (pappus of each disk achene consisting of bristles, pappus of each ray achene absent)

39a Stem with black dots or streaks; pappus bristles whitish to red brown (anthers black or purple; rays 6–18)

40a Rays 3–15 mm long, yellow, sometimes with white tips ...
.... *Layia gaillardioides*
Woodland Layia

40b Rays 1–4 mm long, yellow .... *Layia hieracioides*
Tall Layia; Al, CC (MD)

39b Stem without black dots or streaks; pappus bristles white or off white

41a Anthers usually purple; rays yellow, but usually with white tips (rays 5–18, 3–20 mm long; pappus bristles 14–32; widespread) .... *Layia platyglossa*
[includes *L. platyglossa* ssp. *campestris*] (pl. 13)
Tidytips; Me-s

41b Anthers yellow; rays yellow

42a Rays 5–9, 4–15 mm long; pappus bristles 16–21; often on serpentine soil .... *Layia septentrionalis*
Colusa Layia; Sn; 1b

42b Rays 3–14, 3–22 mm long; pappus bristles 10–15; on sandy soil, inland .... *Layia glandulosa*
White Layia; CC-s

38b Achene of each ray flower partly or not at all, enclosed by the adjacent phyllary (rays orange or yellow, sometimes with purple veins)

  43a Phyllaries in 1 series

    44a Leaves toothed; rays orange; a phyllary not enclosing each ray achene; achene prickly; pappus none (plant glandular hairy; annual; up to 15 cm tall) . . . . *Calendula arvensis*
Field-marigold; eu

    44b At least lower leaves lobed; rays yellow, sometimes with purple veins; a phyllary partly enclosing each ray achene; achene not prickly; pappus of each disk achene absent or consisting of 6–12 scales, pappus of each ray achene absent

      45a Leaves with spinelike tips or lobes (anthers black; pappus of each disk achene consisting of 8–12 scales; rays 10–20, 3–4 mm long, 2 lobed) . . . . . . *Hemizonia fitchii*
Fitch Spikeweed; Me-SLO

      45b Leaves without spinelike tips

        46a Anthers yellow; pappus of each disk achene consisting of 6–12 scales (rays 5, 4–8 mm long) . . . . . .
. . . . *Hemizonia kelloggii*
Kellogg Spikeweed; SFBR-s

        46b Anthers black; pappus absent

          47a Rays 18–24, 4–7 mm long, without purple veins . . . . *Hemizonia corymbosa*
Coast Spikeweed, Coast Tarweed; Me-Mo

          47b Rays 5–13, 5–11 mm long, usually with purple veins . . . . . *Hemizonia congesta* ssp. *congesta*
. . . [includes *H. lutescens* and *H. multicaulis* sspp. *multicaulis* and *vernalis*] (pl. 12)
Yellow Hayfield Tarweed; Me-Ma, Al, Sn

  43b Phyllaries in more than 1 series (lower leaves usually toothed, but not lobed; pappus consisting of 1–6 bristles)

    48a Tips of outer phyllaries curled back, sometimes so much that they form loops

      49a Upper stem somewhat hairy; plant slightly succulent; rays 12–25 mm long; in coastal areas, including borders of salt marshes (rays 20–60; sometimes prostrate; widespread) . . . . *Grindelia stricta* var. *platyphylla*
[includes *G. stricta* ssp. *venulosa*]
Pacific Gumplant

      49b Upper stem not hairy; plant not succulent; rays 5–11 mm long; inner Coast Ranges

        50a Lower leaves up to 15 cm long, not sessile; rays 11–28 . . . . *Grindelia nana*
Idaho Gumplant; Na-n

50b Lower leaves up to 8 cm long, sessile; rays 32–39
... *Grindelia camporum* [includes *G. camporum*
var. *parviflora* and *G. procera*] (pl. 11)
Great Valley Gumplant

48b Tips of outer phyllaries straight, spreading outward or
curved, but not curled back

51a Upper stem mostly hairy (rays 20–60, 14–20 mm long;
in Coast Ranges and salt marshes) ....
.... *Grindelia hirsutula* var. *hirsutula*
[includes *G. humilis*]
Hairy Gumplant; Na-Mo

51b Upper stem usually not hairy

52a Lower leaves up to 18 cm long, not sessile; rays
10–40, 11–15 mm long (on ocean bluffs) ....
.... *Grindelia hirsutula* var. *maritima*
[*G. maritima*]
San Francisco Gumplant; SF; 1b

52b Lower leaves up to 8 cm long, sessile; rays 32–39,
8–11 mm long ... *Grindelia camporum* [includes
G. camporum var. *parviflora*
and *G. procera*] (pl. 11)
Great Valley Gumplant

14b Plant not glandular, but sometimes hairy

53a Leaves smooth margined, but species of *Helianthus* sometimes have toothed or
lobed leaves

54a Leaves rough to the touch because of bumps or short hairs; sometimes lower
leaves opposite (most leaves 4 times as long as wide; disk flowers yellow or
brown)

55a Lower phyllaries shorter than the combined height of the involucre and
disk; mostly in dry places (pappus consisting of 2 scales)

56a Phyllaries with long hairs on the margins; leaf with 1 prominent vein
from the base .... *Helianthus ciliaris*
Blueweed; na

56b Phyllaries with short hairs; leaf with 3 prominent veins from the
base .... *Helianthus gracilentus* (pl. 12)
Slender Sunflower; CC-s

55b Lower phyllaries longer than the combined height of the involucre and
disk; mostly in wet places

57a Flower heads 2.5–3.5 cm wide, solitary on each stem; pappus absent
or consisting of 2 scales .... *Helianthella castanea*
Diablo Helianthella; 1b

57b Flower heads 1.8–2.5 cm wide, in panicles on each stem; pappus
consisting of 2 scales .... *Helianthus californicus*
California Sunflower; Na-s

54b Leaves not rough to the touch because of bumps or short hairs; all leaves alternate (flower head less than 2.5 cm wide)

    58a Rays 2–3 mm long

        59a Leaves not often more than 1 mm wide, without dark, glandular pits; flower heads usually not more than 10 on each plant, solitary at the tips of branches; rays fewer than 15; pappus scales 1–5 or absent; annual; not more than 30 cm tall . . . . *Rigiopappus leptocladus*

        59b Leaves usually at least 3 mm wide, with dark, glandular pits; flower heads numerous, in cymes; rays 15–25; pappus bristles 25–45; perennial; often more than 100 cm tall . . . . *Euthamia occidentalis [Solidago occidentalis]* (fig.)

                                      Western Goldenrod

    58b Rays at least 5 mm long

        60a Leaves with dark glandular pits; pappus consisting of 5–10 scales

            61a Leaves about 2 mm wide, clustered; rays 7–10 (rays 6–12 mm long) . . . . *Helenium amarum*

                                  Fineleaf Sneezeweed; na

            61b Leaves 10–40 mm wide, not clustered; rays 13–20

                62a Rays 13–15, 4–10 mm long; inflorescence usually with more than 10 flower heads; annual . . . . *Helenium puberulum* (pl. 11)

                                  Rosilla, Sneezeweed

                62b Rays 14–20, 13–25 mm long; inflorescence with fewer than 10 flower heads; perennial . . . . *Helenium bigelovii*

                                  Bigelow Sneezeweed

        60b Leaves without dark glandular pits; pappus consisting of scales and bristles (rays 3–30, 3–10 mm long; in sandy soil)

            63a Leafy bracts not present beneath the flower head; leaves up to 5 cm long; involucre 7–10 mm high . . . . *Heterotheca sessiliflora* ssp. *echioides* [*Chrysopsis villosa* vars. *camphorata* and *echioides*] (pl. 12)

                          Bristly Golden Aster; Sn, Al, SCl(MH)-s

            63b Leafy bracts present beneath the flower head; leaves not more than 4 cm long; involucre 10–15 mm high (usually coastal)

                64a Leaves and stem densely hairy, the upper leaves more than 1 cm long, with petioles; plant branching extensively from the base . . . . *Heterotheca sessiliflora* ssp. *bolanderi* [*Chrysopsis villosa* var. *bolanderi*]

                          Bolander Golden Aster; SFBR-Me

                64b Leaves and stem not densely hairy, the upper leaves not more than 1 cm long, sessile; plant usually not branching from the base but sometimes branching above . . . . *Heterotheca sessiliflora* ssp. *sessiliflora* [*Chrysopsis villosa* var. *sessiliflora*]

                  Sessile Golden Aster, Hairy Golden Aster; Me-s

53b At least some leaves toothed or lobed, except *Layia chrysanthemoides* sometimes has all smooth-margined leaves

65a Achene of each ray flower enclosed by a phyllary (lower leaves lobed, upper toothed, or smooth margined; pappus of each ray achene absent)

66a Leaves without spinelike tips; rays 6–16, 3 lobed, yellow with white tips (rays 3–18 mm long; phyllaries with stout hairs on the margins; pappus of each disk achene absent or consisting of 2–18 bristles; involucre 6–11 mm high) . . . . *Layia chrysanthemoides*

[includes *L. chrysanthemoides* ssp. *maritima*]

Smooth Layia; Me-Mo

66b Leaves with spinelike tips; rays 9–30 or more, 2 lobed, yellow

67a Involucre 5–10 mm high; pappus of each disk achene consisting of 3–5 scales

68a Leaves and stem glandular and with long hairs; rays 3–6 mm long (often near marshes) . . . . *Hemizonia parryi* ssp. *parryi*

Pappose Spikeweed; SM-n

68b Leaves and stem not glandular and without long hairs; rays 2.5–3 mm long . . . . *Hemizonia parryi* ssp. *congdonii*

Congdon Tarplant; Al, CC-s

67b Involucre 3–6 mm high; pappus absent (rays 3–5 mm long)

69a Leaves rough to the touch; usually much more than 10 cm tall; inland, in dry habitats . . . . *Hemizonia pungens* ssp. *pungens*

Common Spikeweed; Ma

69b Leaves not rough to the touch; less than 10 cm tall; around San Francisco Bay, in salt marshes as well as in dry habitats . . . .
. . . . *Hemizonia pungens* ssp. *maritima*

Salt Marsh Spikeweed; SFBR-s

65b Achenes not enclosed by phyllaries

70a Flower head more than 1.5 cm wide (lower leaves sometimes lobed, upper toothed or smooth margined)

71a Lower leaves opposite, upper alternate

72a Leaves rough to the touch because of bumps or short hairs; pappus scales 2

73a Leaves rarely so much as twice as long as wide; disk flowers reddish (flower head 2.5–3.5 cm wide; widespread) . . . .
. . . . *Helianthus annuus*

[includes *H. annuus* ssp. *lenticularis*] (pl. 11)

Common Sunflower

73b Most leaves 4 times as long as wide; disk flowers reddish purple (phyllaries with long hairs on the margins) . . . .
. . . . *Helianthus ciliaris*

Blueweed; na

72b Leaves not rough to the touch, but covered with soft hairs; pappus absent

74a Involucre 8–13 mm high; phyllaries united, forming a cup; in grassland . . . . *Monolopia major*
Cupped Monolopia; SF-Mo

74b Involucre 5–7 mm high; phyllaries separate; in wooded or shrubby areas . . . . *Monolopia gracilens*
Woodland Monolopia; CC (MD)-s

71b All leaves alternate

75a Rays curled at the tips (pappus of each disk flower consisting of bristles and sometimes scales; rays 25–40, yellow)

76a Stem and leaves hairy; rays 5–8 mm long; pappus bristles 3–5 mm long, red (widespread) . . . .
. . . . *Heterotheca grandiflora* (pl. 12)
Telegraphweed

76b Stem and leaves not hairy; rays 12–18 mm long; pappus bristles about 7 mm long, yellow to red brown
Prionopsis ciliata [Haplopappus ciliatus]; na

75b Rays not curled at the tips

77a Rays 20, golden yellow, 3–4 mm wide, each with a deep notch at the tip; pappus absent; leaves mostly twice lobed
. . . . *Chrysanthemum segetum*
Corn Daisy, Corn Chrysanthemum; me

77b Rays 30–60, reddish yellow or purple, 1 mm wide, each usually somewhat frayed at the tip but without a distinct notch; pappus scales 4, 1–3 mm long; leaves coarsely toothed . . . . *Hulsea heterochroma*
Red-rayed Hulsea; SCl-s

70b Flower head less than 1.5 cm wide

78a Leaves lobed

79a Stem and sometimes also the leaves and phyllaries with woolly hairs; rays 10–20, 7–15 mm long; pappus a crown of scales (lobes of leaves regularly toothed or divided again into small secondary lobes) . . . . *Anthemis tinctoria*
Golden Marguerite; me

79b Stem, phyllaries, and leaves without woolly hairs; rays 6–13, 5–7 mm long; pappus absent

80a Lower leaves smooth margined, the upper ones with 1–3 lobes, these 10–30 mm long (stigmas red; in hard-packed soil of dried vernal pools) . . . . *Blennosperma bakeri*
Baker Blennosperma; Sn; 1b

80b All leaves with 3–15 lobes, these not more than 5 mm long

81a Disk flowers 60–100; pollen yellow; achene 3–4.5 mm long; in sandy soil on Pt. Reyes . . . .
. . . . *Blennosperma nanum* var. *robustum*
Point Reyes Blennosperma; Ma (PR); 1b

81b Disk flowers 20–70; pollen white; achene 2.5–3 mm long; on wet clay soil (widespread) ....
.... *Blennosperma nanum* var. *nanum* (pl. 9)
Common Blennosperma, Glueseed

78b Leaves not lobed but some usually toothed

82a Stem and leaves not hairy (involucre 5–6 mm high; pappus bristles 25–45; mostly along the coast) ....
.... *Solidago spathulata* (pl. 14)
Coast Goldenrod, Dune Goldenrod; Mo-n

82b Stem hairy, leaves sometimes hairy at least on 1 surface

83a Phyllaries not obscured by dense hairs, but some hairs may be present; pappus consisting of 25–45 bristles

84a Inflorescence usually 2–4 times as long as wide; leaves hairy on both surfaces; basal and lower leaves toothed, upper leaves smooth margined, much reduced (widespread) ..... *Solidago californica* (pl. 14)
California Goldenrod

84b Inflorescence usually not more than 2 times as long as wide; leaves either not hairy or hairy only on the undersides; all leaves either toothed or smooth margined, similar in size (widespread) ....
.... *Solidago canadensis* ssp. *elongata* (pl. 14)
Canada Goldenrod

83b Phyllaries obscured by dense hairs; pappus absent (rays 5, often with red veins; lower leaves toothed, the upper smooth margined; anthers black)

85a Rays 8–13 mm long, bright yellow; involucre 4–5 mm high; longest leaves 5.5 cm .... *Lagophylla minor*
Lesser Hareleaf; Na-n

85b Rays 3–5.5 mm long, pale yellow; involucre 4–7 mm high; longest leaves 12 cm

86a Flower heads in dense clusters, these 1.5–6 cm wide .... *Lagophylla ramosissima* ssp. *congesta* [*L. congesta*]
Rabbitfoot; SCr-n

86b Flower heads single or in clusters not more than 1.5 cm wide ....
.... *Lagophylla ramosissima* ssp. *ramosissima*
Common Hareleaf

BERBERIDACEAE (BARBERRY FAMILY) The Barberry Family consists of shrubs and perennial herbs, usually with alternate leaves. The basic flower formula is as follows: six sepals, often petal-like, in two circles; six petals, also in two circles; six stamens; fruit dry or fleshy, with only one compartment. But there are exceptions, such as *Achlys triphylla* (Vanilla-leaf), whose flower lack sepals and petals and usually has nine stamens. The ovary is superior.

Many exotic barberries have long been in cultivation, and some of our native species are now available, including *Berberis aquifolium* (Oregon-grape), the Oregon state flower, which is not sharply distinct from the local *B. pinnata*. Also, *Vancouveria planipetala* (Inside-out-flower) is a useful herb for shaded gardens. It spreads by underground stems and makes an attractive ground cover.

1a   Shrubs; leaves not basal, pinnately compound; marginal teeth of leaflets spine tipped (fruit a blue purple berry)

     2a   Stem usually not branched; leaves 25–45 cm long; leaflets usually 11–21, each with 12–24 teeth; usually in shaded habitats . . . . *Berberis nervosa*
                              Longleaf Oregon-grape; Mo-n

     2b   Stem branched; leaves 5–15 cm long; leaflets usually 5–11, each with 16–40 teeth; usually in sunny habitats . . . . *Berberis pinnata* (pl. 15)
                     Shinyleaf Oregon-grape, California Barberry

1b   Herbs; leaves basal, compound, but not pinnately; marginal teeth, if present on leaflets, not spine tipped

     3a   Leaves with 3 fan-shaped leaflets arising at the top of the petiole, the broad end of the leaflets coarsely toothed; inflorescence dense, narrow, the flower less than 5 mm long; calyx and corolla absent; fruit an achene . . . . *Achlys triphylla* (pl. 15)
                                     Vanilla-leaf; Sn-n

     3b   Leaves with 3 primary divisions, each of these divided again into 3 long-stalked leaflets, these sometimes lobed; inflorescence a loose panicle, the flower 6–8 mm long; sepals 12–15 in 2 whorls, the outer bractlike, petals 6, these turned back; fruit a capsule . . . . . .
        . . . . *Vancouveria planipetala* (pl. 15)
                            Inside-out-flower, Redwood-ivy; Mo-n

BETULACEAE (BIRCH FAMILY)   The trees and shrubs of the Birch Family are represented in our region by a hazelnut and two species of alders, all of which have alternate, deciduous leaves. In alders, both the pistillate and staminate flowers are in catkins; the pistillate catkins, especially after becoming woody, resemble the cones of coniferous trees. In hazelnuts, only the staminate flowers are in catkins; the pistillate flowers are produced singly or in clusters, but as a rule only one flower of a cluster matures as an acornlike nut.

Alders, like many other plants, especially members of the Pea Family, have nitrogen-fixing bacteria in nodules on their roots. These bacteria, with help from their hosts, are able to utilize atmospheric nitrogen gas in synthesis of amino acids and proteins. Much of the nitrogen fixed by the bacteria becomes incorporated into the trees. Thus alder seedlings may flourish on soils that are poor in ammonia and nitrates, which are the usual sources of nitrogen for plants. The symbiotic association accounts, in part, for the success of some species on land that has been deforested. When the leaves decay, their organic nitrogen is converted by various soil bacteria into ammonia and nitrates. By enriching the soil, and also by making the soil more acidic, alders prepare the land for colonization by other trees, such as *Pseudotsuga menziesii* (Douglas-fir). It

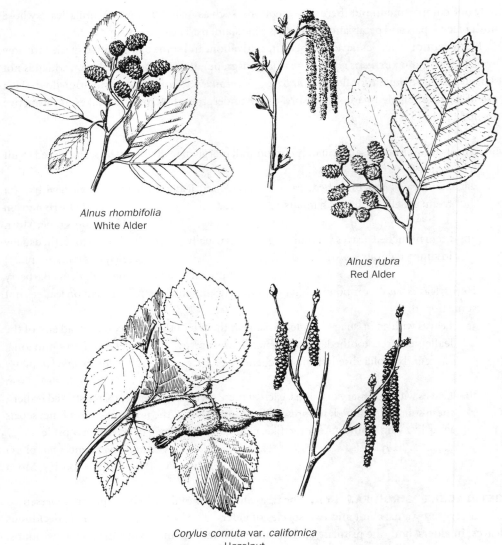

Alnus rhombifolia
White Alder

Alnus rubra
Red Alder

Corylus cornuta var. *californica*
Hazelnut

is a little ironic that the increased acidity makes the soil unsuitable for succeeding generations of alder seedlings.

1a   Leaf slightly notched at the base, the 2 sides of the basal portion often not quite equal; staminate flowers in catkins, pistillate flowers single or in small clusters; fruit not flattened, about 15 mm long, each nut enclosed by 2 fused papery bracts; shrub or small tree (stigmas red) . . . . *Corylus cornuta* var. *californica* (fig.)
Hazelnut

1b   Leaf not notched at the base, and the 2 sides of the basal portion equal; both staminate and pistillate flowers in catkins, the pistillate catkins, when mature, with numerous bracts, and resembling conifer cones; fruit flattened, about 2.5 mm long, each nutlet with a woody bract beneath it; tree

    2a   Margins of leaf blades inrolled and with coarse teeth that usually have small secondary teeth; undersides of leaves slightly rusty gray; fruit with prominent membranous margins that are at least half as wide as the thickened portion; mostly near the coast . . . . . .
. . . . *Alnus rubra [A. oregona]* (fig.)
Red Alder; SCr-n

    2b   Margins of leaf blades not inrolled, the teeth small and more or less the same size; undersides of leaves not at all rusty gray; fruit with thin margins, these not membranous; mostly along streams away from the coast (widespread) . . . . . . *Alnus rhombifolia* (fig.)
White Alder

BORAGINACEAE (BORAGE FAMILY)   Some of the Old World species of the Boraginaceae (Borage Family) have long been used as kitchen flavorings, herbal remedies, and sources of dyes. A few of them, such as *Borago officinalis* and *Symphytum asperum* (comfrey), are found growing as escapes in California, but they are close to civilization and thus not likely to be mistaken for native species. All of our representatives are herbs, except for *Echium candicans* (Viper's-bugloss).

With few exceptions, the foliage and stems of our local representatives have bristly hairs. The leaves are usually smooth margined, and those on the stem are generally alternate, except for a few lower ones that may be opposite. The inflorescences are often one sided and at least slightly coiled, in general form resembling the scroll of a violin. The corolla is generally five lobed, with five stamens attached to the tube. The ovary is superior, and the fruiting portion of the pistil is usually deeply divided into four lobes that typically separate into dry, one-seeded nutlets, as is the case also in Lamiaceae and Verbenaceae. The nutlets may be seen at the base of the pistil, usually without the aid of a hand lens. In some species, however, the nutlets do not develop equally, and not all four will reach maturity. To properly view the distinctive surfaces of the nutlets, a dissecting microscope is required.

1a   Corolla orange or yellow (in *Myosotis discolor,* under choice 1b, the corolla changes from yellow to blue as the flower ages; all leaves alternate or basal; nutlets 4)

    2a   Sepals 2–4; corolla tube with 20 veins near the base

        3a   Corolla yellow, 8–12 mm long and 2–6 mm wide (surface of nutlets resembling a cobblestone pavement; anthers pressed against the stigma; plant bristly; in sandy or gravelly areas, inner Coast Ranges) . . . .
. . . . *Amsinckia tessellata* var. *tessellata* (pl. 16)
Devil's-lettuce; CC-s

        3b   Corolla orange, 12–20 mm long and 6–15 mm wide

            4a   Corolla 12–16 mm long and 6–10 mm wide; surface of nutlets resembling a cobblestone pavement (anthers diverging away from the stigma; plant bristly; in sandy or gravelly areas, inner Coast Ranges) . . . .
. . . . *Amsinckia tessellata* var. *gloriosa [A. gloriosa]*
Largeflower Devil's-lettuce; CC-s

4b   Corolla 14–20 mm long and 10–15 mm wide; surface of nutlets smooth (re-established in some areas) . . . . *Amsinckia grandiflora*

Largeflower Fiddleneck; CC, Al; 1b

2b   Sepals 5; corolla tube with 10 veins near the base (surface of nutlets bumpy, but not re-sembling a cobblestone pavement)

5a   Corolla 4–7 mm long, 2–3 mm wide, the tube barely, if at all, extending beyond the calyx . . . . *Amsinckia menziesii* var. *menziesii* [includes *A. retrorsa*]

Rancher's Fireweed, Harvest Fireweed

5b   Corolla 7–20 mm long, 4–14 mm wide, the tube extending well beyond the calyx

6a   Corolla throat constricted and partly obstructed by somewhat saclike eleva-tions; all stamens attached at the same level below the constriction (usually in moist areas) . . . . *Amsinckia lycopsoides*

Bugloss Fiddleneck; SLO-n

6b   Corolla throat wide open, not obstructed by any elevations; stamens attached somewhat irregularly, not all at the same level

7a   Leaves finely toothed; sometimes 2 or 3 sepals, due to fusion, much shorter than the others, which are separate nearly to their bases; corolla yellow (corolla 7–15 mm long; usually around salt marshes and in coastal sandy habitats) . . . . *Amsinckia spectabilis*

Coast Fiddleneck, Seaside Fiddleneck

7b   Leaves not toothed; sepals separate nearly to their bases; corolla orange

8a   Corolla 10–20 mm long, 8–14 mm wide, the lobes usually without red marks (inland) . . . . *Amsinckia eastwoodiae* [*A. intermedia* var. *eastwoodae*]

Valley Fiddleneck

8b   Corolla 7–11 mm long, 5–12 mm wide, the lobes usually with red marks

9a   Corolla slightly 2 lipped, and its tube bent (Coast Ranges) . . . . *Amsinckia lunaris*

Bentflower Fiddleneck; 1b

9b   Corolla regular, and its tube straight . . . . *Amsinckia menziesii* var. *intermedia* [*A. intermedia*] (pl. 16)

Common Fiddleneck

1b   Corolla blue, purple, white, or white tinged with yellow or pink, sometimes becoming yel-low or pink on drying

10a   Corolla primarily blue or purple (sepals united at their bases; all leaves alternate or basal; nutlets 4)

11a   Blades of longest leaves oval, up to about 15 cm (flowers in a loose raceme; corolla bright blue, the stamens not protruding from them, with a ring of white crests at the throat; nutlets covered with short prickles; generally in shaded habitats) . . . . *Cynoglossum grande* (pl. 16)

Hound's-tongue

11b Blades of longest leaves not more than 10 cm, usually much shorter (sometimes escaping from gardens)

    12a Corolla 20 mm wide (corolla bright blue, the stamens protruding from it, and with a ring of crests at the throat; nutlets roughened) . . . . *Borago officinalis*
Borage; me

    12b Corolla not more than 12 mm wide

        13a Corolla 12 mm wide, the stamens protruding from it, without crests at the throat; inflorescence tightly coiled; nutlets roughened; shrub; 1–2 m tall (corolla purple or dark blue). . . . . *Echium candicans*
Viper's-bugloss; af

        13b Corolla not more than 10 mm wide, the stamens not protruding from it, with a ring of white crests at the throat; inflorescence not tightly coiled; nutlets smooth; herbs; up to 0.5 m tall (in moist areas)

            14a Corolla about 5–8 mm wide, light blue; widest leaves about 2 cm; perennial; up to 50 cm tall (established along the coast) . . . .
. . . . *Myosotis latifolia*
Forget-me-not; af

            14b Corolla about 4 mm wide, at first pale yellow, later blue; widest leaves less than 1 cm; annual; up to 30 cm tall . . . .
. . . . *Myosotis discolor [M. versicolor]*
Yellow-and-blue Scorpion-grass; eu

10b Corolla primarily white or white tinged with yellow, but sometimes becoming yellow or pink on drying

    15a At least lower leaves opposite . . . . BORAGINACEAE, SUBKEY

    15b Leaves all alternate or basal

        16a Upper leaves not markedly reduced; basal cluster of leaves absent or not well developed

            17a Corolla white to bluish, with a purple or purple and yellow ring in the throat; sepals united; leaves slightly succulent, not hairy, with a coating of a bluish wax; perennial; nutlets 4; growing in somewhat saline places, usually at the coast, but sometimes inland . . . .
. . . . *Heliotropium curassavicum [H. curassavicum* var. *oculatum]* (pl. 16)
Seaside Heliotrope, Chinese Pusley

            17b Corolla white, drying to yellow, without a ring in the throat; sepals separate; leaves not succulent, usually at least slightly hairy, without a coating of a bluish wax; annual; nutlets 1–4; in disturbed areas . . . .
. . . . *Lithospermum arvense*
Gromwell; eua

        16b Upper leaves much reduced; basal cluster of leaves well developed, but short lived in *Plagiobothrys infectivus* (leaves hairy; sepals united at their bases; nutlets 1–4)

            18a Corolla 3–9 mm wide; calyx 2–3 mm long; nutlets 1–4 (sap purple; widespread). . . . . *Plagiobothrys nothofulvus* (pl. 16)
Common Popcornflower

18b Corolla not more than 4 mm wide; calyx 3–7 mm long; nutlets 2–4

    19a Calyx 3–5 mm long; corolla 1–3 mm wide; sap usually not purple (widespread) .... *Plagiobothrys tenellus*

        Delicate Popcornflower, Slender Popcornflower

    19b Calyx 5–6 mm long; corolla 3–4 mm wide; sap purple

        20a Inflorescence without bracts; basal leaves persisting; widespread .... *Plagiobothrys fulvus* [includes *P. fulvus* var. *campestris*]

            Hairy Popcornflower; SCl-n

        20b Inflorescence with bracts below each of several flowers; basal leaves short lived; in clay soils .... *Plagiobothrys infectivus*

            Dye Popcornflower; SLO-n

Boraginaceae, Subkey: Corolla white or white tinged with yellow; at least some leaves opposite

1a   Sepals with hooked bristles; mature nutlets spreading, with hooked bristles (sepals separate; corolla not extending beyond the calyx; lower leaves fused at their bases, the upper ones alternate; mostly less than 20 cm tall)........................... *Pectocarya pusilla*

        Little Pectocarya; Mo-n

1b   Sepals usually without hooked bristles; mature nutlets erect, without hooked bristles

    2a   Stem and leaves with densely crowded long hairs; corolla usually with 5 crests in the tube; sepals usually separate, with white or straw-colored hairs, these usually longer than the sepals; stem usually branched only near the tip into 2–5 racemes; side of each nutlet on which the attachment scar is located without a keel but with a distinct lengthwise groove

        3a   Bristles on sepals distinctly curved, sometimes hooked (sepals 3–4 mm long in the fruiting stage; hairs on stem lying flat, pointing upward; nutlet 1, smooth; mostly on dry slopes) .... *Cryptantha flaccida*

            Flaccid Cryptantha

        3b   Bristles on sepals not distinctly curved or hooked

            4a   Sepals usually less than 2 mm long in the fruiting stage

                5a   Nutlets 4, 1 usually noticeably larger than the others and at least 1 roughened (often on burned areas)........... *Cryptantha micromeres* (pl. 16)

                    Minuteflower Cryptantha; Ma-s

                5b   Nutlet solitary, smooth (on dry slopes) .... *Cryptantha microstachys*

                    Hairy Cryptantha, Tejon Cryptantha

            4b   Sepals at least 2 mm long in the fruiting stage

                6a   Many hairs on stem pointing outward rather than lying flat against the stem

                7a   Corolla not more than 1.5 mm wide (sepals 3–6 mm long in the fruiting stage, its bristles 2–3 mm long; flowers in the fruiting stage not touching each other; nutlets 4, smooth; mostly on hillsides near the coast) ....

                  ....*Cryptantha torreyana* [includes *C. torreyana* var. *pumila*]

                    Torrey Cryptantha; Ma-n

                7b   Corolla 2–7 mm wide

8a Each sepal not wider at the base than near the middle; flowers in the fruiting stage touching each other; nutlets smooth (nutlet usually 1; sepals 3–6 mm long in the fruiting stage) . . . .
. . . . *Cryptantha clevelandii* [includes *C. clevelandii* var. *florosa*]
Cleveland Cryptantha; Ma (PR)

8b Each sepal widest at or near the base; flowers in the fruiting stage not touching each other; at least 1 nutlet roughened

9a Style of pistil not extending beyond tip of nutlets; sepals 3–5 mm long in the fruiting stage; nutlets 1–4, elongated . . .
. . . . *Cryptantha intermedia*
Common Cryptantha; La, Na, SLO

9b Style of pistil extending beyond tip of nutlets; sepals 2–4 mm long in the fruiting stage; nutlets 4, rounded (mostly in gravelly or rocky habitats; widespread) . . . .
. . . . *Cryptantha muricata* (pl. 16)
Prickly Cryptantha; CC-s

6b Most hairs on stem pointing upward and lying flat against stem

10a At least 1 nutlet roughened (nutlets usually 4)

11a Corolla less than 1 mm long; sepals 4–6 mm long in the fruiting stage; flowers in the fruiting stage touching each other; up to 15 cm tall, the stem sometimes falling; in open sandy habitats at elevations below 2,500 ft . . . .*Cryptantha hooveri*
Hoover Cryptantha; CC-s; 1b

11b Corolla 1–1.5 mm long; sepals 3–6 mm long in the fruiting stage; flowers in the fruiting stage not touching each other; up to 40 cm tall, the stem not falling; in woods at elevations above 2,500 ft . . . .*Cryptantha simulans*
Pine Cryptantha

10b Nutlets smooth

12a Corolla 2–3 mm wide; on serpentine soil (sepals 2–3 mm long in the fruiting stage; flowers in the fruiting stage not touching each other; nutlet usually solitary; inner Coast Ranges) . . . .
. . . . *Cryptantha hispidula*
Napa Cryptantha; La, Na

12b Corolla less than 2 mm wide; not on serpentine soil

13a Sepals 2–3 mm long in the fruiting stage; flowers in the fruiting stage touching each other; nutlets 3 or 4; corolla less than 2 mm wide; in sandy habitats near the shore . . . .
. . . . *Cryptantha leiocarpa*
Coast Cryptantha

13b Sepals 3–6 mm long in the fruiting stage; flowers in the fruiting stage not touching each other; nutlets usually 1 or 2; corolla less than 1 mm wide; on slopes, mostly away from the coast . . . . *Cryptantha nemaclada*
Colusa Cryptantha; SLO-n

2b   Stem and leaves without densely crowded long hairs, often scarcely hairy; corolla without crests; sepals united at least at their bases, with rust-colored hairs at the tips and sometimes with white hairs on the lower parts, the hairs very much shorter than the sepals; stem branched near the base, each branch with some leaves in the lower part and a raceme above; side of each nutlet on which the attachment scar is located with a prominent lengthwise keel that may extend from the tip to the attachment scar

14a Corolla 5–12 mm wide

15a Leaves up to 5 mm wide (corolla 5–10 mm wide; plant with long hairs, those on the calyx lobes yellow; bracts absent; perennial; up to 30 cm tall, the stem often falling) . . . . *Plagiobothrys mollis* var. *vestitus*

Petaluma Popcornflower; Sn; 1a (1888)

15b Leaves generally much less than 5 mm wide

16a Stem and leaves not hairy; known only from around sulfur springs near Calistoga, Napa County (corolla 5–6 mm wide; bracts absent or a few near base of inflorescence; attachment scar of each nutlet deeply concave and about half the length of the nutlet) . . . . *Plagiobothrys strictus*

Calistoga Popcornflower; Na; 1b

16b Stem and leaves at least slightly hairy; not confined to sulfur springs

17a Inflorescence with bracts in the lower half; attachment scar of each nutlet at its base; in alkaline habitats or vernal pools (corolla 5–12 mm wide) . . . . *Plagiobothrys stipitatus*

Valley Popcornflower; Na, Sn-n

17b Inflorescence with bracts near the base of inflorescence; attachment scar of each nutlet not at its base; in moist areas

18a Corolla 6–10 mm wide; pedicel usually longer than the calyx (keel on each nutlet within a groove; stem either prostrate with tip rising or erect; mostly coastal) . . . .*Plagiobothrys chorisianus* var. *chorisianus*

Choris Popcornflower; SF-SLO; 1b

18b Corolla 5–7 mm wide; pedicel usually shorter than the calyx

19a Stem prostrate; corolla 5–6 mm wide; keel on each nutlet within a groove (mostly coastal) . . . . *Plagiobothrys chorisianus* var. *hickmanii*

Hickman Popcornflower; 4

19b Stem usually erect; corolla 5–7 mm wide; keel on each nutlet not in a groove . . . . *Plagiobothrys tener*

Inland Popcornflower; Na-n

14b Corolla less than 5 mm wide

20a Stem and leaves not hairy, at least in the upper half of the plant

21a Stem succulent, somewhat inflated and hollow; corolla about 3 mm wide; bracts present at the base of inflorescence; around salt marshes or saline habitats inland . . . . *Plagiobothrys glaber*

Hairless Popcornflower; SF; 1a (1954)

21b Stem not succulent, not inflated or hollow; corolla 4–5 mm wide; bracts absent or a few at the base of inflorescence; known only from around sulfur springs near Calistoga, Napa County....... *Plagiobothrys strictus*

Calistoga Popcornflower; Na; 1b

20b Stem and leaves at least slightly hairy throughout

22a Inflorescence with a few bracts at the base

23a Corolla 2–4 mm wide; nutlets with prominent spines about 1 mm long, these covered with small barbs; mostly on clay or dried mud of vernal pools .... *Plagiobothrys greenei*

Green Popcornflower; Sn-n

23b Corolla 3–7 mm wide; nutlets without spines; in mud flats or vernal pools, mostly away from the coast .... *Plagiobothrys tener*

Inland Popcornflower; Na-n

22b Inflorescence with bracts either in the lower half or throughout

24a Inflorescence with bracts throughout (corolla 1–2 mm wide; calyx 3–6 mm long in the fruiting stage)

25a Nutlets with prominent spines, these covered with small barbs; attachment scar of each nutlet at least one-fourth the length of the nutlet .... *Plagiobothrys hystriculus*

Bearded Popcornflower; Sl; 1a (1892)

25b Nutlets without spines; attachment scar of each nutlet less than one-fourth the length of the nutlet (in mud flats or vernal pools) .... *Plagiobothrys trachycarpus*

Roughseed Popcornflower; CC-s

24b Inflorescence with bracts in the lower half (nutlets without spines; attachment scar of each nutlet less than one-fourth the length of the nutlet)

26a Keel on each nutlet not in a groove (corolla 1–3 mm wide; calyx 2–4 mm long in the fruiting stage; in mud flats or vernal pools) .... *Plagiobothrys bracteatus*

Bracted Popcornflower

26b Keel on each nutlet within a groove

27a Attachment scar of each nutlet about as wide as long; corolla 2–4 mm wide; calyx 2–4 mm long in the fruiting stage; in coastal locations .... *Plagiobothrys reticulatus* var. *rossianorum* [includes *P. diffusus*]

Northern Popcornflower; Ma, SF

27b Attachment scar of each nutlet narrow; corolla 1–2 mm wide; calyx 2 mm long in the fruiting stage; in mud flats and vernal pools .... *Plagiobothrys undulatus*

Southern Popcornflower, Coast Popcornflower; Me-s

BRASSICACEAE (MUSTARD FAMILY)   Plants of Brassicaceae are easily recognized when in flower, for there are few exceptions to the following formula: four separate sepals, four separate petals, six stamens (four long, two short), and a pistil that is usually partitioned lengthwise into two divisions, each with its own row of seeds. The ovary is superior. The mature fruit, which is crucial for identification of mustards, may be either short and broad or long and narrow. Sometimes the upper portion of the pistil persists on the fruit as a "beak." The leaves are alternate. Some species of *Lepidium* lack petals.

This large family of mainly herbaceous plants, which used to be called Cruciferae, provides us not only with mustard, but also with radishes, cabbage, cauliflower, broccoli, bok choy, kohlrabi, watercress, and other food plants, the flavors of which are due to characteristic oils. It also gives us numerous valuable ornamentals such as *Aubrieta deltoidea* (rock cress) and *Erysimum capitatum* (wallflower).

Relatively few California representatives are showy enough to be cultivated. Certain species of *Erysimum, Arabis,* and *Draba* are, however, grown widely in rock gardens.

1a   Fruit usually not more than twice as long as wide but never more than 4 times as long as wide (fruit flattened, except in *Cardaria*)

    2a   Seam either on margins of flattened fruit or not visible (fruit either flattened parallel to the partition that separates the 2 halves or not divided into 2 halves)

        3a   Fruit 3–4 times as long as wide (leaves not lobed, all basal; petals white, deeply notched, about 2.5 mm long; blooms in late winter; fruit 3–10 mm long, the pedicel turned up; rarely more than 10 cm tall; widespread) . . . . *Draba verna* (fig.)
Vernal Whitlowgrass; SM, SCl-n

        3b   Fruit less than twice as long as wide

            4a   Petals 1.5–2 cm long, deep purple; fruit 3–4 cm long and nearly as wide, the pedicel 1–1.5 cm long (pedicel turned up; leaves toothed, the upper clasping the stem) . . . . *Lunaria annua*
Honesty, Matrimony-plant; eu

            4b   Petals less than 1 cm long, not deep purple, but sometimes tinged with purple; fruit much less than 2 cm long and wide, the pedicel not more than 1 cm long, usually much less

                5a   Fruit without an abrupt swelling in the central portion

                    6a   Fruit with crowded hooked hairs (fruit 2–3 mm long, the pedicel turned down; petals 1.5 mm long; leaves basal and also on the stem, the longest 3 cm, sometimes toothed; inflorescence about 8 cm long, 1 sided) . . . . *Athysanus pusillus*
Dwarf Athysanus, Sandweed

                    6b   Fruit, if hairy, without hooked hairs

                        7a   Leaves entirely basal, some of them lobed or toothed; flower solitary on a peduncle 3–13 cm long; fruit up to 12 mm long; in moist, gravelly habitats, mostly at elevations above 2,000 ft . . . .*Idahoa scapigera*
Flatpod; SCl (MH)

                      7b   Leaves scattered along the stem, none of them lobed or toothed; flowers in dense racemes; fruit less than 3 mm long; widespread in disturbed lowlands (flowers sweet smelling, produced throughout the year) . . . . *Lobularia maritima*
Sweet Alyssum; me

Arabis breweri
Brewer Rock Cress

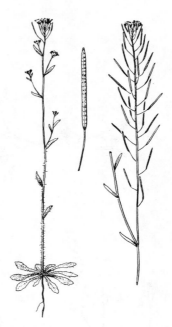

Arabidopsis thaliana
Thale Cress

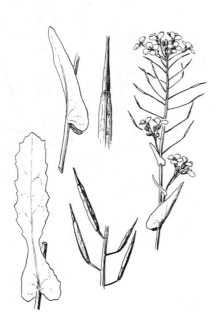

Brassica rapa
Field Mustard

Cardamine oligosperma
Bitter Cress

Coronopus didymus
Lesser Wartcress

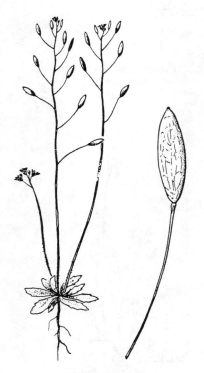

Draba verna
Vernal Whitlowgrass

Lepidium nitidum
Common Peppergrass

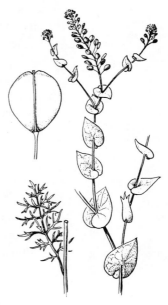

Lepidium oxycarpum
Sharp-pod Peppergrass

Lepidium perfoliatum
Shield Cress

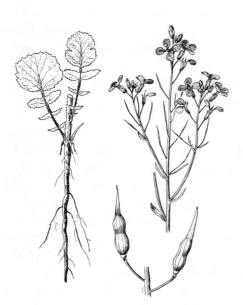

Raphanus sativus
Wild Radish

Thlaspi arvense
Fanweed

5b Fruit with an abrupt swelling in the central portion where the single seed is located

    8a Leaves not toothed or lobed; flowers crowded in the upper part of the raceme; petals pale yellow or cream colored, fading to white; fruit divided into 2 halves, without perforations (pedicel turned up) . . . . *Alyssum alyssoides*

        Yellow Alyssum, Pale Alyssum; eua

    8b Some leaves toothed or lobed; flowers not crowded; petals white or purple tinged; fruit not divided into 2 halves, the outer portion sometimes perforated

        9a Fruit 8–10 mm wide, the margins not perforated; outer portion of fruit with distinct, slender ridges that extend outward from the swollen central portion; pedicel turned up (in moist habitats) . . . . *Thysanocarpus radians*

           Ribbed Fringepod; SCl-n

        9b Fruit 3–8 mm wide, the margins sometimes perforated; outer portion of fruit without distinct ridges, but there may be indistinct, broad ridges; pedicel turned down

           10a Leaves hairy, clasping the stem and the lower ones forming a basal cluster; fruit 5–8 mm wide, the outer portion sometimes perforated (widespread) . . . . *Thysanocarpus curvipes* [includes *T. curvipes* var. *elegans*] (pl. 17)

              Hairy Fringepod

           10b Leaves not hairy, not clasping the stem or forming a basal cluster; fruit 3–6 mm wide, the outer portion often perforated . . . . *Thysanocarpus laciniatus* [includes *T. laciniatus* var. *crenatus*]

              Narrowleaf Fringepod; CC-s

2b Seam on faces of flattened fruit (fruit flattened at right angles to the partition that separates the 2 halves; pedicel spreading or turned up)

    11a Upper leaves clasping the stem

        12a Fruit nearly triangular, widest at the top where there is a shallow notch, narrowing evenly to the pedicel (leaves hairy, sometimes toothed or lobed; fruit 4–8 mm long; petals white; widespread) . . . . *Capsella bursa-pastoris*

           Shepherd's-purse, Medic; eu

        12b Fruit not nearly triangular, both margins mostly rounded

           13a Fruit 10–15 mm long; pedicel 7–15 mm long (fruit notched; leaves hairy, sometimes toothed; petals white) . . . . *Thlaspi arvense* (fig.)

              Fanweed, Field Penny Cress; eu

           13b Fruit 4–6 mm long; pedicel less than 7 mm long

              14a Fruit not notched at the tip (petals white; fruit 3–5 mm wide, swollen; some leaves usually toothed)

                  15a Plant hairy throughout; fruit obviously hairy . . . . *Cardaria pubescens*

                    Whitetop; as

15b Plant usually not hairy in the upper half; fruit not obviously hairy . . . . *Cardaria draba*
Heartpod Hoary Cress; eua

14b Fruit slightly notched at the tip

16a Upper leaves usually less than twice as long as wide, the basal ones 2–3 times lobed or compound; petals yellow; fruit 4 mm long, not swollen; plant not hairy in upper half . . . .
. . . . *Lepidium perfoliatum* (fig.)
Shield Cress, Perfoliate Peppergrass; eua

16b Upper leaves 2–3 times as long as wide, the basal ones sometimes toothed or lobed; petals white; fruit 5–6 mm long, swollen; plant hairy throughout (mostly at elevations above 2,000 ft) . . . . *Lepidium campestre*
Field Cress, English Peppergrass; eu

11b Leaves not clasping the stem

17a Fruit so deeply notched that the lobes on both sides of the notch are at least one-half the length of rest of the fruit; petals green (leaves toothed, lobed, or compound; fruit 5–7 mm long; racemes very dense; plant hairy, low, branching from the base; in vernal pools and alkaline soils) . . . . *Lepidium latipes*
Dwarf Peppergrass

17b Fruit not so deeply notched that the lobes on both sides of the notch are more than one-third the length of the rest of fruit; petals white, sometimes absent

18a Both halves of fruit bulging, with either a networklike surface pattern or conspicuous pointed bumps, the beak scarcely evident (leaves deeply lobed, sometimes hairy; fruit less than 4 mm long)

19a Fruit notched at the tip; surface of fruit with a networklike pattern (widespread) . . . . *Coronopus didymus* (fig.)
Lesser Wartcress; eu

19b Fruit not notched at the tip; surface of fruit with pointed bumps
. . . . *Coronopus squamatus*
Swine Cress; eu

18b Halves of fruit not bulging, usually without a networklike surface pattern or without conspicuous pointed bumps, the beak evident

20a Pedicel not flattened to the extent that it is twice as wide as thick

21a Lower leaves more than 10 cm long, usually evenly toothed; fruit not notched at the tip; often more than 1 m tall (plant sometimes not hairy; fruit hairy; widespread) . . . .
. . . . *Lepidium latifolium*
Slender Perennial Peppergrass; eua

21b Lower leaves not more than 8 cm long, irregularly toothed, lobed, or compound; fruit notched at the tip; usually less than 0.5 m tall

22a Fruit 1.5 mm long, hairy; plant not hairy, at least in the lower half; petals, if present, shorter than the sepals . . . . . .
. . . . *Lepidium pinnatifidum*
Eurasian Cress; eua

22b Fruit 2–4 mm long, not hairy; plant at least somewhat hairy throughout; petals as long as, or longer than, the sepals .... *Lepidium virginicum* var. *virginicum*
Tonguegrass, Tall Peppergrass; na

20b Pedicel distinctly flattened, at least twice as wide as thick

23a Sepals persisting at least until the fruit is nearly mature; plant prostrate, with tip of stem rising (plant hairy; petals none or inconspicuous; some leaves lobed or compound; fruit sometimes hairy, the lobes sharp pointed) .... *Lepidium strictum*
Prostrate Cress, Wayside Peppergrass

23b Sepals not persisting until the fruit is nearly mature; plant erect

24a Fruit at least slightly hairy; some leaves toothed or lobed, but not compound (plant hairy; fruit hairy, the lobes not sharp pointed)

25a Fruit with a pronounced networklike pattern; upper leaves 1–3 mm wide, usually smooth margined; prostrate with tip of stem rising (in alkaline soil) ....
.... *Lepidium dictyotum*
Alkali Peppergrass; CC

25b Fruit without a pronounced networklike pattern; upper leaves at least 4 mm wide, usually toothed; prostrate or erect .... *Lepidium lasiocarpum*
Hairypod Peppergrass; Ma

24b Fruit not hairy; leaves often compound (pedicel sometimes hairy)

26a Lobes of fruit sharp pointed (plant erect, sometimes hairy; petals sometimes absent; leaves sometimes lobed or compound; in alkaline soils) ....
.... *Lepidium oxycarpum* (fig.)
Sharp-pod Peppergrass; Sn-SB

26b Lobes of the fruit not sharp pointed

27a Plant hairier in the upper half; pedicel not winged; longest leaves 10 cm, the basal ones compound; up to 40 cm tall; in alkaline areas, vernal pools .... *Lepidium nitidum* (fig.)
Common Peppergrass

27b Plant equally hairy throughout; pedicel winged; longest leaves 15 cm, the basal ones lobed or compound; up to 70 cm tall; common in disturbed areas .... *Lepidium virginicum* var. *pubescens*
Hairy Tonguegrass

1b Fruit more than 4 times as long as wide

28a Fruit divided into 2 units, the lower of these shorter than the upper one (fruit not flattened, beak conspicuous)

    29a Petals yellow; upper portion of the fruit nearly globular and with 5 lengthwise ridges, tapering abruptly to a slender beak that is about as long as the upper portion (lower leaves lobed or compound) .... *Rapistrum rugosum*
                                                     Wild Turnip; eu

    29b Petals pale purple to nearly white; upper portion of fruit flattened or ovoid, somewhat angular, tapering gradually to a short, stout beak (on backshores of sandy beaches)

        30a Leaves usually with wavy margins and toothed, or sometimes lobed; fruit widest in the upper unit; petals about 6 mm long ....
        .... *Cakile edentula* [includes *C. edentula* ssp. *californica*]
                                                     Searocket; na

        30b Leaves without wavy margins, but lobed, the lobes sometimes separated nearly to the midrib; fruit widest where the 2 units are joined; petals about 10 mm long...................................... *Cakile maritima* (pl. 17)
                                       Horned Searocket; eu

28b Fruit not divided into 2 slightly unequal units, but there may be constrictions between the seeds

    31a Upper portion of pistil either not developing into a beak, or the beak less than one-sixth the length of the body of the fruit

        32a Calyx usually not green, usually pouched at the base; petals often ruffled, the upper portion often not wider than the lower portion (fruit usually flattened; petals usually at least partly purple, or with some purple veins) ....
        .... BRASSICACEAE, SUBKEY 1

        32b Calyx green, not pouched at the base; petals not ruffled, the upper portion usually wider than the lower portion .... BRASSICACEAE, SUBKEY 2

    31b Upper portion of pistil, usually just the style, developing into a conspicuous beak that is at least one-sixth the length of the body of the fruit (usually some leaves lobed, with the end lobe longer than the side lobes; fruit not flattened)

        33a Beak of fruit more than half as long as body of fruit

            34a Fruit covered with conspicuous, flattened hairs; leaves not compound; petals 8–11 mm long (plant usually bristly; longest leaves 20 cm; petals pale yellow to white; body and beak of fruit each 2–3 cm long, the beak flattened) .... *Sinapis alba [Brassica hirta]*
                                  White Mustard; eua

            34b Fruit scarcely if at all hairy; some leaves often compound; petals 15–25 mm long

                35a Fruit not at all constricted between seeds, divided lengthwise by a partition; beak flattened; longest leaves 15 cm (petals white to yellow, with red purple veins) ... *Eruca vesicaria* ssp. *sativa [E. sativa]*
                                    Garden-rocket; eu

                35b Fruit constricted between seeds, not divided lengthwise by a partition; beak conical; longest leaves 20 cm

36a Petals usually yellow, fading to white; fruit constricted about halfway to the midline, with obvious grooves and ridges; seeds 2–12 .... *Raphanus raphanistrum*
Jointed Charlock; me

36b Petals usually purple with darker veins, sometimes white; fruit constricted less than halfway to the midline, the grooves and ridges absent or not obvious; seeds 1–5 (widespread) ....
.... *Raphanus sativus* (fig.)
Wild Radish; me

33b Beak of fruit less than half the length of body of fruit (leaves usually toothed or lobed, but not compound; fruit scarcely if at all hairy)

37a Upper leaves sessile and clasping the stem; body of fruit 30–70 mm long, the beak 10–15 mm long (plant scarcely if at all hairy; petals 6–11 mm long, yellow; annual; common)...... *Brassica rapa* [*B. campestris*] (fig.)
Field Mustard, Wild Turnip; eu

37b Upper leaves, if sessile, not clasping the stem; body of fruit 10–40 mm long, the beak 2–10 mm long

38a Plant not hairy (body of fruit 20–40 mm long, the beak 5–10 mm long; basal leaves soon deciduous; petals 7–10 mm long, pale yellow; annual) .... *Brassica juncea*
Indian Mustard; eua

38b Plant at least sparsely hairy

39a Fruit not pressed against the stem; plant bristly at least in the lower half; body of fruit 20–35 mm long, the beak 5 mm long, 4 sided (petals 8–10 mm long, yellow; leaves 5–15 cm long; annual) .... *Sinapis arvensis* [includes *Brassica kaber* vars. *kaber* and *pinnatifida*]
Charlock; eu

39b Fruit pressed against the stem; plant sparsely to densely hairy, but not bristly; body of fruit 10–20 mm long, the beak 2–6 mm long, not 4 sided (widespread)

40a Longest leaves 10–20 cm, sparsely to densely hairy; petals 7–11 mm long, pale yellow; beak of fruit almost uniformly slender; annual .... *Brassica nigra*
Black Mustard; eu

40b Longest leaves 4–10 cm, densely hairy; petals 5–6 mm long, pale yellow to white; lower part of beak much thicker than the tip; biennial or perennial ....
.... *Hirschfeldia incana* [*Brassica geniculata*] (pl. 17)
Summer Mustard; eu

Brassicaceae, Subkey 1: Fruit more than 4 times as long as wide, the beak less than one-sixth the length of the body; calyx usually not green, usually pouched at the base; petals often ruffled, the upper portion usually not wider than the lower portion; seam on margins of flat-tened fruit, but fruit of *Caulanthus coulteri* var. *lemmonii* is usually not flattened

1a   Stigma obviously 2 lobed, the lobes 1–3 mm long; fruit usually not flattened (leaves sessile, clasping the stem; sepals 7–18 mm long, unequal, dark purple before flower opens, becoming paler; petals 8–31 mm long, whitish, cream colored, or purple) . . . .
. . . . *Caulanthus coulteri* var. *lemmonii*
Lemmon Caulanthus; Al-s; 1b

1b   Stigma not 2 lobed, except in *Streptanthus callistus* and *S. hispidus*, but these lobes are less than 1 mm long; fruit flattened

2a   Calyx slightly irregular, 1 sepal extending outward farther than the other 3, and somewhat separate from them (sepals 4–10 mm long; petals 6–15 mm long)

3a   Stem and leaves with dense bristles, especially on margins and midribs of leaves of lower half of plant; petals equal, the upper often darker than the lower ones (fruit 3–11 cm long, sometimes bristly; sometimes on serpentine)

4a   Flowers not mostly on 1 side of the inflorescence (sepals dark purple; petals purple) . . . . *Streptanthus glandulosus* ssp. *glandulosus*
Common Jewelflower; Sn-SLO

4b   Flowers mostly on 1 side of the inflorescence

5a   Sepals red purple; petals purple; inflorescence crowded; usually less than 30 cm tall . . . . *Streptanthus glandulosus* ssp. *pulchellus*
Mount Tamalpais Jewelflower; Ma (MT); 1b

5b   Sepals purple, rose, white, or yellow green; petals white to purple; inflorescence not crowded; usually more than 30 cm tall . . . .
. . . . *Streptanthus glandulosus* ssp. *secundus*
One-sided Jewelflower; Me-Ma

3b   Stem and leaves usually scarcely hairy; upper petals usually longer than the lower ones, similar in intensity of color shade

6a   Sepals 5–7 mm long, very dark purple; fruit 2–7 mm long (petals dark purple; on serpentine) . . . . *Streptanthus niger*
Tiburon Jewelflower; Ma; 1b

6b   Sepals 5–10 mm long, white to pale lavender; fruit 4–12 mm long

7a   Sepals pale green; petals white, with purple or brown veins; on serpentine . . . . *Streptanthus albidus* ssp. *albidus* [*S. glandulosus* var. *albidus*]
Metcalf Canyon Jewelflower; SCl (MH); 1b

7b   Sepals pale lavender; petals usually purple; sometimes on serpentine
. . . . *Streptanthus albidus* ssp. *peramoenus*
[included in *S. glandulosus* ssp. *glandulosus*]
Most Beautiful Jewelflower; Al, CC, SCl; 1b

2b   Calyx regular (petals similar in size)

8a   Leaves and stem with bristly hairs, especially near base of plant; fruit bristly; stigma 2 lobed; sepals purple, but turning brown or green; usually less than 20 cm tall (sepals 3–8 mm long; in rocky areas)

9a   Fruit 3–7 cm long; petals 6–10 mm long, purple, with white margins; bristly hairs crowded . . . . *Streptanthus hispidus*
Mount Diablo Jewelflower; CC (MD); 1b

9b Fruit 1.5–2 cm long; petals 8–11 mm long, red purple; bristly hairs scattered, usually separated by a distance equal to their length ... *Streptanthus callistus*
Mount Hamilton Jewelflower; SCl (MH); 1b

8b Leaves and stem without bristly hairs; fruit usually not bristly; stigma not lobed; sepals purple; generally at least 30 cm tall

    10a Lower portion of inflorescence with a few broad, leaflike bracts that clasp the stem (lower leaves usually less than 3 times as long as wide; petals yellow, with purple veins; up to 100 cm tall)

        11a Sepals 5–8 mm long; petals 6–8 mm long; usually biennial . . . .
. . . . *Streptanthus tortuosus* var. *tortuosus*
Mountain Jewelflower; Sn-n

        11b Sepals 8–10 mm long; petals 11–13 mm long; perennial (sometimes on serpentine) . . . . *Streptanthus tortuosus* var. *suffrutescens*
Shrubby Mountain Jewelflower

    10b Inflorescence without broad, leaflike bracts (often on serpentine)

        12a Lower leaves, including the basal ones, usually more than 3 times as long as wide (annual; up to 80 cm tall)

            13a Lower leaves 5–22 cm long, toothed or lobed; sepals 5–11 mm long, cream colored, pale yellow, or sometimes purple; petals 8–15 mm long, ruffled, white, sometimes with purple veins (inner Coast Ranges) . . . . *Guillenia flavescens* [*Thelypodium flavescens*]
Yellowflower Mustard; Sl-SB

            13b Lower leaves 6–11 cm long, usually smooth margined; sepals 3–6 mm long, yellow green, or sometimes purple; petals 5–10 mm long, not ruffled, upper pair white, lower pair purple with white margins ...
. . . . *Streptanthus barbiger*
Bearded Streptanthus; Me-Na; 4

        12b Lower leaves not often more than 3 times as long as wide (upper petals often white with purple veins, the lower ones purple)

            14a Lower leaves clasping the stem (sepals 3–9 mm long, purple or pale green, not hairy; fruit 2–11 cm long; up to 60 cm tall)

                15a Flowers almost entirely on 1 side of the inflorescence; petals 6–8 mm long; leaves yellow green, up to 5 cm long; fruit curved, approximately C shaped ... *Streptanthus breweri* var. *hesperidis*
Yellow-green Streptanthus; Na; 1b

                15b Flowers not just on 1 side of the inflorescence; petals 5–13 mm long; leaves blue green, up to 12 cm long; fruit only slightly curved . . . . *Streptanthus breweri* var. *breweri*
Brewer Streptanthus; SB-n

            14b Leaves not clasping the stem

                16a Lower leaves up to 2.5 cm long, mostly 2–3 times as long as wide; sepals 3–4 mm long; fruit 2–3 cm long; annual (petals 5–7 mm long, white with purple veins; sepals green or purple with white tips, not hairy; up to 18 cm tall) . . . . . . *Streptanthus batrachopus*
Tamalpais Jewelflower; Ma (MT); 1b

16b Lower leaves often more than 3 cm long, rarely so much as twice as long as wide; sepals 5–9 mm long; fruit 4–8 cm long; biennial

    17a Petals 6–10 mm long, cream to salmon colored, with dark veins; sepals 5–7 mm long, often at least slightly hairy, yellow to red purple; up to 1 m tall .... *Streptanthus morrisonii*
[includes *S. morrisonii* ssp. *hirtiflorus*]
Morrison Jewelflower; Sn; 1b

    17b Petals 7–9 mm long, white; sepals 6–9 mm long, not hairy, rose purple, with yellow bases; up to 0.5 m tall .... *Streptanthus brachiatus*
Socrates Mine Jewelflower; Sn; 1b

Brassicaceae, Subkey 2:   Fruit more than 4 times as long as wide, the beak less than one-sixth the length of the body; calyx usually green; upper portions of petals usually wider than the lower portions

1a Petals white, pink, or purple (*Arabis glabra*, in choice 1b, is sometimes purple)

  2a Tip of fruit with a pair of hornlike lobes (biennial or short-lived perennial; flower very fragrant; petals white, pink, or purple; leaves grayish, with dense, often branched hairs) .... *Matthiola incana*
Stock; eu

  2b Tip of fruit without a pair of hornlike lobes

    3a Plant growing in shallow water in streams, ditches, and at the edges of ponds (petals 3–4 mm long, white; fruit 10–15 mm long; leaves compound) .... *Rorippa nasturtium-aquaticum*
Watercress; eua

    3b Plant of terrestrial habitats

      4a Basal leaves all originating at about the same level, thus forming a compact cluster (fruit scarcely if at all hairy)

        5a Leaves compound (petals 2–4 mm long, white; fruit 15–25 mm long, not flattened; widespread) .... *Cardamine oligosperma* (fig.)
Bitter Cress

        5b Leaves not compound, sometimes toothed

          6a Petals white, 3–4 mm long; fruit 10–15 mm long, sometimes flattened; annual; in disturbed areas .... *Arabidopsis thaliana* (fig.)
Thale Cress, Mouse-ear Cress; eu

          6b Petals rose purple, up to about 15 mm long; fruit 20–40 mm long, flattened; perennial; in rocky habitats .... *Arabis blepharophylla* (pl. 17)
Coast Rock Cress; Sn-SCr; 4

      4b Basal leaves not all originating at about the same level (fruit flattened)

7a   Petals pink, rose, or purple, sometimes nearly white in *Arabis breweri;* leaves not lobed or compound, but sometimes toothed (in rocky habitats)

     8a   Fruit 3–7 cm long; petals 6–9 mm long; lower leaves not often more than 3 cm long; somewhat woody at the base ... *Arabis breweri* (fig.)

                                                                    Brewer Rock Cress; Mo-n

     8b   Fruit 6–12 cm long; petals 9–12 mm long; lower leaves usually more than 3 cm long; not woody .... *Arabis sparsiflora* var. *arcuata*

                                                                    Elegant Rock Cress; SCl-s

7b   Petals white, sometimes becoming pale rose; leaves sometimes lobed or compound (petals 9–14 mm long; fruit 2–5 cm long)

     9a   Stem leaves smooth margined (basal leaves smooth margined; in moist, shaded areas) ...... *Cardamine californica* var. *cardiophylla*

                                   [Dentaria californica var. *cardiophylla]*

                                           Undivided Milkmaids; Sl-n

     9b   Stem leaves toothed, lobed, or compound

         10a   Stem leaves lobed or compound; lobes or leaflets usually with obvious sharp teeth (basal leaves compound; in shaded areas; widespread) .... *Cardamine californica* var. *californica*

                                  [Dentaria californica ssp. *californica]* (pl. 17)

                                Common Milkmaids, California Toothwort

         10b   Stem leaves sometimes lobed or compound; lobes or leaflets if present, either not toothed or the teeth not obvious

             11a   Basal leaves usually compound; stem leaves lobed or compound; lobes or leaflets either not toothed or the teeth not obvious; in open areas (coastal) ....

                     .... *Cardamine californica* var. *integrifolia*

                                   [Dentaria californica var. *integrifolia]*

                                       Coast Milkmaids; Mo-n

             11b   Basal leaves not compound; stem leaves sometimes lobed or compound; lobes or leaflets if present, not toothed; in shaded areas .... *Cardamine californica* var. *sinuata*

                                     [Dentaria californica var. *sinuata]*

                                      Dividedleaf Milkmaids; Sn-n

1b   Petals cream colored, yellow, orange, orange brown, or red brown

     12a   Stem somewhat angled or prominently ridged (petals yellow; at least lower leaves lobed; fruit not flattened)

         13a   Fruit 1–3 cm long, the beak 2–3 mm long (petals 6–8 mm long, bright yellow; basal leaves usually with fewer than 6 pairs of side lobes)............ *Barbarea vulgaris*

                                       Common Winter Cress; eua

         13b   Fruit 2.5–8 cm long, the beak usually less than 2 mm long

             14a   Fruit 4.5–8 cm long; basal leaves usually with at least 5 pairs of side lobes; petals 5–7 mm long, bright yellow .... *Barbarea verna*

                                   Early Winter Cress, Scurvy-grass; eua

14b Fruit 2.5–5 cm long; basal leaves with fewer than 5 pairs of side lobes; petals 4–6 mm long, pale yellow (in moist areas). . . . .
. . . . *Barbarea orthoceras* [includes *B. orthoceras* var. *dolichocarpa*]
Erectpod Winter Cress, American Winter Cress

12b Stem not at all angled or ridged

15a Fruit less than 3 cm long (fruit not flattened)

16a In wet meadows and wet areas adjacent to aquatic habitats: stream banks, ditch banks, shores of ponds (petals 1–2 mm long, yellow; fruit 6–15 mm long, curved like a banana) . . . . *Rorippa curvisiliqua*
Yellow Cress

16b Not in wet meadows or wet areas adjacent to aquatic habitats

17a Longest pedicel of fruit 5–20 mm long (pedicel spreading to turned up)

18a Fruit 15–25 mm long, the pedicel 15–20 mm long; petals 4–5 mm long, yellow; hairs not branched; in alkaline soil (fruit separates into 4 sections) . . . .*Tropidocarpum capparideum*
Caperfruit Tropidocarpum; CC (MD); 1a (1957)

18b Fruit 5–12 mm long, the pedicel 5–10 mm long; petals less than 4 mm long, bright yellow; hairs branching; in sandy soil . . . .
. . . . *Descurainia pinnata* ssp. *menziesii*
Tansy Mustard; CC-s

17b Pedicel of fruit 1–4 mm long

19a Fruit 8–15 mm long, pressed against the stem, the pedicel turned up; petals pale yellow; leaves lobed . . . . *Sisymbrium officinale*
Hedge Mustard; eu

19b Fruit 10–25 mm long, not pressed against the stem, the pedicel usually turned down; petals white to pale yellow; leaves sometimes lobed . . . . *Guillenia lasiophylla*
[includes *Thelypodium lasiophyllum* vars. *nalienum*, *lasiophyllum*, and *rigidum*]
California Mustard

15b Fruit at least 3 cm long

20a Upper leaves clasping the stem (petals 3 mm long, cream colored or pale yellow, rarely purple; fruit 4–10 cm long, not flattened, the pedicel turned up; leaves toothed or lobed, with branched hairs; annual; up to 20 cm tall; in shaded areas) . . . . *Arabis glabra*
Tower Mustard

20b Leaves not clasping the stem

21a Leaves lobed or compound

22a Petals 3–6 mm long

23a Pedicel of fruit 10–15 mm long, turned up; petals 4 mm long, yellow . . . . *Tropidocarpum gracile*
Lacepod, Dobiepod

23b Pedicel of fruit 1–4 mm long, turned down; petals 3–6 mm long, white or pale yellow . . . . *Guillenia lasiophylla* [includes *Thelypodium lasiophyllum* vars. *inalienum, lasiophyllum,* and *rigidum*] California Mustard

22b Petals 6–32 mm long (pedicel spreading or turned up)

24a Petals 14–32 mm long, deep yellow; leaves somewhat succulent, lobed, but not compound; stigma not 2 lobed; fruit cylindrical, becoming flattened; hairs branched; rarely so much as 15 cm tall . . . . *Erysimum menziesii* ssp. *menziesii* Menzies Wallflower; Mo-n; 1b

24b Petals 6–10 mm long, pale yellow; leaves not succulent, often compound; stigma 2 lobed; fruit never flattened; hairs not branched; up to at least 30 cm tall

25a Petals 8–10 mm long; fruit 3–10 cm long, the pedicel 3–6 mm long, turned up; upper leaves usually not lobed except for 2 lobes at their bases, these not very narrow; up to 30 cm tall . . . . *Sisymbrium orientale* Eastern Rocket, Oriental Mustard; eu

25b Petals 6–8 mm long; fruit 5–10 cm long, the pedicel 4–10 mm long, spreading; upper leaves with more than 2 lobes, these very narrow; up to 150 cm tall . . . . *Sisymbrium altissimum* Tumble Mustard, Jim Hill Mustard; eu

21b Leaves smooth margined or toothed

26a Pedicel of fruit 1–4 mm long, turned down; fruit 3–7 cm long; petals 3–6 mm long; hairs not branched (petals white to pale yellow) . . . . *Guillenia lasiophylla* [includes *Thelypodium lasiophyllum* vars. . . . . *inalienum, lasiophyllum,* and *rigidum*] California Mustard

26b Pedicel of fruit 3–14 mm long, usually turned up; fruit 3–18 cm long; petals 12–32 mm long; hairs branched

27a Fruit 4 sided (fruit 3–15 cm long; petals yellow or orange; up to 80 cm tall)

28a Lower leaves usually less than 10 cm long; in dry places away from the coast (widespread) . . . . *Erysimum capitatum* ssp. *capitatum* (pl. 17) Western Wallflower

28b Lower leaves usually more than 10 cm long; on sand dunes near Antioch, Contra Costa County . . . . *Erysimum capitatum* ssp. *angustatum* Contra Costa Wallflower; CC; 1b

27b Fruit flattened

    29a Basal leaves at least 10 times as long as wide (petals cream colored to yellow; fruit 4–13 cm long; biennial or perennial; up to 40 cm tall; coastal, sometimes on serpentine, but mostly on other sandy or rocky soils) .... *Erysimum franciscanum*

        San Francisco Wallflower; SM-n; 4

    29b Basal leaves not more than 5 times as long as wide

        30a Petals deep yellow; leaves somewhat succulent, the lower about twice as long as wide; fruit 3–8 cm long, the pedicel less than 1 cm long; very bushy, branching at the base; rarely so much as 15 cm tall .... *Erysimum menziesii* ssp. *menziesii*

        Menzies Wallflower; Mo-n; 1b

        30b Petals cream colored to yellow; leaves not succulent, the lower about 5 times as long as wide; fruit 3–13 cm long, the pedicel 1.5–2 cm long; not branching at the base; usually more than 15 cm tall .... *Erysimum menziesii* ssp. *concinnum*

        Coast Wallflower; Ma (PR)-n

BUDDLEJACEAE (BUDDLEJA FAMILY) Plants belonging to the Buddlejaceae are mostly shrubs or trees with opposite leaves. They were formerly included in Loganiaceae. Their flowers have a bell-shaped, four-lobed calyx, a tubular, four-lobed corolla, and four stamens that are attached to the corolla tube. A native American species, *Buddleja utahensis,* occurs in deserts from Utah to southern California. A Chinese species, *B. davidii* (Summer Lilac or Butterfly-bush), is widely cultivated for its attractive purple or lilac flowers, grouped in dense, pointed inflorescences up to 40 cm long. The leaf blades are up to 25 cm long, have toothed margins, and are dark green above and white woolly on the undersides. This large shrub, sometimes 5 m tall, has become thoroughly naturalized in some sunny habitats.

CALLITRICHACEAE (WATER-STARWORT FAMILY) Water-starworts are rooted in mud and grow in shallow water and at the edges of ponds or vernal pools. The stems of these plants are very slender and weak, and the leaves are opposite and smooth margined. The pistillate and staminate flowers, in the leaf axils, are extremely small and lack sepals and petals. They have either a single stamen or a four-lobed pistil with two styles. The ovary is superior. As the fruit matures and dries, the four lobes, each with just one seed, separate.

1a   Bases of pairs of leaves not joined by a membrane; leaves all submerged and all the same type, usually 6–10 mm long and about 1 mm wide .... *Callitriche hermaphroditica*

        Hermaphroditic Water-starwort

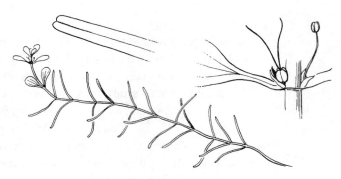

Callitriche heterophylla var. *heterophylla*
Variedleaf Water-starwort

1b   Bases of pairs of leaves joined by a narrow, thin membrane; leaves sometimes of varying form, depending on their location on the stem

     2a   Pistillate flower and fruit on a pedicel 1–25 mm long (fruit less than 2 mm long, slightly longer than wide, sometimes with prominent wings) . . . . *Callitriche marginata* [includes *C. longipedunculata*]

                                                      California Water-starwort

     2b   Pistillate flower and fruit sessile or nearly so

         3a   Fruit with prominent wings that extend from top to bottom along the narrow edges (fruit about as long as wide). . . . . *Callitriche trochlearis*

                                                      Northern Water-starwort; Sn-n

         3b   Fruit either without wings, or with wings that are prominent only near the tip, becoming narrower or disappearing below

             4a   Pitlike markings on broad face of fruit forming perceptible vertical rows. . . . . *Callitriche verna*

                                                      Vernal Water-starwort

             4b   Pitlike markings on broad face of fruit distributed so irregularly that they do not form vertical rows

                 5a   Fruit up to 0.8 mm wide. . . . . *Callitriche heterophylla* var. *heterophylla* (fig.)

                                                      Variedleaf Water-starwort

             5b   Fruit up to 1.4 mm wide. . . . . *Callitriche heterophylla* var. *bolanderi*

                                                      Bolander Water-starwort

CALYCANTHACEAE (SPICEBUSH FAMILY)   Members of the Spicebush or Sweetshrub Family have smooth-margined, opposite leaves. Their flowers have numerous similar sepals and petals that collectively form several circles. There are many pistils and stamens; the stamens nearest the pistils lack anthers. The pistils ripen to form one-seeded achenes that are almost completely enclosed by the cup-shaped receptacle, which in turn is covered tightly by bracts.

Our only representative is *Calycanthus occidentalis* (pl. 17), (Spicebush or Sweetshrub) which grows in Napa County northward. It is generally found close to streams, especially in canyons, and sometimes around ponds. The flower, about 5 cm wide, has deep-red perianth segments that are slightly fleshy. The leaves, commonly 10–12 cm long, yield a spicy aroma when rubbed or bruised. This attractive and unusual shrub is now frequently cultivated in California.

CAMPANULACEAE (BLUEBELL FAMILY)   In members of the Bluebell or Bellflower Family, the leaves are usually alternate, sessile, and not lobed or compound. The corolla has five lobes, and there are five stamens. The ovary is usually inferior or half-inferior. Typically there is a single style with two to five lobes. In *Downingia* and some species of *Nemacladus,* the corolla is strongly two lipped and the stamens are fused together, but in most of our genera, the corolla is regular and the stamens are not fused. Even if the stamens are not fused, however, before the flower opens, the anthers form a tube through which the style grows and collects pollen. In a newly opened flower, the lobes of the style may be obscured by a dense coating of pollen. The mature fruit is a dry capsule that opens by pores or by splitting lengthwise or crosswise, yielding numerous small seeds.

1a   Corolla irregular, 2 lipped; stem not 4 sided; filaments of stamens, and sometimes also the anthers, fused (leaves usually sessile)

    2a   Leaves basal; flower not sessile; corolla less than 2 mm long, the upper lip 3 lobed and the lower lip 2 lobed, not spotted; anthers not fused; in dry areas; stem erect, up to 18 cm tall

        3a   Pedicel 10–15 mm long, the tip turned up; corolla white or purple; leaves 6–18 mm long, sometimes toothed and hairy; base of fruit rounded; on serpentine . . . . *Nemacladus montanus*
Mountain Nemacladus; Na, La

        3b   Pedicel 8–12 mm long, the tip not turned up; corolla white; leaves 3–15 mm long, not toothed, sometimes hairy; base of fruit shaped like an inverted cone; usually not on serpentine . . . . *Nemacladus capillaris*
Common Nemacladus, Common Threadplant; SCl-n, w

    2b   Leaves on stem, sometimes absent at flowering; flower sessile; corolla usually much more than 2 mm long, the upper lip usually 2 lobed and the lower 3 lobed, usually with a white or yellow spot; anthers fused; in wet areas such as the drying mud of vernal pools and the edges of ponds and ditches; stem falling or erect, up to 40 cm tall

        4a   Corolla 2–4 mm long, about as long as, or shorter than, the calyx; upper lip of corolla 3 lobed; fruit less than 3 cm long (corolla either with white upper lobes and tricolored [blue, white, yellow] lower lobes, or entirely white; anthers protruding from corolla) . . . . *Downingia pusilla*
Dwarf Downingia; SN-s; 2

        4b   Corolla 7–19 mm long, longer than the calyx; upper lip of corolla 2 lobed; fruit 3–7.5 cm long

            5a   Lower lip of corolla with a white blotch within which are 3 purple spots alternating with 2 larger yellow spots; anthers protruding from corolla (upper lip of corolla, and tips of lobes of lower lip, purple; lobes of corolla without a fringe of hairs; fruit up to 7.5 cm long) . . . . *Downingia pulchella*
Flat-faced Downingia; Mo-n

Campanula prenanthoides
California Harebell

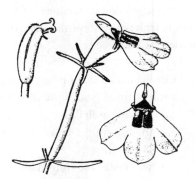

Downingia concolor
Maroon-spotted Dowingia

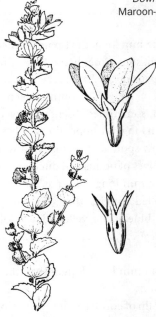

Triodanis biflora
Venus-looking-glass

5b   Lower lip of corolla with a white blotch within which are either 1 purple spot or 1 or 2 yellow spots; anthers not protruding from corolla

        6a   Margins of upper lobes of corolla usually with a fringe of hairs; lower lip of corolla with a large purple spot within the basal portion of a white blotch; rest of corolla blue; fruit up to 5 cm long . . . . *Downingia concolor* (fig.)

                                              Maroon-spotted Downingia; La-Mo

        6b   Margins of upper lobes of corolla without a fringe of hairs; lower lip of corolla with a yellow spot, or 2 somewhat separate yellow spots that are within a white blotch; rest of corolla pale blue or lavender, sometimes white; fruit up to 7 cm long . . . . *Downingia cuspidata*

                                           Cuspidate Downingia; SLO-n

1b   Corolla regular; stem 4 sided; stamens not fused but appearing so when flower first opens (leaves toothed and sessile except in *Campanula prenanthoides* and sometimes in *C. californica*)

    7a   Calyx definitely shorter than corolla (flower not sessile; corolla 7–18 mm long)

        8a   Style protruding from corolla; corolla bright blue; sometimes more than 50 cm tall (leaves 10–60 mm long, not sessile; stem erect or reclining; in dry, wooded areas) . . . . *Campanula prenanthoides* (fig.)

                                             California Harebell

        8b   Style not protruding from corolla; corolla pale blue or deep purple; not often more than 30 cm tall

            9a   Longest leaves 15–20 mm and about twice as long as wide, sometimes sessile; perennial; stem often sprawling; limited to swampy habitats near the coast (corolla pale blue; pedicel 1–20 mm long) . . . . *Campanula californica*

                                         Swamp Harebell; Me-Ma; 1b

            9b   Leaves not more than 11 mm long, mostly more than twice as long as wide, sessile; annual; stem erect; on rocky slopes

                10a Lower and upper leaves several times as long as wide; corolla pale blue; pedicel 3–20 mm long (often on serpentine) . . . . *Campanula exigua*

                              Chaparral Harebell; CC (MD), SCl (MH); 1b

                10b Lower leaves scarcely twice as long as wide, the upper leaves usually 3–4 times as long as wide; corolla deep purple; pedicel 1–3 mm long . . . . *Campanula sharsmithiae*

                              Sharsmith Harebell; Al, SCl (MH); 1b

    7b   Calyx about as long as or longer than corolla (erect; up to 40 cm tall, but often much shorter)

        11a Corolla 4–14 mm long (flower solitary; corolla deep blue, usually with a pedicel; in chaparral and oak woodland) . . . . *Githopsis specularioides*

                                        Common Bluecup

        11b Corolla 3–9 mm long

            12a Corolla 5–9 mm long; (flowers sessile, in clusters of 1–4, the lower ones not opening; corolla blue, lilac, or violet; in disturbed areas) . . . . *Triodanis biflora* (fig.)

                                      Venus-looking-glass

12b Corolla 2–6 mm long
    13a Flowers sessile, the lower ones not opening; corolla blue; in moist, shaded areas. . . . . *Heterocodon rariflorum*
    13b Flowers not sessile, all of them opening; corolla pale blue to white; on chaparral, often on serpentine
        14a Most leaves more than 3 times as long as wide; pedicel 1–6 mm long; fruit distinctly longer than wide; plant sometimes hairy . . . . *Campanula griffinii [C. angustiflora* var. *exilis]*
                  Griffin Harebell; SN, Ma, SCr
        14b Most leaves only 2–3 times as long as wide; pedicel 3–20 mm long; fruit nearly spherical; plant hairy . . . . *Campanula angustiflora*
                  Eastwood Campanula; SN, Ma, SCr

CAPRIFOLIACEAE (HONEYSUCKLE FAMILY) All members of the Honeysuckle Family are perennial, and most are woody vines, shrubs, or small trees. The leaves are opposite, and in *Lonicera*, they are sometimes joined at their bases in such a way that the pairs form disks around the stem. The corolla, often with a substantial tube, may have four or five essentially equal lobes, or it may be markedly two lipped, with the upper lip four lobed and the lower lip consisting of a single large lobe. The ovary is inferior, and the fruit, fleshy in all our species, has two to five seed-forming divisions.

This family has contributed many garden plants, mostly European, Asiatic, and North American species of *Lonicera* and *Viburnum*. Two of our native species, *Lonicera involucrata* var. *ledebourii* and *Sambucus racemosa,* do well if planted in moist soil. Species of *Symphoricarpos* (snowberries), especially *S. albus* var. *laevigatus,* spread so aggressively by vegetative means that they may need to be controlled.

1a   Leaves compound (corolla regular, cream colored or white; large shrubs up to 6 m tall)
    2a   Leaflets 3–5; inflorescence not as tall as wide; fruit bluish or black if the powdery coating is rubbed off . . . . *Sambucus mexicana* [includes *S. caerulea*] (pl. 18)
                  Blue Elderberry
    2b   Leaflets usually 7; inflorescence at least as tall as wide; fruit red (in moist areas) . . . . *Sambucus racemosa [S. callicarpa]* (pl. 18)
                  Red Elderberry; SM-n
1b   Leaves not compound but sometimes toothed or lobed
    3a   Leaves with coarse, evenly spaced teeth in the upper half (corolla white; fruit red; shrub; up to 4 m tall; in chaparral and pine forests). . . . . *Viburnum ellipticum*
                  Oval-leaf Viburnum; CC, Sn-n; 2
    3b   Leaves without regularly spaced teeth, but in species of *Symphoricarpos* they may have shallow, irregular lobes
        4a   Flowers borne in pairs above broad bracts that become red; leaves usually pointed at the tips, 3–7 cm long (corolla regular, yellow; fruit black; shrub, up to 3 m tall; in moist habitats) . . . . *Lonicera involucrata* var. *ledebourii* (pl. 18)
                  Black Twinberry; La-SLO

4b Flowers not in pairs above bracts; leaves rounded, not more than 4 cm long, except in *Lonicera hispidula* var. *vacillans*

5a Corolla cream colored or yellowish (corolla irregular; fruit red; leaves 1–3 cm long; stipules absent; usually sprawling on other plants)

6a Plant woody throughout; leaves of 1 or more upper pairs united around the stem; corolla not hairy . . . . *Lonicera interrupta*
Chaparral Honeysuckle

6b Plant woody only at the base; leaves not united around the stem; corolla often hairy . . . .
. . . . *Lonicera subspicata* var. *denudata* [*L. subspicata* var. *johnstonii*]
Southern Honeysuckle; CC (MD), SCl (MH)

5b Corolla pink or purplish (corolla hairy)

7a Leaves 3.5–8 cm long, those of 1 or more upper pairs united around the stem; stipules present; corolla irregular; fruit red (sprawling on other plants; widespread) . . . . *Lonicera hispidula* var. *vacillans* (pl. 18)
Hairy Honeysuckle

7b Leaves 1–4 cm long, not united around the stem; stipules absent; corolla regular; fruit white

8a Plant often more than 100 cm tall, upright; flowers 8–16 in each cluster; leaves not hairy on the undersides; corolla with a nectar gland below only 1 lobe; fruit 8–12 mm long (widespread) . . . .
. . . .*Symphoricarpos albus* var. *laevigatus* [*S. rivularis*] (pl. 18)
Common Snowberry

8b Plant rarely more than 50 cm tall, usually sprawling; flowers 2–8 in each cluster; leaves usually hairy on the undersides; corolla with nectar glands below all 5 lobes; fruit about 8 mm long . . . .
. . . .*Symphoricarpos mollis* (pl. 18)
Creeping Snowberry, Tripvine; Me-s

CARYOPHYLLACEAE (PINK FAMILY)   We are indebted to Caryophyllaceae not only for cultivated pinks, carnations, and baby's-breath, but also for some attractive wildflowers. The most striking species native to our region is *Silene californica* (Indian Pink), which is usually bright red. The family is a large one, and it has given us many weeds, some of which have become well established in habitats where they may seem to be part of the original flora.

In plants of the Pink Family, the leaves are opposite, and the nodes where they originate are usually swollen. The flower typically has five sepals, which may be separate or united, and five separate petals, often deeply lobed. In a few species, some flowers may lack petals. There are generally ten or five stamens. The pistil has either two to five styles or one style with two or three lobes. The ovary is superior, and the fruit, which is dry when mature, has from one to many seeds.

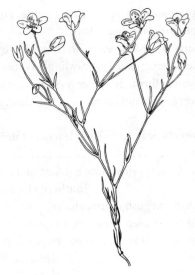

Minuartia douglasii
Douglas Sandwort

1a   Sepals with a bristle at the tips; petals either absent or scalelike (sepals separate, hairy; styles either 2, or 1 with 2 lobes; leaves narrow, stiff, up to about 1 cm long; stipules present; perennial; forming low, tufted growths)

    2a   Sepals all alike, each one narrowing to a stiff bristle; petals absent; stamens 5; on grassy hillsides . . . . *Paronychia franciscana*
                 California Whitlow-wort; sa

    2b   Sepals unequal, the 3 outer ones larger than the 2 inner ones, and also tipped with stouter and slightly longer bristles; petals scalelike; stamens 3–5; backshores of sandy beaches . . . . *Cardionema ramosissimum*
                 Sandmat

1b   Sepals without bristles; petals usually present and not scalelike, except sometimes in species of *Cerastium, Minuartia, Sagina,* and *Stellaria,* which are otherwise not as described in choice 1a

    3a   Sepals united, forming a tube; each petal with a narrow, stalklike base and often with a pair of scalelike outgrowths on the inner face (stamens 10; styles usually 3; calyx usually glandular hairy; often on fire-burned areas)

        4a   Petals bright red or sometimes deep pink; styles usually protruding from the corolla (calyx 15–25 mm long, with 10 ribs; each petal with 4–6 lobes and outgrowths on inner face) . . . . *Silene californica* (pl. 19)
                 Indian Pink

        4b   Petals white, pink, or lavender; styles not protruding from the corolla

           5a   Calyx with 16–30 prominent ribs (calyx 8–12 mm long; each petal notched at the tip, outgrowths on inner face absent) . . . . *Silene multinervia*
                 Many-nerved Catchfly; Sn-s

5b Calyx with 10 ribs, these not always prominent
    6a Calyx 4–10 mm long
        7a Calyx usually not hairy; each petal with 2 lobes, the exposed portion about twice the length of the calyx lobes; petals with no outgrowths on their inner faces; inflorescence not 1 sided . . . . .*Silene antirrhina*
                Sleepy Catchfly, Snapdragon Catchfly
        7b Calyx hairy; each petal sometimes notched but not lobed, the exposed portion at least 3 times the length of the calyx lobes; petals with outgrowths on their inner faces; inflorescence nearly 1 sided (widespread) . . . . *Silene gallica* (pl. 19)
                Windmill Pink; eu
    6b Calyx 10–16 mm long (calyx hairy; each petal with 2–4 lobes and outgrowths on its inner face; inflorescence not 1 sided)
        8a Lower leaves 12–20 mm wide; petals with 2–4 lobes (on hills and bluffs close to the coast) . . . . *Silene scouleri* ssp. *grandis* (pl. 19)
                Scouler Catchfly; SM-n
        8b Lower leaves 2–9 mm wide; petals with 2 lobes
            9a Longest leaves 3–6 cm; calyx sparsely hairy, usually purple; coastal, in sandy areas . . . .*Silene verecunda* ssp. *verecunda*
                San Francisco Campion; SF-SCr; 1b
            9b Longest leaves 5–9 cm; calyx densely hairy, usually green; inland, at elevations above 2,000 ft . . . .
        . . . . *Silene verecunda* ssp. *platyota*
                Cuyamaca Catchfly; La, CC (MD)-s
3b Sepals separate; petals without stalklike bases or outgrowths
  10a Leaves with membranous stipules (at least the inflorescence usually glandular hairy; not more than 40 cm tall)
    11a Leaves usually 2 or 4 in each axil, if 4, 2 of the leaves definitely longer (corolla usually shorter than calyx; styles 3; in salt marshes, sand dunes, and saline or alkaline habitats)
        12a Stamens 2–5 (leaves 2–4 cm long; stipules 2–4 mm long; petals white or pink; fruit up to 6 mm long). . . . . *Spergularia marina*
                Salt Marsh Sand-spurrey
    12b Stamens 8–10
        13a Leaves up to 2 cm long; stipules 2–4 mm long; petals white or pink; fruit 3–5 mm long, the seeds slightly roughened; erect . . . .
        . . . . *Spergularia bocconei*
                Boccone Sand-spurrey; me
        13b Leaves up to 5 cm long; stipules 3–6 mm long; petals white; fruit 5–8 mm long, the seeds smooth; erect or prostrate . . . . *Spergularia media*
                Middle-sized Sand-spurrey; eu

11b Leaves often more than 4 in each axil, all similar in size

  14a Styles 5 (petals white, 2–4 mm long; leaves 1–5 cm long; stipules 1–2 mm long) . . . . *Spergula arvensis*

    Stickwort, Starwort; eu

  14b Styles 3

    15a Petals 4–7 mm long; sepals 5–10 mm long; stamens 9 or 10 (leaves 10–35 mm long; stipules 5–10 mm long)

      16a Petals pink or lavender; on seaside bluffs and around salt marshes . . . . . . *Spergularia macrotheca* var. *macrotheca* (pl. 19)

        Perennial Sand-spurrey, Largeflower Sand-spurrey

      16b Petals white; around inland freshwater marshes and in alkaline habitats . . . . *Spergularia macrotheca* var. *longistyla*

        Whiteflower Sand-spurrey; Na-Al

    15b Petals 2–5 mm long; sepals 3–5 mm long; stamens 6–10

      17a Petals pink; leaves 6–12 mm long; stipules 4–5 mm long; stem below inflorescence not glandular hairy; in gravelly areas and mud flats . . . . *Spergularia rubra* (pl. 19)

        Ruby Sand-spurrey, Purple Sand-spurrey; eu

      17b Petals white; leaves 10–40 mm long; stipules 3–8 mm long; stem below inflorescence glandular hairy; in sandy soil, usually near the coast . . . . *Spergularia villosa*

        Villous Sand-spurrey; sa

10b Leaves without stipules (petals white)

  18a Each petal either not lobed or with only a shallow notch, but the petals sometimes absent in species of *Minuartia* and in *Sagina procumbens* (leaves sessile)

    19a Styles 4 or 5, alternating with sepals (plant not hairy)

      20a Stamens 4; petals if present, less than 2 mm long, shorter than sepals; leaves 3–10 mm long, rarely 20 mm; stem prostrate, with tip rising; in disturbed areas . . . . *Sagina procumbens*

        Arctic Pearlwort, Procumbent Pearlwort; Ma-n

      20b Stamens 10; petals 2–4 mm long, about as long as sepals; leaves 7–22 mm long; up to 18 cm tall; on ocean bluffs and sand dunes . . . . *Sagina maxima* ssp. *crassicaulis* [*S. crassicaulis*]

        Beach Pearlwort; Mo-n

    19b Styles 3, opposite sepals (stamens 10)

      21a Leaves 2–5 mm long; petals if present, 3–4 mm long; plant not hairy (flower sometimes solitary in leaf axils; annual; up to 12 cm tall; widespread, sometimes on serpentine) . . . . *Minuartia californica* [*Arenaria californica*]

        California Sandwort

      21b Leaves 5–55 mm long; petals if present, 2–8 mm long; plant somewhat hairy

22a Plant glandular hairy, at least in upper part; leaves 5–40 mm long; annual; erect, up to 30 cm tall (petals if present, 3–6 mm long; flower sometimes solitary in leaf axils; on dry, rocky, or gravelly hillsides, often on serpentine) ....
.... *Minuartia douglasii* [includes *Arenaria douglasii* vars. *douglasii* and *emarginata*] (fig.)
Douglas Sandwort

22b Plant not glandular hairy, but with some hairs; leaves 15–55 mm long; perennial; sprawling, rooting at the nodes

    23a Flower solitary in each leaf axil; petals 5–6 mm long; plant hairy at stem nodes; in swampy habitats ....
.... *Arenaria paludicola*
Marsh Sandwort; SF; 1b

    23b Flowers 2–5 in each leaf axil; petals 2–8 mm long; plant with minute hairs throughout; on shaded slopes at elevations above 1,500 ft, sometimes on serpentine ....
.... *Moehringia macrophylla* [*Arenaria macrophylla*]
Largeleaf Sandwort; SCl-n

18b Each petal deeply divided into 2 lobes, but the petals may be absent in *Cerastium glomeratum*, *Stellaria borealis* ssp. *sitchana*, and *S. crispa*

24a Styles 5

    25a Most leaves at least 6 times as long as wide; corolla 7–12 mm long, longer than the calyx; in rocky or grassy habitats (pedicel 1–4 times as long as calyx; perennial; widespread) .... *Cerastium arvense* (pl. 19)
Field Chickweed; Mo-n

    25b Leaves rarely more than 3 times as long as wide; corolla 5–8 mm long, sometimes longer than the calyx; in disturbed areas

        26a Pedicel 1–4 times as long as calyx; corolla 6–8 mm long, present on each flower; uppermost leaves usually with membranous margins; short-lived perennial; forming mats, but with flowering stem rising ....*Cerastium fontanum* ssp. *vulgare*
Large Mouse-ear Chickweed; eu

        26b Pedicel usually no longer than the calyx; corolla 5–6 mm long, absent on some flowers; uppermost leaves entirely green; annual; erect ....*Cerastium glomeratum*
Mouse-ear Chickweed; eu

24b Styles 3 (petals 1–7 mm long, often absent in *Stellaria borealis* ssp. *sitchana* and *S. crispa*)

    27a Stem with a single lengthwise row of hairs (leaves not clasping the stem, the blades oval, 8–45 mm long; usually in somewhat shaded places; annual; widespread) .... *Stellaria media*
Common Chickweed; me

    27b Stem, if hairy, without a single lengthwise row of hairs

        28a Plant with long hairs; leaves sessile and usually clasping the stem (leaf blades 10–45 mm long, oval; in moist habitats among sand dunes; perennial) .... *Stellaria littoralis*
Beach Chickweed; SF-n; 4

28b Plant scarcely if at all hairy; leaves sometimes sessile, but not clasping the stem

29a Most leaves several to many times as long as wide; annual (leaf blades 5–15 mm long; lower leaves not sessile; in grassy habitats and among sand dunes) . . . *Stellaria nitens*
Shiny Chickweed

29b Leaves usually less than twice as long as wide; perennial

30a Margins of leaves often wavy; flower solitary in each leaf axil (leaf blades 8–20 mm long; petals usually absent; in moist areas) . . . . *Stellaria crispa*
Chamisso Chickweed; Ma-n

30b Margins of leaves not wavy; flowers more than 1 in each leaf axil

31a Corolla if present, 2–4 mm long, shorter than the calyx; leaf blades 15–45 mm long; in moist places up to 6,000 ft . . . . *Stellaria borealis*
ssp. *sitchana* [*S. sitchana* var. *bongardiana*]
Northern Chickweed; Ma-n

31b Corolla 4–5 mm long, as long as or longer than the calyx; leaf blades 10–35 mm long; established in lawns . . . . *Stellaria graminea*
Lesser Chickweed; eu

CELASTRACEAE (STAFF-TREE FAMILY) Celastraceae has two uncommon representatives in our region. Both are attractive and interesting shrubs that merit a place in gardens. They both have opposite, toothed leaves and angled branchlets but are otherwise very different.

Euonymus occidentalis (pl. 19) (Western Burningbush) is deciduous and grows from Monterey County northward. Its leaves are 4–20 cm long, with obvious petioles. The five rounded petals, about 4–7 mm long, are brownish purple, dotted delicately with white. When the fruit splits open, some red tissue from the inside adheres to the seeds, as it does in European and Asiatic relatives that are widely cultivated in gardens. Western Burningbush grows tall, sometimes reaching a height of more than 5 m. It is found primarily in wooded ravines and requires considerable moisture.

Paxistima myrsinites (pl. 19) (Oregon Boxwood) is evergreen and grows from Marin County northward. Its nearly sessile, oval leaves are generally about 18–22 mm long. The four petals are only about 1 mm long, but they are of an unusual reddish brown color. Oregon Boxwood grows slowly and compactly, reaching a height of about 1 m. It is found mostly in partly shaded habitats above 2,000 ft. In cultivation, it requires good drainage and can stand considerable drought once it is well established.

CERATOPHYLLACEAE (HORNWORT FAMILY) Members of Ceratophyllaceae may be confused with two other families of aquatic plants: Lentibulariaceae (bladderworts) and Haloragaceae (water-milfoils). Plants of all three families are of about the same size and of similar form, with submerged or floating stems often more than 1 m long. The leaves of hornworts, however, are stiff and somewhat rough to the touch, owing to small teeth. The only species in all of California is *Ceratophyllum demersum* (fig.) (Hornwort) which grows in quiet freshwater.

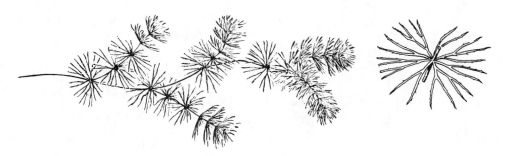

Ceratophyllum demersum
Hornwort

CHENOPODIACEAE (GOOSEFOOT FAMILY) The Goosefoot Family consists largely of herbaceous plants, although it does include some shrubs and near shrubs. The leaves and stems are often powdery, owing to the presence of small scales, and some species are decidedly succulent. The Latin and common names of the family refer to the fact that the leaf blades are often triangular in shape, like a goose's foot. The flowers, usually clustered, are small and without petals. In certain genera staminate and pistillate flowers are separate and may even be on separate plants. The ovary is usually superior. Each fruit, generally dry at maturity, contains a single seed.

This family has given us the beet, whose cultivated forms include the sugar beet, Swiss chard, and spinach. Various other members of the group have been used as food or medicines.

Many representatives of the family live in saline or alkaline habitats, and certain of them—notably species of *Salicornia* and *Salsola*—have been important sources of sodium salts. Huge piles of these plants were formerly incinerated to produce ash needed for making soap and glass.

1a Stem, at least those of recent growth, succulent and jointed; leaves scalelike
  2a Branches alternate; shrub, often more than 100 cm tall; limited to inland alkaline soils (inflorescence 5–25 mm long, the flowers in spirals; calyx 1–2 mm long, green) . . . .
. . . . *Allenrolfea occidentalis*
Iodinebush; Al, CC
  2b Branches opposite; herbs, rarely so much as 50 cm tall; in coastal salt marshes and saline habitats inland
    3a Inflorescence without flowers in the portion near the tip (inflorescence 10–40 mm long and 2–3 mm wide; perennial; up to 30 cm tall) . . . . *Salicornia subterminalis*
Glasswort; SF-s
    3b Inflorescence with flowers throughout its length
      4a Inflorescence usually slightly thinner than most stems; plant branching from the base and sometime woody there; perennial (inflorescence 15–60 mm long and 2–7 mm wide) . . . . *Salicornia virginica* (pl. 20)
Virginia Pickleweed
      4b Inflorescence usually slightly thicker than most stems; plant branching well above the base, not woody; annual

Atriplex hortensis
Garden Orache

Chenopodium album
Lamb's-quarters

Chenopodium foliosum
Strawberry-blite

Chenopodium murale
Wall Goosefoot

5a   Inflorescence 2–4 mm wide and 25–70 mm long; plant branching throughout (usually turning bright red in autumn) . . . .
. . . . *Salicornia europaea*
Slender Glasswort; Ma, Sn

5b   Inflorescence 4–6 mm wide and 15–90 mm long; plant branching above the middle . . . . *Salicornia bigelovii*
Bigelow Pickleweed

1b   Stem neither especially succulent nor jointed; leaves not scalelike, although they may be slender

6a   Leaves spine tipped (leaves alternate, succulent, smooth margined, the lower ones slender, up to 5 cm long, the upper ones much shorter and ending in a stout spine; plant very branched, up to 1 m tall; annual; widespread) . . . . *Salsola tragus* (pl. 20)
Russian-thistle, Tumbleweed; eua

6b   Leaves not spine tipped

7a   Leaves succulent, cylindrical, the blades usually at least 5 times as long as wide, not scaly on the surfaces (leaves alternate, smooth margined, 5–35 mm long; woody, at least at the base; in alkaline and saline habitats)

8a   Flowers 2–3 mm wide, 1–5 in clusters scattered throughout the plant; scars of fallen leaves generally persisting as obvious bumps; up to 80 cm tall, often forming a mound; in or close to salt marshes . . . . *Suaeda californica*
California Seablite; SF-s

8b   Flowers 1–1.5 mm wide, 1–12 in clusters usually restricted to upper branches; scars of fallen leaves generally not persisting as obvious bumps; up to 15 cm tall; in coastal salt marshes and inland alkaline habitats . . . . *Suaeda moquinii*
Bush Seepweed, Shrubby Seablite; SCl-s

7b   Leaves not especially succulent, the blades not often so much as 4 times as long as wide, often scaly on the surfaces

9a   Shrub (leaves alternate, the blades smooth margined, oval to arrowhead shaped, up to 5 cm long and wide, grayish, powdery on both surfaces; pistillate flowers sandwiched between a pair of bracts; often more than 1.5 m tall; in coastal and inland saline habitats) . . . .
. . . . *Atriplex lentiformis* [includes *A. lentiformis* ssp. *breweri*]
Big Saltbush, Quailbush; SF-s, e

9b   Herbs, but *Atriplex semibaccata* is woody at the base . . . .
. . . . CHENOPODIACEAE, SUBKEY

Chenopodiaceae, Subkey: Stems not jointed; leaves neither scalelike, spine tipped, nor succulent, the blades rarely so much as 4 times as long as wide

1a   Flowers pistillate or staminate, sometimes on separate plants; pistillate flowers sandwiched between a pair of bracts (lower leaves opposite in some species)

2a   Bracts enclosing fruit usually at least 10 mm wide (bracts often nearly circular; leaf blades sparsely or not at all scaly, 1.5–12 cm long, the lower ones opposite; annual; often more than 1.5 m tall) . . . . *Atriplex hortensis* (fig.)
Garden Orache; eua

2b   Bracts enclosing fruit rarely more than 8 mm wide

3a   Bracts enclosing fruit fleshy, red; plant woody at the base (leaf blades 1–3 cm long, usually with a scaly underside; perennial; prostrate, forming a ground cover; in disturbed areas, especially where saline or alkaline) . . *Atriplex semibaccata* (pl. 20)

Australian Saltbush; au

3b   Bracts enclosing fruit not fleshy or red; plant not at all woody

     4a   Longest leaf blade usually decidedly broader near the base than near the middle, thus roughly triangular or arrowhead shaped (annual; often more than 80 cm tall)

         5a   Bracts enclosing fruit united for nearly their entire length; leaf blades gray and scaly on both surfaces, up to 4 cm long, the lower opposite (on saline soils) . . . .

. . . . *Atriplex argentea* var. *mohavensis* [*A. argentea* ssp. *expansa*]

Silver Saltbush; SF-s

         5b   Bracts enclosing fruit united for not more than half their length; leaf blades gray and scaly on the undersides or not at all, up to 6–7 cm long, alternate

             6a   Undersides of leaves sparsely or not at all scaly; usually in moist, saline, or alkaline habitats . . . . *Atriplex triangularis*

[known as *A. patula* ssp. *hastata*]

Spearscale; Al, La

             6b   Undersides of leaves densely scaly; mostly in dry habitats, especially waste places (leaves becoming red; bracts hard when fruit matures) . . . . *Atriplex rosea*

Redscale, Tumbling Orache; eua

     4b   Leaf blade oval or elongated, not obviously broader near the base than near the middle

         7a   Leaves densely scaly; perennial; somewhat mat forming, rarely more than 30 cm tall; on coastal bluffs and backshores of beaches

             8a   Bracts enclosing fruit united for almost half their length; pistillate and staminate flowers usually in separate clusters; leaf blades up to 4 cm long, alternate . . . . *Atriplex leucophylla* (pl. 20)

Beach Saltbush, Seascale

             8b   Bracts enclosing fruit almost completely separate; pistillate and staminate flowers usually in the same cluster; leaf blades up to 2.5 cm long, the lower opposite . . . . *Atriplex californica*

California Saltbush; Ma-s

         7b   Leaves sparsely or not at all scaly; annual; erect, usually more than 40 cm tall; mostly at the edges of salt marshes or on other saline soils (the 2 following varieties intergrade)

             9a   Leaves with 2–4 lobes; bracts enclosing fruit usually with prominent projections . . . . *Atriplex patula* var. *obtusa*

Common Orache; SFBR-n

             9b   Leaves not lobed; bracts enclosing fruit without obvious projections . . . . *Atriplex patula* var. *patula*

Spear Orache

1b   Flowers with stamens and a pistil; none of the flowers sandwiched between a pair of bracts

10a   Leaves alternate or opposite, the blades usually about 10 times as long as wide, rarely so much as 4 mm wide, sometimes with short hairs; inflorescences in the leaf axils, each with not more than 7 flowers (along roadsides and in waste places) . . . .
. . . . *Kochia scoparia* [includes *K. scoparia* var. *subvillosa*]
Summer-cypress; eua

10b   Leaves alternate, the blades not more than 4 times as long as wide, the larger leaves generally at least 10 mm wide, powdery or with glandular hairs, but otherwise not hairy; inflorescences in leaf axils or at the tips of stems, each often with numerous flowers

    11a   Leaves and stem glandular, usually strong scented, the blades often lobed, sometimes shallowly so, usually not triangular (in waste places and along roadsides)

        12a   Calyx with 3–5 small teeth, but not lobed; leaves not obviously glandular (leaf blades lobed, up to 2 cm long; up to 30 cm tall) . . . . *Chenopodium multifidum*
Cutleaf Goosefoot; sa

        12b   Calyx 5 lobed, the lobes separate nearly to their bases; leaves obviously glandular

            13a   Leaf blades up to 2 cm long; up to 25 cm tall (leaves toothed or lobed)
. . . . *Chenopodium pumilio*
Small Goosefoot; au

            13b   Longest leaf blades 6.5–10 cm; often more than 65 cm tall

                14a   Flowers on short but distinct pedicels; branches of inflorescence noticeably curved; leaves lobed; up to 65 cm tall . . . .
. . . . *Chenopodium botrys*
Jerusalem-oak; eu

                14b   Flowers sessile; branches of inflorescence straight; leaves toothed or lobed; up to 130 cm tall . . . . *Chenopodium ambrosioides*
Mexican-tea; sa

    11b   Leaves and stem not glandular, sometimes strong scented, the blades sometimes toothed but not lobed, often triangular (leaves sometimes powdery)

        15a   Most flower clusters on the leafless end portions of main stems, with some clusters in the leaf axils lower on the stem; perennial (leaf blades up to 9 cm long, not powdery; calyx lobes not ridged; in open areas, especially in sand or clay) . . . . *Chenopodium californicum*
California Goosefoot

        15b   Most flower clusters in the leaf axils; annual

            16a   Most or all fruit wider than long, somewhat pumpkin shaped, with the calyx 5 lobed; calyx lobes ridged (mostly in waste places)

                17a   Upper surfaces of leaves shiny, dark green, the undersides sparsely powdery (leaf blades up to 6 cm long, unpleasantly scented) . . . . . .
. . . . *Chenopodium murale* (fig.)
Wall Goosefoot, Nettleleaf Goosefoot; eu

17b Upper surfaces of leaves dull green, the undersides densely powdery

    18a Leaves unpleasantly scented, the blades up to 5 cm long; ridge on each calyx lobe about one-fourth as wide as the lobe .... *Chenopodium berlandieri*

        [includes *C. berlandieri* var. *zschackei*]

        Pitseed Goosefoot; Ma

    18b Leaves not unpleasantly scented, the blades up to 7 cm long; ridge on each calyx lobe much less than one-fourth as wide as the lobe (common) .... *Chenopodium album* (fig.)

        Lamb's-quarters, Pigweed; eu?

16b Most or all fruit longer than wide, egg shaped, with calyx 3 lobed, but usually some fruit pumpkin shaped and the calyx with 4 or 5 lobes; calyx lobes not ridged

    19a Flower clusters rounded, usually 3–5 mm wide, but sometimes 10 mm; calyx fleshy (calyx becoming red; leaf blades up to 4 cm long, not powdery; in open places, especially in sand or gravel) .... *Chenopodium foliosum [C. capitatum]* (fig.)

        Strawberry-blite; eu

    19b Flower clusters longer than wide, often more than 10 mm long; calyx not fleshy

        20a Young leaves densely covered with white powder, but becoming green (leaf blades up to 5 cm long; calyx lobes rounded at the tips, completely enclosing the ripe egg-shaped fruit; in wet habitats, including marshes) .... *Chenopodium macrosperum* var. *halophilum*

        [*C. macrosperum* var. *farinosum*]

        Coast Goosefoot; sa?

        20b Leaves with little or no white powder (mostly in saline habitats, including the edges of salt marshes)

            21a Leaf blades up to 9 cm long, sometimes sparsely powdery on the undersides; calyx lobes red, not pointed at the tips, scarcely enclosing the ripe egg-shaped fruit .... *Chenopodium rubrum*

            Red Goosefoot

            21b Leaf blades up to 3 cm long, not powdery; calyx lobes not red, pointed at the tips, completely enclosing the ripe egg-shaped fruit .... *Chenopodium chenopodioides*

            South American Goosefoot; sa

CISTACEAE (ROCKROSE FAMILY)   In Cistaceae, the flower has three or five sepals, five petals, at least 12 stamens, and a pistil whose fruiting portion becomes a three-lobed, dry capsule. The ovary is superior. The petals often fall away early. The leaves, alternate or opposite, are smooth margined. Most members of the family are at least somewhat shrubby. We have only one native genus, but a few of the cultivated rockroses from the Mediterranean region have become naturalized.

Helianthemum scoparium
Peak Rushrose

1a   Leaves alternate, shed early; sepals 5, the outer 2 narrower; petals yellow, up to 1 cm long; up to 45 cm tall (leaves up to 4 cm long and up to 6 mm wide; in sandy or gravelly areas) . . . . *Helianthemum scoparium* [includes *H. scoparium* var. *vulgare*] (fig.)
Peak Rushrose; Me-s

1b   Leaves opposite, persistent; sepals 3 or 5, equal; petals not yellow, 1–8 cm long; up to 100 cm tall (naturalized in a few areas)

     2a   Petals mauve, pale purple, or other nonwhite colors, not spotted; leaf not sessile, with 1 prominent vein from the base (flowers in clusters of 1–7; sepals 5; petals 2–3 cm long; leaves 3–7 cm long) . . . . *Cistus creticus* [*C. villosus* var. *corsicus*]
Crete Rockrose; me

     2b   Petals white, sometimes with a red or yellow spot; leaf sessile, with 3 prominent veins from the base

         3a   Flower solitary; sepals 3; each petal 3–5 cm long, with a red or yellow spot at the base; leaves 4–8 cm long . . . . *Cistus ladanifer*
Gum Cistus; me

         3b   Flowers in clusters of 2–8; sepals 5; petals 1–1.5 cm long, not spotted; leaves 1.5–5 cm long . . . . *Cistus monspeliensis*
Montpellier Rockrose; me

CONVOLVULACEAE (MORNING-GLORY FAMILY)   Plants of Convolvulaceae are mostly climbing or trailing plants. The leaves are alternate, although the flower pedicels sometimes have opposite leaflike bracts. The flower has five separate or nearly separate sepals, five united petals that usually form a funnel-shaped corolla, and five stamens that are attached to the corolla tube. The corolla is usually twisted when in the bud stage and is often pleated. The ovary

Convolvulus arvensis
Field Bindweed

is superior. In our species, the pistil has either two styles or a stigma with two or three lobes. The fruit is a capsule.

Various morning-glories of the genera *Ipomoea* and *Convolvulus* are grown commercially, and *Dichondra micrantha* is used for lawns. Also in this family is sweet potato, a species of *Ipomoea*, as well as *Convolvulus arvensis*, a widespread weed in orchards, pastures, and gardens.

1a   Corolla less than 1 cm long, not much longer than the calyx, deeply lobed
    2a   Leaf blades kidney shaped; corolla about 3 mm long, the lobes rounded; bracts absent beneath each flower (leaf blades 1–2 cm wide; corolla whitish; styles 2; small ground cover, native, but widely cultivated and escaping) . . . . *Dichondra donelliana*
            California Dichondra; Mo-n
    2b   Leaf blades elliptical or elongated; corolla 6–8 mm long, the lobes pointed; bracts present beneath each flower
        3a   Leaf blades oblong, up to 5 cm long; corolla pink or bluish; stigma 2 lobed; plant with trailing stems; annual; in wet clay, sometimes on serpentine. . . . .
        . . . . *Convolvulus simulans*
            Smallflower Morning-glory; CC-s; 4
        3b   Leaf blades elliptical, about 1 cm long; corolla white; styles 2; plant tufted; perennial; in alkaline soil. . . . . . *Cressa truxillensis* [includes *C. truxillensis* var. *vallicola*]
            Alkali-weed
1b   Corolla more than 1 cm long, extended well beyond the calyx, not deeply lobed
    4a   Each flower without a pair of bracts below it; stigma usually with 2 or 3 lobes; annual (corolla white, pink, purple, or blue, 5–6 cm long; leaf blades heart shaped, up to 12 cm long, hairy; widely cultivated and frequently escaping). . . . . *Ipomoea purpurea*
            Garden Morning-glory; sa
    4b   Each flower with a pair of bracts below it; stigma 2 lobed; perennial
        5a   Bracts usually not directly beneath each flower, but if so, not concealing the calyx
            6a   Calyx 5 mm long; corolla 20–25 mm long; leaf blades up to 25 mm long (leaves with minute hairs or none; corolla white or pink; arising from a deep root; widespread). . . . . *Convolvulus arvensis* (fig.)
            Field Bindweed, Orchard Morning-glory; eu

6b   Calyx 7–15 mm long; corolla 20–52 mm long; leaf blades up to 50 mm long

7a   Leaf blades hairy, up to 4 cm long; bracts 1–7 mm beneath the flower; corolla white to pale yellow. . . . . *Calystegia occidentalis* [includes *C. polymorpha*]
Western Morning-glory; Na-n

7b   Leaf blades not hairy, up to 5 cm long; bracts 3–16 mm beneath the flower; corolla white, cream colored, or purple

8a   Leaves with rounded lobes; bracts alternate, usually lobed . . . . *Calystegia purpurata* ssp. *saxicola*
Marin Morning-glory; Me-Ma; 1b

8b   Leaves with pointed, angular lobes; bracts opposite, smooth margined . . . . *Calystegia purpurata* ssp. *purpurata*
[includes *C. purpurata* ssp. *solanensis*] (pl. 20)
Climbing Morning-glory

5b   Bracts directly beneath each flower, concealing the calyx or nearly so

9a   Leaf blades 4–8 cm long, much longer than wide; bracts 13–28 mm long; corolla 35–70 mm long (leaves usually not hairy; corolla white or pink; growing in salt marshes and other moist, saline habitats) . . . . *Calystegia sepium* ssp. *limnophila*
Hedge Bindweed; SF

9b   Leaf blades 1–4.5 cm long, about as long as wide; bracts 7–20 mm long; corolla 30–62 mm long

10a   Leaf blades not hairy, kidney shaped; corolla pink; coastal, growing on backshores of sandy beaches (leaf blades 1–3 cm long). . . . . *Calystegia soldanella* (pl. 20)
Beach Morning-glory

10b   Leaf blades hairy, more or less triangular; corolla white or cream colored; inland, growing in dry, gravelly habitats

11a   Leaves hairy, but not woolly; bracts 4–9 mm wide (leaf blades 3–4 cm long; corolla 33–62 mm long, white or cream colored; stem up to 20 cm long) . . . . *Calystegia subacaulis*
Hill Morning-glory; Sn, Na-SLO

11b   Leaves woolly; bracts 5–14 mm wide

12a   Leaf blades less than 3 cm long; corolla 30–54 mm long; stems up to 15 cm long, not radiating away from the rootstock; bracts 8–15 mm long, rounded at the tips (sometimes on serpentine) . . . . *Calystegia collina*
Woolly Morning-glory; Me-SCl

12b   Leaf blades 3–4.5 cm long; corolla 20–45 mm long; stems up to 100 cm long, radiating away from the rootstock; bracts 7–20 mm long, each usually tapering to a point . . . . *Calystegia malacophylla* ssp. *pedicellata* (pl. 20)
Hairy Morning-glory; Al (MH)

CORNACEAE (DOGWOOD FAMILY)   All members of the Dogwood Family native to North America have deciduous, opposite leaves. In certain species, however, successive pairs of leaves may be so close together that they appear to form a whorl. The flower is small. Each has a four-lobed calyx, four separate petals, and four stamens. The ovary is inferior, and the fruit is fleshy and two seeded.

This family has given us some attractive garden subjects, including *Cornus florida* of eastern North America, *Cornus kousa* of Asia, and some Asiatic species of *Aucuba.* Of the three species native to our region, *Cornus sericea* ssp. *occidentalis* is perhaps the most likely candidate for a garden here, although it requires considerable moisture. *Cornus nuttallii* is a spectacular tree when in flower and when its foliage turns red in autumn. Although it grows well in the mountains and in the Pacific Northwest, it is not likely to succeed in lowland gardens. Futhermore, it is subject to an extremely debilitating disease caused by a fungus.

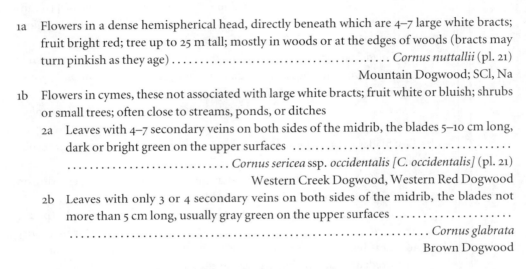

1a   Flowers in a dense hemispherical head, directly beneath which are 4–7 large white bracts; fruit bright red; tree up to 25 m tall; mostly in woods or at the edges of woods (bracts may turn pinkish as they age) . . . . *Cornus nuttallii* (pl. 21)
    Mountain Dogwood; SCl, Na

1b   Flowers in cymes, these not associated with large white bracts; fruit white or bluish; shrubs or small trees; often close to streams, ponds, or ditches

    2a   Leaves with 4–7 secondary veins on both sides of the midrib, the blades 5–10 cm long, dark or bright green on the upper surfaces . . . . *Cornus sericea* ssp. *occidentalis* [*C. occidentalis*] (pl. 21)
        Western Creek Dogwood, Western Red Dogwood

    2b   Leaves with only 3 or 4 secondary veins on both sides of the midrib, the blades not more than 5 cm long, usually gray green on the upper surfaces . . . . *Cornus glabrata*
        Brown Dogwood

CRASSULACEAE (STONECROP FAMILY)   Crassulaceae provides us with numerous plants that are easily grown in rockeries, dish gardens, and pots. Various species of *Sedum* (stonecrops) and *Sempervivum* (houseleeks) are especially popular. The nearly ubiquitous *Sempervivum tectorum,* a European species called hen-and-chickens because of its habit of growth, used to be planted on tile roofs because it was thought to provide protection from lightning. Although it may not help in that regard, its juice may be useful for curing ringworm and for removing warts.

The members of this family in our area have four or five sepals and petals. There are 5, 8, or 10 stamens and three to five pistils, which are sometimes fused at their bases. The ovary is superior. The alternate or opposite leaves are smooth margined and often somewhat succulent.

1a   At least lower leaves opposite, less than 1 cm long; stamens 4 or 5; petals not more than 2 mm long, greenish (petals separate or united at their bases; annual; up to 8 cm tall)

    2a   Flowers in cymes at ends of branches; upper leaves alternate, lower leaves opposite; petals 5; stamens 5; pistils 5; up to 8 cm tall (leaves 4–5 mm long, the lower shed early, but the places where they were attached will be visible; in rocky habitats, including serpentine) . . . . *Parvisedum pentandrum*
        Mount Hamilton Parvisedum; La-SB

2b Flowers 1 or 2 in leaf axils; all leaves opposite; petals 4; stamens 4; pistils 3–5; rarely more than 6 cm tall

    3a Flowers usually 2 in leaf axils; leaves 1–3 mm long; upright; in open fields and on burned areas (often forming dense masses, becoming reddish; widespread) . . . . *Crassula connata [C. erecta]* (pl. 21)
        Sand Pygmyweed

    3b Flower single in leaf axils; leaves 2–6 mm long; sprawling, with stem tips rising; in muddy areas that dry out in late spring and summer . . . . *Crassula aquatica*
        Pygmyweed, Water Pygmy

1b Leaves all alternate or basal, more than 1 cm long except *Sedum radiatum;* stamens 8 or 10; petals 4–14 mm long, white, yellow, or red

    4a Leaves 5–22 mm long; petals 4 or 5, 5–10 mm long, separate or united at their bases; pistils 4 or 5; stamens 8 or 10 (in rocky areas)

        5a Leaves 5–11 mm long, those on stem nearly cylindrical, widest near their bases, not whitish; petals yellow to white; annual or biennial; sometimes on serpentine . . . . *Sedum radiatum [S. stenopetalum* ssp. *radiatum]*
            Narrow-petaled Stonecrop; Mo-n

        5b Leaves 11–22 mm long, those on stem flattened, widest near their tips, usually whitish; petals yellow; perennial; widespread . . . . . . *Sedum spathulifolium* (pl. 21)
            Broadleaf Stonecrop, Pacific Stonecrop; SCr-n

    4b Leaves at least 25 mm long; petals 5, 7–14 mm long, united at their bases; pistils 5; stamens 10 (perennial)

        6a Upper reduced leaves nearly triangular, about as long as wide; petals 10–14 mm long; strictly limited to coastal cliffs (leaves 2.5–6 cm long; petals pale yellow) . . . . *Dudleya farinosa*
            Powdery Dudleya, Bluff-lettuce

        6b Upper reduced leaves slender, mostly 2–3 times as long as wide; petals 7–14 mm long; inland as well as near the coast

            7a Leaves 3–17 cm long, the basal ones often more than 2 cm wide; petals bright yellow or red; widespread . . . . *Dudleya cymosa* (pl. 21)
                Common Dudleya, Spreading Dudleya; SCl-n

            7b Leaves 3–8 cm long, the basal ones rarely so much as 2 cm wide; petals pale yellow; in rocky areas including serpentine . . . . *Dudleya setchellii [D. cymosa* ssp. *setchellii]*
                Santa Clara Valley Dudleya; CC-SB; 1b

CUCURBITACEAE (GOURD FAMILY) The Gourd Family is a generous one, for it has given us cucumbers, watermelons, cantaloupes, squashes, and pumpkins. Important medicinal substances are also derived from certain species. Plants of this group are mostly climbing or trailing; climbing types have tendrils originating at the leaf axils. The alternate leaves are usually palmately lobed.

    The flowers are pistillate or staminate. The more or less cup-shaped corolla has five lobes, and the calyx may seem to be absent when it has no lobes. Each staminate flower has five stamens, but because these are partly joined to one another, they appear as three or four. The ovary is more or less inferior, and the fruit has a few large seeds.

Marah fabaceus
California Manroot

1a   Tendrils not branched; calyx usually obvious, with 5 lobes; staminate and pistillate flowers on different plants; stigmas 3, each 2 lobed; fruit a red or orange berry (occasionally naturalized). . . . . *Bryonia dioica*

White Bryony; eu

1b   Tendrils branched; calyx lacking lobes, therefore appearing to be absent; staminate flowers and single pistillate flower in same inflorescence on same plant; stigma 1, not lobed; fruit a dry gourdlike capsule

    2a   Leaves mostly more than 10 cm wide; pistillate flower 15–17 mm wide; staminate flower 12–15 mm wide (corolla white; fruit 2–3.5 cm wide, the spines, if present, about 6 mm long, weak; Coast Ranges) . . . . *Marah oreganus*

Coast Manroot; SCl-n

    2b   Leaves mostly less than 10 cm wide; pistillate flower 8–15 mm wide; staminate flower 5–12 mm wide

        3a   Staminate flower 5–7 mm wide; corolla white; fruit 2–3.5 cm wide, the spines 1–2 mm long, weak; inland . . . . *Marah watsonii*

Taw Manroot; Sn, Sl

        3b   Staminate flower 7–12 mm wide; corolla yellow green, cream colored, or white; fruit 4–5 cm wide, the spines 5–25 mm long, rigid; widespread . . . . *Marah fabaceus* [includes *M. fabaceus* var. *agrestis*] (fig.)

California Manroot, Wild Cucumber; Ma-Mo

CUSCUTACEAE (DODDER FAMILY)   Members of Cuscutaceae, often included in Convolvulaceae (Morning-glory Family), are rootless and have almost no chlorophyll. The stems are orange or yellow, and the leaves are reduced to tiny scales. All dodders parasitize other flowering plants, whose tissues they penetrate by specialized branches of their slender, twining stems. Each small flower has four or five calyx and corolla lobes and four or five stamens. The corolla is typically white and usually has outgrowths on the inner faces that are about half the length of the corolla. The ovary is superior. The pistil, with two styles, develops into a small dry fruit with one to four seeds.

1a   Stem bright orange; restricted to salt marshes and alkaline areas (outgrowths on inner face of corolla present)

     2a   Flower 2–3 mm long; parasitic on plants of Chenopodiaceae in inland saline habitats . . . . *Cuscuta salina* var. *salina*
Inland Dodder, Alkali Dodder

     2b   Flower 3–4.5 mm long; parasitic on *Salicornia* (Chenopodiaceae) and some other plants in coastal salt marshes . . . . *Cuscuta salina* var. *major* (pl. 21)
Salt Marsh Dodder

1b   Stem yellow, or orange in *Cuscuta subinclusa;* in many habitats, both moist and dry, but rarely in salt marshes or alkaline areas

     3a   Stem orange; flower 5–6 mm long (calyx lobes pointed; outgrowths on corolla inner face present; pedicel, if present, shorter than the flower; parasitic on various trees [*Quercus, Salix, Prunus*] and shrubs [*Ceanothus, Rhus*]) . . . . *Cuscuta subinclusa*
Canyon Dodder

     3b   Stem yellow; flower mostly less than 5 mm long

         4a   Calyx lobes pointed; outgrowths on corolla inner face absent or scarcely visible (flower 2–4 mm long)

             5a   Flower sessile; coastal (often parasitic on species of *Grindelia* and *Solanum,* but not confined to these) . . . . *Cuscuta californica* var. *breviflora* [*C. occidentalis*]
Western Dodder

             5b   Flower with a short pedicel; not restricted to the coast (parasitic on many native shrubs and herbs) . . . . *Cuscuta californica* var. *californica*
California Dodder, Chaparral Dodder

         4b   Calyx lobes blunt; outgrowths on corolla inner face present

             6a   Flower 3–5 mm long; calyx definitely shorter than corolla (parasitic on *Medicago sativa* and various other plants) . . . . *Cuscuta indecora*
Common Dodder

             6b   Flower 2–3 mm long; calyx nearly as long as corolla (parasitic on species of *Trifolium, Medicago sativa,* and various members of Asteraceae) . . . . *Cuscuta pentagona* [includes *C. campestris*]
Western Field Dodder

DATISCACEAE (DATISCA FAMILY)   *Datisca glomerata* (fig.) (Durango-root), an herbaceous perennial, is the only species of Datiscaceae in California. It reaches a height of about 1 m. The leaves, about 15 cm long, are alternate, but some, especially lower ones, are nearly opposite or even whorled. They have toothed margins, and most are raggedly lobed in a pinnate pattern. The flowers, clustered in the leaf axils, are small and lack petals. They are, moreover, of two types, these on separate plants. Strictly staminate flowers, with 8 to 12 stamens, have three to nine small calyx lobes; flowers that have a pistil (this with three forked styles) and sometimes also two to four stamens, have three calyx lobes. The ovary is inferior, and the fruit is a some-

Datisca glomerata
Durango-root

what three-sided capsule about 8 mm long. Durango-root is most commonly found in stream beds that are dry in summer.

DIPSACACEAE (TEASEL FAMILY)   In our region, Dipsacaceae is represented by only three species, all introduced from Europe. These herbaceous plants have opposite leaves. Their small flowers are concentrated in dense heads whose bases are surrounded by bracts, which are inconspicouous in *Scabiosa* but spinelike in *Dipsacus*. These spinelike inflorescences of *D. sativus* (Fuller's Teasel) were formerly used in textile mills for raising the nap on woolen cloth. A unique feature of the flower of this family is a cuplike or funnel-like bractlet that encloses the calyx. There are usually four stamens. The ovary is inferior, and the dry fruit is one seeded. In *Dipsacus,* the corolla is irregular with two larger lobes and two smaller lobes. In *Scabiosa,* some flowers have a regular five-lobed corolla, but others near the base of the head are irregular, having two smaller lobes and three larger ones.

1a Plant not especially stiff, bushy, without prickles, mostly less than 50 cm tall; calyx 5 lobed, the lobes bristlelike; flowers in heads 3 cm wide and high; bracts and bractlets not spinelike; corolla 5 lobed, pink, rose, purple, or white . . . . *Scabiosa atropurpurea* (pl. 22)
Pincushion, Morning-bride; me

1b Plant stiff, usually with a single main stem, this prickly, often more than 100 cm tall; calyx 4 lobed, not bristlelike; flowers in heads 5–10 cm high and about half as wide; bracts and bractlets spinelike; corolla 4 lobed, lavender or white

2a Bracts beneath the inflorescence curving upward, some of them longer than the inflorescence; bracts beneath each flower ending in a straight spine . . . . *Dipsacus fullonum* (pl. 22)
Wild Teasel; eu

2b Bracts beneath the inflorescence directed outward, generally shorter than the inflorescence; bracts beneath each flower ending in a downcurved spine . . . . . *Dipsacus sativus*

Fuller's Teasel; eu

DROSERACEAE (SUNDEW FAMILY) Sundews live in moist habitats, especially sphagnum bogs. Their leaves are covered with sticky, red hairs whose glandular tips trap small insects. The insects are then slowly digested, and the soluble products are absorbed by the plant. Our only species is *Drosera rotundifolia* (Roundleaf Sundew), which grows from Sonoma County northward. Its basal leaves, resembling flattened spoons 4–12 mm wide, form a tight cluster. The small flowers, with five white or pinkish petals 4–6 mm long, are on one or more upright peduncles 5–25 cm long. The ovary is superior.

ELAEAGNACEAE (OLEASTER FAMILY) *Elaeagnus angustifolius [E. angustifolia]* (Oleaster or Russian-olive), is probably native to Asia but is common in Europe and North America. Through self-seeding, it sometimes becomes so well established that it may appear to be native. It is a shrub or tree up to about 7 m tall with alternate, smooth-margined leaves. The twigs and leaves are silvery, owing to the presence of scalelike hairs. The flowers, in clusters, lack petals but have four yellowish, calyx lobes that are sweet scented. The ovary is superior. The oval fruit, 1–2 cm long, are yellowish or brownish and are covered with silvery scales similar to those on other parts of the plant. In Europe, the fruit is dried and used to flavor cakes.

A species native to California, east of the Sierra Nevada, is *Shepherdia argentea* (Buffaloberry). The silvery leaves and red berries of this shrub or small tree make it an attractive subject for some gardens.

ERICACEAE (HEATH FAMILY) The Heath Family consists of trees, shrubs, and perennial herbs. In some floras, it is separated into several families. Familiar garden plants belonging to this large group are rhododendrons, azaleas, heaths, heathers, and blueberries. The flower is regular or nearly so and usually has five separate sepals and four or five separate or united petals. There are usually 8–10, sometimes 5, stamens, which are interesting because their anthers release pollen through a neat pore or slit instead of simply splitting open. The fruit is typically partitioned into five seed-producing divisions. Generally, the ovary is superior, but in a few genera, such as *Vaccinium,* it is inferior.

Probably all plants of this family have fungi associated with their roots. The fungi contribute to the symbiotic relationship by absorbing nutrients and making these available to their plant hosts. Phosphorus is a particularly important element that enters the plants by way of the fungi. Several of our representatives of Ericaceae lack chlorophyll and are saprophytic.

Many shrubs of this family are excellent subjects for gardens. For our region, species of *Arctostaphylos* (manzanitas) are especially useful in drought-tolerant gardens. Some of them, such as *A. uva-ursi,* are low ground covers, but others are more than 1 m tall. A wide variety of manzanitas are now commercially available, and they are easily propagated from cuttings. *Rhododendron occidentale* (Western Azalea), *Vaccinium ovatum* (Evergreen Huckleberry), and *Gaultheria shallon* (Salal), are attractive shrubs for somewhat shaded situations where soil moisture is maintained throughout summer.

Arbutus menziesii (Pacific Madrone) is much admired but difficult to transplant successfully, even when small. Perhaps this is due, in part, to its root fungi not taking kindly to disturbance. You may be successful, however, in growing it from seed. Collect the fleshy fruit in autumn, crush them to squeeze out the seeds, and press these into the soil where you want a tree to grow. A year or two later, thin the crop so that the strongest seedling will have no competition. Remember that the Pacific Madrone might become 40 m tall, its leathery leaves may cause problems on a roof or in the garden, and it is susceptible to some diseases caused by fungi. Maybe you'll be better off without it.

1a   Plant with scalelike leaves, neither these nor the rest of the plant green (in shaded habitats)
    2a   Stem with red and white stripes (inflorescence 5–40 cm long; petals 5, separate; stamens 10; up to 40 cm tall; at elevations above 2,000 ft) . . . . *Allotropa virgata*
                               Sugarstick; Na, Sn-n

    2b   Stem without red and white stripes
       3a   Flowers in loose racemes at least 6 cm long; style protruding out of the flower; petals 5; stamens 10 (petals separate) . . . . *Pyrola picta* [*P. picta* forma *aphylla*]
                        Whitevein Shinleaf (nongreen form); Na, Sn-n
       3b   Flowers in congested heads less than 6 cm long; style not protruding out of the flower; petals usually 4; stamens usually 8
          4a   Inside of corolla not hairy; leaves often fringed with hairs (plant yellowish or cream colored; petals separate; up to 20 cm tall) . . . . *Pleuricospora fimbriolata*
                                 Fringed Pinesap; SM-n
          4b   Inside of corolla hairy; leaves not fringed with hairs
              5a   Plant pink or cream colored, up to 12 cm tall; anthers straight; petals united at their bases . . . . *Hemitomes congestum*
                                     Gnomeplant; Mo-n
              5b   Plant cream colored to yellowish, up to 20 cm tall; anthers bent; petals completely separate . . . . *Pityopus californicus* (fig.)
                                California Pinefoot; La, Ma; 4

1b   Plant with typical leaves, these green
    6a   Herbs, sometimes woody at the base, not more than 40 cm tall (petals 5, separate; stamens 10; in shaded habitats)
       7a   Leaves usually not toothed, basal; petioles 2–7 cm long; inflorescence at least 6 cm long (leaves with white or yellowish veins; petals white, yellow, green, or purple) . . . . *Pyrola picta* (fig.)
                        Whitevein Shinleaf (green form); Na, Sn-n
       7b   Leaves usually toothed, on the stem and usually in whorls; petioles absent or not more than 0.5 cm long; inflorescence not more than 5 cm long
          8a   Leaves not more than 3.5 cm long, sometimes with white veins; flowers 1–3 in each inflorescence; petals white, turning pink . . . . *Chimaphila menziesii*
                             Pipsissewa, Little Prince's-pine
          8b   Leaves 3–7 cm long, the veins not white; flowers 3–7 in each inflorescence; petals pink . . . .
         . . . . *Chimaphila umbellata* [includes *C. umbellata* var. *occidentalis*] (fig.)
                                        Prince's-pine

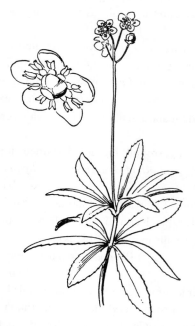

Ledum glandulosum
Western Labrador-tea

Chimaphila umbellata
Prince's-pine

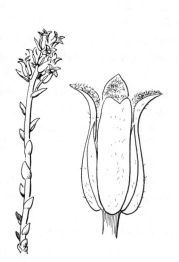

Pityopus californicus
California Pinefoot

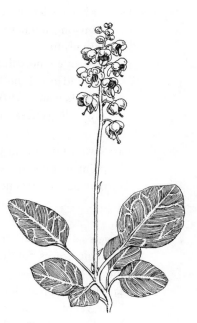

Pyrola picta
Whitevein Shinleaf

6b   Trees or shrubs, woody throughout, often more than 100 cm tall, but some species low and creeping

9a   Petals either separate, or united for about half their length, in which case the lower portion of the corolla is funnel shaped and the corolla as a whole is at least 2 cm long (petals 5)

10a   Petals 5–8 mm long, separate or nearly so, white to yellowish; up to 1.5 m tall (stamens 8–10; leaves 1–3.5 cm long, evergreen, leathery; in swamps and bogs) ..... *Ledum glandulosum* [includes *L. glandulosum* ssp. *columbianum*] (fig.)
Western Labrador-tea; SCr-n

10b   Petals at least 25 mm long, united for about half their length, white, cream, or rose; often more than 2 m tall

11a   Leaves evergreen, 6–20 cm long, smooth margined; stamens 10; corolla mostly pink or rose (coastal) ..... *Rhododendron macrophyllum* (pl. 23)
Rosebay; Mo-n

11b   Leaves deciduous, 3–8 cm long, finely toothed; stamens 5; corolla white or cream, often pink tinged and with considerable yellow on the uppermost lobe (flower fragrant; usually in moist habitats) ....
.... *Rhododendron occidentale* (pl. 23)
Western Azalea; SCr-n

9b   Petals united for more than half their length, the portion of the corolla below the lobes shaped like a cup or urn, and the corolla as a whole not more than 1 cm long

12a   Flowers in panicles or racemes that are at least 5 cm long; leaves usually more than 7 cm long, not hairy; corolla 5 lobed (leaves evergreen and leathery; stamens usually 10)

13a   Tree, often more than 20 m tall; leaves elliptical, usually rounded at the tips, and usually not toothed; flowers in panicles; corolla yellowish white or pink; fruit red to orange; bark reddish brown, the outer layer generally peeling off, exposing a smooth surface. ... *Arbutus menziesii* (pl. 22)
Pacific Madrone

13b   Shrub, up to 2 m tall, the stem sometimes vinelike; leaves abruptly pointed at the tips and toothed; flowers in racemes; corolla white; fruit dark purple; bark not obviously peeling (in shaded habitats) ....
.... *Gaultheria shallon* (pl. 23)
Salal

12b   Flowers solitary or in panicles, corymbs, or racemes that are rarely more than 3 cm long; leaves not more than 7 cm long, often hairy; corolla with 4 or 5 lobes

14a   Flowers in panicles or racemes at the ends of branches; ovary superior; leaves often more than 2 cm long; bracts several on pedicel; stamens usually 10; bark of trunk and main stem usually smooth, reddish brown, and peeling away; in open areas, including chaparral (leaves leathery, evergreen, usually smooth margined)................ ERICACEAE, SUBKEY
Arctostaphylos

14b Flowers solitary or in corymbs or racemes in the leaf axils; ovary inferior; leaves about 2 cm long; bracts absent or not more than 2 on pedicel; stamens 8 or 10; bark not reddish brown or peeling; in shaded areas

    15a Leaves finely toothed, firm, evergreen; new branches without prominent ridges; flowers in corymbs or racemes; corolla white; fruit dark blue black (coastal) . . . . *Vaccinium ovatum* (pl. 23)

        Evergreen Huckleberry, California Huckleberry

    15b Leaves usually not toothed, thin, deciduous; new branches with prominent ridges; flower solitary; corolla pink or green; fruit red . . . . *Vaccinium parvifolium* (pl. 23)

        Red Huckleberry; SCl-n

Ericaceae, Subkey: Evergreen trees or shrubs; flowers in a panicle or raceme rarely more than 3 cm long, with some bracts on the peduncle and usually 1 or 2 on each pedicel; petals not more than 1 cm long, united for more than half their length, usually urn shaped; leaves not more than 7 cm long, leathery, usually smooth margined; bark of trunk and main stems usually smooth, reddish brown, and peeling away; base of *Arctostaphylos glandulosa, A. tomentosa,* and sometimes *A. uva-ursi* enlarged as a burl . . . . *Arctostaphylos*

1a Bracts on each pedicel up to 6 mm long, usually scalelike and shorter than the pedicel, but sometimes the lowest bract of the inflorescence is longer and leaflike (leaves not overlapping or clasping the stem)

  2a Mature plant often less than 1 m tall, the stem prostrate or with the tip rising (leaves sparsely if at all hairy; pedicel not hairy and usually not glandular)

    3a Lowest bract of inflorescence 5–10 mm long, leaflike (leaves up to 3 cm long, both surfaces bright green; pedicel 4–5 mm long; corolla 4–5 mm long; fruit 5–6 mm wide). . . . . *Arctostaphylos densiflora*

        Vine Hill Manzanita; Sn; 1b

    3b Lowest bract of inflorescence similar to other bracts, not leaflike

      4a Upper surfaces of leaves darker green than the undersides (leaves up to 3 cm long; pedicel 2–4 mm long; base of plant sometimes enlarged) . . . . *Arctostaphylos uva-ursi* [includes *A. uva-ursi* vars. *coactilis* and *marinensis*] (pl. 23)

        Bearberry, Kinnikinnick; SM-n

    4b Both surfaces of leaves bright green

      5a Leaves 3.5–6 cm long; inflorescence a panicle; corolla bright pink; not restricted to serpentine . . . . *Arctostaphylos stanfordiana* ssp. *decumbens* [*A. stanfordiana* var. *repens*]

        Rincon Manzanita; Sn; 1b

      5b Leaves up to 3.5 cm long; inflorescence a raceme; corolla white to light pink; restricted to serpentine

        6a Corolla 4–5 mm long; petioles 2–4 mm long; fruit 4–5 mm wide (leaf blades 1–1.5 cm wide; branchlets hairy, but not densely so) . . . . *Arctostaphylos hookeri* ssp. *ravenii*

        Presidio Manzanita; SF; 1b

6b Corolla 5–7 mm long; petioles 3–5 mm long; fruit 6–8 mm wide

7a Leaf blades up to 1 cm wide; branchlets hairy, but not densely so
.... *Arctostaphylos hookeri* ssp. *franciscana*
Franciscan Manzanita; SF; 1a (1942)

7b Leaf blades 1–1.5 cm wide; branchlets densely hairy ....
.... *Arctostaphylos hookeri* ssp. *montana*
[*A. pungens* var. *montana*]
Mount Tamalpais Manzanita; Ma (MT); 1b

2b Mature plant at least 1 m tall, often much more, the stem erect

8a Lowest bract of inflorescence 5–10 mm long, leaflike

9a Pedicel bristly; leaf blades 2.5–5 cm long, both surfaces with a whitish coating, not hairy; petioles 7–15 mm long; sometimes treelike; not on serpentine ....
.... *Arctostaphylos glauca*
Bigberry Manzanita; CC (MD)-s

9b Pedicel not hairy; leaf blades 1–3 cm long, both surfaces dark green, hairy or glandular hairy; petioles 3–6 mm long; shrubs; often on serpentine

10a Fruit red brown, 8–10 mm wide, its pedicel 6–8 mm long.............
.... *Arctostaphylos bakeri* [*A. stanfordiana* ssp. *bakeri*]
Baker Manzanita; 1b

10b Fruit tan, 5–7 mm wide, its pedicel 5–7 mm long......................
.... *Arctostaphylos hispidula* [*A. stanfordiana* ssp. *hispidula*]
Howell Manzanita; Sn; 4

8b Lowest bract of inflorescence similar to other bracts, not leaflike

11a Bracts on each pedicel less than 2 mm long; fruit 3–8 mm wide

12a Corolla 4 lobed, white to pink; petioles 1–3 mm long; leaves up to 2.5 cm long, the upper surfaces darker green than the undersides; fruit 3–4 mm wide ....
.... *Arctostaphylos nummularia* [includes *A. nummularia* var. *sensitiva*]
Fort Bragg Manzanita; Ma-SCl

12b Corolla 5 lobed, bright pink; petioles 8–12 mm long; leaves 3.5–6 cm long, both surfaces bright green; fruit 6–8 mm wide ....
.... *Arctostaphylos stanfordiana* ssp. *stanfordiana* (pl. 23)
Stanford Manzanita; Na, Sn

11b Bracts on each pedicel 2–4 mm long; fruit 5–12 mm wide (corolla 5 lobed, white or pink; leaves bright green on both surfaces)

13a Corolla 7–8 mm long, white or pink; leaves up to 6 cm long; fruit 8–12 mm long, red; often treelike (widespread)...............................
.... *Arctostaphylos manzanita* ssp. *manzanita*
Parry Manzanita; CC-n

13b Corolla 4–7 mm long, white; leaves about 3 cm long; fruit 5–8 mm long, dark brown; shrub............ *Arctostaphylos manzanita* ssp. *laevigata*
Contra Costa Manzanita; CC (MD); 1b

1b   Bracts on each pedicel 5–20 mm long, leaflike and often longer than the pedicel (pedicel hairy, the hairs sometimes glandular)

14a   Leaves overlapping, their bases clasping the stem; petioles up to 4 mm long

15a   Leaf surfaces not alike, the upper ones not hairy, the undersides bristly (leaf blades 4–7 cm long, often with some teeth; often over 2 m tall) . . . . *Arctostaphylos andersonii*
Santa Cruz Manzanita; SF-SCr; 1b

15b   Leaf surfaces similar

16a   Leaf surfaces with a whitish coating, this sometimes obscured by dense hairs (leaf blades 1.5–4.5 cm long)

17a   At least young leaves hairy; pedicel 4–10 mm long, hairy, but not glandular, becoming curved in fruit; bracts 5–15 mm long; fruit hairy; often over 1 m tall . . . . *Arctostaphylos auriculata* (pl. 22)
Mount Diablo Manzanita; Al, CC (MD); 1b

17b   Leaves not hairy; pedicel 8–12 mm long, glandular hairy, not curved in fruit; bracts 5–9 mm long; fruit glandular hairy; often over 2 m tall . . . . *Arctostaphylos pallida [A. andersonii* var. *pallida]* (pl. 22)
Pallid Manzanita; Al, CC; 1b

16b   Leaf surfaces pale green, without a whitish coating (pedicel glandular hairy)

18a   Leaf blades 3–6 cm long, glandular hairy at least when young; pedicel 6–10 mm long (often more than 2 m tall) . . . . *Arctostaphylos regismontana*
King Mountain Manzanita; SM (MM); 1b

18b   Leaf blades 2.5–4 cm long, sparsely hairy, not glandular; pedicel 3–6 mm long

19a   Stem prostrate or with tip rising, less than 1 m tall; corolla 3–5 mm long; pedicel 3–5 mm long . . . . *Arctostaphylos imbricata [A. andersonii* var. *imbricata]*
San Bruno Mountain Manzanita; SM (SBM); 1b

19b   Stem erect, often much more than 1 m tall; corolla 6–9 mm long; pedicel 5–6 mm long . . . . *Arctostaphylos montaraensis*
Montara Manzanita; SM (MM); 1b

14b   Leaves usually not overlapping, their bases not clasping the stem; petioles 2–10 mm long

20a   Mature plant treelike, usually at least 2 m tall (branchlets glandular hairy)

21a   Petioles 4–10 mm long; leaves not overlapping, the blades 4–6 cm long, dark green, densely hairy at least when young; pedicel 2–4 mm long; bracts 10–18 mm long . . . . *Arctostaphylos columbiana*
Hairy Manzanita, Douglas-fir Manzanita; Sn-n

21b   Petioles 2–4 mm long; leaves somewhat overlapping, the blades 3–5 cm long, bright green, sparsely hairy; pedicel 3–8 mm long; bracts 8–20 mm long . . . . *Arctostaphylos virgata*
Marin Manzanita; Ma; 1b

20b Mature plant not treelike, but usually at least 1 m tall

22a Leaves scarcely if at all hairy (base of plant enlarged; leaf blades 2–5 cm long)

23a Petioles 5–10 mm long; leaf blades flat, bright green, sometimes with a whitish coating, hairy or not; pedicel hairy or glandular hairy. . . . .
. . . . *Arctostaphylos glandulosa*
[includes *A. glandulosa* var. *cushingiana*] (pl. 22)
Eastwood Manzanita; Ma, Na, Sn

23b Petioles 2–5 mm long; leaf blades curved, dark or bright green, without a whitish coating, usually not hairy; pedicel hairy

24a Branchlets with long, white bristles . . . .
. . . . *Arctostaphylos tomentosa* ssp. *crustacea* [*A. crustacea* var. *crustacea* and *A. glandulosa* var. *campbellae*] (pl. 23)
Brittleleaf Manzanita; CC, SF-s

24b Branchlets hairy, but without long, white bristles . . . .
. . . . *Arctostaphylos tomentosa* ssp. *rosei* [*A. crustacea* var. *rosei*]
Lake Merced Manzanita, Rose Manzanita; SF

22b At least undersides of leaves densely hairy

25a Leaves not hairy on the upper surfaces; pedicel 2–5 mm long, not curved in fruit; base of plant enlarged (pedicel hairy) . . . .
. . . . *Arctostaphylos tomentosa* ssp. *crinita*
[*A. tomentosa* var. *tomentosiformis*]
Hairy Manzanita; SM-SLO

25b Leaves densely hairy on both surfaces; pedicel 5–9 mm long, curved in fruit; base of plant not enlarged

26a Branchlets and pedicel hairy, but not glandular; bracts up to 20 mm long . . . . *Arctostaphylos canescens* ssp. *canescens*
[includes *A. canescens* var. *candidissima*]
Hoary Manzanita; SCr-n

26b Branchlets and pedicel glandular hairy; bracts up to 12 mm long
. . . . *Arctostaphylos canescens* ssp. *sonomensis*
Sonoma Manzanita; Sn; 1b

EUPHORBIACEAE (SPURGE FAMILY)   The Spurge Family, with both shrubby and herbaceous representatives, has given us some interesting garden and greenhouse subjects, including poinsettias. Certain species from southern Africa resemble cacti and are popular with collectors of succulents. There are a few weeds in the group, as well as plants that have long been exploited as sources of dyes, oils, and other substances. One of the more valuable species is *Ricinus communis* (Castor-bean), whose seeds yield castor oil, widely used in medicine; this oil is also used as fuel for lamps and in the manufacture of soap, candles, varnish, and polishes. The plant as a whole is very poisonous, but when seeds are pressed, the toxic substance remains in the pulp. Species of *Euphorbia* and *Chamaesyce* have a milky sap that may irritate the skin, and the seeds of some species of these genera are so toxic that their extracts have been used for coating arrowheads and for killing fish.

The flower of a spurge is small and staminate or pistillate. None of ours have petals, and in *Euphorbia* and *Chamaesyce* there are no sepals either, although some bracts below a group of flowers may form what looks like a calyx. In these two genera, several staminate flowers, each consisting of one stamen, surround a pistillate flower that has a single pistil. The ovary is superior, and the fruit has three divisions that separate, each releasing one or two seeds.

1a   Flower usually with sepals but no petals; pistillate and staminate flowers not in the same cluster and sometimes on separate plants

    2a   Leaves opposite (leaf blades 2–5 cm long, with fine teeth; staminate flowers in dense inflorescences at stem tips, pistillate flowers in the leaf axils of separate plants; up to 30 cm tall) . . . . *Mercurialis annua*
Mercury; eu

    2b   Leaves alternate, but sometimes crowded at the tips of branches in *Eremocarpus setigerus*

        3a   Leaves 10–40 cm long, with 5–11 palmate lobes (pistillate and staminate flowers on same plant; shrub, up to 3 m tall) . . . . *Ricinus communis* (pl. 24)
Castor-bean; eu

        3b   Leaves not more than 6 cm long, not lobed (widespread)

            4a   Leaves and stem covered with long, stiff hairs; pistillate and staminate flowers on same plant; stem branching in twos; annual; forming mats or nearly hemispherical masses up to 20 cm tall (leaves with 3 prominent veins) . . . . *Eremocarpus setigerus* (pl. 24)
Turkey-mullein, Doveweed

            4b   Leaves and stem covered with short, matted hairs; pistillate and staminate flowers on separate plants; branching, but not in twos; perennial, sometimes woody at the base; forming a spreading mass or upright, up to 100 cm tall (leaves grayer on the undersides than above; in sandy soil) . . . . *Croton californicus* (pl. 24); SF-s

1b   Flower without sepals or petals; pistillate and staminate flowers in the same cluster, within a toothed or lobed cup (there may be 4 glands alternating with the teeth or lobes of the cup, or there may be leaves that resemble petals below the cup)

    5a   Leaves generally alternate

        6a   Leaves of flowering stem wedge shaped; plant upright or low with stem tips rising (leaves finely toothed; annual) . . . . *Euphorbia helioscopia*
Wartweed; eu

        6b   Leaves of flowering stem more or less oval or oblong; plant upright

            7a   Leaves below the flowering stem with petioles about half the length of the leaves (annual; up to 35 cm tall) . . . . *Euphorbia peplus* (pl. 24)
Petty Spurge; eu

            7b   Leaves below the flowering stem without distinct petioles

                8a   Leaves without fine teeth; up to 60 cm tall (leaves up to 3.5 cm long, the tips sometimes pointed; usually annual) . . . . *Euphorbia crenulata*
Chinese-caps

                8b   Leaves with fine teeth; up to 45 cm tall

                    9a   Stem hairy; stem leaves 4–6 cm long; perennial . . . . *Euphorbia oblongata*
European Euphorbia; eu

9b   Stem not hairy; stem leaves 1–3 cm long; annual . . . .
. . . . *Euphorbia spathulata*
Spatulateleaf Spurge, Reticulate-seed Spurge

5b   All leaves opposite (annual or biennial)
10a Leaves 5–14 cm long (leaves toothed, sessile, not hairy, those beneath flowers 2–6 cm long; up to 1 m tall). . . . . *Euphorbia lathyris*
Caper Spurge, Gopher-plant; eu

10b Leaves not more than 3 cm long
11a Tips of leaves with 1 or more teeth
12a Fruit hairy only on the angles (prostrate, the stem up to 25 cm long)
. . . . *Chamaesyce prostrata [Euphorbia prostrata]*
Prostrate Spurge; sa

12b Fruit either not hairy or rather uniformly hairy
13a Plant not hairy; fruit not hairy; prostrate or up to 35 cm tall (widespread) . . . .*Chamaesyce serpyllifolia [Euphorbia serpyllifolia]*
Thymeleaf Spurge

13b Plant hairy, but sometimes only on the tips of young branches; fruit hairy; prostrate . . . .*Chamaesyce maculata* [includes *Euphorbia maculata* and *E. supina*] (pl. 24)
Spotted Spurge, Large Spurge; na

11b Tips of leaves not toothed (prostrate, the stem up to 25 cm long)
14a Leaves up to 7 mm long, rounded . . . .
. . . . *Chamaesyce serpens [Euphorbia serpens]*
Serpent Spurge; sa

14b Leaves up to 11 mm long, oblong
15a Staminate flowers 4 in each very small cup . . . .
. . . . *Chamaesyce prostrata [Euphorbia prostrata]*
Prostrate Spurge; sa

15b Staminate flowers many in each cup . . . .
. . . . *Chamaesyce ocellata [Euphorbia ocellata]*
Valley Spurge

FABACEAE (PEA FAMILY)   Fabaceae (formerly called Leguminosae), as represented in our area, consists mostly of herbaceous plants with compound, alternate leaves. It does, however, include some shrubs and small trees. Many of our species are introduced, either intentionally or accidentally. All of the medicks and sweet clovers, and a large proportion of the true clovers and vetches, are of exotic origin. So are some shrubby species such as *Cytisus scoparius* (Scotch Broom), *Genista monspessulana* (French Broom), *Ulex europaea* (Gorse), and *Spartium junceum* (Spanish Broom) that have colonized extensive coastal areas where the native vegetation has been disturbed. Although these are attractive in one way or another, they are so aggressive that they prohibit the recovery of natural vegetation. Elimination of all of them is advised.

The flower typically has a distinctive irregular corolla. The two lower petals are fused, along the edges where they touch, into a structure called the keel; the two petals on the side, which often stand out at a 90 degree angle to the keel, are called wings, and the upper petal, usually the largest, forms the banner. In most of our genera, the filaments of nine of the stamens are fused into a tube, but the uppermost one is free. The ovary is superior, and the pistil is not divided into compartments, so it has a single row of seeds. When the fruit dries, it usually splits open rather forcefully, thus effectively scattering its seeds. In some genera the flowers deviate markedly from the pattern typical of the family. *Acacia*, for instance, has a regular, inconspicuous corolla and numerous stamens, and *Amorpha* has a single petal, the banner.

This family provides some of the world's most important food plants: peas, beans, peanuts, and soybeans are just a few of them. Futhermore, certain bacteria that live in nodules on the roots of nearly all plants in Fabaceae are capable of utilizing atmospheric nitrogen. These bacteria convert atmospheric nitrogen into organic compounds that they share with their host plants, which, if growing in nitrogen-poor soils, must depend to a considerable extent on their bacterial symbionts. When the plants die and are decomposed by other bacteria, a net gain in the nitrogen content of the soil results. This is one reason why so many members of Fabaceae are grown as cover crops.

Particularly good subjects for gardens are some of the annual and perennial lupines, which can be grown from seed. In the category of large shrubs or small trees, *Cercis occidentalis* (Western Redbud), should be considered. Its flowers, produced freely in late winter or early spring, are of a beautiful reddish purple color.

1a Trees or large shrubs, generally over 2 m tall (plant without thorns or prickles, but *Robinia* has stipules that are spines)

    2a Leaves not compound, except sometimes the young leaves of *Acacia longifolia* and *A. melanoxylon*

        3a Leaf blade rounded but with a notch at the base and sometimes indented at the tip; corolla 8–12 mm long, irregular; stamens 10, not protruding out of corolla (corolla reddish purple). . . . . *Cercis occidentalis* (pl. 24)

                                        Western Redbud; Sl-n

        3b "Leaf" (actually a flattened petiole) oblong and usually curved, without a notch or indentation; corolla less than 3 mm long, regular; stamens many, protruding out of corolla

            4a Protruding stamens cream colored, the flowers in bunches along a raceme; fruit curved or twisted; tree. . . . . *Acacia melanoxylon*

                                         Blackwood Acacia; au

            4b Protruding stamens yellow, the flower solitary along a raceme; fruit straight; shrub or small tree. . . . . *Acacia longifolia*

                                    Sydney Golden Wattle; au

  2b Leaves compound

    5a Leaves once pinnately compound, the leaflets usually at least 2 cm long

        6a Leaflets an odd number, usually 11–17, oval, rounded at the tips, not hairy; base of petiole with a pair of spines; corolla whitish, irregular; deciduous; tree. . . . . *Robinia pseudoacacia*

                                  Black Locust, Desert Locust; na

        6b Leaflets an even number, 12–16, about 3 times as long as wide, pointed at the tips, hairy; petiole without spines; corolla yellow, only slightly irregular; evergreen; shrub or small tree. . . . *Senna multiglandulosa [Cassia tomentosa]*; mx

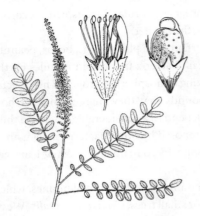

Amorpha californica var. *napensis*
False Indigo

Astragalus nuttallii var. *virgatus*
Nuttall Milkvetch

Glycyrrhiza lepidota
Wild Licorice

Lathyrus latifolius
Everlasting Pea

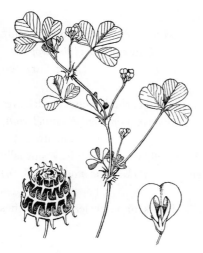

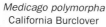

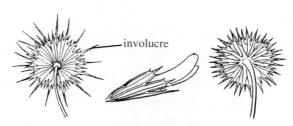

Medicago polymorpha
California Burclover

Trifolium willdenovii
Tomcat Clover

5b   Leaves twice pinnately compound, the leaflets not more than 1.5 cm long
 7a Leaves with 3–6 pairs of primary leaflets (primary leaflets with 12–20 pairs of secondary leaflets, these 5–7 mm long; corolla regular, less than 5 mm long; stamens many; tree) . . . . *Acacia baileyana*
                    Cootamundra Wattle; au
 7b Leaves usually with more than 7 pairs of primary leaflets
  8a Primary leaflets with 15–35 pairs of secondary leaflets, these 5–15 mm long; flowers in several rounded clusters; corolla regular, less than 5 mm long; stamens many; fruit with conspicuous constrictions; tree . . . .
   . . . . *Acacia decurrens*
                       Green Wattle; au
  8b Primary leaflets with 7–11 pairs of secondary leaflets, these less than 8 mm long; flowers not in clusters; corolla irregular, over 35 mm long; stamens 10; fruit without obvious constrictions; shrub (corolla yellow, orange, and red; stamens red, 8–9 cm long). . . . . *Caesalpinia gilliesii*
                     Bird-of-paradise; sa
1b   Shrubs or herbs, usually not more than 2 m tall (corolla irregular)
 9a Shrubs with thorny branches, spinelike leaves, or nearly leafless branches
  10a Branches without thorns; leaves not spiny, when present, much reduced (corolla yellow; branches green) . . . . *Spartium junceum* (pl. 27)
                     Spanish Broom; me

10b Branches with thorns; leaves becoming spiny

    11a Corolla yellow (leaves not compound; flowers about 25 mm long, usually more than 1 in each cluster)................. *Ulex europaea [U. europaeus]*

        Gorse, Furze; eu

    11b Corolla not yellow

        12a Flower 8–9 mm long, 4–6 in each cluster, the corolla lavender red; leaves not compound.................... *Alhagi pseudalhagi [A. camelorum]*

        Camelthorn; as

        12b Flower 16–19 mm long, solitary, the corolla rose purple; leaves usually palmately compound, with 3 leaflets ...... *Pickeringia montana* (pl. 26)

        Chaparral Pea; Me–s

9b Shrubs or herbs without thorny branches, spinelike leaves, or nearly leafless branches (leaves compound, with 3 or many leaflets)

    13a Leaves palmately compound

        14a Leaflets 4 or many (flowers in showy racemes; fruit hairy)

            15a Calyx with prominent glands; leaflets 4 or 5 (corolla 8–12 mm long, blue or purple; herb) ....

                .... *Pediomelum californicum [Psoralea californica]* (pl. 27)

            Indian Breadroot

            15b Calyx without glands; leaflets 4 or many .... FABACEAE, SUBKEY 1

            Lupinus

        14b Leaflets 3 (herbs)

            16a Flowers 17–19 mm long, in racemes 15–25 cm long; filaments of stamens separate (corolla 15–25 mm long, yellow) ....

                .... *Thermopsis macrophylla* (pl. 27)

            False Lupine

            16b Flowers not more than 12 mm long, in heads or umbels; 9 filaments of stamens united................................. FABACEAE, SUBKEY 2

            Trifolium

    13b Leaves pinnately compound, but leaflets sometimes only 3, each with a stalk

        17a Each leaf with a tendril or short projection about 3 mm long at the tip (leaflets either 2 or more than 3; herbs) .... FABACEAE, SUBKEY 3

        17b Leaves without tendrils or other projections at the tips

            18a Prominent glands present on the leaves (sometimes not visible on young leaves) and often on the calyx

                19a Leaflets 9–27

                      20a Flower with only 1 petal, the banner; fruit without hooked bristles; shrub; often more than 2 m tall ....

                        ....*Amorpha californica* var. *napensis* (fig.)

                    False Indigo; Ma; 1b

                    20b Flower with 5 petals; fruit with hooked bristles; herb; up to 1 m tall (in moist areas) .... *Glycyrrhiza lepidota* (fig.)

                    Wild Licorice

19b Leaflets 3 (herbs)

    21a Calyx lobes up to 5 mm long

        22a Calyx not hairy, the lobes more or less equal in length; corolla white or yellow, the banner 10–14 mm long . . . . *Rupertia physodes [Psoralea physodes]* (pl. 27)

                                              California-tea

        22b Calyx hairy, with 1 lobe 2–3 mm longer than the others; corolla blue or purple, the banner 5–40 mm long (in moist areas) . . . . . . *Hoita macrostachya [Psoralea macrostachya]*

                                              Leather-root

    21b Calyx lobes 10–15 mm long (corolla blue or purple)

        23a Calyx lobes about 1 cm long, equal in length; banner 10–16 mm long (moist areas) . . . . *Hoita orbicularis [Psoralea orbicularis]*

                                              Roundleaf Hoita

        23b One calyx lobe about 1.5 cm long, longer than the others; banner 12–13 mm long . . . . *Hoita strobilina [Psoralea strobilina]*

                              Loma Prieta Hoita; CC-SCl; 1b

18b Prominent glands not present on the leaves or calyx

    24a Leaflets toothed, at least at the tips (herbs)

        25a Corolla white or blue

            26a Corolla white, 4–5 mm long; teeth not confined to the upper half of each leaflet . . . . *Melilotus alba [M. albus]*

                                        White Sweetclover; eua

            26b Corolla blue, 8–10 mm long; teeth present only in the upper half of each leaflet . . . . *Medicago sativa*

                                      Alfalfa, Lucerne; eu

        25b Corolla yellow

            27a Flowers 2–5 in each cluster; corolla 3–6 mm long (fruit coiled, sometimes spiny) . . . .*Medicago polymorpha* [includes *M. polymorpha* var. *confina*] (fig.)

                                  California Burclover; me

            27b Flowers numerous in each cluster; corolla 2–3 mm long

                28a Leaflets fan shaped, 1 cm long; fruit coiled . . . . *Medicago lupulina*

                                Black Medick, Yellow Trefoil; eu

                28b Leaflets elongate, 2–2.5 cm long; fruit not coiled . . . . *Melilotus indica [M. indicus]* (pl. 26)

                            Sourclover, Indian Melilot; me

    24b Leaflets not toothed

        29a Herbs . . . . FABACEAE, SUBKEY 4

                                    Astragalus, Lotus

29b Shrubs (corolla yellow; leaflets 3–6; widespread)

    30a Flowers in clusters at the ends of branches; calyx hairy; fruit densely hairy . . . . *Genista monspessulana* [*Cytisus monspessulanus*] (pl. 24)

        French Broom; me

    30b Flowers single or in clusters in the leaf axils; calyx not hairy; fruit not hairy or hairy only on margins

        31a Flower almost sessile, the banner less than 10 mm long; stem not angled, usually sparingly leafy; fruit not hairy . . . . *Lotus scoparius* (pl. 25)

            Deerweed, California Broom

        31b Flower not sessile, the banner 15–18 mm long; stem obviously angled, usually densely leafy; fruit hairy on margins (extremely invasive) . . . . *Cytisus scoparius* (pl. 24)

            Scotch Broom; eu

Fabaceae, Subkey 1: Shrubs or herbs, without spiny branches; leaves palmately compound, with 4 to many leaflets; flowers in showy racemes; fruit hairy . . . . *Lupinus*

1a Plant woody, at least at the base

    2a Lower keel margin hairy; upper keel margins not hairy (banner hairy on back; corolla light violet to blue; leaflets 1–2.5 cm long; coastal) . . . . *Lupinus chamissonis*

        Chamisso Bush Lupine, Silver Bush Lupine; Ma-s

    2b Lower keel margin not hairy; upper keel margins usually hairy

        3a Flower 14–18 mm long, the petals usually yellow, sometimes lilac or mixed colors; stipules 8–12 mm long; leaflets 2–6 cm long (banner not hairy on back; coastal) . . . . *Lupinus arboreus* [includes *L. arboreus* var. *eximius*] (pl. 26)

            Yellow Bush Lupine, Tree Lupine

        3b Flower 8–16 mm long, the petals not yellow, except sometimes in *Lupinus variicolor;* stipules 6–8 mm long; leaflets 1–3.5 cm long

            4a Banner not hairy on back; petioles less than 3 cm long; petals white, yellow, pink, or purple; prostrate or with stem tip rising; coastal . . . . *Lupinus variicolor*

                Varied Lupine, Lindley Varied Lupine; SLO-n

            4b Banner usually hairy on back; petioles 2–8 cm long; petals violet to lavender; upright; not coastal

                5a Plant usually not woody throughout, less than 50 cm tall; inflorescence 4–14 cm long; upper keel margins sometimes not hairy . . . . *Lupinus albifrons* var. *collinus*

                    Bay Area Silver Lupine, Collin Bush Lupine; SFBR

                5b Plant woody throughout, usually more than 50 cm tall; inflorescence 8–30 cm long; upper keel margins hairy . . . . *Lupinus albifrons* var. *albifrons* (pl. 26)

                    Silver Lupine, Silver Bush Lupine

1b Plant not woody, even at the base

6a Longest leaflets not more than 4 cm

    7a Leaflets not more than 2.5 cm long (petals blue or lavender; banner not hairy, with a spot)

        8a Upper keel margins hairy their entire length; petioles 1–3 cm long; leaflets 3–5; flower 11–13 mm long; perennial; in sand dunes (lower keel margin not hairy) . . . . *Lupinus tidestromii* [includes *L. tidestromii* var. *layneae*]
> Tidestrom Lupine; Ma-Mo; 1b

        8b Keel not hairy; petioles 4–8 cm long; leaflets usually 7; flower 7–9 mm long; annual; in disturbed areas . . . . *Lupinus pachylobus*
> Bigpod Lupine, Mt. Diablo Annual Lupine; SCl-n

    7b Leaflets 1–4 cm long (upper keel margins hairy)

        9a Inflorescence 1–8 cm long; flower 4–10 mm long (leaflets 5–7; petioles 1–7 cm long; banner not hairy, with a white spot; upper keel margins hairy near tips; lower keel margin not hairy; petals blue, rarely pink or white; fruit 1–3 cm long) . . . . *Lupinus bicolor* [includes *L. bicolor* sspp. *tridentatus* and
> *umbellatus, L. bicolor* var. *trifidus,* and *L. polycarpus*]
> Miniature Lupine

        9b Inflorescence up to 22 cm long; flower 6–16 mm long

            10a Lower keel margin hairy; fruit 1–1.5 cm long; leaflets 7–9 (petioles up to 5 cm long; flower 10–16 mm long; banner not hairy; petals pale yellow, rarely pink or blue) . . . . *Lupinus luteolus*
> Butter Lupine, Kellogg Lupine

            10b Lower keel margin not hairy; fruit 2–4 cm long; leaflets 5–9

                11a Petioles 2–8.5 cm long; flower 6–15 mm long; petals blue, rarely lavender, pink, or white; banner sometimes hairy on back, with a white spot; annual (upper keel margins hairy near tips) . . . .
> . . . . *Lupinus nanus* [includes *L. nanus* ssp.
> *latifolius* and *L. vallicola* ssp. *apricus*]
> Douglas Lupine, Sky Lupine

                11b Petioles less than 3 cm long; flower 11–16 mm long; petals white, yellow, rose, or purple; banner not hairy on back, without a spot; perennial . . . . *Lupinus variicolor*
> Varied Lupine, Lindley Varied Lupine; SLO-n

6b Longest leaflets more than 4 cm

    12a Longest leaflets at least 7 cm (flower 9–16 mm long; banner not hairy on back; perennial)

        13a Leaflets 7–9, 2.5–7 cm long; petioles up to 7 cm long; fruit 3–4.5 cm long (stipules 4–15 mm long; petals purple; keel not hairy; in dry areas) . . . .
> . . . . *Lupinus formosus*
> Summer Lupine

        13b Leaflets 5–17, 4–15 cm long; petioles up to 20 cm or more long; fruit 2–4.5 cm long

14a Leaflets 9–17, up to 15 cm long; petioles up to 45 cm long; stipules 5–30 mm long; petals violet, lavender, pink, or white; keel usually not hairy (in moist areas) .... *Lupinus polyphyllus*
Swamp Lupine; SCr-n

14b Leaflets 5–11, up to 10 cm long; petioles up to 20 cm long; stipules 5–10 mm long; petals blue, purple, or white; each upper keel margin hairy from middle to base, lower margin usually hairy

15a Flower 13–16 mm long; stem densely hairy; in chaparral ....
.... *Lupinus latifolius* var. *dudleyi*
Dudley Lupine; SM (MM)

15b Flower 10–14 mm long; stem not densely hairy; in moist areas .....
.... *Lupinus latifolius* var. *latifolius*
Broadleaf Lupine

12b Longest leaflets not more than 6 cm

16a Leaves with stinging hairs, these arising out of small blisters; petals bright pink or magenta; upper keel margins not hairy (leaflets 5–8; petioles 4–9 cm long; flower 12–18 mm long; banner not hairy; lower keel margin hairy from middle to base; fruit 2–4 cm long; often on fire burns) ....
.... *Lupinus hirsutissimus*
Stinging Lupine; SM-s

16b Leaves without stinging hairs; petals usually not bright pink or magenta; upper keel margins hairy, except in *Lupinus adsurgens*

17a Upper keel margin toothed; banner hairy on back (leaflets 5–8; petioles 3–10 cm long; flower 8–12 mm long; each upper keel margin hairy from middle to tip; lower keel margin not hairy; petals blue; fruit 3–5 cm long; annual) .... *Lupinus affinis*
Sky Lupine; SCr-n

17b Upper keel margin not toothed; banner usually not hairy

18a Fruit 1–1.5 cm long; leaflets 5–11 (petioles 3–15 cm long; flower 8–18 mm long; lower keel margin sometimes hairy near base; annual)

19a Petals white or yellow, rarely rose or purple; fruit usually on 1 side of flowering stem .... *Lupinus microcarpus* var. *densiflorus* [includes *L. densiflorus* vars. *aureus* and *densiflorus*] (pl. 26)
Gully Lupine

19b Petals pink or purple, rarely yellowish or white; fruit usually not just on 1 side of flowering stem ....
.... *Lupinus microcarpus* var. *microcarpus* [*L. subvexus*]
Chick Lupine, Redflower Lupine; La-s

18b Fruit 2–4 cm long; leaflets 4–9

20a Petioles 2–6 cm long; stipules 5–17 mm long; flower 9–14 mm long; keel not hairy (petals yellowish, lavender, or violet; banner not hairy; perennial) .... *Lupinus adsurgens*
Silky Lupine; SCl-n

20b Petioles 5–15 cm long; stipules less than 8 mm long; flower 12–18 mm long; keel hairy, at least on upper margins

    21a Leaflets 4–7; fruit 2–3 cm long; inflorescence 10–30 cm long; petals purple or violet; banner sometimes hairy on back; upper keel margins hairy from bases to tips; lower keel margin usually not hairy; perennial . . . . *Lupinus sericatus*
    Cobb Mountain Lupine; Na, Sn, La; 1b

    21b Leaflets 7–9; fruit 3.5–5 cm long; inflorescence 9–15 cm long; petals blue purple, rarely white, pink, or lavender; banner not hairy; both upper and lower margins of keel hairy at the bases; annual . . . . *Lupinus succulentus* (pl. 26)
    Arroyo Lupine, Succulent Annual Lupine; Me-s

Fabaceae, Subkey 2:   Herbs, without spiny branches; leaves palmately compound, with 3 leaflets; flowers not more than 12 mm long, in heads or umbels . . . . *Trifolium*

1a Flower head without bracts at the base, but there may be stem leaves or a ringlike structure near the head

    2a Each flower in each head with a pedicel 1–2 mm long

        3a Corolla yellow or cream colored (annual)

            4a Flower head often with more than 20 flowers, these all similar; corolla yellow, 3–4 mm long; petioles usually not longer than leaf blades . . . . *Trifolium dubium*
    Shamrock, Little Hop Clover; eu

            4b Flower head with fewer than 10 flowers, some with corollas and some without; corolla cream colored, 8–12 mm long; petioles usually longer than leaf blades (flowers without corollas are sterile and grow up over developing fruit, transforming the head into a bur that may be buried as the peduncle bends) . . . . *Trifolium subterraneum*
    Subterranean Clover; eu

        3b Corolla white, yellow, pink, or purple

            5a Peduncle and calyx hairy (leaflets notched at the tips; corolla 6–9 mm long, yellow to purple; annual)

                6a Each leaflet up to 5 times as long as wide, not toothed, the tip notched up to one-half its entire length . . . . *Trifolium bifidum* var. *bifidum*
    Notchleaf Pinole Clover; Mo-n

                6b Each leaflet not more than 3 times as long as wide, toothed or not, the tip either without a notch or this less than one-half its entire length . . . . *Trifolium bifidum* var. *decipiens*
    Deceiving Clover

            5b Peduncle and calyx not hairy

                7a Pedicel about 5 mm long; corolla 6–10 mm long (in moist areas)

                    8a Stipules 10–25 mm long; corolla pale pink; annual or perennial . . . . *Trifolium hybridum* (pl. 27)
    Alsike Clover; eu

8b   Stipules 4–10 mm long; corolla white to pink; perennial . . . .
. . . . *Trifolium repens*
White Clover; eua

7b   Pedicel 1–3 mm long, corolla 5–7 mm long (corolla pink to purple; annual)

9a   Stipules less than 10 mm long; peduncle 2–6 cm long . . . .
. . . .*Trifolium gracilentum*
Pinpoint Clover

9b   Stipules 15–30 mm long; peduncle 5–15 cm long . . . .
. . . . *Trifolium ciliolatum*
Tree Clover

2b   Each flower in each head sessile

10a   Flower head sessile, with some leaves and stipules directly below it

11a   Flower heads often in pairs on stem or branches; corolla purple or of 2 colors, 6 mm long, the calyx about as long; prostrate, with stem tip rising; annual . . .
. . . . *Trifolium macraei*
Double-headed Clover

11b   Flower heads solitary on stem or branches; corolla red purple, 10–20 mm long, the calyx half as long; erect, up to 60 cm tall; perennial . . . .
. . . . *Trifolium pratense* (pl. 27)
Red Clover; eu

10b   Flower head on a peduncle arising from the uppermost group of leaves (flower head solitary on stem or branches; annual)

12a   Corolla hidden by the calyx (calyx up to 10 mm long; corolla purple and white) . . . . *Trifolium albopurpureum* var. *olivaceum* [*T. olivaceum*]
Olive Clover

12b   Corolla at least as long as the calyx

13a   Corolla 10–16 mm long; calyx 7–10 mm long

14a   Corolla 12–16 mm long, purple, the petals with white tips; calyx 10–12 mm long, the lobes feathery; flower head 2.5 cm long and about as wide . . . . *Trifolium amoenum*
Showy Indian Clover; Ma–Sl; 1b

14b   Corolla 10–14 mm long, crimson; calyx 7–10 mm long, the lobes densely hairy and with bristlelike tips; flower head about 4 cm long and half as wide . . . . *Trifolium incarnatum*
Crimson Clover; eu

13b   Corolla 5–12 mm long; calyx 3–8 mm long

15a   Corolla 5–8 mm long, about as long as the calyx (calyx lobes feathery; corolla purple and white) . . . .
. . . .*Trifolium albopurpureum* var. *albopurpureum*
Rancheria Clover

15b   Corolla 6–12 mm long, longer than the calyx

16a Flower head narrow, with more than 10 flowers; calyx 4–8 mm long, the lobes feathery; corolla purple and white . . . .
. . . *Trifolium albopurpureum* var. *dichotomum* [*T. dichotomum*]
Branched Indian Clover; SCl-n

16b Flower head globular, with fewer than 10 flowers; calyx 3–4 mm long, the lobes not feathery; corolla pink purple, the petals with white tips (ringlike structure present at base of the head) . . . . .
. . . . *Trifolium depauperatum* var. *depauperatum*
Dwarfsack Clover; Al-n

1b Flower head with bracts at the base

17a Bracts of flower head completely separate (banner of corolla becoming inflated in fruit; annual)

18a Corolla 10–20 mm long, cream or yellow, turning pink with age (in moist areas)
. . . . . *Trifolium fucatum* [includes *T. fucatum* vars. *gambelii* and *virescens*] (pl. 27)
Bull Clover, Sour Clover

18b Corolla 4–9 mm long, pink purple, the petals with white tips

19a Bracts about as long as the calyx (in grassy areas) . . . .
. . . . . *Trifolium depauperatum* var. *truncatum* [*T. amplectens* var. *truncatum*]
Common Palesack Clover; La-s

19b Bracts shorter than the calyx

20a Fruit up to 4 mm long, without a distinct stalklike base; in dry, grassy areas . . . . *Trifolium depauperatum* var. *amplectens*
[*T. amplectens* var. *amplectens*]
Palesack Clover; La-s

20b Fruit up to 3 mm long, with a distinct stalklike base; in saline and alkaline areas . . . . *Trifolium depauperatum* var. *hydrophilum*
[*T. amplectens* var. *hydrophilum*]
Saline Clover; SLO-n; 1b

17b Bracts of flower head at least partly joined, the lobes often toothed

21a United bracts forming a shallow cup-shaped structure that covers much of the flowers when the head is pressed (annual)

22a Lobes of bracts not toothed (corolla 4–7 mm long, pink to lavender . . . .
. . . . *Trifolium microcephalum* (pl. 28)
Small-headed Clover, Maiden Clover

22b Lobes of bracts toothed

23a Corolla 8–16 mm long, definitely longer than the calyx; bract cup 1.5–2 cm wide (corolla purple; in moist areas) . . . .
. . . . *Trifolium barbigerum* var. *andrewsii* [*T. grayi*]
Gray Clover; Me-Mo

23b Corolla 4–10 mm long, about as long as the calyx; bract cup about 1 cm wide

24a Corolla 5–10 mm long, purple; prostrate (coastal, in moist areas)
. . . . *Trifolium barbigerum* var. *barbigerum*
Bearded Clover

24b Corolla 4–6 mm long, white or pink; often more than 10 cm tall (in grassland) . . . . *Trifolium microdon*
Valparaiso Clover; SLO-n

21b United bracts forming a flat disk that does not cover the flowers when the head is pressed

25a Corolla 8–18 mm long (tips of petals white)

26a Plant hairy and glandular (corolla 14–18 mm long, pale lavender to purple; annual; in moist areas) . . . . *Trifolium obtusiflorum*
Creek Clover

26b Plant not hairy or glandular (common)

27a Disk of united bracts less than 2 cm wide; leaves not mainly basal; corolla 8–15 mm long, lavender to purple; annual (in grassland) . . .
. . . . *Trifolium willdenovii [T. tridentatum]* (pl. 28; fig.)
Tomcat Clover

27b Disk of united bracts usually more than 2 cm wide; leaves mainly basal; corolla 12–16 mm long, pink purple or magenta; perennial (in moist areas; widespread) . . . . *Trifolium wormskioldii*
Cow Clover

25b Corolla 5–10 mm long

28a Corolla definitely longer than the calyx; annual or perennial (calyx lobes not hairy; corolla lavender to purple, the petals usually white at the tips; in moist areas) . . . . *Trifolium variegatum* [includes *T. appendiculatum*]
Whitetip Clover

28b Corolla about as long as the calyx; annual

29a Corolla dark purple; calyx hairy (in moist areas) . . . .
. . . . *Trifolium barbigerum* var. *barbigerum*
Bearded Clover

29b Corolla mostly lavender, the petals with white tips; calyx not hairy
. . . . *Trifolium oliganthum*
Fewflower Clover; SLO-n

Fabaceae, Subkey 3: Herbs, without spiny branches; leaves pinnately compound, with either 2 leaflets or more than 3; each leaf with a tendril or short projection about 3 mm long at the tip

1a Flower 1–3 cm long; style with hairs only on 1 side; wings free from keel except at the bases; each wing with a crescent-shaped ridge at the base

2a Leaflets 2

3a Flower solitary, 10–13 mm long; corolla red . . . . *Lathyrus cicera*
Flatpod Peavine; eu

3b   Flowers not solitary, 20–30 mm long; corolla pink, purple, or red, sometimes white in *Lathyrus latifolius*

       4a   Inflorescence with 1–3 flowers (coastal) .... *Lathyrus tingitanus*
                 Tangier Pea; eu

       4b   Inflorescence with 5–15 flowers .... *Lathyrus latifolius* (fig.)
                 Everlasting Pea, Perennial Sweet Pea; eu

2b   Leaflets more than 8

    5a   Tendrils reduced to a projection about 3 mm long; stem not winged; flowers less than 2 cm long, not more than 10 in each raceme

       6a   Flowers 8–13 mm long, 1 or 2 in each raceme; corolla lilac to pale purple blue; in open woods... *Lathyrus torreyi*
                 Redwood Pea; SCr-n

       6b   Flowers 15–18 mm long, 2–10 in each raceme; corolla pink purple and white; on backshores of sandy beaches .... *Lathyrus littoralis* (pl. 25)
                 Silky Beach Pea; Mo-n

    5b   Tendrils much more than 3 mm long, coiling; stem winged; flowers about 2 cm long, 15–20 in each raceme

       7a   Corolla white to lavender, fading to yellow (in woods or under shrubs) .....
        .... *Lathyrus vestitus* [includes *L. vestitus* ssp. *bolanderi*] (pl. 25)
                 Woodland Pea, Common Pacific Pea; SFBR-n

       7b   Corolla crimson, the banner sometimes paler (near water)

          8a   Plant not hairy; corolla uniformly bright pink ....
          .... *Lathyrus jepsonii* var. *jepsonii*
                 Delta Tule Pea, Jepson Pea; SFBR; 1b

          8b   Plant hairy; banner usually paler than the wings and keel ....
          .... *Lathyrus jepsonii* var. *californicus*
                 Bluff Pea; SLO-n

1b   Flower often less than 1 cm long; style usually with hairs all around tip; wings joined at least partly to keel; wings without a ridge at the bases

    9a   Raceme with a peduncle not more than 2 cm long, or the flowers sometimes sessile on the stem

       10a   Corolla white or yellow and white

          11a   Flower 6 mm long; corolla white .... *Lens culinaris*
                 Lentil; eua

          11b   Flower over 15 mm long; corolla yellow and white

            12a   Flower 15–18 mm long; leaflets 14–18 .... *Vicia pannonica*
                  Hungarian Vetch; eu

            12b   Flower 20–25 mm long; leaflets 8–16 .... *Vicia lutea*
                  Yellow Vetch; eu

       10b   Corolla violet, purple, or white with a purple spot

          13a   Flower 10–18 mm long (corolla uniformly purple; raceme with 1–3 flowers)
          .... *Vicia sativa* ssp. *nigra* [*V. angustifolia*]
                 Narrowleaf Vetch, Common Vetch; eu

13b Flower 18–35 mm long

    14a Raceme with 3–6 flowers; corolla white, with purple on the wings; leaflets 6 or 7, 15 mm wide; tendrils less than 1.5 cm long . . . . *Vicia faba*
        Horsebean, Broadbean; eua

    14b Raceme with 1–3 flowers; corolla purple, with violet on the wings; leaflets 8–16, 2–3 mm wide; tendrils 2–3 cm long . . . .
    . . . . *Vicia sativa* ssp. *sativa* (pl. 28)
        Spring Vetch, Common Vetch; eu

9b Raceme with a peduncle at least 2.5 cm long

  15a Raceme with 1 or 2 flowers, these 2–7 mm long

    16a Calyx lobes about as long as the calyx tube (corolla pale lavender to light purple) . . . . *Vicia tetrasperma*
        Tare; eu

    16b Calyx lobes much shorter than the calyx tube

      17a Leaflets 2–3 cm long; corolla pale blue . . . .
      . . . . *Vicia ludoviciana* [*V. exigua* var. *exigua*]
        Slender Vetch, California Vetch

      17b Leaflets 1–2 cm long; corolla lavender to white . . . .
      . . . . *Vicia hassei* [*V. exigua* var. *hassei*]
        Hasse Vetch; Ma, Al-s

  15b Raceme with more than 3 flowers, these 10–18 mm long

    18a Raceme with 3–10 flowers, these not grouped on 1 side; leaflets 8–16 (corolla blue purple to lavender; widespread) . . . .
    . . . . *Vicia americana* [includes *V. americana* ssp. *oregana*,
    V. americana vars. *linearis* and *truncata*, and *V. californica*]
        American Vetch

    18b Raceme with more than 10 flowers, sometimes fewer in *Vicia benghalensis*, the flowers grouped on 1 side; leaflets usually 16–24

      19a Upper leaflets of each leaf usually considerably smaller than the lower ones

        20a Stem and leaves obviously hairy; tip of keel dark blue or violet, rest of keel whitish . . . .*Vicia villosa* ssp. *villosa*
          Hairy Vetch, Woolly Vetch; eu

        20b Stem and leaves scarcely hairy if at all; tip of keel not markedly different from rest of keel, this dull red or orange red, sometimes yellow or purple (in moist areas) . . . . *Vicia gigantea* (pl. 28)
          Giant Vetch; SLO-n

      19b All leaflets more or less equal

        21a Leaflets up to 35 mm long and 10 mm wide; raceme with 3–10 flowers; corolla rose purple; fruit hairy . . . . *Vicia benghalensis*
          Purple Vetch; eu

        21b Leaflets up to 15 mm long and 4 mm wide; raceme usually with more than 10 flowers; corolla light blue to purple; fruit not hairy . . . .
        . . . . *Vicia villosa* ssp. *varia* [*V. dasycarpa*]
          Winter Vetch; eu

Fabaceae, Subkey 4:　Herbs, without spiny branches; leaves pinnately compound and without tendrils or prominent glands; leaflets not toothed; *Astragalus, Lotus*

1a　Leaflets 7–40 (flowers in racemes; fruit of some species inflated)

　2a　Leaflets 17–40 (flowers 15–125 in each raceme; perennial)

　　3a　Flower usually less than 1 cm long; fruit less than 1 cm long

　　　4a　Raceme dense, 6 cm long and 2 cm wide; leaflets very hairy (near salt marshes, sand dunes) . . . . *Astragalus pycnostachyus*
　　　　　　　Marsh Milkvetch; Me-SM

　　　4b　Raceme open, 10–15 cm long and 1 cm wide; leaflets only slightly hairy (in moist areas, often on serpentine) . . . . *Astragalus clevelandii*
　　　　　　　Cleveland Milkvetch; Na; 4

　　3b　Flower 1–1.5 cm long; fruit 2–3 cm long

　　　5a　Pedicel of fruit 2–4 cm long; flowers 15–45 in each raceme; inland . . . .
　　　　　. . . . . *Astragalus asymmetricus*
　　　　　　　San Joaquin Milkvetch; Sl, CC-s

　　　5b　Pedicel of fruit less than 0.5 cm long; flowers 40–125 in each raceme; coastal
　　　　　. . . . . *Astragalus nuttallii* var. *virgatus* [*A. nuttalli* var. *virgatus*] (fig.)
　　　　　　　Nuttall Milkvetch; Ma-SM, Al

　2b　Leaflets 7–17

　　6a　Flowers 20–95 in each raceme (perennial; in moist areas, often on serpentine) . . .
　　　　. . . . . *Astragalus clevelandii*
　　　　　　Cleveland Milkvetch; Na; 4

　　6b　Flowers 4–15 in each raceme

　　　7a　Flower 3–6 mm long; raceme not more than 1 cm long; corolla white, tinged with violet (annual)

　　　　8a　Raceme 8–10 mm long; fruit 3 mm long, roundish, not twisted at the tip; sometimes prostrate. . . . . *Astragalus didymocarpus*
　　　　　　　Twoseed Milkvetch, Common Dwarf Locoweed; CC-s

　　　　8b　Raceme 4–8 mm long; fruit 4–5 mm long, flattened, twisted at the tip; upright . . . . *Astragalus gambelianus*
　　　　　　　Gambel Milkvetch, Gambel Dwarf Locoweed

　　　7b　Flower 7–12 mm long; raceme over 1 cm long; corolla lilac or purple, sometimes white with a purple spot at the tip of each petal

　　　　9a　Leaflets about 2 mm wide (leaflets 10–14 mm long; fruit 10–15 mm long, somewhat hairy; annual; in moist, alkaline soil) . . . . *Astragalus tener*
　　　　　　　Alkali Milkvetch, Slender Rattleweed; Sl, SF, SB; 1b

　　　　9b　Leaflets more than 2 mm wide

　　　　　10a　Leaflets 10 mm long; fruit 15–30 mm long (fruit somewhat hairy; annual; often on serpentine) . . . .
　　　　　　. . . *Astragalus rattanii* var. *jepsonianus* [*A. rattani* var. *jepsonianus*]
　　　　　　　Jepson Milkvetch; Na; 1b

　　　　　10b　Leaflets 6 mm long; fruit 6–15 mm long

　　　　　　11a　Fruit densely hairy; raceme umbel-like, with 6–9 flowers; annual (often on serpentine) . . . . *Astragalus breweri*
　　　　　　　　Brewer Milkvetch; Ma; 4

        11b Fruit somewhat hairy; raceme not umbel-like, with more than
                11 flowers; perennial . . . . . . *Astragalus lentiginosus* var. *idriensis*
                Freckled Milkvetch, Idria Locoweed; SCl (MH)

1b   Leaflets 3–7 (flowers in leaf axils, in umbels or only 1 or 2)
    12a Stipules obvious, similar in size to the leaflets, but not always green
       13a Stipules green and leaflike; leaflets 9–19 (corolla red and white, about 1 cm long)
          . . . . *Lotus stipularis*
                Stipulate Lotus; Mo-n

       13b Stipules not green, not leaflike; leaflets 4–7
         14a Corolla white or pink, 8–9 mm long (at edges of woods). . . . .
            . . . . *Lotus aboriginus [L. aboriginum]*
                Roseflower Lotus; Sn-n

         14b Corolla with some yellow, 9–14 mm long
           15a Stipules 3 mm long; corolla mostly yellow green, with some red; in dry
               areas. . . . . *Lotus crassifolius* (pl. 25)
                   Broadleaf Lotus; SLO-n

           15b Stipules 8 mm long; banner yellow, wings and keel rose; in moist coastal
               areas. . . . . *Lotus formosissimus*
                  Witch's-teeth, Coast Lotus; Mo-n

    12b Stipules not obvious, not similar in size to the leaflets, sometimes represented by small
       red or black glands
       16a Peduncle of inflorescence up to 5 mm long, sometimes absent
         17a Peduncle 2–5 mm long
            18a Plant densely hairy, at least on the new stem and leaves; flowers 4–10 in
               each leaf axil; corolla 5–6 mm long, the wings longer than the other
               petals; leaflets 4–6, 4–16 mm long; mostly prostrate, forming mats . . . . .
               . . . . *Lotus heermanii* var. *orbicularis [L. heermanii* var. *eriophorus]*
                   Southern Lotus, Woolly Lotus; SCr-s

           18b Plant scarcely hairy; flowers 2–8 in each leaf axil; corolla 6–8 mm long,
               the wings not longer than other petals; leaflets 3–5, 5–10 mm long; mostly
               upright, or at least the stem tip rising (sometimes on serpentine) . . . . . .
               . . . . *Lotus junceus*
                   Rush Lotus; Me-SLO

       17b Peduncle nearly absent (in dry areas; widespread)
         19a Flowers 2–7 in each leaf axil; corolla 7–12 mm long; leaflets 3–6, 6–15 mm
            long; perennial; often more than 1 m tall (plant scarcely hairy, some-
            times woody at the base) . . . . *Lotus scoparius* (pl. 25)
                 Deerweed, California Broom

         19b Flowers 1–4 in each leaf axil; corolla 5–9 mm long; leaflets 4, 4–15 mm
            long; annual; prostrate, with stem tip rising
            20a Flower solitary in each leaf axil; calyx lobes not twice as long as calyx
               tube; plant scarcely hairy . . . . *Lotus wrangelianus* (pl. 25)
                 California Lotus, Chile Lotus

           20b Flowers 1–4 in each leaf axil; calyx lobes up to twice as long as calyx
               tube; plant obviously hairy . . . . *Lotus humistratus* (pl. 25)
                 Colchita, Shortpod Lotus

16b Peduncle of inflorescence at least 10 mm long (peduncle sometimes with a leaflike bract just below the flower cluster)

21a Peduncle 4–15 cm long

22a Flowers 3–9 in each umbel; peduncle 4–8 cm long; corolla 15–25 mm long, pale green or yellow .... *Lotus grandiflorus*

Largeflower Lotus; Me-s

22b Flowers 8–12 in each umbel; peduncle 5–15 cm long; corolla 11–12 mm long, mainly yellow, but with some red .... *Lotus uliginosus*

Wetland Deerweed; eu

21b Peduncle 1–4 cm long

23a Corolla 8–14 mm long (flowers 3–8 in each umbel; corolla mostly yellow, but the banner sometimes red; leaflets 5, 5–20 mm long; calyx 2–4 mm long; annual)................................... *Lotus corniculatus*

Birdsfoot Deerweed, Birdsfoot Trefoil; eu

23b Corolla 3–10 mm long

24a Corolla white, pink, or salmon colored, rarely yellow

25a Flowers 3–10 on each peduncle; corolla 7–9 mm long, white or pink; calyx 3–6 mm long; perennial .... *Lotus benthamii*

Bentham Lotus; Sn-s

25b Flower solitary on each peduncle; corolla 5–6 mm long, pink or salmon colored; calyx 2–3 mm long; annual .... *Lotus micranthus*

Hill Lotus, Smallflower Lotus

24b Corolla yellow, rarely pink

26a Wings longer than keel (leaflets 4–9, 3–10 mm long; flowers 1 or 2 on each peduncle; corolla 5–10 mm long, yellow, turning orange or red; calyx 3–6 mm long; annual; prostrate) .... *Lotus strigosus* [includes *L. strigosus* var. *hirtellus*]

Hairy Lotus, Bishop Lotus; Ma-s

26b Wings not longer than keel

27a Flower solitary on each peduncle; leaflets usually 3, up to 20 mm long (corolla 5–9 mm long, yellow or pink; calyx 3–7 mm long; annual) .... *Lotus purshianus* (pl. 25)

Spanish Lotus

27b Flowers 1–8 on each peduncle; leaflets 3–7, not more than 15 mm long

28a Calyx 6–7 mm long; leaflets 5; flowers 1–3 (corolla 5–8 mm long; annual) .... *Lotus angustissimus*

Slenderpod Deerweed; eu

28b Calyx 2–5 mm long; leaflets 3–7; flowers 2–8

29a Flowers 2–8; corolla 6–8 mm long, yellow with red tinges; leaflets 3–5, up to 10 mm long; perennial .... *Lotus junceus*

Rush Lotus; Me-SLO

29b Flowers 2–4; corolla 3–10 mm long, yellow; leaf-
       lets 3–7, up to 15 mm long; annual . . . .
       . . . . *Lotus salsuginosus*
       Hooked-beak Lotus, Coastal Lotus; SCl-s

FAGACEAE (OAK FAMILY)   Besides oaks, Fagaceae includes chestnuts, chinquapins, and beeches. All of these are trees or shrubs with alternate leaves and separate staminate and pistillate flowers. Both types of flowers are on the same plant. Staminate flowers, each with 4–12 stamens, are in catkins. Pistillate flowers, each with a single pistil, may be solitary and separated from the staminate catkins, or 1–3 at the base of each staminate catkin. The ovary is inferior and develops into a nut with a single seed. Neither the staminate nor the pistillate flowers have petals, but there are a few very small sepals.

Two genera of our region, *Quercus* and *Lithocarpus*, produce acorns. These consist partly of the nut derived from the pistil and partly of a scaly cup formed by closely associated bracts that were below the flower. In *Chrysolepis*, comparable bracts contribute to a spiny bur that encloses 1–3 nuts.

Worldwide, trees of Fagaceae are of great economic importance. They provide lumber for furniture, flooring, and other uses, as well as cork, food, and fuel. An extract from our native *Lithocarpus densiflorus* (Tanbark Oak), has long been used for tanning leather.

The most conspicuous oak in the Bay Area is *Quercus agrifolia* (Coast Live Oak). In some years, its leaves are severely damaged by caterpillars of *Phryganidia californica* (California Oak Moth).

1a   Leaves with golden scales on the undersides, at least when young; nuts 1–3, enclosed by a
      spiny bur; staminate catkins upright; pistillate flowers at the base of each staminate catkin
      (leaves smooth margined, the margins sometimes wavy, the tips tapered, the older ones
      sometimes olive colored on the undersides)
        2a   Leaf blades flat; tree, sometimes more than 30 m tall; at elevations up to 1,500 ft
             . . . . *Chrysolepis chrysophylla* var. *chrysophylla*
             Giant Chinquapin; Ma-n
        2b   Leaf blades often folded along the midrib or with wavy margins; shrub or small tree,
             rarely more than 8 m tall; at elevations up to 6,000 ft . . . .
             . . . . *Chrysolepis chrysophylla* var. *minor* (pl. 28)
             Golden Chinquapin
1b   Leaves not golden on the undersides, except in *Quercus chrysolepis;* nut 1, partly enclosed
      within a cup; staminate catkins hanging down, except in *Lithocarpus densiflorus*
        3a   Shrubs up to about 3 m high, but usually less (leaves often with spiny teeth; ever-
             green)
               4a   Both leaf surfaces hairy (leaves up to 3 cm long; petioles less than 5 mm long; often
                    on serpentine) . . . . *Quercus durata*
                    Leather Oak
        4b   Upper leaf surface not hairy

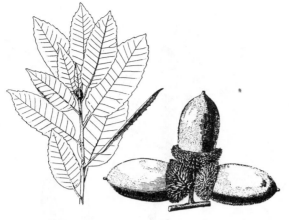
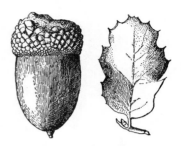

Lithocarpus densiflorus
Tanbark Oak

Quercus berberidifolia
California Scrub Oak

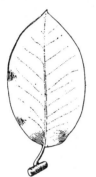

Quercus chrysolepis
Canyon Live Oak

Quercus douglasii
Blue Oak

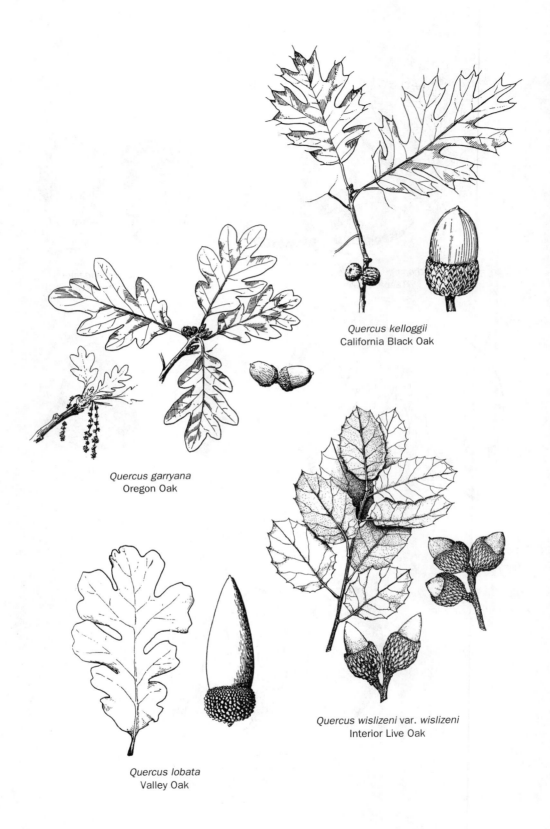

Quercus kelloggii
California Black Oak

Quercus garryana
Oregon Oak

Quercus lobata
Valley Oak

Quercus wislizeni var. *wislizeni*
Interior Live Oak

5a   Leaves 2–6 cm long, the undersides not hairy; petioles 3–15 mm long; acorns at least twice as long as wide; scales of acorn cup overlapping . . . . *Quercus wislizeni* var. *frutescens*

Dwarf Interior Live Oak; La-s

5b   Leaves not more than 3 cm long, the undersides hairy; petioles 1–4 mm long; acorns less than twice as long as wide; scales of acorn cup like bumps, not overlapping . . . . *Quercus berberidifolia* (fig.)

California Scrub Oak

3b Trees often more than 20 m tall, but sometimes shorter and shrublike

6a   Leaves deeply lobed; deciduous

7a   Lobes of leaves with sharp tips and sometimes with a few teeth; depth of acorn cup nearly or fully half the length of the nut . . . . *Quercus kelloggii* (fig.)

California Black Oak, Kellogg Oak

7b   Lobes of leaves more or less rounded, smooth margined; depth of acorn cup usually much less than half the length of the nut

8a   Leaves 5–11 cm long, not leathery; petioles less than 1.5 cm long; nuts commonly 3–4 cm long, about 3 times as long as wide . . . . *Quercus lobata* (fig.)

Valley Oak, Roble

8b   Leaves 9–17 cm long, leathery; petioles 1.5–2.5 cm long; nuts not more than 2.5 cm long, not more than twice as long as wide . . . *Quercus garryana* (fig.)

Oregon Oak, Garry Oak; SCl, Ma-n

6b Leaves not deeply lobed, but sometimes toothed; evergreen, except *Quercus douglasii*

9a   Longest leaves often more than 12 cm; petioles 10–20 mm long; catkins mostly upright; scales of acorn cup drawn out into slender projections (leaf tips blunt, usually hairy on the undersides, with more than 8 prominent secondary veins, each one ending at a marginal tooth; catkins with an unpleasant odor; cup of acorn one-fourth as long as the nut; in coastal forests) . . . . *Lithocarpus densiflorus* [*L. densiflora*] (fig.)

Tanbark Oak, Tan Oak

9b   Longest leaves usually not more than 9 cm; petioles 2–15 mm long; catkins hanging down; scales of acorn cup not drawn out into slender projections

10a Leaves bluish green on the upper surfaces, sometimes shallowly lobed; deciduous; cup of acorn about one-fourth as long as the nut (leaves paler and slightly hairy on the undersides; diameter of cup of acorn less than that of the widest part of the nut; mostly away from the coast; widespread) . . . . *Quercus douglasii* (fig.)

Blue Oak

10b Leaves dark green on the upper surfaces, the margins toothed or smooth; evergreen; cup of acorn at least one-third as long as the nut

11a Undersides of young leaves with golden hairs, these generally soon shed and the surface becoming grayish; acorn cup 17–30 mm wide (acorn cup densely hairy on the scales and inside the cup; leaves flat or wavy, usually smooth margined, the tips tapered) . . . . *Quercus chrysolepis* (fig.)

Canyon Live Oak, Maul Oak

11b Undersides of leaves neither with golden hairs, nor becoming grayish; acorn cup not more than 18 mm wide

    12a Leaves usually smooth margined, the tips tapered (undersides
of leaves not hairy) . . . . *Quercus parvula* var. *shrevei*
Shreve Oak, Oracle Oak; SFBR

    12b Leaves usually with small teeth, but sometimes smooth mar-
gined, the tips blunt (widespread)

        13a Leaves flat, the undersides not hairy; nut usually about 3
times as long as wide . . . .
. . . .*Quercus wislizeni* var. *wislizeni* (fig.)
Interior Live Oak

        13b Leaves generally convex, the undersides often with tufts of
hair on the veins; nut usually about twice as long as wide
(the most common oak in our region) . . . .
. . . . *Quercus agrifolia* (pl. 28)
Coast Live Oak, Encina

FRANKENIACEAE (FRANKENIA FAMILY) Plants belonging to the Frankeniaceae have opposite, sessile or nearly sessile leaves. Sometimes two pairs of leaves may be so close together
that they appear to form a whorl. The four to seven sepals are united into a tube with short lobes,
and the four to seven petals are attached to the inside of the calyx tube. Each petal has a small
scalelike structure on its inner face. The ovary is superior, and the fruit is a dry capsule. Our only
representative is *Frankenia salina [F. grandifolia]* (pl. 28) (Alkali-heath), a much-branched
shrubby plant that grows at the edges of salt marshes from Marin and Solano counties south to
Baja California. Its petals are pink or blue purple, and there are usually six stamens. The style is
generally three lobed. *Frankenia* reaches a height of about 30 cm but may spread out to a width
of 1 m.

GARRYACEAE (SILK-TASSEL FAMILY) Silk-tassels, belonging to the genus *Garrya*, are
shrubs or small trees. They have opposite, rather tough, evergreen leaves. The staminate and pistillate inflorescences resemble catkins and are on separate plants. Within an inflorescence, each
group of one to four flowers has a little bract below it. Each staminate flower has four stamens
and four united sepals. Staminate inflorescenses may be as long as 25 cm. Pistillate inflorescences
are eye catching, too, especially because the black or purple fruit persist for much of the year. The
ovary is inferior. Silk-tassels, easily propagated from cuttings, are valuable garden subjects because they are attractive, fast growing, and drought tolerant.

1a Leaf margins wavy (fruit densely hairy)

    2a Leaf blades 6–8 cm long, so densely hairy on the undersides that individual hairs are
scarcely distinguishable with a hand lens; petioles 6–12 mm long (widespread) . . . .
. . . . *Garrya elliptica* (fig.)
Coast Silk-tassel

Garrya elliptica
Coast Silk-tassel

2b Leaf blades 2.5–6 cm long, not so densely hairy on the undersides that individual hairs are not distinguishable with a hand lens; petioles 4–8 mm long . . . .
. . . . *Garrya congdonii* [*G. congdoni*]
Congdon Silk-tassel; SB-n

1b Leaf margins flat (leaf blades 2–5 cm long)

3a Fruit scarcely or not at all hairy; undersides of leaves scarcely hairy . . . *Garrya fremontii*
Fremont Silk-tassel; Mo-n

3b Fruit densely hairy; undersides of young leaves densely hairy, becoming much less hairy in time . . . . *Garrya flavescens* [includes *G. flavescens* var. *pallida*]
Ashy Silk-tassel; Al-s

GENTIANACEAE (GENTIAN FAMILY)   Gentians and their relatives have opposite, smooth-margined leaves. Both the calyx and corolla have four or five lobes, and the number of stamens is the same. The stamens are attached to the corolla tube. In the bud stage, the corolla lobes overlap one another with a slight twist. The ovary, usually with a two-lobed stigma, is superior, and the dry fruit splits lengthwise into two halves.

Various exotic species—mostly European and Asiatic species of the genus *Gentiana*—have long been in cultivation. Most highly prized are certain alpine species grown in rock gardens. Some of the native montane gentians from California and other western states are strikingly beautiful, but they demand very special conditions. The genus *Centaurium* is represented by several rather attractive pink-flowered species, all annuals that are easily grown from seed. One of them, *C. erythraea,* was introduced from Europe and has become weedy in the Pacific Northwest. It should perhaps be avoided.

1a   Corolla blue, sometimes with greenish dots or streaks within (flower parts in fours or fives)

    2a   Leaves less than twice as long as wide; folds within the corolla, located at the bases of the clefts between lobes, fringed with a few teeth; all flowers sessile .... *Gentiana affinis* var. *ovata [G. oregana]*
Prairie Gentian; Ma-n

    2b   Most leaves about 3 times as long as wide; folds within the corolla not fringed; some flowers with pedicels (mostly in sphagnum bogs; coastal) ... *Gentiana sceptrum* (pl. 29)
King's Gentian; Sn-n

1b   Corolla yellow, pink, rose purple, or sometimes white, without dots or streaks

    3a   Corolla deep yellow; flower parts in fours (corolla about 7 mm long, narrower than the cup-shaped calyx, closing in the afternoon; less than 5 cm tall; in grassy places) .... *Cicendia quadrangularis*
American Cicendia

    3b   Corolla pink, rose purple, or sometimes white; flower parts usually in fives

        4a   Corolla lobes 8–15 mm long, about as long as or longer than the corolla tube; anthers 3–6 mm long (pedicel absent or sometimes very short; the following 2 species hybridize)

            5a   Corolla lobes up to 10 mm long, pink; corolla throat usually pink; anthers 3–4 mm long; leaves 10–30 mm long; in moist, alkaline soil .... *Centaurium trichanthum*
Alkali Centaury; SM-n

            5b   Corolla lobes up to 15 mm long, rose purple, sometimes white; corolla throat white; anthers 4–6 mm long; leaves 5–25 mm long; in dry areas............ .... *Centaurium venustum*
Canchalagua; Na, Sn

        4b   Corolla lobes 3–6 mm long, shorter than the corolla tube; anthers 1–2 mm long (corolla pink; leaves 3–25 mm long; in moist areas)

            6a   Pedicel less than 1 mm long; corolla tube 8–12 mm long; corolla lobes 3–4 mm long (common) .... *Centaurium muehlenbergii* [includes *C. floribundum*] (pl. 29)
Monterey Centaury; Mo-n

            6b   Pedicel up to 10 mm long, but absent on some flowers; corolla tube 8–10 mm long; corolla lobes 4–6 mm long (coastal) .... *Centaurium davyi*
Davy Centaury; SLO-n

GERANIACEAE (GERANIUM FAMILY)   Geraniums and their relatives usually have regular flowers, although the various cultivated pelargoniums, native to southern Africa, have a slightly irregular corolla. There are five separate sepals and petals and 5 or 10 stamens, some of which may not have anthers. The basal portions of the filaments of the stamens are more or less joined to one another. The pistil, with a superior ovary, is deeply five lobed; the five styles are partly fused together to form a long beak. When the fruit matures, the lobes separate, and each one, containing a single seed, gets one of the styles. In *Erodium,* the styles become coiled like corkscrews as they dry. When wet, however, they straighten out, and the action of uncoiling literally drills the seeds into the soil.

This family has contributed numerous weeds to our flora. Some of the species of *Erodium* are so well established in otherwise nearly wild areas that they may seem to be native.

1a Anther-bearing stamens 10; antherless stamens absent; stem leaves alternate or opposite

    2a Leaves compound, with 3 leaflets, these deeply lobed (petals 8–13 mm long, pink to red purple; pedicel 3–8 mm long; longest leaves 5–15 cm, with a strong odor; in open or shaded areas) . . . . *Geranium robertianum* (pl. 29)

Herb-Robert; eu

    2b Leaves lobed

        3a Sepals not bristle tipped; petals 3–4 mm long; fruit not hairy, but often transversely wrinkled (sepals 3–4 mm long; petals deep pink, notched; pedicel 5–20 mm long; longest leaves 4–12 cm; in open or slightly shaded areas) . . . . *Geranium molle*

Dove's-foot Geranium; eu

        3b Sepals bristle tipped; petals 4–8 mm long; fruit slightly hairy, but not transversely wrinkled

            4a Petals not notched (petals 4–6 mm long, purple; sepals 3–4 mm long; pedicel 5–15 mm long; longest leaves 4–15 cm; perennial) . . . . . . *Geranium retrorsum*

New Zealand Geranium; au

            4b Petals notched

                5a Longest leaves 2–6 cm (petals 5–7 mm long, pale purple; sepals 4–8 mm long; fruit beak 3–5 mm long; pedicel 10–30 mm long; annual or biennial; in open or shaded areas) . . . . *Geranium bicknellii* [includes *G. bicknellii* var. *longipes*] (pl. 29)

Bicknell Geranium; Mo, Ma-n

                5b Longest leaves 5–15 cm

                    6a Pedicel 10–20 mm long; sepals 4–5 mm long, petals 6–8 mm long; longest leaves 5–15 cm; fruit beak less than 1 mm long; perennial; in moist, shaded areas (petals pink to red) . . . . *Geranium potentilloides* [*G. microphyllum* and *G. pilosum*]

Cinquefoil Geranium; au

                    6b Pedicel 2–10 mm long; sepals and petals 4–7 mm long; longest leaves 8–15 cm; fruit beak 2–3 mm long; annual; in open or shaded areas

                        7a Pedicel 2–7 mm long, without glandular hairs; petals usually rose purple, sometimes white or of intermediate colors; in open or shaded areas . . . . *Geranium carolinianum*

Carolina Geranium

                        7b Pedicel 6–10 mm long, with glandular hairs; petals rose purple; in open areas . . . . *Geranium dissectum* (pl. 29)

Cutleaf Geranium; eu

1b Anther-bearing stamens 5; scalelike antherless stamens 5; stem leaves opposite (annual or biennial)

    8a Leaf blades toothed or shallowly lobed, somewhat heart shaped or kidney shaped

        9a Sepals and petals 4–6 mm long, the petals purple to lavender; longest leaves 4–15 cm, scattered along the stem; stem sometimes more than 20 cm tall . . . . *Erodium malacoides*

Soft Stork's-bill; me

9b   Sepals 8–10 mm long, petals 10–15 mm long, white with a purple tinge; longest leaves 10–15 cm, mostly basal; stem less than 5 cm tall . . . . *Erodium macrophyllum*
Largeleaf Filaree; Ma; 2

8b   Leaf blades deeply lobed or compound, not heart shaped or kidney shaped (petals lavender or red lavender)

10a   Sepals 3–5 mm long, equal to the petals (sepals bristle tipped; longest leaves 3–10 cm, compound with 9–13 leaflets; fruit not hairy; mature styles 2–5 cm long) . . . . *Erodium cicutarium* (pl. 29)
Redstem Filaree; eua

10b   Sepals 6–13 mm long, shorter than the petals

11a   Sepals 10–13 mm long; fruit not hairy or glandular; mature styles 5–12 cm long (sepals sometimes bristle tipped; petals 11–14 mm long; longest leaves 3–15 cm, lobed or compound). . . . . *Erodium botrys* (pl. 29)
Broadleaf Filaree, Longbeak Filaree; me

11b   Sepals 6–10 mm long; fruit hairy or glandular; mature styles 2–8 cm long

12a   Longest leaves 15 cm, compound, with 11–15 leaflets; sepals not bristle tipped; petals 10–15 mm long; fruit glandular; mature styles 2–4 cm long . . . . *Erodium moschatum*
Whitestem Filaree; eu

12b   Longest leaves 10 cm, lobed; sepals bristle tipped; petals 8–11 mm long; fruit hairy but not glandular; mature styles 5–8 cm long . . . . *Erodium brachycarpum*
Southern European Stork's-bill; me

GROSSULARIACEAE (GOOSEBERRY FAMILY)   Our representatives of the Gooseberry Family are currants and gooseberries, all of which belong to the genus *Ribes,* which is sometimes included in Saxifragaceae. Species of *Ribes* are woody shrubs with alternate and often palmately lobed leaves that are usually deciduous. The small flowers are borne in racemes that generally hang down from the leaf axils. There are five stamens and five sepals that are united and longer than the five separate petals. The ovary is inferior, and the fruit becomes a fleshy berry. The two styles of the pistil are fused except at the tips.

Several native species, especially *Ribes sanguineum* var. *glutinosum, R. speciosum, R. malvaceum,* and *R. aureum* var. *gracillimum* are excellent subjects for gardens. They are easily propagated from cuttings and can also be grown from seed.

1a   Branches without spines at the nodes (fruit 4–8 mm wide)

2a   Calyx lobes and petals yellow, sometimes tinged with red; lobes of leaves usually not toothed; fruit red, not glandular (leaves glandular when young; style not hairy) . . . . *Ribes aureum* var. *gracillimum* (pl. 30)
Golden Currant; Al-s

2b   Calyx lobes and petals pink or rose, rarely white; lobes of leaves usually toothed; fruit not red, glandular

3a   Leaves dull green, glandular and hairy; style hairy at the base; calyx lobes and petals pale pink to bright rose; fruit purplish with white hairs; in chaparral or oak woodland . . . . *Ribes malvaceum*
Chaparral Currant; CC, Ma-s

3b Leaves bright green, slightly hairy, but not glandular; style not hairy; calyx lobes and petals pale pink, rarely white; fruit blue black; in sunny or somewhat shaded habitats (widespread) . . . . *Ribes sanguineum* var. *glutinosum* (pl. 30)
Pinkflower Currant

1b Branches usually with stout spines at the nodes, but the spines sometimes lacking in *Ribes divaricatum* var. *pubiflorum*

4a Petals and sepals 4, scarlet, the calyx lobes equal in length to the petals; stamens protruding well beyond the corolla (fruit 10–12 mm wide, with gland-tipped bristles) . . . .
. . . . *Ribes speciosum* (pl. 30)
Fuchsia-flower Gooseberry; SCl-s

4b Petals and sepals 5, not scarlet, the calyx lobes longer than the petals; stamens shorter than the corolla, or only the anthers protruding

5a Fruit not bristly (fruit black; petals white)

6a Leaves 2–5 cm wide; calyx lobes green or purplish; style hairy at the base; fruit 6–10 mm wide . . . . *Ribes divaricatum* var. *pubiflorum* [*R. divaricatum*]
Straggly Gooseberry

6b Leaves 1–2 cm wide; calyx lobes yellow; style not hairy; fruit 7–8 mm wide
. . . . *Ribes quercetorum*
Oak Gooseberry; Al-s

5b Fruit bristly (style not hairy)

7a Fruit yellow or purple, the bristles gland tipped; leaves 15–50 mm wide, glandular (fruit 8–10 mm wide)

8a Calyx lobes green to purplish, petals white or greenish white; leaves up to 5 cm wide; petioles as long as the width of the leaves . . . . . . *Ribes victoris*
Victor Gooseberry; Ma, Sn, Na, Sl; 4

8b Calyx lobes and petals reddish; leaves up to 4 cm wide; petioles 1–2.5 cm long . . . . *Ribes menziesii* [includes *R. menziesii* vars. *leptosmum* and *senile*] (pl. 30)
Canyon Gooseberry; SLO-n

7b Fruit red, the bristles sometimes not gland tipped; leaves 12–30 mm wide, the undersides not glandular and often without any hairs (petals white or whitish)

9a Calyx lobes purplish red; fruit 14–16 mm wide . . . .
. . . . *Ribes roezlii* var. *cruentum*
Sierra Gooseberry; Na, Sn-n

9b Calyx lobes greenish, whitish, or purplish; fruit 9–10 mm wide . . . .
. . . . *Ribes californicum* (pl. 30)
Hillside Gooseberry; Mo-Me

GUNNERACEAE (GUNNERA FAMILY)   The Gunnera Family consists of some large-leaved waterside plants native to South America. It is sometimes included in Haloragaceae. *Gunnera tinctoria* [*G. chilensis*] is from Chile and has become naturalized in some damp, shaded places. It is huge for an herbaceous plant. The leaves have somewhat heart-shaped, palmately lobed blades 1–2 m wide, and the succulent petioles are often more than 1 m long. The small flowers, in dense panicles, have two tiny sepals, usually two hoodlike petals, two stamens, and a pistil that develops into a small red fruit.

HALORAGACEAE (WATER-MILFOIL FAMILY)   Water-milfoils in our area are mostly perennials of the genus *Myriophyllum,* which are rooted in mud of freshwater lakes and ponds. The leaves are arranged in whorls around the flexible stems and are divided into slender, almost hairlike lobes. The inconspicuous staminate and pistillate flowers, with a cuplike, obscurely four-lobed calyx and sometimes with four tiny petals, are borne in the axils of the submerged leaves or in an inflorescence that is raised up out of the water. Staminate flowers, with four or eight stamens, are generally above the pistillate flowers, which have two to four styles, usually with feathery stigmas. The ovary is inferior. The small fruit splits apart into four hard, one-seeded nutlets.

Also included in the family is *Haloragis erecta* (Seaberry), a terrestrial shrub that has become established in a few places. Its flowers have stamens and a pistil but are otherwise somewhat similar to those of *Myriophyllum.*

1a   Plant terrestrial; shrub; leaves opposite, toothed; all flowers with stamens and a pistil (stem 4 sided; leaves 2–5 cm long). . . . . *Haloragis erecta*
Seaberry; au

1b   Plant aquatic; herbs; leaves whorled, most lobed; flowers pistillate or staminate

  2a   All leaves similar, lobed, and submerged, 15–35 mm long; inflorescence usually submerged . . . . *Myriophyllum aquaticum [M. brasiliense]*
Parrot's-feather; sa

  2b   Leaves of 2 types: submerged leaves lobed, 10–30 mm long, leaves beneath flower clusters 1–3 mm long, toothed or smooth margined; inflorescence raised above the water

    3a   Submerged leaves with fewer than 26 lobes, the lobes usually not opposite one another. . . . . *Myriophyllum sibiricum*
Water-milfoil

    3b   Submerged leaves usually with more than 28 lobes, the lobes opposite one another . . . . *Myriophyllum spicatum*
Eurasian Water-milfoil, Eurasian Milfoil; eua

HIPPOCASTANACEAE (BUCKEYE FAMILY)   The Buckeye Family consists of deciduous trees and shrubs that have opposite, palmately compound leaves. Several species of the principal genus, *Aesculus,* are native to eastern North America, and one of them, *A. glabra,* is the state flower of Ohio. *A. hippocastanum* (horse chestnut), a native of Europe, is widely cultivated in the United States and Canada.

The only species native to our region is *Aesculus californica* (pl. 30) (California Buckeye), a large shrub or tree that is sometimes more than 8 m tall. It is common on hillsides and in ravines. The flowers, with a rather disagreeable odor, are produced in showy, dense panicles. They are slightly irregular because the five lobes of the calyx and also the four or five white or pale pink petals are unequal. All the flowers have five to seven stamens, but only those in the lower part of each panicle have a pistil. The tough-skinned fruit, 5–8 cm wide, remains on the leafless trees long into fall and usually produces one or two seeds, each 2–3 cm wide. The seeds are very poisonous and have been used by Native Americans to stupefy fish.

HIPPURIDACEAE (MARE'S-TAIL FAMILY)   The Mare's-tail Family, sometimes included in Haloragaceae, is one of the smallest. Some authorities recognize only one species, *Hippuris*

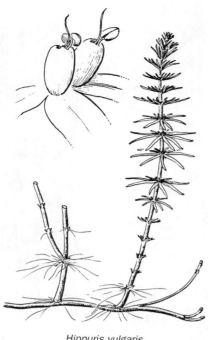

Hippuris vulgaris
Mare's-tail

vulgaris (fig.)(Mare's-tail), a perennial that grows partly submerged at the edges of lakes, ponds, and sluggish streams. It propagates vegetatively by creeping stems rooted in the mud. The narrow leaves, 1–4 cm long, are in whorls and stand out stiffly from the stems, which are up to almost 30 cm tall. Each flower, less than 2 mm long, is sessile in a leaf axil and has a minute calyx, but no petals. Most flowers have one stamen and a pistil, but some are either pistillate or staminate. The ovary is inferior, and the fruit, about 2 mm long, is a one-seeded nutlike structure.

HYDROPHYLLACEAE (WATERLEAF FAMILY)   Hydrophyllaceae is mostly restricted to the western part of North America, and it gives us no bothersome weeds. A few of the phacelias, including *Phacelia nemoralis* (Bristly Phacelia), however, have stinging hairs. Contact with them may lead to an unpleasant rash.

Except for one shrub, *Eriodictyon californicum* (Yerba-santa), all of the numerous local representatives of the family are herbaceous. The inflorescence is often a tightly coiled, one-sided cluster. The flower has five calyx and corolla lobes, five stamens, and a pistil that usually has two partly or completely partitioned divisions. Thus, some of our species have two separate styles; others have a single style, which is either two lobed or has two stigmas. The ovary is usually superior.

Nemophila menziesii var. *menziesii* (Baby-blue-eyes), *Emmenanthe penduliflora* var. *penduliflora* (Whispering-bells), and some species of *Phacelia* are attractive plants for gardens. *Eriodictyon* is a possible choice for gardens, but it suffers from a fungus that causes its leaves to look as though they had been dusted with soot.

Phacelia bolanderi
Bolander Phacelia

Phacelia nemoralis
Bristly Phacelia

1a Shrub (leaves evergreen, 4–15 cm long, usually toothed, often sticky, usually hairy; young leaves dark green, shiny, the older ones often blackened by a fungus; styles 2; up to 2 m tall; widespread) .... *Eriodictyon californicum* (pl. 31)
      Yerba-santa; SB-n

1b Herbs, not at all woody
    2a Leaf blades nearly circular, about as long as wide, toothed or shallowly lobed (leaves mainly basal, the upper ones much reduced; corolla 5–12 mm long, white, with some yellow inside; in moist, rocky habitats near the coast)...............................
      .... *Romanzoffia californica [R. suksdorfii]*
      Mistmaiden; SCr-n

    2b Leaf blades not circular, definitely longer than wide, usually deeply lobed
      3a All flowers hanging downward when open; corolla yellow, pink, cream colored, or white (inflorescences at stem ends; leaves 1–12 cm long, with 10–20 short lobes; corolla 6–15 mm long; style 2 lobed; in dry, rocky areas, including those with serpentine)
        4a Corolla yellow to cream colored..
        .... *Emmenanthe penduliflora* var. *penduliflora* (pl. 31)
          Whispering-bells

        4b Corolla pink, drying white .... *Emmenanthe penduliflora* var. *rosea*
          Rose Whispering-bells; SCl-s

3b Most flowers not hanging downward when open; corolla mostly white, blue, or violet, sometimes tinged with yellow

    5a Flowers not in a coiled, 1-sided inflorescence; often some leaves opposite . . . . HYDROPHYLLACEAE, SUBKEY 1

    5b Flowers in a coiled, 1-sided inflorescence that resembles the scroll of a violin; leaves usually alternate (style 2 lobed) . . . . HYDROPHYLLACEAE, SUBKEY 2

                                                     Phacelia

Hydrophyllaceae, Subkey 1: Flowers not in a coiled, 1-sided inflorescence; leaf blades longer than wide, usually deeply lobed; corolla mostly white, blue, or violet; often some leaves opposite

1a Flowers in tight clusters at stem ends; all leaves alternate

    2a Leaves lobed or compound, 5–40 cm long, often mainly basal, usually with light blotches; corolla 6–10 mm long, white to violet; stigma 2 lobed; perennial; up to 60 cm tall; usually in moist, shaded areas. . . . . *Hydrophyllum occidentale* (pl. 31)

                                 Heliotrope, California Waterleaf; Mo-n

    2b Leaves smooth margined, up to 1.5 cm long, not mainly basal, without blotches; corolla 1–3 mm long, white to pale pink; style deeply 2 lobed; annual; up to 10 cm tall; in dry gravelly soil. . . . . *Nama californicum [Lemmonia californica]* La-s

1b Flowers either solitary in the leaf axils, or in loose racemes at stem ends or in leaf axils; at least some leaves opposite (leaves lobed)

    3a Upper stem and flower peduncle either glandular or with downcurved prickles (lower leaves opposite, upper ones alternate)

        4a Upper stem and flower peduncle glandular, without downcurved prickles; primary lobes of leaf blades again deeply lobed; flowers 8–15 in each cluster; stigma 2 lobed; on fire burns, but also on coastal bluffs and in shaded areas (corolla 4–8 mm wide, white, yellow tinged) . . . . *Eucrypta chrysanthemifolia*

                                      Common Eucrypta

        4b Upper stem and flower peduncle not glandular but with downcurved prickles; primary lobes of leaf blades not lobed again; flowers 1–10 in each cluster; style 2 lobed; in shaded habitats

            5a Corolla 10–30 mm wide and 7–15 mm long, blue, lavender, or purple, with darker markings; prickles usually visible without a hand lens . . . . *Pholistoma auritum*

                                   Fiesta-flower; La-S

            5b Corolla up to 10 mm wide and 3–6 mm long, white, with a purple spot on each lobe; prickles visible only with a hand lens. . . . . *Pholistoma membranaceum*

                              White Fiesta-flower; CC-s

    3b Plant neither glandular nor with downcurved prickles (style deeply 2 lobed)

        6a Corolla 10–40 mm wide (leaves all opposite; in moist areas)

            7a Corolla 5–20 mm long, bright blue, with a light center; widespread . . . . *Nemophila menziesii* var. *menziesii* (pl. 31)

                                 Baby-blue-eyes

7b Corolla 6–12 mm long, white, or white with blue lines, and with very small black spots; coastal . . . . *Nemophila menziesii* var. *atomaria*
White Baby-blue-eyes; SCl-n

6b Corolla 1–12 mm wide

    8a Corolla 4–12 mm wide and 3–10 mm long, decidedly longer than calyx (corolla white or blue, without veins or spots; usually some upper leaves alternate; in partly shaded areas; common) . . . . *Nemophila heterophylla*
Variableleaf Nemophila; SB-n

    8b Corolla 1–8 mm wide and 2–5 mm long, only slightly longer than the calyx (in moist, often shaded areas)

        9a Corolla 1–5 mm wide, the lobes without dark markings; usually some upper leaves alternate, the lobes usually 5 . . . . *Nemophila parviflora*
Smallflower Nemophila; Mo-n

        9b Corolla 2–8 mm wide, the lobes with dark veins, spots, or a blotch; leaves all opposite, the lobes 5–9 . . . . *Nemophila pedunculata*
Meadow Nemophila

Hydrophyllaceae, Subkey 2: Flowers in a coiled, 1-sided inflorescence; leaf blades longer than wide, usually deeply lobed; corolla mostly white, blue, or violet; style 2 lobed; leaves usually alternate . . . . *Phacelia*

1a Larger leaves not compound and either not lobed, or with only 1 or 2 lobes at the bases (leaves sometimes toothed)

    2a Leaves toothed (larger teeth sometimes with smaller teeth on the margins)

        3a Stamens protruding out of corolla; perennial (corolla 10–12 mm long, lavender; restricted to coastal areas) . . . . *Phacelia bolanderi* (fig.)
Bolander Phacelia; Sn-n

        3b Stamens not protruding out of corolla; annual

            4a Coarse teeth of leaves without smaller teeth; corolla 7–11 mm long, the lobes lavender to purple, the tube yellow; leaves and stem not especially hairy . . . .
. . . . *Phacelia suaveolens*
Sweet-scented Phacelia; SCl-n

            4b Coarse teeth of leaves with smaller teeth; corolla 4–5 mm long, white or pale blue; leaves and stem usually very hairy . . . . *Phacelia rattanii*
Rattan Phacelia; SLO-n

    2b Leaves not toothed

        5a Corolla 10–15 mm long; leaves sometimes with 1 or 2 lobes at the bases (corolla lavender to violet, the stamens not protruding; stem not glandular hairy; annual)
. . . . *Phacelia divaricata*
Divaricate Phacelia; Me-Mo, SB

        5b Corolla 4–7 mm long; leaves without lobes

            6a Stem glandular hairy; corolla white, the stamens protruding; perennial; on serpentine . . . . *Phacelia corymbosa*
Serpentine Phacelia; Sn-n

6b   Stem not glandular hairy; corolla white to lavender, the stamens not protruding; annual; not on serpentine (in shade of chaparral, near summits of peaks) . . . . *Phacelia phacelioides*

Mount Diablo Phacelia; CC (MD)–SCl (MH); 1b

1b   Larger leaves either compound or deeply lobed (lobes or leaflets usually also toothed or divided)

   7a   Leaves lobed, the primary lobes smooth margined

      8a   Leaves not mainly basal, 1–4 cm long; corolla pale blue, the stamens not protruding; annual (in rocky soil) . . . . *Phacelia breweri*

Brewer Phacelia; CC-SB

      8b   Leaves mainly basal, 5–15 cm long; corolla white or lavender, the stamens protruding; perennial

         9a   Each calyx lobe about twice as long as wide, overlapping one another, especially as fruit matures; leaves with 3–15 lobes; corolla white or lavender; not on serpentine. . . . . *Phacelia imbricata*

Rock Phacelia, Imbricate Phacelia

         9b   Each calyx lobe usually at least 3 times as long as wide, not overlapping one another; leaves with 3–7 lobes; corolla white; on serpentine . . . . *Phacelia corymbosa*

Serpentine Phacelia; Sn-n

   7b   Leaves lobed or compound, the primary lobes or leaflets deeply divided or toothed

      10a   Leaves with stinging hairs (basal leaves compound with 3–7 leaflets; corolla pale green, the stamens protruding; usually in moist, wooded areas) . . . . *Phacelia nemoralis* (fig.)

Bristly Phacelia, Shade Phacelia; SB-n

      10b   Leaves without stinging hairs

         11a   Corolla bright blue, with a paler center; each calyx lobe not more than 3 times as long as wide, and widest below the middle (corolla up to 2 cm wide; annual) . . . . *Phacelia ciliata* (pl. 31)

Field Phacelia, Great Valley Phacelia

         11b   Corolla, if blue, without an obviously paler center; each calyx lobe more than 3 times as long as wide, and not widest below the middle

            12a   Stamens not protruding beyond corolla; flowers not crowded in the inflorescence (corolla about 2 cm wide, bluish purple; in sandy soil; annual) . . . . *Phacelia douglasii*

Douglas Phacelia; SF-s

            12b   Stamens protruding beyond corolla; flowers crowded in the inflorescence

               13a   Calyx lobes decidedly widest above the middle (stem glandular hairy; corolla white to lavender; perennial; usually spreading outward from the base)

                  14a   Stem below the inflorescence with soft, spreading hairs . . . . *Phacelia ramosissima* var. *ramosissima* (pl. 31)

Branched Phacelia; SCL-n

14b Stem below the inflorescence with coarse, erect hairs . . . .
. . . . *Phacelia ramosissima* var. *latifolia*
[*P. ramosissima* var. *suffrutescens*]
Inland Phacelia; SCl-s

13b Calyx lobes not widest above the middle

15a Plant densely glandular hairy (hairs yellow; corolla white; fruit rounded, hairy; annual) . . . . *Phacelia malvifolia*
Stinging Phacelia; SLO-n

15b Plant, if densely hairy, not glandular hairy

16a Plant densely hairy, but not glandular; corolla pale lavender or nearly white; perennial (stem sometimes upright, sometimes sprawling, branching near the base) . . . .
. . . . *Phacelia californica* (pl. 31)
California Phacelia; SCl-n

16b Plant usually not densely hairy, but the hairs glandular; corolla blue purple, sometimes lavender, cream colored or white; annual

17a Inflorescence usually without small leaves where clusters of flowers branch off; each calyx lobe usually at least 4–5 times as long as wide; fruit hairy only near the tip, longer than wide; in sandy or gravelly soil . . . .
. . . . *Phacelia tanacetifolia*
Tansy Phacelia; La-s

17b Inflorescence typically with small leaves where clusters of flowers branch off; each calyx lobe usually about 3 times as long as wide; fruit hairy to below the middle, nearly round; in clay or rocky soil (common)
. . . . *Phacelia distans*
Wild Heliotrope, Common Phacelia

HYPERICACEAE (ST. JOHN'S-WORT FAMILY) All of our species of Hypericaceae (St. John's-wort Family) have opposite leaves. Their flowers have five slightly united sepals, five separate yellow petals, and usually many stamens, which are often in three clusters. The ovary is superior, and the fruit becomes a dry capsule that cracks open to release its small seeds.

Several exotic species of *Hypericum* are widely cultivated. The best known—and perhaps too commonly used in landscaping—is *H. calycinum,* a low, creeping shrub native to southeastern Europe.

1a Petals 2–3 mm long; leaves usually not more than 1 cm long; plant with creeping stems, forming mats; flowering stem rarely more than 7 cm tall (in wet places, including sphagnum bogs) . . . . *Hypericum anagalloides* (pl. 32)
Tinker's-penny; SLO-n

1b Petals at least 10 mm long; leaves usually more than 1 cm long; plant not forming mats; flowering stem more than 12 cm tall

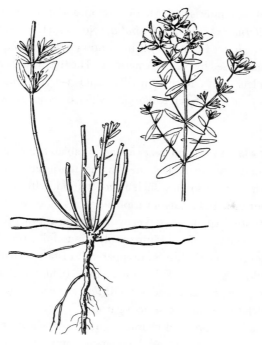

Hypericum perforatum
Klamathweed

2a Leaves folded, usually at least 5 times as long as wide, tapering to pointed tips; flowering stem not often more than 20 cm tall (on dry, brushy hillsides) . . . .
. . . . *Hypericum concinnum* (pl. 32)
Goldwire; Me-Ma

2b Leaves not folded, usually less than 4 times as long as wide, not tapering to pointed tips; flowering stem commonly more than 25 cm tall

3a Sepals slender, 4–5 mm long; numerous short branches on each plant not bearing flowers; in disturbed areas (very noxious weed) . . . . . *Hypericum perforatum* (fig.)
Klamathweed; eu

3b Sepals oval, 3 mm long; most branches on each plant bearing flowers; in wet meadows and ditches . . . . *Hypericum formosum* var. *scouleri*
Scouler St. John's-wort; Mo-n

JUGLANDACEAE (WALNUT FAMILY)   The two principal genera of the Walnut Family are *Juglans* (walnuts and butternuts) and *Carya* (hickories and pecans). These are deciduous trees with alternate, pinnately compound leaves. The staminate and pistillate flowers are on the same tree and are both associated with some bracts. Each has a four-lobed calyx but no petals. The staminate flowers are in drooping catkins and have three to many stamens. The pistillate flowers are borne singly or in clusters at the ends of branches. The ovary is inferior. The fruit has a double wall: an outer husk that is fleshy at first but later dries out, and an inner hard nutshell, familiar to everyone who has cracked open a walnut or pecan to get at the large seed.

Only one member of the family is native to our region, and it is considered to be endangered (1b). This is *Juglans californica* var. *hindsii [J. hindsii]* (Northern California Black Walnut), limited to the foothills of the inner Coast Ranges in Contra Costa and Napa counties. It is found around campsites formerly used by Native Americans. The tree reaches a height of about 25 m. Its leaves have 11–19 leaflets, each 5–10 cm long. The fruit is 3–3.5 cm wide. Farther south in California, it is replaced by *J. californica* var. *californica*, which does not grow so tall and has smaller fruit.

LAMIACEAE (MINT FAMILY)   Lamiaceae (formerly Labiatae) has given us many herbs used in cooking and perfumery. Rosemary, lavender, thyme, marjoram, and sage are just a few of them. Some so-called medicinal species, such as horehound, selfheal, and motherwort, have provided remedies reported to be useful for various ailments. And even the sleepiest cats usually become animated when stimulated by a packet of dry *Nepeta cataria* (Catnip).

Many of our native members of the Mint Family are useful garden plants. The flowers of some attract hummingbirds, and the leaves are apparently distasteful to deer. A few species with vibrant purplish blue flowers, such as *Salvia columbariae* (Chia), are easily grown from seed. Others, such as woody species of *Salvia* and *Monardella*, can be propagated from cuttings. They are fast growing and each species has a distinctive aroma.

Mints have a distinctive complex of characters, many of which are present in any given species. The stem is usually four sided, and the leaves are opposite and glandular, often with a minty or otherwise pungent aroma. The flowers, in whorls in the axils of the upper leaves, may be so crowded that the inflorescence seems continuous. Each whorl often has bracts and a pair of leaves directly beneath it. The irregular corolla is two lipped, with two lobes generally forming the upper lip and three forming the lower lip. The five-lobed calyx may also show some tendency toward being two lipped. There are four stamens protruding from the corolla. One of the two pairs may be antherless. The ovary is superior. The fruiting part of the pistil is four lobed, each lobe maturing into a nutlet that encloses a single seed. This is such a large group that deviations from the typical formula may be expected. But a mint is a mint, and one will not need much experience to place an unfamiliar plant correctly in this family.

1a   Bracts with spines usually at least 5 mm long, sometimes 10 mm long (corolla white, sometimes pink or lavender tinged; bracts broad, with prominent veins; on serpentine)

    2a   Corolla 20–25 mm long, the upper lip 2 lobed, about equal to the lower lip; style and anthers hairy .... *Acanthomintha lanceolata*
                                         Santa Clara Thornmint; Al, SCl-Mo; 4

    2b   Corolla 12–16 mm long, the upper lip not lobed, decidedly smaller than the lower lip; style and anthers not hairy (stem usually not branching; inflorescence at stem ends; anthers pinkish red) .... *Acanthomintha duttonii [A. obovata* ssp. *duttonii]*
                                         San Mateo Thornmint; SM; 1b

1b   Bracts, if present, either not spiny or the spines less than 3 mm long

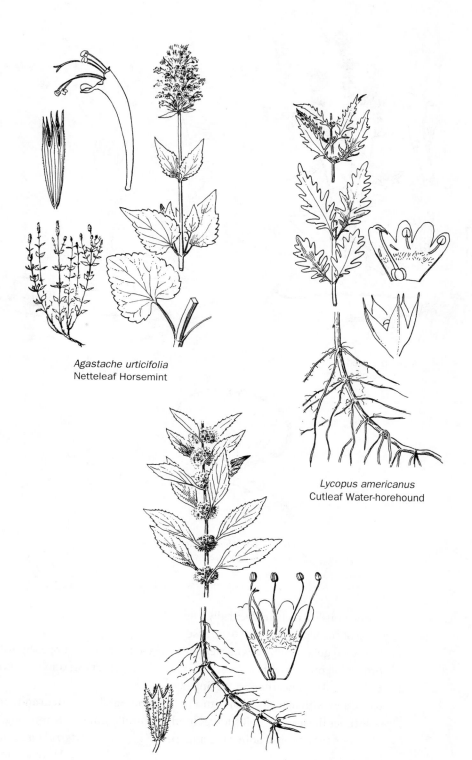

Agastache urticifolia
Netteleaf Horsemint

Lycopus americanus
Cutleaf Water-horehound

Mentha arvensis
Field mint

Mentha × piperita
Peppermint

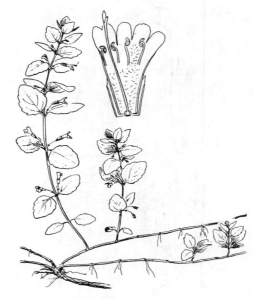

Satureja douglasii
Yerba-buena

Trichostema lanceolatum
Vinegarweed

3a   Corolla conspicuously irregular, decidedly 2 lipped

    4a   Upper lip of corolla not divided to its base

        5a   Calyx irregular, 2 lipped (calyx either with 3–6 teeth or the upper lip with a projection on its back) . . . . LAMIACEAE, SUBKEY 1

        5b   Calyx regular or nearly so, not 2 lipped

            6a   Calyx with 10 teeth, each with a hooked spine at the tip (stem and sometimes the leaves woolly; leaves toothed; corolla white, 5–6 mm long, the stamens not protruding; perennial) . . . . *Marrubium vulgare*
Horehound; eu

            6b   Calyx with not more than 6 teeth, if any, these without spines, or the spines not hooked . . . . LAMIACEAE, SUBKEY 2

4b   Upper lip of corolla divided to its base (corolla blue or lavender; calyx regular, but 1 lobe sometimes narrower; stamens markedly curved and protruding well beyond the corolla; leaves smooth margined; annual)

       7a   Stamens 3–7 mm long (petioles less than 5 mm long; corolla tube 2–4 mm long, gradually curved upward; in dry margins of streams and ponds) . . . . *Trichostema oblongum*

                                                                   Mountain Bluecurls; Na-n

       7b   Stamens 7–20 mm long

          8a   Corolla tube 5–10 mm long, bent upward at nearly a 90 degree angle; leaves with sharp tips; petioles of lower leaves not more than 4 mm long; in dry, open fields (plant with a strong odor of vinegar) . . . . *Trichostema lanceolatum* (fig.)

                                                                      Vinegarweed

          8b   Corolla tube 4–8 mm long, not bent upward; leaves without sharp tips; petioles of lower leaves usually at least 10 mm long; mostly in damp, gravelly habitats . . . . *Trichostema laxum*

                                                         Turpentine-weed; Na, Sn-n

3b   Corolla only slightly irregular, with 4 or 5 lobes almost equal in size and shape

       9a   Flowers in heads at stem or branch ends, each head with some wide bracts beneath; corolla 12–20 mm long, 5 lobed; in dry places (very aromatic)

          10a   Margins of leaves wavy (leaves slightly or not at all hairy; corolla 14–20 mm long, purple; annual; in sandy soils) . . . . *Monardella undulata*

                                                Curlyleaf Monardella; Ma-s; 4

          10b   Margins of leaves not wavy

             11a   Outer bracts below each flower head not bent down, not leaflike (leaves hairy; in chaparral and oak woodland)

                12a   Leaf margins inrolled; corolla 14–17 mm long, the stamens not protruding; perennial (flower head 10–15 mm wide; corolla lavender, rose, or purple) . . . . *Monardella viridis*

                                            Green Monardella; Na; 4

                12b   Leaf margins not inrolled; corolla 11–14 mm long, the stamens protruding; annual

                    13a   Flower head 10–15 mm wide; corolla deep purple; bracts silvery, with darker margins and veins (sometimes on serpentine) . . . . *Monardella douglasii*

                          Douglas Monardella, Fenestra Monardella; CC-Mo

                13b   Flower head 20–30 mm wide; corolla rose; bracts papery throughout (in sandy areas) . . . . *Monardella breweri*

                                          Brewer Monardella; Al-s

          11b   Outer bracts below each flower head bent down, leaflike (perennial)

             14a   Leaves not hairy, sometimes glandular; stem sometimes purple; bracts sometimes fringed; corolla 12–14 mm long, rose or purple (leaf blades 15–30 mm long; often on serpentine) . . . . *Monardella purpurea*

                          [includes *M. subglabra* and *M. villosa* ssp. *neglecta*]

                                          Siskiyou Monardella

14b Leaves at least sparsely hairy; stem not purple; bracts not fringed; corolla 10–18 mm long, purple, pink, or white

    15a Leaf blades woolly, especially on the undersides, 22–50 mm long; flower head 20–40 mm wide; in coastal scrub . . . . *Monardella villosa* ssp. *franciscana*

        Coyotemint; Ma-s

    15b Leaf blades usually not woolly, 10–22 mm long; flower head 10–30 mm wide; in chaparral and oak woodland (widespread) . . . . *Monardella villosa* ssp. *villosa*

        [includes *M. villosa* ssp. *subserrata*] (pl. 32)

        Common Coyotemint

9b Flowers either in clusters in the leaf axils, or in elongated inflorescences at stem ends, without wide bracts; corolla 2.5–7 mm long, with 4 or 5 lobes; in moist places

  16a Flowers at the ends of leafless stems (corolla white, pink, lavender, or violet)

    17a Leaves 3–6 cm long, with petioles 3–8 mm long; stamens not protruding from corolla (leaves not hairy; corolla 3–6 mm long) . . . . *Mentha* × *piperita* [includes *M. citrata* and *M. piperita*] (fig.)

        Peppermint; eu

    17b Leaves 1–6 cm long, more or less sessile; stamens protruding from corolla

      18a Leaves mostly more than twice as long as wide, not hairy; corolla 3–4 mm long . . . . *Mentha spicata*

        Spearmint; eu

      18b Leaves not more than twice as long as wide, woolly, especially on the undersides; corolla 2–3 mm long . . . . *Mentha suaveolens*

        Sweet Mentha; me

  16b Flowers in clusters in the leaf axils (stamens protruding from corolla)

    19a Anther-bearing stamens 2; corolla 2–5 mm long, white; leaves 2–10 cm long, sometimes hairy, not aromatic

      20a Leaves lobed, at least some with petioles; stem usually not hairy . . . . *Lycopus americanus* (fig.)

        Cutleaf Water-horehound

      20b Leaves toothed, sessile; stem hairy . . . . *Lycopus asper*

        Bugleweed; Sl

    19b Anther-bearing stamens 4; corolla 4–8 mm long, usually not white; leaves 2–5 cm long, hairy, at least on the undersides, aromatic (stem hairy; lower leaves with petioles)

      21a Leaves sometimes toothed, up to 2.5 cm long, the upper ones reduced; pedicel at most only slightly colored; corolla lavender or violet . . . . *Mentha pulegium* (pl. 32)

        Pennyroyal; eu

21b Leaves toothed, up to 5 cm long, the upper ones not reduced; pedicel purple; corolla light purple, pink, or white . . . .
. . . . *Mentha arvensis* [includes *M. arvensis* var. *villosa*] (fig.)
Field Mint

Lamiaceae, Subkey 1:   Corolla conspicuously irregular, 2 lipped; calyx irregular, 2 lipped and with either 3–6 teeth or the upper lip with a projection on its back

1a   Upper lip of calyx with a projection on its back, but neither lip divided into teeth (stamens not protruding from corolla)

2a   Corolla white, yellow tinged (corolla 16–19 mm long; lower leaves reddish; longest petioles 5–10 mm) . . . . *Scutellaria californica*
California Skullcap; Al-n

2b   Corolla blue, violet, or purple, the lower lip often with a white patch

3a   Corolla 25–35 mm long; most leaves at least 5 times as long as wide; upper leaves sessile (longest petioles 10–20 mm) . . . . *Scutellaria siphocampyloides*
Grayleaf Skullcap; Al-SB

3b   Corolla 13–20 mm long; most leaves much less than 5 times as long as wide; all leaves with petioles

4a   Longest petioles 5–10 mm; stem erect, with short, tightly curled hairs . . . .
. . . . *Scutellaria antirrhinoides*
Snapdragon Skullcap; Sn-n

4b   Longest petioles 5–20 mm; stem falling down to some extent, with long, straggly hairs . . . .
. . . . *Scutellaria tuberosa* [includes *S. tuberosa* var. *similis*] (pl. 33)
Blue Skullcap, Dannie Skullcap; Ma-n

1b   Upper lip of calyx without a projection on its back, but the upper lip with 3 teeth and the lower lip with 2 larger teeth

5a   Anther-bearing stamens 4; upper lip of calyx fanlike; stamens not protruding from corolla

6a   Flowers in leaf axils; leaves toothed (corolla 8–15 mm long, white; leaf blades 2–14 cm long; fragrant; stem much branched; often more than 1 m tall) . . . .
. . . . *Melissa officinalis*
Beebalm, Lemon Balm; me

6b   Flowers crowded at ends of stems; leaves usually not toothed (in moist areas)

7a   Corolla blue or violet, less than 10 mm long; calyx up to 5 mm long; leaves up to 3 cm long . . . .
. . . *Prunella vulgaris* var. *vulgaris* [includes *P. vulgaris* var. *parviflora*] (pl. 33)
European Selfheal; eu

7b   Corolla dark violet, 10–20 mm long; calyx 5–10 mm long; leaves up to 5 cm long (Coast Ranges) . . . .
. . . . . *Prunella vulgaris* var. *lanceolata* [includes *P. vulgaris* var. *atropurpurea*]
Narrowleaf Selfheal

5b   Anther-bearing stamens 2; upper lip of calyx not fanlike; stamens usually protruding from corolla

    8a   Shrub (leaves toothed, 2–7 cm long, hairy on the undersides; corolla blue, white, or lavender, the tube 5–9 mm long; 1–2 m tall; widespread)... *Salvia mellifera* (pl. 33)

                  Black Sage; CC-s

    8b   Herbs, but sometimes woody at the base

        9a   Leaves with small, regular teeth; plant forming mats (leaves densely hairy on the undersides; perennial)

            10a   Corolla red purple, the tube 25–35 mm long; leaves 8–20 cm long; plant not woody .... *Salvia spathacea* (pl. 33)

                  Hummingbird Sage, Pitcher Sage; Sl-s

            10b   Corolla blue or blue violet, the tube 6–15 mm long; leaves 3–6 cm long; plant woody at the base .... *Salvia sonomensis*

                  Sonoma Sage, Creeping Sage; Na-n

        9b   Leaves lobed or with large, irregular teeth; plant upright

            11a   Leaves woolly and spiny; corolla lavender, the tube 15–25 mm long, the lower lip ragged; up to 100 cm tall (leaves 3–10 cm long; annual; in sandy soil)... *Salvia carduacea*

                  Thistle Sage; CC-s

            11b   Leaves hairy, but not woolly and not spiny; corolla blue or blue violet, the tube 6–15 mm long, the lower lip not ragged; up to 60 cm tall

                12a   Leaves 2–10 cm long, many of the lobes divided again; corolla tube 6–8 mm long; annual .... *Salvia columbariae* (pl. 33)

                  Chia

                12b   Leaves 5–19 cm long, the lobes or large teeth not divided again; corolla tube 6–15 mm long; perennial .... *Salvia verbenacea*

                  Verbena Sage; eu

Lamiaceae, Subkey 2:   Corolla irregular, 2 lipped; calyx not 2 lipped, regular, or nearly so, the lobes sometimes slightly unequal

1a   Calyx lobes nearly as long as the entire calyx, 2 lobes longer than the others

    2a   Flowers readily visible among the leaves and bracts; corolla 1–1.5 cm long............ .... *Pogogyne douglasii* [includes *P. douglasii* ssp. *parviflora*]

                  Douglas Pogogyne; Sn

    2b   Flowers almost completely hidden by the leaves and bracts; corolla less than 1 cm long

        3a   Flowers in the axils of most leaves; corolla without spots; stem prostrate, with tip rising (in moist areas) .... *Pogogyne serpylloides* (pl. 33)

                  Thymelike Pogogyne; SLO-n

        3b   Flowers mostly in the axils of leaves near the tip of the stem; corolla with dark spots on the lower lip; stem mostly erect (in wet places that dry out in spring) .... *Pogogyne zizyphoroides*

                  Sacramento Pogogyne; SCl-n

1b Calyx lobes less than half as long as the entire calyx and mostly equal to one another (slightly unequal in *Glechoma hederacea*)

4a Flowers in the axils of the leaves, solitary or in groups of 2 or 3 (in shaded areas)

5a Flowers in groups of 2 or 3; calyx lobes of slightly unequal length (corolla 10–22 mm, blue or purple; in moist areas) . . . . *Glechoma hederacea*

Ground-ivy; eu

5b Flower solitary; calyx lobes of equal length

6a Shrub; corolla 25–30 mm long, white, pale lavender, or pink; calyx becoming enlarged as fruit develops; 1–2 m tall (widespread) . . . .

. . . . *Lepechinia calycina* (pl. 32)

Pitcher Sage

6b Herb, sometimes woody at the base; corolla 3–8 mm long, white to purple; calyx not becoming enlarged as fruit develops; forming mats (fragrance sweet) . . . . *Satureja douglasii* (fig.)

Yerba-buena

4b Flowers at the tip of the stem, in whorls that may be distinct and well spaced or continuous and crowded

7a Whorls of flowers continuous on upper stem, crowded together for a length of about 6 cm

8a Corolla rose or violet; stamens protruding 6–8 mm beyond corolla; up to 2 m tall (in moist habitats). . . . . *Agastache urticifolia* (fig.)

Nettleleaf Horsemint; SLO-n

8b Corolla white with purple marks; stamens protruding only slightly, if at all, beyond corolla; less than 1 m tall

9a Corolla white with purple veins, the tube 6–7 mm long, the upper lip 3–4 mm long (in moist habitats) . . . . *Stachys pycnantha*

Short-spiked Hedgenettle; Ma, CC-s

9b Corolla white with purple dots, the tube 8–10 mm long, the upper lip 2 mm long . . . . *Nepeta cataria*

Catnip; eua

7b Whorls of flowers distinct on upper stem, spaced clearly apart

10a Inflorescence with 1–4 whorls, each with more than 20 flowers (corolla 6–7 mm long, white; calyx hairy; in moist areas) . . . .

. . . . *Pycnanthemum californicum*

Mountain Mint

10b Inflorescence with about 7 whorls, each with up to 10 flowers

11a Tube of corolla 18–20 mm long, the upper lip 7–9 mm long; stamens protruding 4–5 mm beyond corolla (in moist areas) . . . . *Stachys chamissonis*

Chamisso Hedgenettle; SLO-n

11b Tube of corolla 6–16 mm long, the upper lip 2–6 mm long; stamens protruding 1–3 mm beyond corolla

12a Leaves narrowed at their bases, with long matted hairs (corolla tube 10–15 mm long; in moist areas) . . . . *Stachys ajugoides* var. *ajugoides*

Bugle Hedgenettle; Sn-s

12b Leaves not obviously narrowed at their bases, without matted hairs, but there may be separate long hairs

 13a Corolla white, the tube 6 mm long (in moist areas) . . . . *Stachys stricta*
              Sonoma Hedgenettle; Sn-n

 13b Corolla rose or purple, the tube more than 6 mm long

  14a Stem only slightly hairy; longest leaves 1–3 cm; petioles up to 3 cm long or absent

   15a Upper leaves sessile . . . *Lamium amplexicaule* (pl. 32)
        Clasping Henbit, Giraffehead; eua

   15b Upper leaves with petioles up to 3 cm long . . . . *Lamium purpureum* (pl. 32)
            Red Henbit; eu

  14b Stem obviously hairy; longest leaves 4–18 cm; petioles up to 6 cm long

   16a Stem branched, the hairs on angles stiffer than those in between; leaf blades 3–18 cm long; flower tube with a ring of hairs inside less than 2 mm from the base; in dry areas . . . . *Stachys bullata* (pl. 34)
      California Hedgenettle, Wood Mint; SF-s

   16b Stem not branched, the hairs similar; leaf blades 5–9 cm long; flower tube with a ring of hairs inside at least 2 mm from the base; in moist or dry areas . . . . *Stachys ajugoides* var. *rigida* [includes *S. rigida* sspp. *quercetorum, rigida,* and *rivularis*]
            Rigid Hedgenettle

LAURACEAE (LAUREL FAMILY)

The Laurel Family is a primarily tropical group that includes avocado, cinnamon, and camphor. The European *Laurus nobilis* is famous as a symbol of glory. From its branches were made the laurel wreaths that crowned heroes and poets of centuries past. The bay leaves used for seasoning soups and stews also come from this tree.

Our only representative of the family is *Umbellularia californica* (pl. 34) (California Bay). This tree, often more than 30 m tall, grows from southern Oregon to San Diego County, mostly within about 75 miles of the coast. Its olive green leaves, 3–10 cm long and 1.5–3 cm wide, are alternate, smooth margined, rather tough, and aromatic. They can be used for seasoning but have a stronger flavor than those of *L. nobilis*. The small flowers, produced in umbels in some leaf axils, have a six-lobed yellow green calyx 6–8 mm long, a pistil, and nine stamens. There are no petals. The ovary is superior. The fruit, about 2.5 cm long, is fleshy and has the shape of a plump olive, and it becomes purplish as it ripens and has a single large seed.

Umbellularia californica is not difficult to grow from seed. It deserves to be planted in places where it can be given the room it needs to form its broad crown. Once it has reached a large size, however, the accumulation of its fallen leaves on the ground discourage the growth of many other plants.

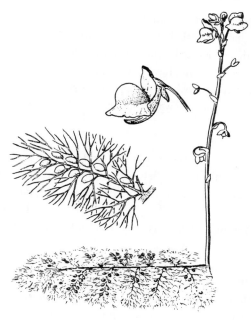

Utricularia vulgaris
Common Bladderwort

LENTIBULARIACEAE (BLADDERWORT FAMILY) Bladderworts are carnivorous aquatic plants, totally submerged except for their inflorescences. The so-called leaves, divided into very slender lobes, are specialized branches of the stems. The irregular flower, mostly yellow, has a two-lipped corolla, with a saclike spur on the lower lip. Small bladders, located on what appear to be leaves or on other side branches of some species, function as traps for microscopic organisms that wander into them, fail to escape, and eventually are digested.

1a "Leaves" up to about 4 cm long, the 2 primary lobes dividing by twos, but unequally and several times, so that there are numerous ultimate lobes; racemes with 6–20 flowers; lower lip of corolla 12–15 mm long, the spur slightly longer . . . . *Utricularia vulgaris* (fig.) Common Bladderwort
1b "Leaves" not more than 1 cm long, usually with only 2 lobes; racemes with 1–3 flowers; lower lip of corolla 6–10 mm long, the spur shorter . . . . *Utricularia gibba* Swollenspur Bladderwort; ?

LIMNANTHACEAE (MEADOWFOAM FAMILY) Limnanthaceae consists of only two genera, both restricted to North America and mostly found in the Pacific states. They are annuals that grow in moist habitats, especially around vernal pools. One species, *Limnanthes douglasii*, is native to our region. It was collected over 150 years ago by David Douglas, an intrepid botanical explorer of our west coast, and has been grown in English and European gardens ever since.

The flowers of *L. douglasii* have parts in fives. The sepals and petals are separate, the petals have a U-shaped band of hairs near their bases. The ovary is superior. There are generally 10 stamens and five pistils with the five styles fused except at the tips. Each pistil produces a nutlet with a single seed. The leaves are pinnately compound.

Limnanthes macounii, with 4 sepals, 4 white petals, 8 stamens, and 4 pistils has been found near the coast of San Mateo County. It was probably introduced from near Victoria, British Columbia.

1a   Petals at least partly yellow (leaves once compound)

    2a   Petals yellow with white tips; leaflets 5–11, smooth margined to lobed; Coast Ranges . . . . *Limnanthes douglasii* ssp. *douglasii* (pl. 34)
                        Douglas Meadowfoam; SB-n

    2b   Petals entirely yellow; leaflets 7–13, toothed or lobed; coastal . . . . *Limnanthes douglasii* ssp. *sulphurea*
                        Point Reyes Meadowfoam; Ma (PR); 1b

1b   Petals white, sometimes aging pink (Coast Ranges)

    3a   Leaves sometimes twice compound, the leaflets very narrow, with only a few teeth, if any; petals, sometimes aging pink, with pink or cream-colored veins . . . . *Limnanthes douglasii* ssp. *rosea*
                        Pale Meadowfoam

    3b   Leaves once compound, the leaflets not especially narrow, with teeth or lobes; petals with purple veins . . . . *Limnanthes douglasii* ssp. *nivea*
                        Snow Meadowfoam; SLO-n

LINACEAE (FLAX FAMILY)   The flowers of our flaxes have five separate sepals, five separate petals that often fall away rather early, and five stamens. The stamens are usually joined together at their bases, forming a low collar, and sometimes there are five antherless stamens. The ovary is superior. The pistil is partitioned lengthwise into two to five divisions, each of which produces two flattened seeds. There are two to five styles. After the fruit has ripened and dried, its wall splits apart into 4, 6, or 10 sections. The leaves are usually narrow and smooth margined and often have glandular stipules.

Linum usitatissimum (Common Flax), whose species name means "of maximum usefulness," has been associated with humans for a long time. Linen, made from its long fibers, is believed to be the oldest known textile. In the past, flax fibers were also used for making nets and ropes. The seeds are pressed to obtain linseed oil, which is incorporated into paints, inks, and varnishes. The material left after the oil has been squeezed out is fed to cattle.

One of our perennial native species, *Linum lewisii* (Western Blue Flax), grows throughout the west and is available as seed for gardens. Although its bright blue petals do not last very long, a colony of Western Blue Flax usually produces a good display of color.

1a   Petals blue, rarely white, 8–15 mm long; antherless stamens usually 5; styles 5 (leaves smooth margined, usually some alternate)

    2a   Petals 8–10 mm long; pedicel 5–18 mm long (leaves up to 2.5 cm long; stigmas elongate . . . . *Linum bienne* (pl. 34)
                        Narrowleaf Flax; me

2b   Petals 10–15 mm long; pedicel 10–30 mm long

    3a   Leaves 1–3.5 cm long, with 3 prominent veins; inner surface of sepals hairy near the margins; stigmas elongate . . . . *Linum usitatissimum*

                Common Flax; eu

    3b   Leaves 1–2 cm long, with 1 prominent vein; inner surface of sepals not hairy near the margins; stigmas nearly round (widespread) . . . .
. . . . *Linum lewisii [L. perenne* ssp. *lewisii]* (pl. 34)

               Western Blue Flax

1b Petals yellow, white, pink, or rose, not more than 8 mm long; antherless stamens absent; styles 2 or 3

    4a   Petals white to pink, sometimes streaked with rose; anthers white, pink, or purple (leaves smooth margined, alternate)

        5a   Petals 2–3 mm long; styles usually 3, sometimes 2 (petals white to pink, sometimes streaked with rose; pedicel 5–15 mm long; sometimes on serpentine) . . . .
. . . . *Hesperolinon micranthum*

               Smallflower Dwarf Flax

        5b   Petals 6–8 mm long; styles 3

            6a   Pedicel 1–8 mm long; petals pink to rose; anther margins not white; on serpentine (glandular stipules red) . . . . *Hesperolinon congestum*

               Marin Dwarf Flax; Ma-SM; 1b

            6b   Pedicel 6–20 mm long; petals white to pale pink; anther margins white; sometimes on serpentine . . . . *Hesperolinon spergulinum*

               Slender Dwarf Flax; Sn, Na

    4b   Petals yellow; anthers usually yellow

        7a   Petals 4–10 mm long (styles 3; pedicel less than 3 mm long; leaves alternate, smooth margined; sometimes on serpentine) . . . . *Hesperolinon breweri*

               Brewer Dwarf Flax; CC (MD); 1b

        7b   Petals 3–4 mm long

            8a   Leaves opposite, the upper ones toothed (styles 2; pedicel less than 3 mm long) . . . . *Sclerolinon digynum*

               Yellow Flax

        8b   Leaves alternate, not toothed

            9a   Styles 2; pedicel 2–12 mm long. . . . . *Hesperolinon bicarpellatum*

               Two-carpellate Dwarf Flax; Na; 1b

            9b   Styles 3; pedicel 5–25 mm long . . . . *Hesperolinon clevelandii*

               Cleveland Dwarf Flax; Na, SCl

LOASACEAE (BLAZING-STAR FAMILY)   All of our species of the Blazing-star or Loasa Family belong to a single genus, *Mentzelia,* and all are natives with alternate leaves. The plants are often bristly hairy. The flower has five separate yellow petals that have a silken sheen. The bases of the petals, and also the numerous stamens, are attached to a five-lobed calyx. The stigma is three lobed, and the ovary is inferior. The calyx lobes generally persist on the fruit as this ripens, dries out, and opens at the top to release its seeds. A particularly beautiful species, easily grown from seed, is *Mentzelia lindleyi.*

1a   Petals 3–8 cm long; perennial; up to 100 cm tall (leaves usually lobed, the basal ones 9–24 cm long; petals pale yellow, at least 3 times as long as wide, tapering to a point) . . . . . .
. . . . *Mentzelia laevicaulis*
Giant Blazing-star

1b   Petals either 2–4 cm long or less than 1 cm long; annual; rarely more than 50 cm tall

     2a   Petals 20–40 mm long; all leaves deeply lobed (each petal golden yellow, with an orange red base, rounded, with a tiny projection at the tip) . . . *Mentzelia lindleyi* (pl. 34)
Lindley Blazing-star; Al-Mo

     2b   Petals 3–6 mm long; at least stem leaves not lobed, sometimes toothed

         3a   Each petal pale yellow, without an orange spot at the base; basal leaves not lobed, up to 18 cm long . . . . *Mentzelia micrantha* (pl. 34)
Golden Blazing-star, San Luis Stickleaf

         3b   Each petal yellow, with an orange spot at the base; basal leaves sometimes lobed, less than 10 cm long . . . . *Mentzelia dispersa*
Nevada Stickleaf, Nada Stickleaf

LYTHRACEAE (LOOSESTRIFE FAMILY)

The flowers of loosestrifes are produced singly or in clusters in the axils of the upper leaves, and the inflorescence as a whole is elongated. At least the lower leaves are opposite or whorled, and all of them are smooth margined. In our species, the calyx is tubular, ribbed lengthwise, and has five to seven lobes. Alternating with the calyx lobes are some other flattened structures that are often conspicuous. There are four to six separate petals and at least as many stamens. The ovary is superior. The elongated fruit, tightly enclosed by the calyx but free from it, is partitioned into two divisions.

1a   Flowers 3–5 in leaf axils; calyx not much longer than wide; petals 4; all leaves opposite and somewhat clasping the stem (corolla 3–4 mm long, rose purple; stamens 4; in wet places)
. . . . *Ammannia coccinea*
Longleaf Ammannia; Ma

1b   Flowers usually 1 or 2 in leaf axils; calyx at least twice as long as wide; petals 4–6; lower leaves opposite or whorled, the upper ones usually alternate, not clasping the stem

     2a   Corolla 4–14 mm long, red purple or purple

         3a   Stem not hairy; corolla 4–8 mm long; stamens usually 6; up to 60 cm tall; lower leaves opposite (near marshes, ponds, and streams) . . . .
. . . . *Lythrum californicum* (pl. 35)
California Loosestrife; Ma, Sl

         3b   Stem hairy; corolla 8–14 mm long; stamens usually 12; up to 150 cm tall; lower leaves opposite or whorled (very invasive in wet areas) . . . . *Lythrum salicaria*
Purple Loosestrife; eu

     2b   Corolla 2–5 mm long, lavender, pink, rose, or white (stem usually not hairy; lower leaves opposite; stamens 4–6)

         4a   Corolla 2–5 mm long, pink to rose, sometimes white, the style protruding; up to 60 cm tall . . . . *Lythrum hyssopifolium* [*L. hyssopifolia*]
Grasspoly; eu

         4b   Corolla less than 3 mm long, lavender, the style not protruding; usually prostrate, rarely more than 30 cm tall (in wet areas, including drying pools) . . . .
. . . . *Lythrum tribracteatum*
Three-bracted Loosestrife; me

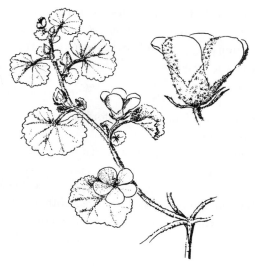

Malvella leprosa
Alkali Mallow

MALVACEAE (MALLOW FAMILY)   Mallows are familiar to gardeners because of the holly-
hock, believed to come from China, and varous species of *Hibiscus*. The characteristic way the
filaments of the numerous stamens are joined to form a tube around the pistil is familiar to any-
one who has looked inside these exotic flowers. The anthers may be concentrated at the top of
the tube or scattered along part of its length. There are five separate petals that are usually rolled
up together in the bud stage and a five-lobed calyx. The ovary is superior, and the pistil is partitioned into several seed-producing divisions, which may break apart after the fruit has ripened
and dried. There are usually several stigmas. The alternate leaves may be at least palmately
veined, if not palmately lobed, and have stipules.

Europe has sent us several weedy members belonging to the genus *Malva*. *Lavatera arborea*
(Tree Mallow), native to Europe, is now rather common in coastal areas. *Lavatera assurgen-
tiflora* (Island Mallow), native to the Channel Islands of southern California, has been widely
planted in relatively frost-free situations and sometimes escapes. Among our native mallows,
species of *Malacothamnus* and *Sidalcea* are colorful and decorative candidates for the garden.

1a   Anthers scattered over much of the tube formed by the united filaments of the stamens
      (leaf blades 4–20 cm long; often more than 2 m tall)
   2a   Flower single in leaf axils; petals 6–10 cm long; stigmas more or less globular; leaf
         blades longer than wide, toothed; in wet places along the Sacramento and San Joaquin
         Rivers (petals white or rose and with a dark crimson blotch; leaf blades 6–10 cm long;
         annual) . . . . *Hibiscus lasiocarpus [H. californicus]*
                                          California Hibiscus, Rose Mallow; CC; 2

2b Flowers more than 1 in leaf axils; petals less than 5 cm long; stigmas slender; leaf blades about as long as wide, deeply lobed; not in wet areas; coastal

    3a Petals 25–45 mm long, rose, with darker veins (leaf blades 5–15 cm long; shrub, native to and endangered on Channel Islands, but naturalized elsewhere) . . . . *Lavatera assurgentiflora*
        Island Mallow, Malva Rose; 1b

    3b Each petal 10–20 mm long, either uniformly pink or lilac, or pale purple red with darker veins confined to the base

        4a Petals 10–16 mm long, pink or lilac; leaf blades 4–10 cm long, sparsely hairy, not cottony; annual. . . . . *Lavatera cretica*
            Crete Mallow, Cretan Mallow; me

        4b Petals 15–20 mm long, pale purple red, with darker veins at their bases; leaf blades 5–20 cm long, cottony; shrub (the most common *Lavatera* in our region) . . . . *Lavatera arborea* (pl. 35)
            Tree Mallow; eu

1b Anthers in a single or double ring at the tip of the tube formed by the united filaments of the stamens

    5a Stigmas more or less globular

        6a Shrubs or woody at least at the base, usually more than 100 cm tall (leaf blades 2–6 cm long, usually at least some shallowly lobed)

            7a Calyx white woolly, the hairs almost hiding the calyx lobes of flowers that have not opened (petals pink to rose, 1–2 cm long) . . . . *Malacothamnus fremontii* (pl. 35)
                Fremont Mallow; CC

            7b Calyx hairy, but not white woolly, the hairs not hiding the calyx lobes before the flower opens . . . . *Malacothamnus fasciculatus* [includes *M. arcuatus* and *hallii*]
                Chaparral Mallow; CC-SM, SCl

        6b Herbs, not at all woody, often more than 50 cm tall

            8a Petals 3–8 mm long, dull red; leaf blades 5–8 cm long (leaves usually deeply lobed; stem reclining) . . . . *Modiola caroliniana*
                Wheel Mallow; sa

            8b Petals 10–25 mm long, not dull red; leaf blades 1–5 cm long

                9a Petals cream colored or yellow, 10–15 mm long; leaf blades 1–3 cm long, toothed, usually grayish or whitish, most of the larger hairs in clusters of at least 10; stem reclining; in saline soil. . . . . *Malvella leprosa* [*Sida leprosa* var. *hederacea*] (fig.)
                    Alkali Mallow, Whiteweed

                9b Petals pink lavender to purple, 15–25 mm long; leaf blades 2–5 cm long, deeply lobed, not especially grayish or whitish, the larger hairs in clusters of fewer than 5; stem mostly upright; in grassland and open woods. . . . . *Eremalche parryi*
                    Parry Mallow; Al-s

5b Stigmas slender (herbs)

    10a Each flower with 3 bracts just beneath the calyx (leaves shallowly lobed if at all; up to 80 cm tall)

        11a Petals 12–25 mm long; anthers in a double ring at the top of the tube formed by the united filaments (petals pale pink to lavender; on dry ridges near the coast) . . . . *Sidalcea hickmanii* ssp. *viridis*

                Marin Checkerbloom; Sn-SM; 1b

        11b Petals 4–13 mm long; anthers in a single ring at the top of the tube formed by the united filaments (common weeds)

            12a Leaf blades 3–12 cm long; bracts below calyx nearly as long as wide; petals pink to blue violet (petals 10–12 mm long; stem erect) . . . . *Malva nicaeensis* (pl. 35)

                        Bull Mallow; eua

            12b Leaf blades 2–8 cm long; bracts below calyx at least twice as long as wide; petals white, pink, or lilac

                13a Corolla 8–13 mm long, about twice as long as the calyx; stem reclining . . . . *Malva neglecta* (pl. 35)

                    Common Mallow, Cheeses; eua

                13b Corolla 4–5 mm long, barely longer than the calyx; stem erect . . . . *Malva parviflora*

                    Cheeseweed, Little Mallow; eua

    10b Flower either without bracts beneath the calyx, or with 1 or 2 bracts

        14a Upper leaves compound (plant sometimes sparsely hairy; usually in moist habitats)

            15a Petals pink, each sometimes with a purple center; basal leaves, if present, deeply lobed; leaflets of upper leaves sometimes lobed (petals 18–20 mm long; annual) . . . . *Sidalcea hartwegii*

                    Hartweg Checkerbloom; Me-Na

            15b Petals light purple to white; basal leaves with broad teeth or shallow lobes; leaflets of upper leaves not lobed (stem sometimes reclining)

                16a Petals 10–20 mm long, light purple or sometimes white; leaf blades 2–5 cm long; annual (in moist areas) . . . . *Sidalcea calycosa* ssp. *calycosa* (pl. 35)

                    Annual Checkerbloom; Me, Ma, Na

                16b Petals 20–25 mm long, light purple; leaf blades 2.5–10 cm long; perennial . . . . *Sidalcea calycosa* ssp. *rhizomata*

                    Point Reyes Checkerbloom; Me, Ma (PR); 1b

        14b Leaves not compound, but sometimes deeply lobed

            17a Petals 20–35 mm long; lobes of uppermost leaves divided again; stipules of upper leaves about 10 mm long, often lobed; annual (petals rose to purple, each sometimes with a purple center; leaves hairy) . . . . *Sidalcea diploscypha*

                    Fringed Checkerbloom; SLO-n

17b Petals 7–20 mm long; lobes of uppermost leaves not divided; stipules of upper leaves not more than 5 mm long, not lobed; perennial

    18a Petals 7–15 mm long, white; up to 150 cm tall (leaves hairy, only shallowly lobed; coastal) . . . . *Sidalcea malachroides*

          Mapleleaf Checkerbloom; Mo-n; 1b

    18b Petals 10–20 mm long, light to dark pink, sometimes white, sometimes with white veins; up to 60 cm tall, but the stem sometimes reclining

        19a Base of plant, stipules, and calyx purplish (leaves hairy, the upper ones toothed or shallowly lobed; coastal) . . . .
. . . . *Sidalcea malviflora* ssp. *purpurea*

          Purple Checkerbloom; Me, Sn; 1b

        19b Plant not purplish

            20a Stem leaves divided into narrow lobes; leaves sparsely hairy, especially on upper surfaces . . . .
. . . . *Sidalcea malviflora* ssp. *laciniata*

              Linear-lobed Checkerbloom; Sn-SLO

            20b Stem leaves not divided into narrow lobes; leaves hairy on both surfaces (widespread) . . . .
. . . . *Sidalcea malviflora* ssp. *malviflora* (pl. 35)

              Common Checkerbloom, Checker Mallow; Me-s

Menyanthes trifoliata
Buckbean

MENYANTHACEAE (BUCKBEAN FAMILY) *Menyanthes trifoliata* (fig.) (Buckbean), sometimes placed in Gentianaceae, is the only species of the family occurring within California. It is found from the San Franciso Bay region northward to Alaska. Up to 40 cm tall, it grows in bogs and at the margins of lakes and ponds. The old stems, prostrate and rooted in mud, give rise to shoots that raise the leaves and upright flower racemes out of the water. Each leaf has three leaflets, about 10 cm long, at the top of a succulent petiole. The calyx is usuallyfive lobed, with the tube of the five-lobed corolla extending well above it. The corolla, 10–16 mm long, is white, pink, or purplish, and has crowded, scalelike hairs on the inner surfaces of the lobes. The five stamens are attached to the corolla tube. The ovary is slightly inferior, and the fruit is a capsule.

MOLLUGINACEAE (CARPETWEED FAMILY) Carpetweeds and their relatives form a relatively small family, mostly represented in warmer regions. They are sometimes included in Aizoaceae. In most of them, the flower lacks petals but has four or five small sepals and 3–20 stamens, which sometimes resemble petals. The ovary is superior. The fruit is partitioned lengthwise into three divisions.

1a   Leaves 3–6 in each whorl, not hairy, about 4–5 times as long as wide, nearly sessile; flower pedicel 5–15 mm long; widespread in moist, disturbed areas . . . . . *Mollugo verticillata* (fig.) Carpetweed; sa

1b   Leaves alternate, but sometimes nearly whorled, hairy, not much longer than wide, the petioles as long as the blades; flower more or less sessile; mostly in dried soil bordering ponds . . . . *Glinus lotoides;* sa

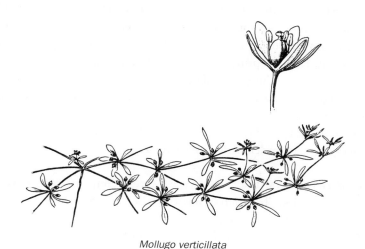

Mollugo verticillata
Carpetweed

Myrica californica
Wax-myrtle

MYRICACEAE (WAX-MYRTLE FAMILY)   Wax-myrtles and their relatives are shrubs or small trees with alternate, aromatic leaves that are often coated with wax. The wax from *Myrica cerifera*, a species of eastern United States, is used to make bayberry candles and soap.

Myrica californica (fig.) (Wax-myrtle) is the only representative native to our region. It grows near the coast from southern California to British Columbia and is commonly about 4 m tall, but occasional specimens are much larger and treelike. The leaves, 5–11 cm long and about 2 cm wide, are glossy on the upper surfaces. The flower lacks petals and sepals but has two to four bracts beneath it. The erect pistillate catkins, 8–12 mm long, are borne near the tips of the branches. Each pistil develops into a globular, one-seeded fruit; this is 2–3 mm long, purplish, and covered with a deposit of whitish wax. Lower down on the branches, drooping staminate catkins, 10–20 mm long, have flowers that consist of 7–16 stamens.

Myrica californica is fast growing, deer resistant, and requires a minimum amount of moisture. It is frequently used in gardens to form a tall hedge.

MYRTACEAE (MYRTLE FAMILY)   Myrtaceae (Myrtle Family) is a large one, with perhaps 100 genera and 3,000 species of trees and shrubs. Only a few are native to temperate regions, and none is native to our area. The most conspicuous introduced member of the family is *Eucalyptus globulus* (Blue Gum), from Tasmania, Australia. This large tree is the main source of oil of eucalyptus and has been widely planted for timber in various parts of the world. In the San Francisco Bay Area it now occupies such large tracts of land, including portions of some of the Regional Parks, that it is often taken for a California native.

Blue Gum has distinctive white, peeling bark and blue gray, sickle-shaped leaves 10–20 cm long. The sepals and petals soon drop off, leaving the many cream-colored stamens protruding from the flower. The ovary is inferior, and the fruit is four ribbed, dry, and about 2 cm long and wide. The juvenile leaves on young plants are opposite, but mature leaves on older plants are alternate. Unfortunately, where *E. globulus* is successful, the soil, covered with its aromatic leaves, can support little other vegetation. Furthermore, its large branches are subject to breaking and dropping unpredictably. Other species of *Eucalyptus,* not discussed in this book, have also become naturalized, but not so extensively as *E. globulus.*

Luma apiculata [Eugenia apiculata] (Temu), a South American shrub or small tree up to 4 m tall, has become established in a few places. Its flowers are similar to those of Blue Gum, but the fruit, about 1 cm wide, is fleshy and black, and the leaves, up to about 2.5 cm long, are opposite and dark green.

NYCTAGINACEAE (FOUR-O'CLOCK FAMILY)   The several genera of Nyctaginaceae found in California are mostly restricted to the deserts, or at least to the southern part of the state. In this family, the leaves are opposite and smooth margined. The flower does not have a corolla, but the bell-shaped or tubular calyx, with four or five lobes, resembles a corolla. There are three to five stamens. There are bracts below each flower or each flower cluster; these are sometimes united to form a cup, and they may also be colored. The ovary is superior. The fruit, frequently with winglike ridges, contains a single seed.

The only representatives native to our region are the sand-verbenas, members of the succulent-leaved genus *Abronia. Abronia umbellata* was the first plant from the west coast of North America to be grown and described in Europe. The seeds were collected by the La Pérouse Expedition in 1786.

1a   Flower solitary in a cup consisting of united bracts; stigma rounded; calyx 30–50 mm long; leaves not succulent (calyx magenta, yellow, or white; leaves sometimes hairy, but not glandular; perennial) .... *Mirabilis jalapa*

Four-o'clock; sa

1b   Each inflorescence with several bracts below it; stigma elongated; calyx 12–32 mm long; leaves succulent (on sandy beaches)

2a   Calyx yellow; inflorescence with 17–34 flowers; leaf blades about as long as wide, very glandular hairy; perennial............................... *Abronia latifolia* (pl. 36)

Yellow Sand-verbena

2b   Calyx mostly magenta, pink, or rose, but the tube sometimes yellow; inflorescence with 8–27 flowers; leaf blades mostly longer than wide, sometimes glandular hairy; annual

3a   Calyx tube 9–13 mm long, green or red; calyx lobes 7–17 mm wide; wings of fruit usually rounded .... *Abronia umbellata* ssp. *umbellata* (pl. 36)

Coast Sand-verbena; Sn-s

3b   Calyx tube 6–8 mm long, green or yellow; calyx lobes 6–8 mm wide; wings of fruit angled................................... *Abronia umbellata* ssp. *breviflora*

Pink Sand-verbena; Ma-n; 1b

NYMPHAEACEAE (WATERLILY FAMILY)   Nymphaeaceae consists of aquatic perennials whose leaves have nearly circular blades. These float or are raised slightly above the water. The stems, rooted in mud, are thick and spread laterally, producing new clusters of leaves and flow-

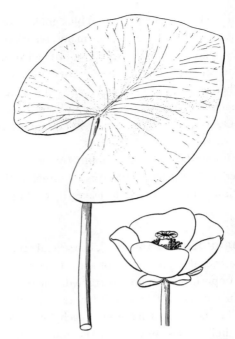

Nuphar lutea ssp. *polysepala*
Yellow Pondlily

ers at intervals. The flower has many stamens arranged in a spiral pattern. The fruit consists of several seed-producing divisions. In *Nuphar lutea* ssp. *polysepala*, whose ovary is superior, the sepals are much larger than the petals and the petals resemble the stamens. In *Nymphaea*, the ovary is half-inferior and the petals are somewhat larger than the sepals.

1a Flower about 5 cm wide, usually raised slightly above the water surface; sepals 5, deep yellow or tinged with red; petals yellow, much smaller than sepals, mostly hidden by the stamens . . . . *Nuphar lutea* ssp. *polysepala* [*N. polysepalum*] (fig.)
Yellow Pondlily; SLO-n

1b Flower up to 10 cm wide, floating on the water surface; sepals 4, green; petals yellow, white, or pink, larger than the sepals (naturalized in some lakes and ponds)
    2a Corolla yellow. . . . . *Nymphaea mexicana*
Yellow Waterlily, Banana Waterlily; na
    2b Corolla white or pink. . . . . *Nymphaea odorata*
Fragrant Waterlily, White Waterlily; na

OLEACEAE (OLIVE FAMILY) *Olea europaea* (olive), is just one of the economically or horticulturally important plants in its family, which also includes *Ligustrum* (privets), *Syringa* (lilacs), *Jasminum* (jasmines), *Forsythia*, and *Fraxinus* (ashes). Collectively these form a diversified assemblage. For instance, although the ovary is superior in all members of the family, some, such as the olive, have fleshy fruit; others have a fruit that splits apart when dry; and the ashes have fruit with a membranous, winglike expansion that aids in dispersal by wind.

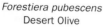

Forestiera pubescens
Desert Olive

Fraxinus latifolia
Oregon Ash

Members of the family in our region have opposite or clustered leaves without stipules. In some species, the staminate and pistillate flowers are on separate plants, but in others the flowers have two stamens and a pistil. Most flowers have a small four-lobed calyx and sometimes two white petals.

1a   Leaves opposite or clustered, not compound; branches ending in thorns; flowers in crowded clusters, these not drooping; fruit fleshy, blue black, 5–7 mm long, not winged (leaves 15–40 mm long, leathery, not hairy; staminate and pistillate flowers on separate plants; petals absent; shrub or small tree up to 3 m tall; along streambanks and in canyons) . . . . *Forestiera pubescens [F. neomexicana]* (fig.)
Desert Olive; CC-s

1b   Leaves opposite, pinnately compound; branches not thorny; flowers in drooping panicles; fruit dry, 2–5 cm long, with a membranous wing

     2a   Leaves 12–33 cm long, the leaflets 2–10 cm long, usually smooth margined and hairy on the undersides; staminate and pistillate flowers on separate plants; petals absent; trees up to 25 m tall; along streams and in moist places . . . . *Fraxinus latifolia* (fig.)
Oregon Ash; SCl-n

     2b   Leaves 5–19 cm long, the leaflets 1–7 cm long, usually toothed and not hairy; most flowers with stamens and a pistil; petals 2; large shrubs or small trees, up to 7 m tall; on hillsides and in canyons . . . . *Fraxinus dipetala*
California Ash

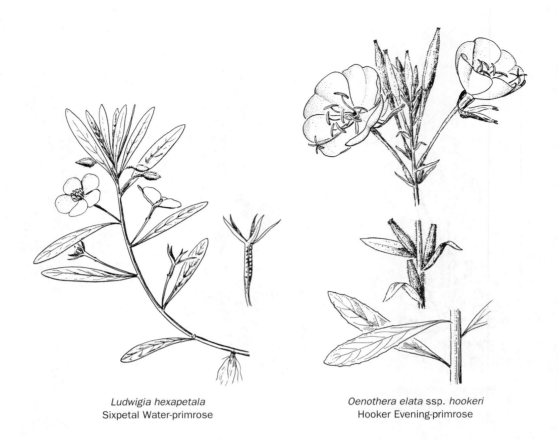

Ludwigia hexapetala
Sixpetal Water-primrose

Oenothera elata ssp. *hookeri*
Hooker Evening-primrose

ONAGRACEAE (EVENING-PRIMROSE FAMILY)   Onagraceae consists mostly of herbaceous plants, although some garden fuchsias, from Central and South America, and also our native *Epilobium canum,* become woody. The family is well represented in western North America, and almost a third of all known species occur in California. Many of them, including a good selection of the beautiful and diverse species of *Clarkia,* are in our region. These plants are easily grown from seed in sunny situations. Another candidate for gardens is *Epilobium canum* (California Fuchsia), whose flowers are usually scarlet. It blooms at the end of summer and well into autumn, but it spreads rampantly by underground stems.

The flowers of this family have an inferior ovary. The flower parts—typically four sepals, four petals, and four or eight stamens—are attached to a floral tube that generally extends at least slightly above the fruit. The stigma is often four lobed. There are exceptions, however, to all of these criteria, other than the position of the developing fruit.

1a   Plant aquatic, rooting in muddy bottoms of lakes, ponds, or marshes, and partly submerged or left behind by receding water; floral tube not present above the fruit-forming part of the pistil, but the sepals persist on the fruit after the rest of the flower has fallen away (perennial)

2a Leaves opposite; petals absent; stamens 4; fruit not more than 5 mm long. . . . .
. . . . *Ludwigia palustris [L. palustris* var. *pacifica]*
Common Water-primrose

2b Leaves alternate; petals 5 or 6, 10–15 mm long, yellow; stamens usually 10; fruit about 20 mm long

    3a Bracts beneath each fruit about as long as wide, broadly triangular; flowering stem usually floating or lying on mud; leaves generally widest near the middle . . .
. . . . *Ludwigia peploides*
Yellow Water-primrose, Yellow Waterweed

    3b Bracts beneath each fruit about twice as long as wide, narrowly triangular; flowering stem usually upright, often more than 1 m tall; leaves generally widest above the middle. . . . . *Ludwigia hexapetala [Jussiaea uruguayensis]* (fig.)
Sixpetal Water-primrose; sa

1b Plant not truly aquatic, but some in very wet habitats; floral tube present above the fruit-forming part of the pistil, and this tube falling with the sepals, petals, and stamens

    4a Sepals 2; petals 2; fruit with hooked hairs (petals white, about 1 mm long, 2 lobed; leaves opposite, 3–11 mm long, the blade usually less than twice as long as wide; perennial; up to 50 cm tall; in moist woodlands) . . . . *Circaea alpina* ssp. *pacifica*
Enchanter's-nightshade; Ma-n

    4b Sepals 4; petals 4; fruit without hooked hairs

        5a Petals yellow, occasionally white in *Camissonia ovata* (seed without a tuft of hairs)
. . . . ONAGRACEAE, SUBKEY 1
Camissonia, Oenothera (in part)

        5b Petals scarlet, purple, red, pink, or white

            6a Petals on 1 side of the floral tube; fruit not splitting into 4 divisions when mature (fruit 7–13 mm long, 4 sided; floral tube 4–14 mm long; petals white or yellow, fading to pink, 6–10 mm long, with slender basal portions; leaves alternate, usually with long hairs; perennial; forming large mats) . . . .
. . . . *Gaura drummondii;* mx

            6b Petals evenly spaced on the floral tube; fruit splitting into 4 divisions when mature

                7a Leaves deeply lobed; petals white, turning pink in age, 2–3 cm long (flower opening in late afternoon; perennial; stem upright or with just the tip rising; restricted to sand dunes near Antioch, Contra Costa County). *Oenothera deltoides* ssp. *howellii* (pl. 38)
Antioch Dunes Evening-primrose; CC; 1b

                7b Leaves not lobed; petals, if white, not more than 1.5 cm long

                    8a Petals with 1 or 2 yellow or green spots at their bases; fruit lumpy, with about 10 seeds in 2 rows; plant with many slender branches; at elevations above 2,300 ft; (floral tube more or less absent; petals 2–3 mm long, white fading to pink; fruit up to 1.5 cm long; about 50 cm tall; annual) . . . . *Gayophytum heterozygum*
Diffuse Gayophytum

8b Petals without yellow or green spots; fruit not lumpy, usually with many seeds in 4 or 8 rows; plant without many slender branches; not restricted to higher elevations; .... ONAGRACEAE, SUBKEY 2
Clarkia, Epilobium

Onagraceae, Subkey 1: Plant not aquatic; sepals 4; petals 4, yellow, occasionally white in *Camissonia ovata;* fruit splitting lengthwise into 4 divisions when mature; seed without a tuft of hairs ....*Camissonia, Oenothera* (in part)

1a Stigma 4 lobed; petals without red dots at their bases; petals at least 2.5 cm long, except in *Oenothera wolfii;* flower not opening before late afternoon (biennial)

    2a Most leaves along the stem not more than 3 times as long as wide (petals 3.5–5 cm long) .... *Oenothera glazioviana [O.* ✕*erythrosepala]* (pl. 38)
Biennial Evening-primrose; eu?

    2b Most leaves along the stem at least 4 times as long as wide

        3a Petals 2.5–5 cm long .... *Oenothera elata* ssp. *hookeri*
[includes *O. hookeri* sspp. *hookeri* and *montereyensis]* (fig.)
Hooker Evening-primrose; SLO-n

        3b Petals 1.5–2.5 cm long .... *Oenothera wolfii [O. hookeri* ssp. *wolfii]*
Wolf Evening-primrose; Ma-n; 1b

1b Stigma disk shaped or nearly globular; petals often with 1 or 2 red dots at their bases; petals less than 2.5 cm long; flower opening in the morning

    4a Stem more or less absent; leaves all basal (petals 8–23 mm long; perennial; in grassland; widespread)......................... *Camissonia ovata [Oenothera ovata]* (pl. 36)
Suncup, Golden-eggs; SLO-n

    4b Stem at least 20 cm long, upright or sprawling; leaves along the stem as well as basal

        5a Leaves up to 4 mm wide; fruit cylindrical (annual)

            6a Petals less than 5 mm long; fruit mostly 1.5–2.5 cm long, without a beak; coastal .... *Camissonia strigulosa*
Hairy Suncup; SLO-n

            6b Petals 5–15 mm long; fruit 2–4 cm long, with a well-defined beak; inland .... *Camissonia campestris [Oenothera campestris]*
Mojave Suncup, Field Primrose; CC-s

        5b Leaves mostly 5–20 mm wide; fruit 4 sided

            7a Leaves 5–50 mm long; petals 6–11 mm long; floral tube 2–5 mm long; perennial; stem prostrate, the tip rising (sandy areas, coastal) .... *Camissonia cheiranthifolia [Oenothera cheiranthifolia]* (pl. 36)
Beach Primrose

            7b Leaves 10–120 mm long; petals 1–9 mm long; floral tube 1–3 mm long; annual; stem upright or falling with the tip rising

                8a Stem upright or falling with the tip rising; leaves up to 12 cm long, the upper ones sessile, but not clasping the stem; petals 1–5 mm long; coastal .... *Camissonia micrantha [Oenothera micrantha]*
Small Suncup, Small Primrose; Ma-s

8b   Stem mostly upright; leaves up to 8 cm long, the upper ones clasping the stem; petals 2–9 mm long; primarily inland . . . .
. . . . *Camissonia hirtella [Oenothera hirtella]*
Self-pollinating Suncup

Onagraceae, Subkey 2:   Plant not aquatic; sepals 4; petals 4, scarlet, purple, red, pink, or white; fruit splitting lengthwise into 4 divisions when mature . . . . *Clarkia, Epilobium*

1a   Sepals, in a flower that has opened, not turned back and not normally adhering to one another

    2a   Floral tube 2–3 cm long; floral tube, sepals, and petals scarlet, rarely pink or white; woody at the base (widespread) . . . .
. . . . *Epilobium canum [Zauschneria californica]* (pl. 38)
California Fuchsia, Zauschneria; Sn-s

    2b   Floral tube, if present, less than 1 cm long; floral tube, sepals, and petals not scarlet; not woody

        3a   Leaves all alternate; raceme usually with at least 15 flowers; leaves sometimes more than 15 cm long; petals pink or lilac pink, rarely white, 1–2.5 cm long, not lobed; seed with a tuft of silky hairs; at least 1 m tall; in disturbed areas, including fire burns. . . . . . . *Epilobium angustifolium* ssp. *circumvagum [E. angustifolium]* (pl. 38)
Fireweed

        3b   Often at least some leaves opposite; plant not conforming in all respects to the other criteria in choice 3a

            4a   Petals usually lobed for at least one-third their length; seed without a tuft of hairs (stem often peeling in the lower portion; annual)

                5a   Fruit nearly 4 sided, with 4 prominent ridges between the slight ridges that mark the 4 partitions (leaves up to 4.5 cm long; petals 2–5 mm long, white to pale pink; fruit 8–12 mm long; seeds in 2 rows in each division of the fruit; mostly in drying mud of vernal pools) . . . .
. . . . *Epilobium cleistogamum [Boisduvalia cleistogama]*
Self-fertilizing Willowherb; Sl

                5b   Fruit nearly cylindrical, without prominent ridges other than those that mark the 4 partitions

                    6a   Petals 3–10 mm long; leaves up to 8.5 cm long (fruit 4–11 mm long, without a beak, the seeds in 1 row in each division) . . . .
. . . . *Epilobium densiflorum [Boisduvalia densiflora]*
Denseflower Willowherb, Denseflower Boisduvalia

                    6b   Petals 1–3 mm long; leaves up to 4.5 cm long

                        7a   Fruit 5–8 mm long, without a beak, the seeds in 2 rows in each division; leaves up to 3.5 cm long . . . .
. . . . *Epilobium pygmaeum [Boisduvalia glabella]*
Smooth Willowherb, Smooth Boisduvalia

                        7b   Fruit 8–13 mm long, with a beak, the seeds in 1 row in each division; leaves up to 4.5 cm long . . . .
. . . . *Epilobium torreyi [Boisduvalia stricta]*
Narrowleaf Willowherb; SCl-n

4b Petals lobed for less than one-third their length; seed with a tuft of silky hairs at one end, this visible, even on seeds from immature fruit

8a Stem peeling, especially in the lower portion, but this sometimes scarcely noticeable in *Epilobium minutum;* annual

9a Lower leaves mostly opposite, the upper ones alternate (floral tube about 1 mm long; petals 2–5 mm long, pink; rarely more than 25 cm tall; often in wet places, but sometimes in dry habitats) . . . . *Epilobium minutum*

Minute Willowherb

9b Leaves all alternate, or some of the upper ones in clusters (in dry habitats)

10a Floral tube less than 1 mm long; petals white; leaves 5–30 mm long; petioles 3–12 mm long; rarely more than 40 cm tall . . . . *Epilobium foliosum*

Manyflower Willowherb

10b Floral tube usually 2–8 mm long; petals white to rose; leaves 10–50 mm long; petioles absent or up to 4 mm long; often more than 100 cm tall . . . . *Epilobium brachycarpum [E. paniculatum]* (pl. 38)

Panicled Willowherb

8b Stem not peeling; perennial (usually in wet habitats)

11a Leaf veins, other than the midrib, inconspicuous; inflorescence often slightly nodding; seeds with small bumps, but without ridges; usually less than 60 cm tall (petals white or pink, 2–5 mm long) . . . . *Epilobium halleanum*

Hall Willowherb; Ma, SCr

11b Leaf veins conspicuous; inflorescence upright; seeds with lengthwise ridges; often more than 100 cm tall

12a Petals rarely more than 4 mm long, pale pink or nearly white . . . . . . *Epilobium ciliatum* ssp. *ciliatum* [includes *E. adenocaulon* vars. *adenocaulon, holosericeum, occidentale,* and *parishii*]

Common Willowherb, California Willowherb

12b Petals 5–14 mm long, deep pink or purple, but sometimes pale pink . . . . *Epilobium ciliatum* ssp. *watsonii* [includes *E. watsonii* vars. *franciscanum* and *watsonii*]

Watson Willowherb, San Francisco Willowherb; SLO-n

1b Sepals, in a flower that has opened, turned back, and sometimes adhering by the tips to form pairs or a single group of 4 (petals more than 5 mm long; annual)

13a Each petal with 2 or 3 lobes (petals 1–3 cm long, pink or purplish pink)

14a Petals 2 lobed . . . . *Clarkia biloba* (pl. 36)

Bilobe Clarkia; CC

14b Petals 3 lobed

15a Each petal about as long as wide, the middle lobe narrower than the 2 outer lobes . . . . *Clarkia breweri*

Brewer Clarkia; Al-s; 4

15b Each petal about twice as long as wide, the middle lobe about equal to the outer lobes . . . . *Clarkia concinna* (pl. 37)

Redribbons; SCl-n

13b Petals not divided into lobes, but the margins may be slightly ragged

16a Each petal widening abruptly from a narrow stalklike region, this at least one-third the entire length of the petal

17a Stalklike portion of each petal with a pair of teeth or small, rounded side lobes . . . . *Clarkia rhomboidea* (pl. 37)

Tongue Clarkia, Rhomboid Clarkia

17b Stalklike portion of each petal without side teeth or lobes

18a Developing fruit, floral tube, and sepals with some long hairs; leaves green; Coast Ranges and coastal locations . . . . *Clarkia unguiculata* (pl. 38)

Elegant Clarkia, Canyon Clarkia; Me-s

18b Developing fruit, floral tube, and sepals without long hairs; leaves gray green; inner Coast Ranges. . . . . *Clarkia tembloriensis*

Temblor Clarkia; Al-s

16b Each petal widening gradually, either without a narrow stalklike region or this much less than one-fourth the entire length of the petal

19a Each petal with a narrow basal portion, this less than one-fourth the entire length of the petal; petals white to cream colored, without darker markings, but becoming pink (petals 5–10 mm long; in shaded areas) . . . .
. . . . *Clarkia epilobioides* (pl. 37)

Willowherb Clarkia, Willowherb Godetia; SF-s

19b Petals without narrow basal portions; petals pink, lavender pink, or purple, often with darker markings

20a Immature fruit either with 4 lengthwise grooves or without noticeable grooves, rarely with 8 grooves in *Clarkia gracilis* ssp. *sonomensis*

21a Tip of flowering stem drooping and with unopened buds

22a Each petal 6–23 mm long, uniformly pink, but sometimes a little darker at the base (widespread) . . . .
. . . . *Clarkia gracilis* ssp. *gracilis*

Summer's-darling; SCl-n

22b Each petal 20–40 mm long, mostly pink, usually with a bright red central spot . . . . *Clarkia gracilis* ssp. *sonomensis* (pl. 37)

Slender Clarkia; Ma, Na-n

21b Tip of flowering stem not drooping

23a Each petal pink or lavender pink, with a red blotch at the base

24a Petals 10–30 mm long; floral tube 4–10 mm long (widespread) . . . . *Clarkia rubicunda*

[includes *C. rubicunda* ssp. *blasdalei*] (pl. 38)

Godetia, Ruby Chalice Clarkia; SFBR

24b Petals 5–13 mm long; floral tube 1–3 mm long (on serpentine) . . . . *Clarkia franciscana* (pl. 37)

Presidio Clarkia; SF; 1b

23b Each petal pink, lavender pink, or white, without a red blotch
at the base, but sometimes with rounded spots or streaks, or the
base may be darker (the following 2 subspecies hybridize)

    25a Stem falling, the tip rising; petals 20–35 mm long; each
flower mostly longer than the internode above the node to
which it is attached . . . . *Clarkia amoena* ssp. *amoena*
Farewell-to-spring, Godetia; Sn, Ma

    25b Stem upright; petals 15–30 mm long; each flower mostly
shorter than the internode above the node to which it is attached . . . . *Clarkia amoena* ssp. *huntiana* (pl. 36)
Hunt Clarkia, Fairy Fans; Ma, Na-n

20b Immature fruit with 8 lengthwise grooves

  26a All 4 sepals usually remaining attached to one another and directed
to 1 side (petals 5–15 mm long, pale pink to dark red, often with purple streaks; in chaparral or foothill woodland; widespread) . . . .
. . . . *Clarkia affinis*
Small Clarkia; Na-s

  26b Sepals usually turned back individually or remaining attached in
pairs

    27a Petals 5–11 mm long, pale lavender or pink, with white or yellow
bases; usually prostrate, the stem tip sometimes rising; tips of
leaf blades blunt (coastal, in sandy soil) . . *Clarkia davyi* (pl. 37)
Davy Clarkia; SM-n

    27b Petals 10–25 mm long, often not pale, without white or yellow
bases; usually upright; tips of leaf blades not blunt

      28a Each petal 20–25 mm long, pale lavender, with a wedge-
shaped purple spot that begins near the middle and widens
toward the tip (in chaparral) . . . . *Clarkia imbricata*
Vine Hill Clarkia; Sn; 1b

      28b Each petal 10–25 mm long, pale pink, pale lavender, deep
purple, or dark red, without a wedge-shaped spot, but
sometimes with a rounded spot or streaks (an extremely
variable and widespread species!)

        29a Leaves mostly less than 5 times as long as wide; flowers
crowded; each flower usually usually longer than the
internode above the node to which it is attached
(petals 10–25 mm long; coastal and inland) . . . .
. . . . *Clarkia purpurea* ssp. *purpurea*
Purple Clarkia, Winecup Clarkia; Ma, Sl-s

        29b Leaves mostly more than 6 times as long as wide;
flowers usually not crowded; each flower shorter than
the internode above the node to which it is attached

30a Petals 15–25 mm long; stamens curving away from the style, and not reaching the stigma (in chaparral or foothill woodland) ....
.... *Clarkia purpurea* ssp. *viminea* (pl. 37)
Large Clarkia

30b Petals less than 15 mm long, usually 10 mm; stamens usually reaching the stigma and clinging to it (the most common subspecies) ....
.... *Clarkia purpurea* ssp. *quadrivulnera* (pl. 37)
Winecup Clarkia, Fourspot

OROBANCHACEAE (BROOMRAPE FAMILY) Broomrapes are parasites, attached to the roots of other flowering plants. They are not green and have no typical leaves, although there are scalelike bracts on the lower portion of the stems and beneath each flower. In Europe, there is a species that parasitizes shrubby members of Fabaceae (Pea Family), including those called brooms. Hence the name broomrape.

The flowers of *Orobanche* have a calyx with four or five lobes and an irregular two-lipped corolla, two lobes of which usually form the upper lip and three lobes form the lower one, similar to the arrangement of most flowers in Scrophulariaceae. The calyx of *Boschniakia*, however, may have one to five very small lobes or none, and the upper lip of its corolla is merely notched instead of two lobed. In both genera, the four stamens are attached to the corolla tube. The stigma is usually two to four lobed. The ovary is superior, and the fruit, dry at maturity, encloses a single seed-forming chamber.

1a Flowers so crowded on the stem that the inflorescence resembles the cone of a pine or fir; upper lip of corolla sometimes notched, but not 2 lobed; bases of stamens hairy; anthers hairy
  2a Widest bract less than 10 mm, the tip usually pointed; corolla 10–15 mm long, usually yellowish; stigma with 2 or 3 lobes; above-ground portion less than 3 cm wide and up to 15 cm tall; parasitic on *Gaultheria shallon* and probably *Arctostaphylos uva-ursi* .... *Boschniakia hookeri*
  Small Groundcone; Ma-n; 2
  2b Widest bract about 10 mm, the tip usually blunt; corolla 15–20 mm long, purplish or reddish brown; stigma usually 4 lobed; above-ground portion more than 3 cm wide and up to 20 cm tall; parasitic on *Arbutus menziesii* and species of *Arctostaphylos* .... *Boschniakia strobilacea* (pl. 39)
  California Groundcone

1b Inflorescence not resembling the cone of a pine or fir; upper lip of corolla 2 lobed; bases of stamens not hairy; anthers sometimes hairy (stigma 2 lobed)
  3a Pedicel 3–15 cm long (corolla 12–35 mm long)
    4a Pedicels 1–3, arising from an underground stem; bracts usually not hairy; corolla usually violet or purple, sometimes yellowish; up to 5 cm tall; parasitic on species of *Saxifraga, Sedum,* and some Asteraceae .... *Orobanche uniflora* [includes *O. uniflora* ssp. *occidentalis* and vars. *purpurea* and *sedi*] (pl. 39)
    Naked Broomrape

4b   Pedicels 5–20, arising from 1 or more stems that are partly above ground; bracts glandular hairy; corolla yellowish to purple tinged; up to 20 cm tall; parasitic on species of *Artemisia, Eriogonum, Eriodictyon,* and other shrubs . . . .
. . . . *Orobanche fasciculata* [includes *O. fasciculata* var. *franciscana*] (pl. 39)
Clustered Broomrape

3b   Pedicel absent or up to 4 cm long on the lower flowers but shorter on the upper ones
    5a   Calyx usually 4 lobed; up to 60 cm tall; parasitic on tomato and on species of *Amaranthus* and *Xanthium* (plant branching from near the base; corolla 10–15 mm long, the lobes blue, the tube yellowish white; pedicel present) . . . .
. . . . *Orobanche ramosa*
Hemp Broomrape; eu

    5b   Calyx 5 lobed; up to 35 cm tall; parasitic on various native plants
       6a   Pedicel mostly absent; corolla 10–18 mm long, yellowish, purplish, or brownish, not obviously streaked; anthers sparsely hairy, if at all (basal portion of stem often somewhat bulbous; parasitic on *Adenostoma fasciculatum* and other chaparral plants) . . . . *Orobanche bulbosa*
Chaparral Broomrape; Sl, Ma-s

       6b   Pedicel up to 4 cm long on the lower flowers and less than 2 cm long on the upper ones; corolla 22–45 mm long, purplish, white, or rose, usually with distinct darker streaks; anthers woolly (mostly parasitic on shrubs of the family Asteraceae)
          7a   Corolla up to 40 mm long and 5–8 mm wide, the lobes usually yellowish or pinkish . . . .
. . . . *Orobanche californica* ssp. *jepsonii* [*O. grayana* var. *jepsonii*]
Jepson Broomrape; Sn-SCr

          7b   Corolla up to 45 mm long and 8–10 mm wide, the lobes usually purple (found on coastal bluffs; usually parasitic on *Grindelia*) . . . .
. . . . *Orobanche californica* ssp. *californica*
[*O. grayana* vars. *nelsonii* and *violacea*]
California Broomrape; Mo-n

OXALIDACEAE (OXALIS FAMILY)   Oxalidaceae is characterized by compound leaves that look like those of clovers and by fleshy underground rhizomes. There are five separate or nearly separate sepals and petals. The petals often overlap one another in bud and have a slight twist. *Oxalis* has 10 stamens of two lengths, and the filaments are united at their bases. The ovary is superior. The fruit-forming part of the pistil is partitioned into five divisions, and there are five styles.

One of the native species of our region, *Oxalis oregana* (Redwood Sorrel), is an excellent ground cover for gardens that are reasonably moist and at least partly shaded. Forms with deep rose pink flowers are especially attractive. Several of our most persistent garden weeds also belong to this family. One of them, *Oxalis pes-caprae* (Bermuda Buttercup), produces little bulblike structures both below and above ground; these are so effective in propagating the plant vegetatively that eradication is very difficult.

1a   Petals white, pale pink, deep pink, purplish red, or reddish violet
    2a   Flower solitary; tips of sepals not orange; petals up to 25 mm long (petals white or pale pink, often with delicate pink streaks)
        3a   Plant nearly stemless; leaves all basal; petioles up to 20 cm long; leaflets up to 4 cm long; in moist coniferous forests . . . . *Oxalis oregana* (pl. 39)

Redwood Sorrel; Mo-n
        3b   Stem up to 30 cm tall; leaves not all basal; petioles up to 7 cm long; leaflets up to 1.5 cm long; in disturbed, shaded sites . . . . *Oxalis incarnata*

Pink Oxalis; af
    2b   Flowers up to 15 in each umbel-like inflorescence; tips of sepals with 2 rows of orange marks; petals up to 15 mm long (plant nearly stemless; leaves basal; petioles up to 30 cm long)
        4a   Leaflets less than 2 cm long; pedicel and sepals hairy; petals white, light to deep pink or purplish pink, with darker streaks . . . . *Oxalis rubra* (pl. 39)

Windowbox Oxalis; sa
        4b   Leaflets about 4.5 cm long; pedicel and sepals usually not hairy; petals purplish red or violet . . . . *Oxalis latifolia [O. martiana]*

Mexican Oxalis; mx
1b   Petals yellow
    5a   Petals 15–20 mm long; tips of sepals often with 2 rows of orange marks; leaflets up to 3.5 cm long, the petioles up to 12 cm long; inflorescence umbel-like, with up to 20 flowers; plant nearly stemless. . . . . *Oxalis pes-caprae*

Bermuda Buttercup, Sourgrass; af
    5b   Petals not more than 12 mm long; tips of sepals not orange; leaflets up to 2 cm long, the petioles up to 7 cm long; inflorescence sometimes umbel-like, with 1–5 flowers; stem up to 40 cm long, but often not upright
        6a   Stem creeping and rooting at the nodes; petals less than 8 mm long; common weed in gardens (leaflets often maroon) . . . . *Oxalis corniculata* (pl. 39)

Creeping Oxalis, Yellow Sorrel; eu
        6b   Stem upright or falling down, but not rooting at the nodes; petals 8–12 mm long; coastal. . . . . *Oxalis albicans* ssp. *pilosa*

Hairy Oxalis, Hairy Wood-sorrel

PAEONIACEAE (PEONY FAMILY)   Peonies form a small family composed of one genus and about 30 speices. They are similar to members of Ranunculaceae (Buttercup Family), but a major point of distinction is that the numerous stamens and two to five pistils of a peony are attached to a fleshy disk, which is not the case in Ranunculaceae. Furthermore, in a peony the anthers on the innermost stamens are the first to produce pollen, whereas in buttercups the anthers of the outermost stamens are the first. Peonies have five separate sepals that persist until the pistils, which are many seeded, have ripened. There are five to ten separate petals. The ovary is superior.

We have one representative, *Paeonia brownii* (fig.) (Western Peony). The flower is more than 3 cm wide and occurs singly at the ends of stems. The rounded petals, 8–13 mm long, are maroon or bronze in the center and yellow or greenish at the edge. The leaves are deeply lobed

Paeonia brownii
Western Peony

to compound. In our region, the Western Peony grows in pine forests and scrubland above 3,000 ft in the Coast Ranges, from Santa Clara County northward.

PAPAVERACEAE (POPPY FAMILY)   The Poppy Family includes *Eschscholzia californica* (California Poppy), the state flower of California, and many other attractive native species that can be grown from seed. Most are herbaceous, but *Dendromecon rigida* (Bush Poppy), is a woody shrub that flowers nearly year-round.

All members of Papaveraceae in our area have two or three sepals that usually fall off after flowering. The leaves are often deeply lobed. Although initially partitioned lengthwise into two or more divisions, the superior ovary usually ripens as a single unit, producing many seeds. An exception is *Platystemon californicus* (Creamcups), in which the several divisions separate as the fruit begins to mature.

Part of the family is often assigned to the Fumariaceae (Fumitory Family). The flowers of this group have six stamens and irregular corollas with two separate outer petals that are different from the two inner petals, which are united at the tips. All other poppies have regular corollas, with four or six separate and similar petals and four to many stamens. The stigma is often lobed.

1a   Corolla irregular, the 2 outer petals separate and different from the 2 inner petals, these united at the tips; stamens 6

2a One outer petal producing a spur at its base; all leaves on the stem; stigma not lobed

    3a Corolla 7–10 mm long including spur, the lower portion usually deep pink or purplish red, the tips purple; leaf lobes sometimes more than 2 mm wide . . . .
. . . . *Fumaria officinalis* (pl. 40)

                                                              Fumitory; eu

    3b Corolla not more than 5 mm long, the lower portion usually pale pink, the tips purple; leaf lobes rarely so much as 1 mm wide . . . . *Fumaria parviflora*

                                            Smallflower Fumitory; eu

2b Each of 2 outer petals producing a pouch at its base; at least some leaves basal; stigma 2 lobed (corolla 12–18 mm long)

    4a Corolla dull rose purple; leaves all basal; mostly less than 40 cm tall; in shady, moist woods. . . . . *Dicentra formosa* (pl. 40)

                                        Bleeding-heart; Mo-n

    4b Corolla yellow, falling away early; leaves basal and also on upper portion of the stem; sometimes more than 100 cm tall; on rather dry ridges . . . .
. . . . *Dicentra chrysantha* (pl. 39)

                                      Golden Eardrops; Me-s

1b Corolla regular, petals 4 or 6 or more, all similar; stamens 12 to many

  5a Stem leaves opposite (petals 6 or more; sepals usually 3; annual)

    6a Stigma 1, 3 lobed; petals 5–10 mm long, white or cream colored, without yellow at their bases; stamens 8–16; most basal leaves broadest above the middle. . . . .
. . . . *Meconella californica*

                                     California Meconella; SFBR

    6b Stigmas many; petals 8–16 mm long, cream colored, with yellow at their bases; stamens more than 12; basal leaves not broadest above the middle (widespread)
. . . . *Platystemon californicus* (pl. 40)

                                          Creamcups

  5b Stem leaves alternate, or all leaves basal

    7a Petals orange, orange red, brick red, or yellow with orange at their bases (sepals 2; petals 4)

      8a Petals 20–60 mm long; sepals joined to form a conical cap that covers the petals before they unfold; flower with a prominent collarlike rim just below the sepals (stigma with 4–8 lobes; stamens 12 to many; widespread) . . . .
. . . . *Eschscholzia californica* (pl. 40)

                                        California Poppy

      8b Petals 10–25 mm long; sepals not forming a conical cap; flower usually without a rim below the sepals

        9a Petals yellow with orange at their bases; stamens 12 to many (stigma with 4–8 lobes; sap not milky) . . . . *Eschscholzia caespitosa* (pl. 40)

                                        Tufted Poppy

        9b Petals orange red or brick red; stamens many

          10a Each petal orange red, with a purple spot at the base; pistil with a distinct style, the headlike stigma with 4–11 lobes; sap yellowish . . . .
. . . . *Stylomecon heterophylla* (pl. 40)

                                       Wind Poppy; La-s

10b Each petal brick red, with a green spot at the base; pistil without a distinct style, but capped by a disklike stigma with 4–20 lobes; sap milky . . . . *Papaver californicum*
Fire Poppy; Ma-s

7b Petals white or cream colored, or if yellow, without orange at their bases
11a Petals white or cream colored; sepals 2 or 3 (stamens many)
12a Petals 4 or 6, 25–40 mm long, white; stem and lobes of leaves with yellow spines; leaves not all basal; stigma with 4–6 lobes (up 1 m tall) . . . . *Argemone munita* [includes *A. munita* ssp. *rotundata*]
Prickly Poppy, Chicalote; CC-s
12b Petals 6, 6–15 mm long, cream colored, their bases yellow; stem and lobes of leaves not spiny; leaves all basal; stigma 3 lobed . . . .
. . . . *Meconella linearis* [*Hesperomecon linearis*]
Narrowleaf Meconella

11b Petals yellow; sepals 2 (petals 4)
13a Shrub; petals 20–30 mm long; stamens many; stigma with 1 or 2 lobes; sepals not forming a conical cap (evergreen) . . . .
. . . . *Dendromecon rigida* (pl. 39)
Bush Poppy; Sn-s
13b Herbs; petals 3–25 mm long; stamens 12 to many; stigma with 4–8 lobes; sepals joined to form a conical cap that covers the petals before they unfold
14a Leaves all basal; seed brown, burlike, with many sharp projections; up to 15 cm tall (petals 7–12 mm long; sepals 6–10 mm long) . . . .
. . . . *Eschscholzia lobbii*
Frying-pan Poppy
14b Leaves not all basal; seed brown or black with a networklike pattern; up to 30 cm tall
15a Sepals 3–4 mm long; petals 3–15 mm long; seeds black . . . .
. . . . *Eschscholzia rhombipetala*
Diamond-petal California Poppy; Al, CC; 1b
15b Sepals 10–18 mm long; petals 10–25 mm long; seeds black or brown . . . . *Eschscholzia caespitosa* (pl. 40)
Tufted Poppy

PHILADELPHACEAE (MOCK-ORANGE FAMILY) Plants making up Philadelphaceae are sometimes included in either Saxifragaceae or Hydrangeaceae. They differ from both of these families, however, in having opposite leaves. In some of the genera, moreover, the flowers have many stamens.

Our only representative is *Whipplea modesta* (fig.) (Yerba-de-selva or Modesty), which grows in coniferous forests from Monterey County northward. It forms a low ground cover and is perhaps underrated as a subject for lightly shaded places in gardens. It is slightly woody at the base and has nearly sessile, oval deciduous leaves. The flowers, in upright inflorescences, have four to six separate sepals and the same number of separate, white petals, 3–6 mm long. There are 8–12 stamens and four or five styles. The ovary is half-inferior , and the globular fruit is dry at maturity.

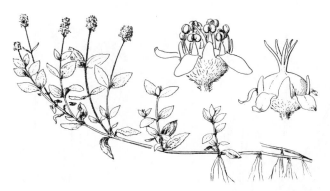

Whipplea modesta
Yerba-de-selva

Two other California natives, which occur outside our region, are widely cultivated and grow well here. One of them, *Carpenteria californica* is an evergreen shrub with large white flowers. *Philadelphus lewisii* is a deciduous shrub and uninteresting when dormant, but its beautiful white flowers have a delightful fragrance.

PHYTOLACCACEAE (POKEWEED FAMILY)   *Phytolacca americana* (Pokeweed or Pigeon-berry) is a perennial native to the eastern United States and Canada. It is an attractive plant and the source of a useful red dye, long used for coloring candy, wine, cloth, and paper. Pokeweed has therefore been cultivated, or at least encouraged, even in areas where it grows wild. It is now well established in southern Europe, and there are a few places in California, including our region, where it has become a moderately common weed.

Phytolacca americana reaches a height of more than 2 m. Its stems are ribbed and often reddish, and they branch in twos. The alternate, smooth-margined leaves are 8–30 cm long. Each small flower has no petals, but there is a five-lobed white or pinkish calyx and 10 stamens. The flowers are borne in dense inflorescences 5–20 cm long. The ovary is superior. The fruit is interesting: a cluster of 5–12 fleshy units, thus something like a blackberry. It starts out reddish, but becomes purplish black as it ripens.

PLANTAGINACEAE (PLANTAIN FAMILY)   All of our plantains belong to a single genus, *Plantago,* characterized by basal leaves and small flowers concentrated in narrow, dense inflorescences at the stem tips. The flower has a four-lobed calyx, a four-lobed, somewhat papery corolla, and two or four stamens. The ovary is superior, and the fruit is partitioned into two major divisions, each producing one or more seeds.

We have an interesting variety of local plantains. Two of them, *Plantago major* and *P. lanceolata,* are aggressive weeds. When in flower they produce much pollen, to which some persons are allergic.

1a   Leaf blades pinnately lobed; inflorescence drooping before fruiting has begun (stamens 4; leaves 4–25 cm long, hairy; annual or biennial; mostly near the coast) . . . . *Plantago coronopus*
Cutleaf Plantain; eu

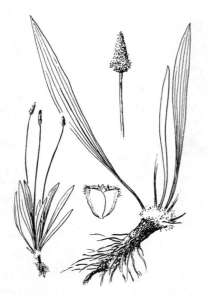

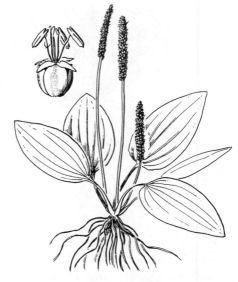

Plantago lanceolata
English Plantain

Plantago major
Common Plantain

1b Leaf blades not pinnately lobed, but sometimes with large, irregular teeth; inflorescence not drooping

  2a Leaves commonly more than 1.5 cm wide and often more than 20 cm long (stamens 4; usually perennial)

    3a Leaves not more than 2.5 cm wide; inflorescence 2–8 cm long (leaves hairy, 5–25 cm long; widespread) .... *Plantago lanceolata* (fig.)
English Plantain, Ribwort; eu

    3b Leaves usually more than 2.5 cm wide; inflorescence 3–30 cm long

      4a Mostly in coastal areas; inflorescence 8–30 cm long; leaves 12–40 cm long; sometimes hairy .... *Plantago subnuda* [*P. hirtella* var. *galeottiana*]
Coast Plantain, Mexican Plantain

      4b Widespread; inflorescence 3–20 cm long; leaves 7–20 cm long, not hairy .... *Plantago major* [includes *P. major* vars. *pilgeri* and *scopulorum*] (fig.)
Common Plantain

  2b Leaves rarely so much as 1 cm wide and not more than 15 cm long

    5a Leaves not hairy or only sparsely hairy

      6a Leaves succulent, not hairy, up to 15 cm long; inflorescence 2–10 cm long; stamens 4; perennial; on sandy beaches, salt marshes, and upper levels of rocky shores .... *Plantago maritima* [includes *P. maritima* ssp. *juncoides* and var. *californica*] (pl. 41)
Sea Plantain

6b Leaves not succulent, sometimes hairy, up to 10 cm long; inflorescence 1–5 cm long; stamens 2; annual; in drying vernal pools and at the borders of salt marshes . . . . *Plantago elongata* [includes *P. elongata* ssp. *pentasperma* and *P. bigelovii*]
Linearleaf Plantain, Annual Coast Plantain; SLO-n

5b Leaves with long hairs (stamens 4; inflorescence 1–3 cm long; annual)

  7a Leaves 3–13 cm long; peduncle up to 30 cm long; corolla lobes spreading outward, exposing the developing fruit (widespread) . . . . *Plantago erecta* [*P. hookeriana* var. *californica*] (pl. 41)
California Plantain

  7b Leaves 1–6 cm long; peduncle up to 6 cm long; corolla lobes forming a beaklike covering over the developing fruit . . . . . *Plantago truncata* ssp. *firma*
Chile Plantain; sa

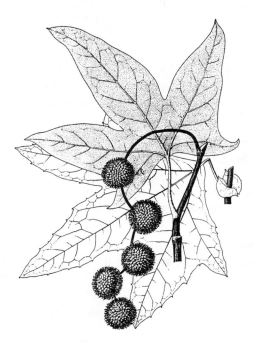

Platanus racemosa
Western Sycamore

PLATANACEAE (SYCAMORE FAMILY) Sycamores and plane trees, all of which belong to a single genus, have alternate, deciduous leaves. The larger blades are palmately lobed, and the bases of the petioles are prominently dilated. The bark flakes off in substantial pieces, and this confers an attractive mottling on the trunk and large branches. The staminate flowers, with three to eight stamens, and the pistillate flowers, with three to nine pistils, are on the same tree and crowded into one to six globular heads attached to a drooping stalk. All of the flowers on a particular stalk are either staminate or pistillate. Both types of flowers have three to eight small sepals and usually three to six very small petals. The ovary is superior.

The only species native to our region and to all of California is *Platanus racemosa* (fig.) (Western Sycamore). It grows mostly along streams in foothills and valleys and reaches a height of about 25 m. The leaf blades, usually with five pointed lobes, are up to 15 cm wide. The heads of the short-lived staminate flowers are about 1 cm wide. Each pistillate flower develops into a hairy, bristly achene. Fruiting heads, about 2.5 cm wide, are called "buttonballs" and persist into autumn.

PLUMBAGINACEAE (LEADWORT FAMILY)   Many plants of the Leadwort Family live in coastal habitats, including salt marshes. A few species of *Limonium* are cultivated, usually under the obsolete genus name *Statice*. Their everlasting flowers, when dried, look much as they did when fresh. The papery texture of the calyx or corolla, or both, is a feature of some members of the family.

The small flower has a five-lobed calyx, which is usually ribbed lengthwise, and five petals that are either nearly separate or quite united. The five stamens, attached to the corolla, are in line with its lobes. The ovary is superior. The pistil, with five styles, develops into a dry fruit containing a single seed. All species in our region are usually perennial, and their leaves are mostly basal.

1a   Flowers in a dense head at the end of an unbranched stem, each head with papery bracts beneath it; petals separated nearly to their bases; leaves up to 3 mm wide; up to 30 cm tall (petals pink; usually on rocky shores above the high-tide line, but sometimes on backshores of sandy beaches or in other open habitats) . . . . *Armeria maritima* ssp. *californica* (pl. 41)
California Thrift, Seapink; SLO-n

1b   Flowers in clusters on a branched panicle, each cluster with scalelike bracts beneath it; petals united into a long tube; most leaves usually at least 30 mm wide; usually more than 30 cm tall

2a   Leaves pinnately lobed (calyx bluish pink, papery; corolla pale yellow, soon withering; stem winged; in disturbed coastal areas) . . . . *Limonium sinuatum*
Winged Sea-lavender; me

2b   Leaves not lobed

3a   Corolla grayish violet; leaves green at flowering time (mostly near the high-tide line in salt marshes; common) . . . . *Limonium californicum* (pl. 41)
California Sea-lavender, Western Marsh-rosemary

3b   Corolla white; leaves withered by flowering time . . . . *Limonium otolepis*
Asian Sea-lavender; as

POLEMONIACEAE (PHLOX FAMILY)   In our region, most members of the Phlox Family are annual, herbaceous plants. The flower generally has a five-lobed calyx and corolla and five stamens that are attached to the corolla tube. The pistil has three stigmas, and its superior ovary has three seed-producing divisions. Certain species, however, deviate from the pattern just described. For instance, flowers with a four-lobed calyx and corolla often predominate in a population or are encountered in inflorescences in which most flowers have a five-lobed calyx and corolla.

All species in the Bay Area are natives. It is not often that we can say that about a family that has so many representatives. *Collomia grandiflora* and species of *Gilia* and *Linanthus* grow well from seed sown in sunny places.

Navarretia intertexta
Needleleaf Navarretia

Polemonium carneum
Jacob's-ladder

1a   Leaves pinnately compound, with 11–21 oval leaflets (leaves alternate; corolla up to 2.5 cm wide, salmon, pink, or purple; perennial; widespread) .... *Polemonium carneum* (fig.)
Jacob's-ladder; SM-n

1b   Leaves not compound but often deeply toothed or lobed

    2a   Leaves smooth margined, toothed, or with a few shallow lobes near the tips

        3a   Leaves alternate (annual)

            4a   Corolla often more than 1 cm wide, salmon, pinkish yellow, or sometimes white; leaves smooth margined; stem usually not branched; often more than 50 cm tall .... *Collomia grandiflora* (pl. 41)
Largeflower Collomia

            4b   Corolla less than 1 cm wide, yellow at the throat, but with purplish lobes; leaves toothed or with 3 shallow lobes near the tips; stem usually branched; rarely more than 10 cm tall (on serpentine soil) .... *Collomia diversifolia*
Serpentine Collomia; Na; 4

        3b   Leaves opposite (leaves smooth margined)

            5a   Corolla lobes 1–2 mm long, usually pink, lavender, or white; leaves not tough; annual; rarely more than 15 cm tall; common in grassy places ....
.... *Phlox gracilis [Microsteris gracilis]* (pl. 42)
Slender Phlox

5b   Corolla lobes about 10 mm long, bright pink; leaves tough; woody-based perennial; sometimes more than 30 cm tall; in rocky habitats . . . .
. . . . *Phlox speciosa* ssp. *occidentalis*
Showy Phlox; Sn-n

2b At least some leaves deeply lobed, the primary lobes sometimes divided again

6a   Flowers in a head above spine-tipped bracts; calyx lobes of distinctly unequal length (leaves alternate; annual)

7a   Inflorescence cobwebby hairy, the hairs intertwined; tips of leaf lobes not spinelike (corolla lobes blue, the tube and throat yellow) . . . .
. . . . *Eriastrum abramsii*
Abrams Eriastrum; La-SB

7b   Inflorescence, if hairy, not cobwebby, the hairs not intertwined; tips of leaf lobes spinelike . . . . POLEMONIACEAE, SUBKEY 1
Navarretia

6b   Flowers, if in a head, without spine-tipped bracts; calyx lobes of equal or nearly equal length

8a   At least some leaves whorled or opposite, with slender palmate lobes (annual) . . . . POLEMONIACEAE, SUBKEY 2
Linanthus

8b   Leaves all alternate, the lower ones, at least, 1–2 times pinnately lobed, sometimes only 1 or 2 lobes on each side . . . . POLEMONIACEAE, SUBKEY 3

Polemoniaceae, Subkey 1:   Flowers in a head above spine-tipped bracts; calyx lobes of distinctly unequal length; leaves alternate, at least some deeply lobed, the tips of the lobes spinelike; annual . . . . *Navarretia*

1a   With 4 corolla lobes, calyx lobes, and stamens (corolla cream or very pale yellow; in moist areas) . . . . *Navarretia cotulifolia [N. cotulaefolia]*
Cotula Navarretia; Sn, Me-SB; 4

1b   With 5 corolla lobes, calyx lobes, and stamens

2a   Each bract beneath the inflorescence usually more than 6 mm wide, not including the lobes (each bract glandular, with 1–3 lobes at the tip) . . . . *Navarretia atractyloides*
Hollyleaf Navarretia

2b   Each bract not more than 4 mm wide, not including the lobes, except in *Navarretia heterodoxa*

3a   Hairs on main stem turned downward, often pressed against the stem (in moist areas, including drying vernal pools)

4a   Stigma deeply divided into 2 or 3 lobes

5a   Stigma 2 lobed; corolla lobes pale blue or white, with 1 prominent vein continuing to the corolla base . . . . *Navarretia intertexta* (fig.)
Needleleaf Navarretia

5b   Stigma 3 lobed; corolla lobes pale blue, with 3 prominent veins continuing to the corolla base . . . . *Navarretia tagetina*
Marigold Navarretia; Na, Sn-n

4b   Stigma rather indistinctly divided into 2 lobes (corolla white, sometimes very compact; mostly growing in drying mud of ponds and lakes)

  6a   Vein on each corolla lobe not branched; corolla 5–7 mm long. . . . .
. . . . *Navarretia leucocephala* ssp. *bakeri* [*N. bakeri*]
Baker Navarretia; Sn-n; 1b

  6b   Vein on each corolla lobe branched; corolla 7–9 mm long . . . .
. . . . *Navarretia leucocephala* ssp. *leucocephala*
Whiteflower Navarretia; SB-n

3b   Hairs on main stem not turned downward

  7a   Corolla yellow (in drying vernal pools) . . . .
. . . . *Navarretia nigelliformis* [*N. nigellaeformis*]
Adobe Navarretia; CC-SLO

  7b   Corolla white, blue, or purple

    8a   Corolla purple, with a dark spot at the base of each lobe . . . .
. . . . *Navarretia jepsonii*
Jepson Navarretia; Na; 4

    8b   Corolla without a dark spot at the base of each lobe

      9a   Corolla tube 1 cm long or longer, the lobes 3 mm long (corolla lobes blue)

        10a   Corolla lobes with 3 dark veins . . . . *Navarretia pubescens* (pl. 42)
Downy Navarretia; SLO-n

        10b   Corolla lobes without dark veins

          11a   Corolla tube up to 12 mm long; primary lobes of leaves irregularly spaced, often divided further; corolla blue, usually pale (plant with a strong skunklike odor; widespread)
. . . . *Navarretia squarrosa*
Skunkweed; Mo-n

          11b   Corolla tube 12–16 mm long; lobes of leaves nearly regularly spaced, not divided further; corolla blue or purple
. . . . *Navarretia viscidula*
Sticky Navarretia; SM-n

      9b   Corolla tube less than 1 cm long, the lobes less than 2 mm long

        12a   Most bracts below the inflorescence with only 1 or 2 lobes on each side (corolla 4–5 mm long, the lobes, pink, purple, or white; at elevations above 1,000 ft) . . . . . . *Navarretia divaricata*
Mountain Navarretia; Na

        12b   Most bracts with several lobes on each side

          13a   Corolla lobes blue, the throat lighter, with purple veins; lobes of each bract evenly distributed along the margins; bract, not including lobes, widest below the middle (corolla 5–7 mm long) . . . . *Navarretia mellita*
Honey-scented Navarretia; SLO-n

13b Corolla uniform in color; lobes of each bract mostly in the lower half; bract, not including lobes, widest above the lobes

    14a Corolla 6–11 mm long, purple, the stamens protruding (on dry rocky slopes) ..... *Navarretia heterodoxa*
    Calistoga Navarretia; Na, Sn-SCl

    14b Corolla 6–7 mm long, lavender to white, the stamens not protruding (on serpentine) ....
    ..... *Navarretia rosulata [N. heterodoxa* ssp. *rosulata]*
    Marin County Navarretia; Ma; 1b

Polemoniaceae, Subkey 2: At least some leaves whorled or opposite, with slender palmate lobes; calyx lobes of equal or nearly equal length; flowers without spine-tipped bracts .... *Linanthus*

1a Corolla tube much longer than the calyx, usually at least twice as long

  2a Corolla lobes 5–10 mm long, usually at least one-third as long as the corolla tube (up to 30 cm tall)

    3a Calyx lobes not hairy on the back, but hairy on the margins; corolla tube 15–28 mm long, pink or lavender, the throat violet at the base; corolla lobes 5–10 mm long, pink, without red spots .... *Linanthus androsaceus*
    Pink-lobed Linanthus; Mo-n

    3b Calyx lobes hairy on the back as well as on the margins; corolla tube 20–35 mm long, maroon, pink, or yellow, the throat yellow at the base; each corolla lobe 5–6 mm long, white, yellow, pink, blue, or purple, often with a red spot at the base (widespread).................... *Linanthus parviflorus* [includes *L. androsaceus* sspp. *croceus* and *luteus*] (pl. 42)
    Common Linanthus; Me-s

  2b Corolla lobes 2–6 mm long, not often more than one-fifth as long as the corolla tube (corolla tube 10–25 mm long)

    4a Corolla lobes yellow; corolla tube yellow; transparent membrane between each calyx lobe scarcely evident or none (up to 15 cm tall) .... *Linanthus acicularis*
    Bristly Linanthus; Al-n; 4

    4b Corolla lobes red, pink, or white; corolla tube white, pink, red, or yellow

      5a Corolla lobes pink or white, usually with a red spot at the base of each lobe; transparent membrane between each calyx lobe almost as wide as the lobe; up to 30 cm tall...................................... *Linanthus ciliatus*
      Whiskerbrush

      5b Corolla lobes red, pink, or white, usually without red spots; transparent membrane scarcely evident; up to 16 cm tall (forming dense colonies; widespread) .... *Linanthus bicolor* (pl. 42)
      Bi-colored Linanthus; SLO-n

1b Corolla tube less than twice as long as the calyx, sometimes not longer than the calyx

  6a Flowers nearly sessile above leaflike bracts, the pedicel less than 5 mm long (corolla throat yellow)

7a   Flower solitary; calyx not hairy; corolla lobes white, tinged with purple (south of San Francisco, the flower usually opens in the evening; north of the Bay, the flower usually opens in the daytime) . . . . *Linanthus dichotomus* [includes *L. dichotomus* ssp. *meridianus*]

Evening-snow; Na

7b   Flowers usually in heads of at least 5 or 6; calyx hairy; corolla lobes white, pale pink, or lilac . . . . *Linanthus grandiflorus* (pl. 42)

Largeflower Linanthus; Sn-s; 4

6b   Flower with a pedicel at least 5 mm long, usually much longer

8a   Corolla without a purple throat, the lobes longer than the corolla tube; calyx lobes united by a transparent membrane for about half their length (corolla 1–2 cm long, white to pale pink or lilac, the tube not extending above the calyx lobes; often more than 40 cm tall; in sandy soil) . . . . *Linanthus liniflorus*

Flaxflower Linanthus; CC-SLO

8b   Corolla usually with a purple throat, the lobes shorter than the corolla tube; calyx lobes united for at least two-thirds their length

9a   Corolla sometimes more than 15 mm long, the lobes usually 5 mm long, pink or blue with yellow at their bases (on serpentine) . . . . . . *Linanthus ambiguus*

Serpentine Linanthus; SM, SCl, SB

9b   Corolla about 10 mm long, the lobes rarely more than 4 mm long, white, pale pink, lilac, or violet . . . . *Linanthus bolanderi* [includes *L. bakeri*]

Bolander Linanthus; CC-Me

Polemoniaceae, Subkey 3:   Leaves all alternate, at least some 1–2 times pinnately lobed, but sometimes with only 1 or 2 lobes on each side; calyx lobes of equal length; flowers without spine-tipped bracts

1a   Corolla tube at least 3 times as long as the calyx

2a   Lower leaves with several lobes on each side, these symmetrically arranged and about the same size, and sometimes subdivided further (corolla lobes pinkish violet, the tube and throat purple) . . . . *Gilia tenuiflora*

Slenderflower Gilia; SCl-s

2b   Lower leaves with lobes usually of unequal size, not often symmetrically arranged, and not subdivided further (annual)

3a   Corolla lobes pink or violet pink, the tube red violet or purple; leaves rarely with more than 1 lobe on each side, these usually much smaller than the end lobe . . . . *Allophyllum divaricatum*

Straggling Gilia; Mo-n

3b   Corolla lobes violet blue, the tube the same color; leaves often with more than 3 lobes on each side, the longest lobes sometimes as long as, or longer than, the end one . . . . *Allophyllum gilioides*

Purple Gilia, Straggling Gilia

1b   Corolla tube not more than twice as long as the calyx

4a   Flowers 50–100 in each dense globular head
> 5a   Corolla 5–8 mm long, the lobes up to 2 mm wide
>> 6a   Inflorescence not hairy or only slightly hairy (widespread) . . . .
>> . . . . *Gilia capitata* ssp. *capitata* (pl. 42)
>> Globe Gilia, Blue Field Gilia; Ma-n
>>
>> 6b   Inflorescence densely hairy at the base . . . . *Gilia capitata* ssp. *tomentosa*
>> Hairy Globe Gilia; Ma, Sn, CC (MD), SCl (MH); 1b
>
> 5b   Corolla 7–13 mm long, the lobes 1–4 mm wide
>> 7a   Corolla dark blue violet; plant without a cluster of leaves at the base . . . .
>> . . . . *Gilia capitata* ssp. *staminea*
>> Pale Gilia; Sn, CC-SB
>>
>> 7b   Corolla light blue violet; plant with a cluster of leaves at the base (coastal)
>> . . . . *Gilia capitata* ssp. *chamissonis*
>> Dune Gilia; SF-Ma; 1b

4b   Flowers up to 50, but usually many fewer, in each inflorescence, this not a dense globular head
> 8a   Deepest portion of each cleft between calyx lobes distended outward, forming a slight lip; inflorescence with a few broad leaves just below it (corolla uniformly rose, pink, or white; leaf lobes with toothed margins; calyx lobes narrowing to slender tips; annual) . . . . *Collomia heterophylla* (pl. 41)
> Variedleaf Collomia; Mo-n
>
> 8b   Deepest portion of each cleft between calyx lobes not distended outward; inflorescence without broad leaves just below it
>> 9a   Corolla uniformly blue, violet blue, or white, without any yellow or orange and without purple spots in the throat
>>> 10a   Flowers 8–25 in each inflorescence; pedicel up to 2 mm long (in sandy soil; widespread) . . . .
>>> . . . . *Gilia achilleifolia* ssp. *achilleifolia* [*G. achilleaefolia*] (pl. 42)
>>> California Gilia; Ma, CC-s
>>>
>>> 10b   Flowers 2–7 in each inflorescence; pedicel 1–30 mm long . . . .
>>> . . . . . *Gilia achilleifolia* ssp. *multicaulis* [*G. achilleaefolia* ssp. *multicaulis*]
>>> Small California Gilia, Many-stemmed Gilia; Ma, CC-s
>>
>> 9b   Corolla not the same color throughout, the tube yellow or orange, the lobes blue or white, and with purple spots in the throat
>>> 11a   Corolla 10–19 mm long, the lobes 4–8 mm wide (plant not glandular)
>>> . . . . *Gilia tricolor* (pl. 42)
>>> Bird's-eyes, Tricolor Gilia
>>>
>>> 11b   Corolla 6–11 mm long, the lobes less than 4 mm wide
>>>> 12a   Corolla 8–11 mm long; plant glandular and with a skunklike odor . .
>>>> . . . .*Gilia millefoliata*
>>>> San Francisco Gilia; SFBR-n; 1b
>>>>
>>>> 12b   Corolla 6–8 mm long; plant not glandular, except perhaps in the inflorescence, usually without a skunklike odor . . . .*Gilia clivorum*
>>>> Grassland Gilia; Sl-s

POLYGALACEAE (MILKWORT FAMILY) The flower of a milkwort is irregular and superficially resembles that of Fabaceae (Pea Family), but it is really very different. Two of the five sepals—those at the sides of a flower—are larger than the others and are usually colored like petals. In the one genus represented in California, *Polygala,* there are three united petals. The lower one, somewhat keel shaped, encloses the six to eight stamens. The stigma is two lobed. A further specialization is that the filaments of the stamens are joined to one another to form an incomplete tube, and this in turn is attached to the base of the side petals. The ovary is superior; the pistil is partitioned into two halves, each producing a single seed.

Our only species, *Polygala californica* (pl. 43) (California Milkwort), is a bushy, slender-stemmed perennial, often somewhat woody at the base. It reaches a height of about 35 cm and has oblong or elliptical, short-petioled leaves 1–6 cm long. The flowers, in short racemes, are pink or rarely white. It is found in the outer Coast Ranges from San Luis Obispo County to southern Oregon. The name *Polygala,* meaning "much milk," is believed to be based on an old idea that some plants of this group, if eaten by cows, stimulate the production of milk. The name "milkwort" is also tied to this notion.

POLYGONACEAE (BUCKWHEAT FAMILY) The Buckwheat Family is well represented in California, especially in dry areas. Cultivated exotic species include *Fagopyrum,* the buckwheat of pancake fame and *Rheum* (rhubarb). Our species are mostly herbs, but a couple are shrubs. Although the inflorescences are often conspicuous, each flower is usually small. In some genera, such as *Eriogonum,* the flowers are grouped in an involucre consisting of fused bracts, which form a cup. Other genera, such as *Polygonum,* are characterized by membranous stipules that are fused around the stem nodes. The flower has no petals, but the five or six sepals, usually at least partly united and sometimes in two whorls, may resemble petals and are sometimes richly colored. The number of stamens ranges from three to nine, and there are generally three styles fused at their bases. The ovary is superior. The fruit, dry at maturity, is an achene.

Numerous species of this family contribute to colorful displays of wildflowers, especially in dry, well-drained habitats. Some of the eriogonums are excellent subjects for gardens. Two genera, *Rumex* and *Polygonum,* include weeds whose rating as nuisances ranges from merely objectionable to pernicious.

1a   Leaves without stipules

     2a   Individual flowers, or groups of flowers, within a cuplike, several-lobed involucre (involucres sometimes in clusters; leaves mostly basal, those on stem below flowers usually opposite or whorled) . . . . POLYGONACEAE, SUBKEY 1

                                                           Chorizanthe, Eriogonum

     2b   Flowers not originating within a cuplike involucre (leaves up to 2 cm long)

         3a   Leaves not basal, those on the stem opposite, the blades fan shaped, with a notch at the tips; each flower with a nonleaflike bract; calyx about 1 mm long, usually 6 lobed, the lobes separate nearly to their bases; growing in the shade of shrubs (calyx cream colored, pink, or rose) . . . . *Pterostegia drymarioides*

         3b   Leaves mostly basal, those on the stem in whorls of 4 or 5, slender, joined together basally and ending in hooked tips; each flower with a whorl of leaves beneath it; calyx about 4 mm long, usually 5 lobed, the lobes separate to about the middle; in sandy, exposed habitats . . . . . *Lastarriaea coriacea [Chorizanthe coriacea];* CC-Mo

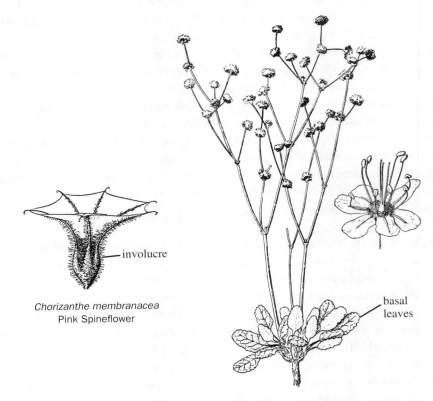

involucre

Chorizanthe membranacea
Pink Spineflower

Eriogonum nudum var. *nudum*
Nakedstem Buckwheat

basal
leaves

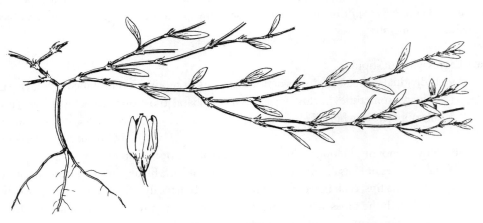

Polygonum arenastrum
Common Knotweed

Polygonum hydropiper
Marshpepper

Rumex conglomeratus
Green Dock

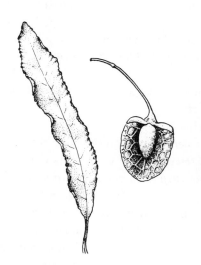

Rumex crispus
Curly Dock

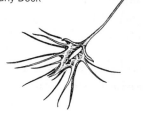

Rumex maritimus
Golden Dock

Rumex obtusifolius
Bitter Dock

1b   Leaves with stipules, these united and forming sheaths that encircle the stem (nodes often swollen)

4a   Calyx 5 lobed, usually not green; leaves usually not mainly basal . . . . POLYGONACEAE, SUBKEY 2
Polygonum

4b   Calyx 6 lobed, often green; leaves often mainly basal

5a   Leaf blades very sour when tasted, some with a pair of basal lobes; pistillate and staminate flowers on separate plants; upright stems arising from underground rhizomes; mostly less than 30 cm tall (widespread) . . . . *Rumex acetosella [R. angiocarpus]* (pl. 44)
Sheep Sorrel; eu

5b   Leaf blades not especially sour when tasted, without basal lobes; most flowers with a pistil and stamens; upright stems arising from a basal rootstock; more than 30 cm tall, sometimes smaller in *Rumex maritimus*

6a   Most basal leaves at least 15 cm long, the margins sometimes wavy

7a   None of the 3 inner calyx lobes with a swelling (in alkaline areas) . . . . *Rumex occidentalis* [includes *R. fenestratus*]
Western Dock, Marsh Dock; SFBR-n

7b   At least 1 of the 3 inner calyx lobes with a swelling

8a   All of the 3 inner calyx lobes with a conspicuous, raised swelling (widespread) . . . . *Rumex crispus* (pl. 44; fig.)
Curly Dock; eua

8b   Only 1 of the 3 inner calyx lobes with a swelling . . . . *Rumex obtusifolius* [includes *R. obtusifolius* ssp. *agrestis*] (fig.)
Bitter Dock; eu

6b   Basal leaves usually less than 15 cm long, the margins not wavy, except sometimes slightly so in *Rumex conglomeratus*

9a   Inner 3 calyx lobes with toothed margins

10a   Inner 3 calyx lobes with 1–3 slender teeth on each side, these teeth longer than the width of the rest of the lobe; annual (in wet areas) . . . . *Rumex maritimus* [includes *R. fueginus* and *R. persicarioides*] (fig.)
Golden Dock; SM, Ma-n

10b   Inner 3 calyx lobes usually with more than 3 teeth on each side, these shorter than the width of the rest of the lobe; perennial . . . . *Rumex pulcher*
Fiddle Dock; me

9b   Inner 3 calyx lobes with smooth margins (in moist areas)

11a   Each leaf blade usually more than 5 times as long as wide, tapering gradually to a petiole . . . . *Rumex salicifolius* (pl. 44)
Willow Dock; Ma-s

11b   Each leaf blade usually less than 5 times as long as wide, tapering abruptly to the base, the base therefore nearly rounded . . . . *Rumex conglomeratus* (fig.)
Green Dock, Clustered Dock; eu

Polygonaceae, Subkey 1: Leaves without stipules, mostly basal, those on stem below flowers opposite or whorled; individual flowers, or groups of flowers, within a cuplike, several-lobed involucre; involucres sometimes in clusters; leaves mostly basal, those on the stem usually opposite or whorled .... *Chorizanthe, Eriogonum*

1a   Each involucre with 1 flower, rarely 2; midrib of each lobe of involucre prolonged as a spine (involucre 6 lobed; calyx white to rose; in sandy or rocky soil)

   2a   Leaves on stem alternate; involucre not hairy in upper half, except for the midribs of the lobes (involucre lobes more or less equal, the spines hooked; calyx densely hairy; stamens 9; leaf blades 1–5 cm long; up to 1 m tall) ....
.... *Chorizanthe membranacea* (pl. 43; fig.)
Pink Spineflower; Me-s

   2b   Leaves all basal, except for a few opposite leaves beneath the flower clusters; involucre not hairy in the upper third or less

      3a   One spine of the involucre much longer than the others; stamens 3 (involucre lobes unequal, the spines hooked; calyx sparsely hairy, 2–3 mm long; leaf blades up to 2 cm long; prostrate, with stem tip rising) .... *Chorizanthe clevelandii*
Cleveland Spineflower

      3b   One spine of the involucre not obviously much longer than the others; stamens 9 (involucre lobes may be unequal)

         4a   Spines of involucre straight; calyx 4–6 mm long; erect, up to 30 cm tall (involucre lobes more or less equal; leaf blades 1–5 cm long) ....
.... *Chorizanthe valida*
Sonoma Spineflower; Sn, Ma; 1b

         4b   Spines of involucre usually hooked, sometimes straight in *Chorizanthe cuspidata;* calyx not more than 4 mm long; usually not erect

            5a   Leaf blades 3–10 mm long; 3 involucre lobes much longer than other 3; calyx densely hairy; prostrate .... *Chorizanthe polygonoides*
Knotweed Spineflower; Ma

            5b   Usually some leaf blades more than 10 mm long; involucre lobes more or less equal or 1 or 2 smaller; calyx hairy, but usually not densely so; erect or prostrate with stem tip rising

               6a   Calyx 2–3 mm long, each lobe with a bristle at the tip; involucre hairy throughout; prostrate, with stem tip rising ....
.... *Chorizanthe cuspidata* [includes
C. cuspidata vars. *marginata* and *villosa*]
San Francisco Spineflower; Sn-SCr; 1b

               6b   Calyx 3–4 mm long, each lobe without a bristle, but sometimes with a tooth; involucre not hairy in upper third; upright, or prostrate with stem tip rising ....
.... *Chorizanthe robusta* [includes *C. pungens* var. *hartwegii*]
Robust Spineflower; SF, Al-Mo; 1b

1b   Each involucre enclosing a group of several to many flowers; midrib of each lobe of the involucre not prolonged as a spine

7a   Leaves along the stem as well as in a basal whorl, the upper leaves sometimes much reduced

    8a   Plant not woody (each peduncle with a single involucre; calyx white to rose; annual)

        9a   Leaves very slender; petioles short or absent . . . . *Eriogonum angulosum*

                  Anglestem Buckwheat; CC-s

        9b   Leaves elliptic; petioles as long as the blades (on clay and serpentine soils) . .
. . . . *Eriogonum argillosum*

                  Clay-loving Buckwheat; SCl-Mo; 4

    8b   Plant woody, at least at the base (leaves either sessile or with short petioles)

        10a   Leaves 15–30 mm long and 5–10 mm wide; clusters of involucres scattered
along the branches (calyx white with green or rose veins) . . . .
. . . . *Eriogonum wrightii* var. *trachygonum*

                  Wright Buckwheat

        10b   Leaves 6–15 mm long and 2–6 mm wide; clusters of involucres at the ends of
branches (flowers usually in umbels)

            11a   Leaf blades narrow, 2–3 mm wide, sessile; calyx white or pink; plant
woody throughout . . . . *Eriogonum fasciculatum* (pl. 43)

                  California Buckwheat; SLO-n

            11b   Leaf blades elliptic, 5–6 mm wide, with petioles; calyx yellow, later red;
plant woody only at the base. . . . .
. . . . *Eriogonum umbellatum* var. *bahiiforme* (pl. 43)

                  Sulfurflower Buckwheat; La-SB; 4

7b   Leaves mainly confined to a basal whorl, sometimes a few on the lower portion of the stem

    12a   Involucres scattered all along the length of the stem

        13a   Involucre less than 2 mm long; calyx white or pink (leaf blades up to 2 cm
long). . . . . *Eriogonum elegans*

                  Elegant Buckwheat; SCl-SLO

        13b   Involucre 2–5 mm long; calyx white, pale yellow, rose, or red

            14a   Stem and involucre densely hairy; leaf blades oblong (leaf blades up to
3 cm long) . . . . *Eriogonum roseum*

                  Rose Buckwheat, Virgate Buckwheat

            14b   Stem and involucre scarcely if at all hairy; leaf blades nearly circular

                15a   Outer and inner lobes of the calyx about the same size and shape, all
lobes slightly hairy near their bases (leaf blades up to 1.5 cm long;
uncommon) . . . . *Eriogonum covilleanum*

                  Coville Buckwheat; Al-SB

                15b   Outer lobes of the calyx distinctly longer and broader than the
inner lobes, none of the lobes hairy (following 2 species difficult to
separate)

                  16a   At least some branches of the inflorescence in whorls of 3 or
more; leaf blades up to 2 cm long . . . .
. . . . *Eriogonum vimineum* (pl. 43)

                  Wicker Buckwheat; Mo-n

16b Branches of the inflorescence rarely in whorls; leaf blades up to 5 cm long (sometimes on serpentine) ....
.... *Eriogonum luteolum* var. *luteolum*
Greene Buckwheat; SFBR-n

12b Involucres concentrated in clusters, mainly at tips of stems or branches, but occasionally also scattered along the upper portions of the stems

17a Involucre hairy

18a Clusters of involucres 1.5–3 cm wide; plant woody at the base (leaves not strictly basal, the blades 1.5–5 cm long) .... *Eriogonum latifolium* (pl. 43)
Coast Buckwheat; SLO-n

18b Clusters of involucres not more than 1.5 cm wide; plant not woody

19a Clusters of involucres less than 5 mm wide; annual; up to 30 cm tall (leaves basal, the blades 2–5 cm long) .... *Eriogonum truncatum*
Mount Diablo Buckwheat; CC (MD); 1a (1940)

19b Clusters of involucres 10–15 mm wide; perennial; up to 100 cm tall

20a Leaves basal, the blades 2–4 cm long, the margins mostly flat ..
.... *Eriogonum nudum* var. *oblongifolium*
Hairy Buckwheat; Na-n

20b Leaves on lower stem, the blades 1–3 cm long, the margins usually wavy .... *Eriogonum nudum* var. *decurrens*
Ben Lomond Buckwheat; CC (MD)–SCr; 1b

17b Involucre not hairy

21a Clusters of involucres up to 6 mm wide; leaves less than 3.5 cm long; not more than 40 cm tall (on shale and serpentine)

22a Leaves up to 3.5 cm long, basal and some usually present at lower stem nodes; calyx rose red ....
.... *Eriogonum luteolum* var. *caninum* [*E. caninum*] (pl. 43)
Tiburon Buckwheat; Ma, Al, CC

22b Leaves up to 2 cm long, strictly basal; calyx white to rose ....
.... *Eriogonum covilleanum*
Coville Buckwheat; Al-SB

21b Clusters of involucres 10–15 mm wide; leaves 1–7 cm long; up to 100 cm tall

23a Leaf blades 3–7 cm long, wavy (calyx white or pink, sometimes yellow, usually not hairy; outer Coast Ranges and near the coast) .....
.... *Eriogonum nudum* var. *auriculatum*
Curledleaf Buckwheat; Sn-Mo

23b Leaf blades 1–5 cm long, usually not wavy

24a Calyx hairy, yellow or white (inner Coast Ranges) ....
.... *Eriogonum nudum* var. *pubiflorum*
Hairyflower Buckwheat

24b Calyx not hairy, mostly white to pink, rarely yellow (Coast Ranges; widespread) ....
.... *Eriogonum nudum* var. *nudum* (pl. 43; fig.)
Nakedstem Buckwheat, Tibinagua; SFBR-n

Polygonaceae, Subkey 2:   Leaves with stipules, these united and forming sheaths that encircle the stem; nodes often swollen; calyx 5 lobed, usually not green; leaves usually not mainly basal . . . . *Polygonum*

1a   Leaf blades triangular or heart shaped (leaf blades 2–6 cm long; annual; stem trailing or clambering over low vegetation; resembling a morning-glory, except for the flower). . . . . .
. . . . *Polygonum convolvulus*
Black Bindweed; eu

1b   Leaf blades not triangular or heart shaped

   2a   Flowers in crowded inflorescences at stem ends, these leafless (in wet places, and sometimes partly submerged)

      3a   Some leaves with petioles at least 1.5 cm long, other leaves on the plant nearly sessile; stem usually with a single raceme, sometimes 2

         4a   Petioles of lower leaves longer than the blades, often more than two-thirds as long as the blades; upper leaves sessile, clasping the stem; in marshes (inflorescence 1–6 cm long; calyx white or pink). . . . . *Polygonum bistortoides*
Western Bistort; Ma-n

         4b   Petioles of lower leaves much shorter than the blades; upper leaves with at least short petioles; in shallow water

            5a   Inflorescence not often more than 3 cm long; peduncle not hairy . . . . . .
. . . . *Polygonum amphibium* var. *stipulaceum*
Shore Knotweed, American Water Persicaria

            5b   Inflorescence usually 3–10 cm long; peduncle glandular hairy. . . . .
. . . . *Polygonum amphibium* var. *emersum* [*P. coccineum*] (pl. 44)
Swamp Knotweed, Kelp

      3b   Leaves either sessile or with petioles less than 1.5 cm long; stem usually with two to several racemes

         6a   Calyx with glandular dots

            7a   Inflorescence nodding at the tip; calyx up to 4 mm long, green with rose or white tips; in wet places; up to 60 cm tall . . . .
. . . . *Polygonum hydropiper* (fig.)
Marshpepper, Waterpepper; eu

            7b   Inflorescence not nodding at the tip; calyx up to 3 mm long, brown with white tips; in shallow water; up to 100 cm tall. . . . *Polygonum punctatum*
Water Smartweed

         6b   Calyx without glandular dots (in moist or wet areas; often more than 80 cm tall)

            8a   Leaves 3–10 cm long, mostly sessile, often with a darker blotch; inflorescence up to 2.5 cm long (calyx pink or purple). . . . .
. . . . *Polygonum persicaria* (pl. 44)
Lady's-thumb; eu

            8b   Leaves 5–20 cm long, mostly with short petioles, without a darker blotch; inflorescence up to 7 cm long

9a   Racemes usually more than 7 at the end of the main stem, this often not branched; raceme very dense, the flowers mostly touching one another; petioles 1–1.5 cm long; annual . . . . *Polygonum lapathifolium*
Willow-weed

9b   Racemes usually fewer than 7 at the end of the main stem or its branches; raceme rather loose, the flowers usually not touching one another; petioles usually less than 1 cm long; perennial . . . .
. . . . *Polygonum hydropiperoides* [includes
P. hydropiperoides var. *asperifolium*]
Waterpepper

2b   Flowers solitary or in small groups in the leaf axils (in *Polygonum paronychia*, the flowers are at stem ends, but they are nevertheless in the leaf axils)

10a   Flowers concentrated at the tip of the stem; midribs of leaf blades raised prominently on the undersides; brownish remains of stipules obvious on the stem; on backshores of sandy beaches (woody at the base; often forming low mats). . . . . . .
. . . . *Polygonum paronychia* (pl. 44)
Beach Knotweed; Mo-n

10b   Flowers scattered along the stem; midribs of leaf blades not raised prominently on the undersides; remains of stipules not obvious; not primarily near sandy beaches, but sometimes around salt marshes

11a   Flowers sessile and usually solitary in the leaf axils (in dry areas)

12a   Flowers rather crowded all along the branches of the inflorescence; leaves usually about 1 cm long; annual; up to 20 cm tall. . . . .
. . . . *Polygonum californicum*
California Knotweed

12b   Flowers not crowded on the branches of the inflorescence, except at the tips; leaves often less than 1 cm long; shrub; up to 60 cm tall . . . .
. . . . *Polygonum bolanderi*
Bolander Knotweed; Na-n

11b   Flowers on short pedicels and in small clusters in the leaf axils

13a   Fruit protruding beyond the calyx lobes (at the edges of salt marshes)
. . . . *Polygonum marinense*
Marin Knotweed

13b   Fruit not protruding beyond the calyx lobes

14a   Calyx 2.5–3 mm long (flowers not hidden among the leaves)

15a   Leaves near the flower head less than 5 mm long; mostly in dry areas, including those with serpentine soils, but sometimes on backshores of sandy beaches . . . .
. . . .*Polygonum douglasii* ssp. *spergulariiforme*
[*P. spergulariaeforme*]
Fall Knotweed

15b   Leaves near the flower head more than 10 mm long; in dry areas . . . . *Polygonum ramosissimum*
Yellowflower Knotweed; na?

14b Calyx 1.5–2 mm long

16a Calyx not hidden among the leaves; upright ....
.... *Polygonum argyrocoleon*

Silversheath Knotweed; as

16b Calyx nearly hidden among the leaves; prostrate ....
.... *Polygonum arenastrum* (fig.)

Common Knotweed, Doorweed; eu

PORTULACACEAE (PURSLANE FAMILY)   The flowers of most purslanes are distinctive in having only two sepals. There are three to many stamens, and either the style or stigma has two to eight lobes. The number of petals varies from 3 to 18, but in *Claytonia, Calandrinia,* and *Montia,* the three genera that account for more than half of our species, there are typically five. The ovary is superior, except in *Portulaca,* and the fruit, dry when mature, encloses several to many seeds. The leaves are usually smooth margined and often somewhat succulent.

Gardeners are familiar with *Portulaca grandiflora,* native to Brazil, an annual grown for the vivid flowers it produces in summer. They also know *Portulaca oleracea,* the Common Purslane, a slightly succulent, mat-forming nuisance that survives being stepped on and that has a knack for colonizing places where it is not wanted. No one, however, can resist the charm of some of the native species, especially the richly colored calandrinias, which grow mostly in sunny grassland habitats, and *Lewisia rediviva.* The latter, called Bitterroot, is the state flower of Montana, but we are privileged to have it near the top of Mount Diablo and in some other rocky places in the inner Coast Ranges.

1a   Petals yellow; ovary half-inferior (petals 4–6; sepals 2, united; stamens 5–20; style with 4–6 lobes; flower opening only in sunshine; leaves up to 3 cm long, whorled at stem tips, opposite or alternate below; annual; stem prostrate; widespread) ...... *Portulaca oleracea* (fig.)

Common Purslane; eu

1b   Petals white, pink, or red; ovary superior

2a   Sepals 6–8 (leaves up to 5 cm long, numerous, basal, narrow; flower solitary on stems, these rarely more than 3 cm long; petals 10–19, 12–25 mm long, pink or white; stamens numerous; style with 6–8 lobes; perennial; in rock crevices or rocky soil) ....
.... *Lewisia rediviva* (pl. 45)

Bitterroot; Mo-n

2b   Sepals 2

3a   Petals 3 or 4; stamens 1–3; sepals sometimes very unequal; stigmas 2 (leaves alternate, at least some of them basal)

4a   Inflorescence a dense umbel; leaves all basal; petals, as they age, twisted; fruit nearly round; perennial (stamens 3; leaves up to 7 cm long; flowering stem prostrate, reddish; petals 4; sepals notched, white or pink, often with a green center)..................................... *Calyptridium umbellatum*

Common Pussypaws; SCl-n

4b   Inflorescence elongated, slightly curved; leaves not all basal; petals, as they age, folding into a cup over the developing fruit; fruit egg shaped to elongated; annual

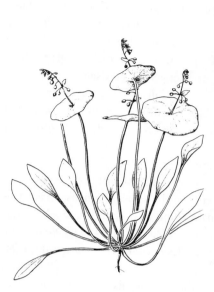

Claytonia perfoliata
Miner's-lettuce

Claytonia sibirica
Candyflower

Montia linearis
Linearleaf Montia

Montia parvifolia
Springbeauty

Portulaca oleracea
Common Purslane

5a   Leaves up to 6 cm long; bases of sepals with a distinct notch; petals 4, white or pink; stamens 1–3; fruit egg shaped, broadest near the base, shorter than the sepals.................. *Calyptridium quadripetalum*

                                              Four-petaled Pussypaws; Na, Sn; 4

5b   Leaves up to 3 cm long; bases of sepals without a notch; petals 3, white; stamens 3; fruit somewhat elongated, longer than the sepals ....
.... *Calyptridium parryi* var. *hesseae*

                                Santa Cruz Mountains Pussypaws; SCl

3b Petals usually 5; stamens 3–15; sepals more or less equal; style 3 lobed or with 3 stigmas

  6a   Stem leaves alternate (stem leaves more than 2)

    7a   Petals red; stamens 3–15 (leaves 1–10 cm long; annual)

      8a   Petals 4–15 mm long, longer than the sepals; stamens 3–15; leaves scattered along the stem; fruit about as long as the sepals (widespread) .... *Calandrinia ciliata*

                        [includes *C. ciliata* var. *menziesii*] (pl. 45)

                                       Redmaids

      8b   Petals 3–5 mm long, shorter than the sepals; stamens 3–6; leaves mostly basal; fruit longer than the sepals ...... *Calandrinia breweri*

                        Brewer Calandrinia, Brewer Redmaids; Sn-s; 4

    7b   Petals white or pink; stamens 3 or 5

      9a   Petals 7–15 mm long; perennial (stamens 5; basal leaves up to 6 cm long; stem leaves much reduced, with small bulbs in axils; in moist, rocky habitats; plants often connected to one another) ....
.... *Montia parvifolia* (fig.)

                        Springbeauty, Small-leaf Montia; Mo-n

      9b   Petals less than 7 mm long, sometimes absent; annual

        10a Leaves basal and on the stem, with petioles, the blades about as long as wide; inflorescence usually with some leaves; stamens 5; petals white or pale pink, 3–5 mm long (in woods) ....
.... *Montia diffusa*

                             Diffuse Montia; Ma-n

        10b Leaves on the stem, sessile, many times as long as wide; inflorescence leafless; stamens 3; petals white, 4–7 mm long (in moist areas) .... *Montia linearis* (fig.)

                            Linearleaf Montia; CC-n

  6b Stem leaves opposite (leaves often joined together by 1 or both edges; petals white or pink; annual)

    11a Stem leaves more than 2

      12a Leaves with petioles, the basal ones 10–25 cm long, usually 3 times longer than the stem leaves; petals 6–12 mm long, pink; stamens 5; in moist woodlands ....*Claytonia sibirica* [*Montia sibirica*] (pl. 45; fig.)

                                Candyflower; SCr-n

12b Leaves sessile, up to 2 cm long, all about the same size, none basal; petals 1–2 mm long, white; stamens 3; in wet places, sometimes floating or left on mud after a drop in the water level ....
.... *Montia fontana* [includes *M. verna*]
Water Chickweed, Blinks; Mo-n

11b Stem leaves 2 (stamens 5)

13a Stem leaves completely united, forming a nearly circular disk (petals 2–6 mm long, white; basal leaves usually 5–15 cm long, but varying greatly; in moist areas; common) ....
...... *Claytonia perfoliata* [*Montia perfoliata* var. *perfoliata*] (fig.)
Miner's-lettuce

13b Stem leaves not completely united (often on serpentine)

14a Inflorescence with 3–15 flowers, the pedicels rarely so much as 3 mm apart on the stalk; petals 2–5 mm long, white or pink; up to 15 cm tall (widespread) .. *Claytonia exigua* [includes *Montia spathulata* vars. *exigua, spathulata, rosulata,* and *tenuifolia*]
Common Claytonia

14b Inflorescence with 3–30 flowers, the pedicels generally at least 5 mm apart on the stalk; petals 5–8 mm long, pinkish; up to 25 cm tall .... *Claytonia gypsophiloides* [includes *Montia gypsophiloides* and *M. perfoliata* var. *nubigena*] (pl. 45)
Santa Lucia Claytonia, Coast Claytonia; Me-SLO

PRIMULACEAE (PRIMROSE FAMILY)   Many members of the herbaceous Primrose Family have basal leaves, but a few have stem leaves, sometimes in what looks like a single whorl. Both the calyx and corolla have a cuplike or tubular lower portion, and generally four or five well-developed lobes. The four or five stamens are attached to the tube of the corolla. In *Trientalis* the number of calyx and corolla lobes varies from five to seven, even on the same plant. The ovary is superior, except in *Samolus.* The pistil develops into a many-seeded dry fruit. An interesting feature of the fruit of some genera is the way it opens by separating into two halves crosswise rather than splitting apart lengthwise.

This family includes the cultivated species of *Primula, Androsace, Soldanella,* and *Cyclamen,* introduced into horticulture from the Old World. The few species of *Primula* native to the mountains of the western states are next to impossible to sustain in gardens. Some species of *Dodecatheon* (shootingstars) can be grown from seed, provided that they have abundant moisture during winter and spring. *Trientalis latifolia* (Starflower) is easy to establish, but its aggressive propagation by underground stems may soon make it a weed. Most gardeners who want a representative of the family in the garden perhaps have one already. This is *Anagallis arvensis* (Scarlet Pimpernel), whose flowers open only when the sky is clear.

1a Some or all leaves in a whorl-like concentration just below the flowers

2a Leaves up to 9 cm long, not basal, all whorled on the stem; corolla about 15 mm wide, pink, with 5–7 lobes; perennial (in coniferous woods) ...... *Trientalis latifolia* (pl. 45)
Starflower; SLO-n

2b   Leaves up to 2 cm long, mostly basal, with 1 whorl on the stem below the flowers; corolla less than 5 mm wide, white, 5 lobed; annual (Coast Ranges, including Mt. Diablo). . . . . *Androsace elongata* ssp. *acuta*
California Androsace; 4

1b   Leaves either all basal, or both basal and scattered along the stem

  3a   Leaves all basal; flowers often in umbels, but sometimes 1–2 in leaf axils (leaves up to 18 cm long; corolla magenta to white, the lobes 6–25 mm long, turned back; perennial)

    4a   Anthers mostly pointed at the tips; tube formed by the filaments of the stamens without a yellow spot at the base of each anther; calyx and corolla lobes 4 or 5 (flowers 3–17 on each stem; in shaded areas; widespread) . . . . *Dodecatheon hendersonii* [includes *D. hendersonii* ssp. *cruciatum*] (pl. 45)
Mosquito-bills, Sailor-caps; SB-n

    4b   Anthers usually rounded or blunt at the tips; tube formed by the filaments of the stamens with a yellow or white spot near the base of each anther; calyx and corolla lobes 5

      5a   Anthers dark; flowers 1–6 on each stem (in moist areas, often on serpentine or alkaline soil) . . . . *Dodecatheon clevelandii* ssp. *patulum*
Padre's Shootingstar; CC, SF-SB

      5b   Anthers usually yellow; flowers 3–7 on each stem; in woods . . . . *Dodecatheon clevelandii* ssp. *sanctarum*
Coastal Shootingstar; CC, SF-s

  3b   Leaves not all basal; flowers in racemes or solitary in leaf axils (corolla less than 3 mm long)

    6a   Leaves alternate, but sometimes the lower ones of *Centunculus* opposite

      7a   Leaves 2–5 cm long, many of them basal; corolla white, in racemes at stem ends; calyx and corolla lobes usually 5; ovary half-inferior; growing along streams and in salt marshes. . . . . *Samolus parviflorus*
Water Pimpernel; CC, Sl

      7b   Leaves less than 1 cm long, none basal; corolla pink, solitary in leaf axils; calyx and corolla lobes usually 4; ovary superior; in moist ground, but not typically along streams or in salt marshes. *Centunculus minimus*
Chaffweed

    6b   Leaves opposite (leaves up to 2 cm long; flower solitary in leaf axils; calyx and corolla lobes 5)

      8a   Corolla usually pinkish orange, sometimes blue, 7–11 mm wide; stem mostly close to the ground (flower opening when the sky is clear; widespread) . . . . *Anagallis arvensis* (pl. 45)
Scarlet Pimpernel, Poor-man's Weatherglass; eu

      8b   Corolla absent, but calyx white or lavender, less than 5 mm wide; stem mostly upright (restricted to coastal salt marshes) . . . . *Glaux maritima*
Sea-milkwort; SLO-n

RANUNCULACEAE (BUTTERCUP FAMILY)   Plants of Ranunculaceae are usually herbaceous and have basal or alternate leaves. Vines of the genus *Clematis*, however, become woody and have opposite leaves. The family includes many widely cultivated garden plants, such as species of *Anemone, Aquilegia, Delphinium, Helleborus* (including Christmas-rose and Lentenrose), and *Ranunculus*. It should be mentioned that nearly all members of the Buttercup Family are poisonous.

The flowers of most species are characterized by numerous stamens. There are commonly several to many closely associated pistils that develop into single-seeded fruit or one or a few separate pistils that develop into many-seeded fruit. In one of our representatives, *Actaea rubra* (Baneberry), the single pistil becomes a fleshy fruit. In all, the ovary is superior.

The flower is usually regular and has five sepals and five petals, but there are many exceptions. When petals are absent, the sepals may be white or brightly colored, thereby resembling petals. Furthermore, the petals may have long, hollow spurs: this is the case in *Aquilegia* (columbines). In *Delphinium* and *Consolida* (larkspurs), the flowers are irregular and have only four petals; the upper two petals have spurs that fit into a spur on one of the sepals. In *Clematis* and *Thalictrum*, the flowers are usually staminate or pistillate, and the two types may be on separate plants.

Clearly this is a remarkably diverse assemblage. But after a little experience, one will generally be able to connect a previously unfamiliar plant with the Ranunculaceae. Some members of Rosaceae (Rose Family) also have numerous stamens and pistils, but their flowers have a floral tube and the sepals are not completely separate as they are in Ranunculaceae.

1a   Flower decidedly irregular, with 4 petals, the 2 upper ones with small spurs that fit into a larger spur on the uppermost sepal
    2a   Flower red or orange red (in moist, shaded areas) ..... *Delphinium nudicaule* (pl. 46)
                                                    Red Larkspur, Orange Larkspur; Mo-n
    2b   Flower not red or orange red
        3a   Leaves with numerous slender lobes, resembling parsley; annual (flower blue, purple, pink, or white) .... *Consolida ambigua*
                                              European Larkspur; eu
        3b   Leaves with a few primary lobes, these again lobed or toothed; perennial
            4a   Lower leaves with primary lobes not separated for more than half the length of the blade (upper petals white; plant scarcely hairy)
                5a   Blades of lower leaves nearly fan shaped, with 3 primary lobes, these usually toothed; sepals bright blue (in wet or dry places, sometimes on serpentine) .... *Delphinium uliginosum*
                                    Swamp Larkspur, Bog Larkspur; Na-n; 4
                5b   Blades of lower leaves not fan shaped, with 5 primary lobes, these toothed; sepals dark blue or purple (coastal) .... *Delphinium bakeri*
                                      Baker Larkspur; Sn, Ma; 1b

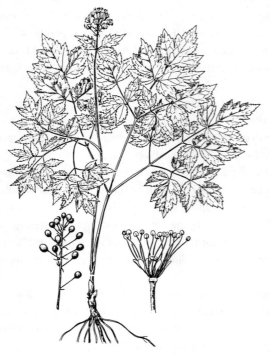

Actaea rubra
Baneberry

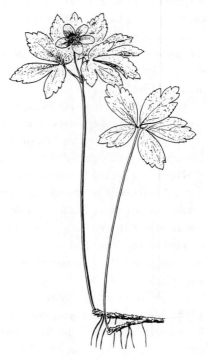

Anemone oregana
Western Wood Anemone

Myosurus minimus
Common Mousetail

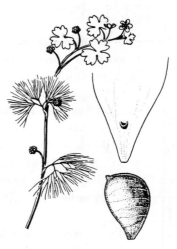

Ranunculus aquatilis var. *hispidulus*
Water Buttercup

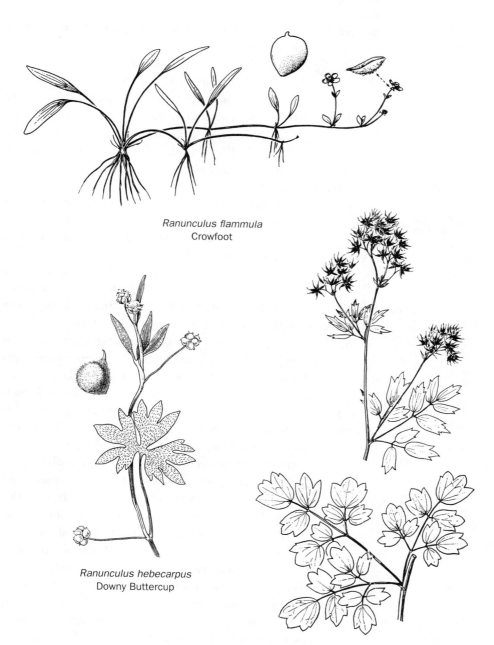

Ranunculus flammula
Crowfoot

Ranunculus hebecarpus
Downy Buttercup

Thalictrum fendleri var. *polycarpum*
Manyfruit Meadowrue

4b   Lower leaves with primary lobes usually separated nearly or fully to the base of each blade

      6a   Sepals green, white, pink, or yellow, but sometimes tinged with purple, or with purple tips

          7a   Leaf blades up to 15 cm wide; flower not opening fully; usually at least 1 m tall (in moist places)

              8a   Sepals green or white, tinged with purple; outer Coast Ranges .... *Delphinium californicum* ssp. *californicum*
California Larkspur, Coast Larkspur; SF-Mo

              8b   Sepals dirty yellow; inner Coast Ranges .... *Delphinium californicum* ssp. *interius*
Hospital Canyon Larkspur; CC-SCl; 1b

          7b   Leaf blades rarely more than 6 cm wide; flower opening fully; rarely so much as 1 m tall

              9a   Sepals and petals yellow, the sepals sometimes with purple tips (on coastal bluffs) .... *Delphinium luteum*
Yellow Larkspur, Golden Larkspur; Sn; 1b

              9b   Sepals and petals white or pink .... *Delphinium hesperium* ssp. *pallescens*
Pale Western Larkspur; SCl

      6b   Sepals blue or purple, but sometimes white or pink in *Delphinium patens*

          10a   Leaves not hairy, or at least not hairy on the upper surfaces

              11a   Leaves hairy on the undersides, somewhat succulent, with 3–15 lobes (sepals blue purple, the side ones 11–24 mm long; upper petals white; mostly on rocky hillsides and coastal bluffs) .... *Delphinium decorum*
Coast Larkspur; SCr-Mo

              11b   Leaves not hairy on either surface, not succulent, with 3–11 lobes

                12a   Sepals bright blue, white, or pink, the side ones 9–20 mm long; upper petals white, usually with blue lines; inland and Coast Ranges (spur curving upward) ... *Delphinium patens*
Spreading Larkspur, Coast Larkspur; La-s

                12b   Sepals light blue, the side ones 10–16 mm long; upper petals white or cream colored; mostly in saline grassland areas .... *Delphinium recurvatum*
Recurved Larkspur, Valley Larkspur; CC-s; 1b

          10b   Leaves hairy, at least on the upper surfaces

              13a   Side sepals 15–20 mm long; inflorescence usually with not more than 12 flowers (sepals deep purple or blue purple; inner lobes of lower petals hairy; in oak woodland) .... *Delphinium variegatum* (pl. 46)
Royal Larkspur; SLO-n

              13b   Side sepals not more than 16 mm long; inflorescence often with more than 12 flowers

14a Sepals dark or light blue; spur up to 15 mm long; upper petals white; veins on undersides of leaves not rust colored; usually in chaparral, sometimes in oak woodland .. .. *Delphinium parryi* [includes *D. parryi* ssp. *seditiosum*]
Parry Larkspur; SCl-s

14b Sepals dark blue or purple; spur up to 18 mm long; upper petals blue with white edges; veins on undersides of leaves usually rust colored; mostly in oak woods ....
.... *Delphinium hesperium* ssp. *hesperium*
Western Larkspur, Coast Larkspur; SCl-n

1b Flower essentially regular, the petals, or sepals that resemble petals, all similar (but sometimes petals fall off early or 1 or more of them are missing)

15a Petals absent, the sepals often petal-like

16a Woody-stemmed vines (leaves opposite, 1 or 2 times compound; sepals usually 4, white; style on fruit feathery, more than 2 cm long; flowers sometimes pistillate or staminate and sometimes on separate plants)

17a Flowers not more than 2 cm wide, in dense panicles; leaflets 5–15, each with irregular lobes or teeth (in moist areas; widespread) ... *Clematis ligusticifolia*
Western Virgin's-bower, Yerba-de-Chiva

17b Flowers up to 5 cm wide, single or in groups of 3; leaflets 3–5, each 3 lobed, the lobes toothed .... *Clematis lasiantha* (pl. 46)
Pipestems, Chaparral Clematis

16b Herbs

18a Leaves lobed or once compound; stem leaves in a single whorl of 3; plant with a single flower (sepals 5–20 mm long, white, blue, purple, or red; basal leaf usually 1; on moist, shaded slopes) ....
.... *Anemone oregana* [*A. quinquefolia* var. *oregana*] (fig.)
Western Wood Anemone, Western Wind-flower; Mo-n

18b Leaves 1–4 times compound; stem leaves numerous; plant usually with several to many flowers

19a Stamens and pistils in separate flowers, these usually on separate plants; flowers in panicles; leaves mostly 15–40 cm long; often more than 100 cm tall; in moist areas (sepals 2–5 mm long, green to purple; widespread) .... *Thalictrum fendleri* var. *polycarpum* [*T. polycarpum*] (fig.)
Manyfruit Meadowrue

19b Stamens and pistils present in each flower; flower solitary; leaves not more than 12 cm long; up to 25 cm tall; in dry, shaded areas

20a Sepals 7–10 mm long, white to pink; stamens 23–27; ultimate leaf lobes 2–9 mm wide .... *Isopyrum occidentale*
Western Rue-anemone; Na, SCl-s

20b Sepals up to 5.5 mm long, white; stamens about 10; ultimate leaf lobes not more than 4 mm wide .... *Isopyrum stipitatum*
Rue-anemone; SCL, Al-n

15b Both petals and sepals usually present (but in *Myosurus minimus* the petals fall early, and in *Actaea rubra* there may be none)

21a Pistil 1; fruit fleshy, shiny, red or white, 5–10 mm wide, poisonous (leaves compound, divided in threes 1–3 times; flowers numerous, in a compact raceme; petals 4–10, white, 2–3 mm long, sometimes absent; in moist, shady habitats)....
.... *Actaea rubra* [includes *A. rubra* ssp. *arguta*] (fig.)
Baneberry; SLO-n

21b Pistil 5 to many; fruit not fleshy

    22a All sepals with short spurs; pistils many, crowded on a slender receptacle that may be more than 3 cm long (leaves smooth margined, all basal; petals 5, whitish, 1–3 mm long, falling off early; mostly around pools that dry out in summer)..................................... *Myosurus minimus* (fig.)
Common Mousetail

    22b Sepals without spurs, but the petals may have spurs; pistils 5 to many, not crowded on an especially long receptacle

        23a Leaves, if compound, not divided in threes, usually not entirely basal; petals without spurs; sepals less than 1 cm long, green, white, or yellow; pistils many, each 1 seeded................ RANUNCULACEAE, SUBKEY
Ranunculus

        23b Leaves 1–3 times compound, divided in threes, all basal; petals with long, hollow spurs; sepals usually more than 1 cm long, orange red, tinted with yellow; pistils usually 5, each many-seeded (petals 5)

            24a Plant sometimes slightly hairy, but not glandular hairy; basal leaves mostly twice compound; spur usually less than 20 mm long ....
..... *Aquilegia formosa* [includes *A. formosa* var. *truncata*] (pl. 46)
Crimson Columbine

            24b Plant glandular hairy; basal leaves mostly 3 times compound; spur 18–30 mm long (in moist serpentine soil) .... *Aquilegia eximia*
Serpentine Columbine, Van Houtte Columbine; Me-s

Ranunculaceae, Subkey: Flower essentially regular, even though some petals may not develop; sepals less than 1 cm long, usually green, sometimes yellowish; pistils several to many, each 1 seeded .... *Ranunculus*

1a Plant aquatic; petals white; submerged leaves divided into slender lobes, the floating leaves, if present, sometimes with 3 broad lobes

    2a Each style, in freshly opened flowers, 2–3 times as long as the portion that develops into a fruit; receptacle of flower not hairy (floating leaves present, with 3 broad lobes)
.... *Ranunculus lobbii*
Lobb Aquatic Buttercup; SCl, Al-Sn, La; 4

    2b Each style, in freshly opened flowers, not longer than the portion that develops into a fruit; receptacle of flower with short hairs

        3a Submerged leaves much shorter than the internodes and usually sessile; pedicel curving back after the fruit has begun to ripen (floating leaves, if present, like submerged leaves)............. *Ranunculus aquatilis* var. *subrigidus* [*R. subrigidus*]
Uncommon Water Buttercup; SF

        3b Submerged leaves about as long as the internodes, and usually with petioles; pedicel not curving back after the fruit has begun to ripen

4a Floating leaves with 3 broad lobes . . . *Ranunculus aquatilis* var. *hispidulus* (fig.)
Water Buttercup; Mo-n

4b Floating leaves, if present, like submerged leaves. . . . .
. . . . *Ranunculus aquatilis* var. *capillaceus*
Inland Water Buttercup

1b Plant terrestrial or growing in marshy places; petals yellow, sometimes becoming white as they age; without distinctly different types of leaves

5a Leaves not compound or deeply lobed (perennial; in marshy places)

6a Petals 1–3, 1–2 mm long; some leaf blades oval; stem not rooting at nodes. . . . . . .
. . . . *Ranunculus pusillus*
Low Buttercup; SCl, Na-n

6b Petals 5 or 10, up to 6 mm long; leaf blades not oval; stem rooting at nodes . . . . . .
. . . . *Ranunculus flammula* [includes *R. flammula* var. *ovalis*] (fig.)
Crowfoot; Ma-n

5b Leaves compound or deeply lobed

7a Main body of fruit spiny or bumpy

8a Petals usually 5, 5–8 mm long; fruit about 5 mm long, with stout, curved bristles; annual or perennial; in wet places . . . . *Ranunculus muricatus* (pl. 46)
Prickleseed Buttercup, Pricklefruit Buttercup; eu

8b Petals 1 or 2, about 1.5 mm long, sometimes absent; fruit about 2 mm long, with slender, hooked bristles; annual; in shaded habitats . . . .
. . . . *Ranunculus hebecarpus* (fig.)
Downy Buttercup

7b Main body of fruit not spiny or bumpy (style on the fruit may be hooked; perennial)

9a Plant with horizontal stems that root at the nodes (stem hairy; leaflets 3; petals often more than 1 cm long; in or near ditches and in wet lawns) . . . . . .
. . . . *Ranunculus repens* (pl. 47)
Creeping Buttercup; eu

9b Plant upright, the stems not rooting at the nodes

10a Petals rarely more than 4 mm long, not all of them developing equally (petals 4 or 5; leaves deeply lobed, hairy; body of fruit, excluding beak, up to 2.5 mm long; in moist shaded areas) . . . .
. . . . *Ranunculus uncinatus* [includes *R. uncinatus* var. *parviflorus*]
Woodland Buttercup; Ma, Sn-n

10b Petals at least 5 mm long, more or less equal in size

11a Petals usually more than 8, but sometimes as few as 5; beak of fruit up to 1 mm long

12a Lower leaves usually compound, sometimes hairy; petioles up to 20 cm long; petals 7–22; beak of fruit usually curved; body of fruit 2–3 mm long, without a ridge (in Coast Ranges and coastal, widespread) . . . .
. . . . *Ranunculus californicus* [includes *R. californicus* vars. *cuneatus* and *gratus*] (pl. 46)
California Buttercup

12b Lower leaves usually deeply lobed, with flattened hairs; petioles up to 12 cm long; petals 5–17; beak of fruit usually straight; body of fruit 3–6 mm long, with a ridge on 1 edge . . . . *Ranunculus canus* [includes *R. canus* var. *laetus*] Sacramento Valley Buttercup; CC

11b Petals 5–8; beak of fruit more than 1 mm long (body of fruit 2–4 mm long, with at least a slight ridge)

13a Lower leaves usually deeply lobed; petals usually 5 or 6, 4–15 mm long; petioles up to 11 cm long; beak of fruit straight or curved (inner Coast Ranges, widespread) . . . . *Ranunculus occidentalis* [includes *R. occidentalis* var. *eisenii*] Western Buttercup; Na-n

13b Lower leaves usually compound, with 3–7 leaflets; petals 5–8, 10–19 mm long; petioles up to 20 cm long; beak of fruit straight

14a Lower leaves hairy, usually with more than 3 leaflets; not confined to moist areas . . . . *Ranunculus orthorhynchus* var. *orthorhynchus* [includes *R. orthorhynchus* var. *platyphyllus*] Bird's-foot Buttercup; Al, Ma

14b Lower leaves not hairy, usually with not more than 3 leaflets; in moist areas . . . . *Ranunculus orthorhynchus* var. *bloomeri* Bloomer Buttercup; SCl-n

RESEDACEAE (MIGNONETTE FAMILY) The garden plants called mignonettes occasionally become established in vacant lots and other places to which they have somehow escaped. They have alternate leaves, and their interesting flowers, noticeably irregular, are concentrated in elongated terminal inflorescences. There are four to six sepals, united at their bases to form a disklike structure, and four to six petals, each of which consists of two portions: a small, flat or concave basal piece and a conspicuously lobed outer part. There is a single pistil, and its superior ovary is divided lengthwise into three or four seed-producing units that separate only partially when the fruit has ripened. The several to many stamens are in a cluster on one side of the flower. *Reseda luteola* (Dyer's-rocket) has been used for centuries as the source of a yellow dye.

1a Petals 4, yellow or yellowish (corolla irregular, the petals 2–4 mm long, of 2 different sizes and with a different number of lobes; stamens 20–25; leaves smooth margined) . . . . *Reseda luteola* Dyer's-rocket; eu

1b Petals 5 or 6, white or greenish white

2a Corolla irregular, the petals 6, 2–5 mm long, of 2 different sizes and with a different number of lobes; at least upper leaves 3 lobed; stamens 20–25 (flower very fragrant) . . . . *Reseda odorata* Garden Mignonette; me

2b Corolla regular, the petals 5 or 6, 5–6 mm long, all about the same size and all 3 lobed; leaves pinnately lobed; stamens 11–13 .... *Reseda alba*
White Mignonette; me

RHAMNACEAE (BUCKTHORN FAMILY) In our region, the Buckthorn Family is represented by shrubs and treelike plants belonging to the genera *Rhamnus* and *Ceanothus*. The flowers are small, but in some species of *Ceanothus*, the crowded inflorescences are extremely showy and may also be deliciously fragrant. The four or five sepals, four or five petals (sometimes absent), and four or five stamens originate at the edge of the cup formed by the floral tube. In *Ceanothus*, the petals have hoodlike tips that may partly enclose the corresponding stamen, and the sepals are usually colored like the petals. There is a single pistil with two to four style lobes; the ovary is superior or half-inferior. In *Rhamnus*, the fruit is fleshy, but in *Ceanothus*, it is dry.

Many species of *Ceanothus* (California lilacs), are widely cultivated. As with the genus *Arctostaphylos* (Ericaceae) and some other shrubby native plants, specimens with special attributes, such as a particular flower color or habit of growth, have been selected for propagation by cuttings. Such horticultural variants are called cultivars. Some nurseries offer a broad selection of cultivars and unselected wild types of *Ceanothus*.

1a Flowers green or yellow green, in clusters in the leaf axils; leaves alternate, with a prominent raised midrib; fruit fleshy, berrylike

    2a Leaf blades 1–4 cm long, the teeth, if present, spine tipped; petals absent; sepals 4; fruit 5–6 mm long

        3a Leaf blades up to 1.5 cm long; petioles 1–4 mm long (branches sometimes thorn tipped)... *Rhamnus crocea* (pl. 47)
Spiny Redberry; La-s

        3b Leaf blades 2–4 cm long; petioles 2–10 mm long............................. .... *Rhamnus ilicifolia* [*R. crocea* ssp. *ilicifolia*]
Hollyleaf Redberry

    2b Leaf blades 3–10 cm long, the teeth if present, not spine tipped; petals 5; sepals 5; fruit 10–12 mm long

        4a Leaves not hairy, or only sparsely hairy, on the undersides (widespread) .... *Rhamnus californica* (pl. 47)
California Coffeeberry

        4b Leaves hairy on the undersides

            5a Leaves not hairy on the upper surfaces .... *Rhamnus tomentella* ssp. *tomentella* [*R. californica* ssp. *tomentella*]
Hoary Coffeeberry

            5b Leaves hairy on the upper surfaces .... *Rhamnus tomentella* ssp. *crassifolia* [*R. californica* ssp. *crassifolia*]
Thickleaf Coffeeberry; Na

1b Flowers white, blue, pink, lavender, purple, or violet, usually in clusters at the ends of branches; leaves alternate or opposite, with 1 or 3 prominent veins; fruit a dry capsule

6a Leaves opposite (leaf blades with only 1 primary vein, the midrib)

    7a Flower white, rarely pale blue in *Ceanothus cuneatus*

        8a Leaves mostly somewhat wedge shaped, each usually with a slight notch at the tip and usually smooth margined (mostly upright, but sometimes prostrate; widespread)...... *Ceanothus cuneatus* [includes *C. ramulosus*] (pl. 47)
Buckbrush

        8b Leaves not wedge shaped, without notches, with 7 or more teeth

            9a Leaves somewhat folded along the midlines, yellow green on the upper surfaces, the margins inrolled and with spinelike teeth; sometimes on serpentine (flower with a musky odor)..... *Ceanothus jepsonii* var. *albiflorus*
White Muskbrush; Na

            9b Leaves usually flat, dark green on the upper surfaces, the margins not inrolled, with teeth, but these not spinelike; restricted to serpentine......
.... *Ceanothus ferrisae*
Coyote Ceanothus; Sl; 1b

    7b Flower blue, lavender, purple, or violet

        10a Margins of leaves inrolled (leaves with sharp or spinelike teeth)

            11a Leaves somewhat folded along the midlines, yellow green on the upper surfaces (flower with a musky odor; sometimes on serpentine; up to 60 cm tall) .... *Ceanothus jepsonii* var. *jepsonii*
Muskbrush; Me-Ma

            11b Leaves usually flat, dark green on the upper surfaces

                12a Plant up to 2 m tall; leaves up to 2.5 cm long, with 5–8 spinelike teeth
.... *Ceanothus divergens*
Calistoga Ceanothus; Na; 1b

                12b Plant prostrate, with stem tip rising, the stem sometimes rooting at the nodes; leaves up to 2 cm long, usually with 3–5 sharp teeth .....
.... *Ceanothus confusus* [*C. divergens* ssp. *confusus*]
Rincon Ridge Ceanothus; Sn; 1b

        10b Margins of leaves not inrolled, except rarely in *Ceanothus prostratus* and *C. sonomensis*

            13a Plant prostrate, sprawling, or with stem tip rising, less than 50 cm tall

                14a Leaves with 3–9 teeth; at elevations of more than 3,000 ft (leaves 1–3 cm long; petioles less than 3 mm long) ....
.....*Ceanothus prostratus* [includes *C. prostratus* var. *occidentalis*]
Mahala-mat; Sn, Na

                14b Leaves usually with many more than 10 teeth; at elevations below 1,000 ft

                    15a Leaves 2–5 cm long; petioles up to 4 mm long; coastal, in sandy soil .... *Ceanothus gloriosus* var. *gloriosus* (pl. 47)
Point Reyes Ceanothus, Glorymat; Me-Ma; 4

                    15b Leaves 1–2 cm long; petioles less than 2 mm long; in forests
.... *Ceanothus gloriosus* var. *porrectus*
Mount Vision Ceanothus; Ma; 1b

13b Plant often more than 100 cm tall

  16a Leaves 1.5–4 cm long, usually with many more than 12 teeth (petioles up to 4 mm long; in chaparral and forests) . . . . *Ceanothus gloriosus* var. *exaltatus*

                                      Glorybrush; Me-Ma; 4

  16b Leaves less than 2.5 cm long, with fewer than 12 teeth (on dry slopes)

    17a Petioles up to 4 mm long (leaves 6–18 mm long, with 9–11 teeth, these may be spinelike) . . . . *Ceanothus masonii*

                                 Mason Ceanothus; Ma; 1b

    17b Petioles less than 2 mm long, sometimes absent (leaves with spinelike teeth)

      18a Leaves up to 1.5 cm long, mostly sessile, with 7 teeth, 3 of these at the tips (sometimes on serpentine) . . . . *Ceanothus sonomensis* (pl. 47)

                               Sonoma Ceanothus; Sn; 1b

      18b Leaves up to 2.5 cm long, on short petioles, with about 10 teeth, these evenly spaced . . . . *Ceanothus purpureus*

                               Hollyleaf Ceanothus; Na; 1b

6b Leaves alternate

  19a Plant with thorny branches, but thorns sometimes absent in *Ceanothus spinosus*

    20a Each leaf blade with 1 primary vein originating at the base; bark olive green (petioles 4–8 mm long; flower pale blue or white; treelike, up to 6 m tall) . . . . *Ceanothus spinosus*

                               Greenbark Ceanothus; SLO-n

    20b Each leaf blade with 3 primary veins originating at the base; bark dark

      21a Petioles 2–3 mm long; leaves 1–4 cm long, with a whitish coating on both surfaces; flower white or blue . . . . *Ceanothus leucodermis*

                               Chaparral Whitethorn; Al-s

      21b Petioles up to 12 mm long; leaves 2–6 cm long, the upper surfaces darker than the undersides; flower white . . . . *Ceanothus incanus*

                               Coast Whitethorn; SCr-n

  19b Plant without spiny branches, although the branches may be very stiff (leaves smooth margined or finely toothed)

    22a Leaves deciduous, either not toothed or with only a few teeth at the tips (inflorescence up to 15 cm long; leaves up to 8 cm long; petioles up to 15 mm long; flower white, blue, or rarely pink; stem usually upright) . . . . *Ceanothus integerrimus* [includes *C. integerrimus* var. *californicus*]

                               Deerbrush

    22b Leaves evergreen, with small glandular teeth around most of the margins

      23a Undersides of leaves either not hairy, or with hairs only on the veins

24a Flower white; leaves up to 8 cm long; petioles up to 18 mm long; inflorescence up to 12 cm long (leaves with 3 prominent veins, not hairy on the upper surfaces, with a sweet, tobaccolike odor; upright, sometimes treelike) .... *Ceanothus velutinus* var. *hookeri*
Tobacco-brush; Ma-n
24b Flower usually blue, rarely white; leaves up to 5 cm long; petioles up to 12 mm long; inflorescence not more than 8 cm long
    25a Leaves sparsely hairy on the upper surfaces (leaves up to 2 cm long; petioles up to 3 mm long; twigs round, gray, sometimes hairy; stem prostrate, usually with tip rising) ....
.... *Ceanothus foliosus* var. *vineatus*
Vine Hill Ceanothus; Sn; 1b
    25b Leaves not hairy on the upper surfaces
        26a Twigs angled, green, not hairy; each leaf up to 5 cm long, with 3 prominent veins; petioles up to 12 mm long; stem prostrate or upright, sometimes treelike (common) .....
.... *Ceanothus thyrsiflorus*
[includes *C. thyrsiflorus* var. *repens*] (pl. 47)
Blue-blossom
        26b Twigs round, becoming gray, often with glandular hairs; each leaf up to 2 cm long, with 1 or 3 prominent veins; petioles less than 3 mm long; stem prostrate with tip rising
.... *Ceanothus foliosus* var. *foliosus*
Wavyleaf Ceanothus; SCr-n
23b Undersides of leaves hairy (flower blue; leaves up to 6 cm long; petioles up to 10 mm long)
    27a Upper surfaces of leaves hairy
        28a Each leaf up to 2 cm long, the margins not inrolled, the upper surface not roughened and glandular, with 1 or 3 prominent veins; petioles less than 3 mm long; inflorescence up to 8 cm long .... *Ceanothus foliosus* var. *medius*
La Cuesta Ceanothus, Wavyleaf Ceanothus; SCl
        28b Each leaf up to 5 cm long, the margins inrolled, the upper surface roughened and glandular, with 1 prominent vein; petioles up to 6 mm long; inflorescence up to 5 cm long ....
.... *Ceanothus papillosus*
Wartleaf Ceanothus; SM-SLO
    27b Upper surfaces of leaves not hairy
        29a Margins of leaves not inrolled; twigs round, sometimes hairy (each leaf not more than 4 cm long, with 3 prominent veins; petioles up to 8 mm long; inflorescence up to 4 cm long; upright, sometimes treelike) ....
.... *Ceanothus oliganthus* var. *sorediatus* [*C. sorediatus*]
Jimbrush

29b Margins of leaves inrolled; twigs angled, hairy

    30a Inflorescence up to 15 cm long, open; each leaf up to 4.5 cm long, with 1 or 3 prominent veins; twigs gray or red brown; upright .... *Ceanothus parryi*
                              Parry Ceanothus; Na, Sn

    30b Inflorescence up to 6 cm long, dense; each leaf up to 6 cm long, with 3 prominent veins; twigs greenish brown; prostrate or upright .... *Ceanothus griseus* (pl. 47)
                              Carmel Ceanothus; Sn

ROSACEAE (ROSE FAMILY)   The Rose Family is a large and diverse group of flowering plants, difficult to define concisely. The arrangement of pistils and structure of fruit are particularly variable features. Nevertheless, after you have learned to recognize some of the common genera, such as *Rubus* (blackberries and raspberries), *Fragaria* (strawberries), *Rosa* (roses), *Prunus* (cherries), and *Potentilla* (cinquefoils), you will probably be able to place relatives of these plants in the same family.

The characteristics given here apply to most members of Rosaceae that grow wild in the region. The leaves are alternate and often have stipules. The flower has a floral tube to which five partly united sepals, five separate petals, and at least 10 stamens are attached. Some species have no petals. Often there are five bractlets alternating with the sepals and attached to the floral tube below them. The ovary is superior or inferior. There may be a single pistil or several to many of them. When there is just one, it may be fused to the floral tube, so the mature fruit, as in an apple

Malus fusca
Oregon Crabapple

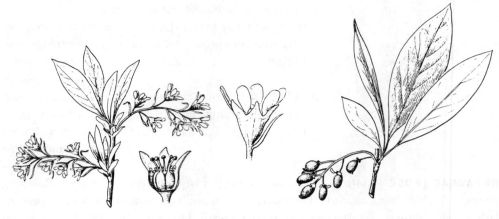

Oemleria cerasiformis
Osoberry

Prunus subcordata
Sierra Plum

or pear, shows the calyx lobes at its free end. When there are many, each pistil may become a one-seeded dry fruit, as in a cinquefoil, or it may become a one-seeded fleshy fruit attached to a conical receptacle, as in a raspberry or blackberry. In a strawberry, the receptacle is fleshy, but each individual fruit embedded in it is of the dry, one-seeded type.

The Rose Family is extremely important because it includes many trees and shrubs that yield edible fruit (and seeds, in the case of almonds), as well as many that are cultivated for their ornamental value.

Native species that are good subjects for wild gardens are available at plant sales and at some nurseries. Valuable herbaceous types are various species of *Fragaria* (strawberries) and *Potentilla* (cinquefoils). Among the shrubs, the following are especially useful.

Amelanchier alnifolia (Serviceberry)
Heteromeles arbutifolia (Toyon)
Holodiscus discolor (Creambush)
Oemleria cerasiformis (Osoberry)
Physocarpus capitatus (Pacific Ninebark)

Prunus ilicifolia (Hollyleaf Cherry)
Rosa californica (California Wild Rose)
Rosa gymnocarpa (Wood Rose)
Rubus spectabilis (Salmonberry)

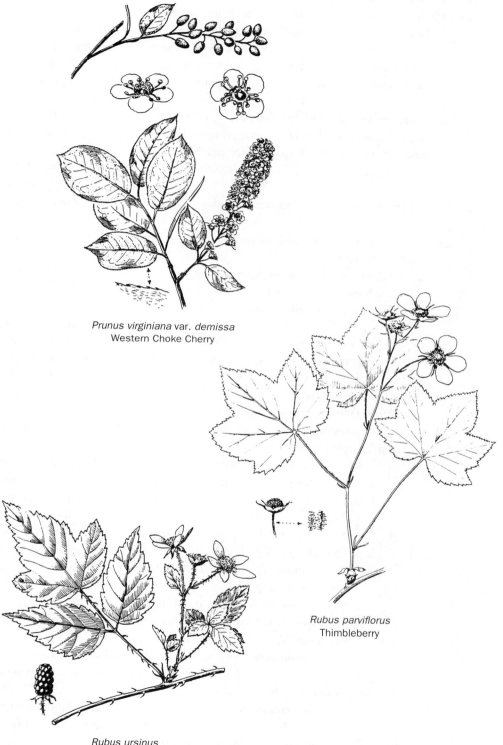

Prunus virginiana var. *demissa*
Western Choke Cherry

Rubus parviflorus
Thimbleberry

Rubus ursinus
California Blackberry

1a Plant not woody, even at the base, and without thorns or prickles (ovary superior)

2a Petals none; sepals 4; pistil 1; stamens 1 or 4; bractlets none or very small

3a Leaves not fan shaped, mostly basal, up to 10 cm long, pinnately compound, with 11–17 leaflets, these with 3–7 lobes; flowers numerous, in nearly globular or elongated clusters at the ends of branches; stamens 4; perennial; up to 25 cm tall (in sandy or rocky areas) . . . . *Acaena pinnatifida* var. *californica* [*A. californica*] California Acaena; Sn-s

3b Leaves fan shaped, scattered along the stem, less than 1 cm long, lobed; flowers in small clusters in the leaf axils; stamen 1; annual; rarely more than 5 cm tall (leaves pale, with stipules that clasp the stem; usually in colonies) . . . . *Aphanes occidentalis* [*Alchemilla occidentalis*] (pl. 48) Western Dewcup, Lady's-mantle

2b Petals 5; sepals 5; pistils at least 10; stamens at least 10; bractlets 5, obvious (leaves compound)

4a All leaves with 3 leaflets; fruit a strawberry, the pistils embedded in a fleshy receptacle; stamens 20–35

5a Leaflets leathery, densely hairy on the undersides; on seacoast bluffs and backshores of sandy beaches . . . . *Fragaria chiloensis* [includes *F. chiloensis* ssp. *pacifica*] (pl. 48) Beach Strawberry; SLO-n

5b Leaflets not leathery, only slightly hairy on the undersides; inland habitats, especially open woods . . . . . *Fragaria vesca* [includes *F. vesca* ssp. *californica*] Wood Strawberry

4b Larger leaves usually with more than 3 leaflets; fruit dry when mature, each developing from a separate pistil; stamens 10–20

6a Petals yellow or cream colored; stamens usually 20, all alike; leaves pinnately or palmately compound

7a Lower leaves palmately compound, with 5–7 leaflets, these deeply toothed (petals pale yellow; leaflets often more than 10 cm long, slightly hairy) . . . . *Potentilla recta* (pl. 49) Pale Cinquefoil; eua

7b Lower leaves pinnately compound (if there are only 3 leaflets, the end leaflet has a long stalk)

8a Flower solitary (petals 8–20 mm long, bright yellow; lower leaves green on the upper surfaces, usually with white hairs on the undersides, and with at least 15 substantial leaflets, as well as some smaller ones; at the margins of coastal salt marshes and in other wet places) . . . . *Potentilla anserina* ssp. *pacifica* [*P. egedei* var. *grandis*] (pl. 49) Pacific Cinquefoil, Pacific Silverweed

8b Flowers mostly in clusters

9a Petals cream colored to pale yellow; stem with glandular hairs (lower leaves mostly with 7–9 leaflets, these much longer than wide; up to 80 cm tall; common) . . . . *Potentilla glandulosa* (pl. 49) Sticky Cinquefoil

9b Petals bright yellow; stem not glandular (usually in wet habitats)

10a Petals less than 4 mm long; basal leaves with 3–5 leaflets, these with large teeth . . . .

. . . . . *Potentilla rivalis* [includes *P. rivalis* var. *millegrana*]

River Cinquefoil

10b Petals 6–10 mm long; basal leaves usually with at least 9 leaflets, these deeply lobed . . . .*Potentilla hickmanii*

Hickman Cinquefoil; Sn-Mo; 1b

6b Petals white; stamens 10, of 2 sizes, or at least 2 forms, these alternating; leaves pinnately compound

11a Pistils rarely more than 30

12a Lower leaves with 7–25 leaflets, these toothed or shallowly lobed; pistils 24–30 (coastal) . . . . *Horkelia marinensis*

Point Reyes Horkelia; Ma-Me; 1b

12b Lower leaves with 21–41 leaflets, these palmately lobed; pistils 9–26 (in sandy soils) . . . . *Horkelia tenuiloba*

Thin-lobed Horkelia; Ma, Sn; 1b

11b Pistils more than 50

13a Bractlets shorter than the sepals (in coastal, sandy areas; distinctions between the 2 subspecies of *H. cuneata* not sharp)

14a Leaves and stem with soft, long hairs, somewhat silky to the touch; stem and leaves not glandular . . . .

. . . . *Horkelia cuneata* ssp. *sericea*

Kellogg Horkelia; Sn-s; 1b

14b Leaves and stem sometimes hairy but not silky to the touch; stem and leaves usually glandular . . . .

. . . .*Horkelia cuneata* ssp. *cuneata*

Wedgeleaf Horkelia; SF-s

13b Bractlets at least as long as the sepals

15a Lower leaves with 7–11 leaflets, these deeply toothed, but rarely truly lobed, except for the end leaflet (pistils 80–220) . . . .

. . . .*Horkelia californica* ssp. *frondosa* [*H. frondosa*]

Leafy Horkelia; Me-Sn

15b Lower leaves generally with 11–21 leaflets, these deeply lobed, the lobes toothed

16a Bractlets usually 3 toothed; pistils 80–200; coastal . . . .

. . . . *Horkelia californica* ssp. *californica* (pl. 49)

California Horkelia; SCr-n

16b Bractlets rarely toothed; pistils 50–100; inland (in moist areas) . . . . *Horkelia californica* ssp. *dissita* [*H. elata*]

Lobed Horkelia; Al-n

1b   Plant woody, and sometimes with thorns or prickles (bractlets none)

    17a   Leaves not compound, but they may be deeply lobed; small trees or shrubs

        18a   Stems not modified to form stout thorns

            19a   Leaves not palmately lobed, but they may be toothed, pinnately lobed, or have 1 or 2 lobes at their bases . . . . ROSACEAE, SUBKEY

            19b   Most leaves palmately lobed, each lobe separated at least one-third the distance to the leaf midrib (petals white; ovary superior; deciduous; shrubs)

                20a   Leaf blades up to about 12 cm wide, usually with 5 distinct lobes; petioles 2–12 cm long; inflorescence open, with rarely more than 7 flowers; petals 15–25 mm long; aggregate of fruit a fleshy raspberry (widespread) . . . . *Rubus parviflorus* [includes *R. parviflorus* var. *velutinus*] (fig.)
Thimbleberry

                20b   Leaf blades usually less than 10 cm wide, usually with 3 lobes; petioles not more than 2 cm long; inflorescence compact, with numerous flowers; petals about 4 mm long; fruit 1–5 on each receptacle, dry when mature (in moist areas; widespread). . . . . *Physocarpus capitatus* (pl. 49)
Pacific Ninebark

        18b   Some short stems modified to form stout thorns (fruit fleshy; petals white)

            21a   Some leaf blades with 1 or 2 large teeth or lobes below the middle, as well as fine teeth in the upper part (leaves deciduous, hairy on the undersides; ovary inferior; fruit applelike, 10–15 mm wide, yellow or reddish; pedicel 2–3 cm long; shrub or tree) . . . . *Malus fusca* (fig.)
Oregon Crab Apple; Sn, Na-n

            21b   None of the leaf blades with 1 or 2 large teeth or lobes below the middle

                22a   Leaves hairy on the undersides, at least twice as long as wide, evergreen; shrub; prostrate or upright (leaves sometimes toothed; ovary half-inferior; fruit 6–8 mm wide, yellow orange or red) . . . . *Pyracantha angustifolia*
Firethorn; as

                22b   Leaves scarcely hairy, if at all, on the undersides, usually less than twice as long as wide, deciduous; shrub or small tree; upright

                    23a   Leaves without teeth in the lower third of the blade; ovary inferior; fruit 7–8 mm wide, blackish . . . . *Crataegus suksdorfii*
Black Hawthorn; Ma-n

                    23b   Leaves with fine teeth around most of the margins; ovary superior; fruit 15–25 mm wide, yellow to dark red (pedicel 10–15 mm long) . . . . *Prunus subcordata* (fig.)
Sierra Plum; Mo-n

    17b   Some leaves compound; shrubs or vines (stem, as a rule, with prickles; ovary superior)

        24a   Leaves pinnately compound, with 5 or more leaflets; fruit single, originating below the sepals; shrubs

            25a   Sepals soon falling away from the ripening fruit (prickles numerous, straight, slender, not broadened at their bases; fruit bright red; in shaded areas) . . . . *Rosa gymnocarpa* (pl. 50)
Wood Rose; Mo-n

25b Sepals persisting on the ripe fruit

    26a Sepals toothed or pinnately lobed (prickles 7–13 mm long, broadened at their bases, often curved; flower 3–5 cm wide) . . . . *Rosa eglanteria*

        Sweetbrier; eu

    26b Sepals not toothed or pinnately lobed

        27a Floral tube covered with gland-tipped bristles; generally less than 30 cm tall (pedicel 5–15 mm long, usually with glandular hairs) . . . .

        . . . .*Rosa spithamea* [includes *R. spithamea* var. *sonomensis*]

        Ground Rose, Sonoma Rose; Me-SLO

        27b Floral tube without gland-tipped bristles, but it may be hairy; usually more than 50 cm tall

            28a Prickles straight, slender, not broadened at their bases; pedicel 10–20 mm long, sometimes with glandular hairs (in woodlands or canyons) . . . . *Rosa pinetorum*

            Pine Rose; Mo-n; 1b

            28b Prickles curved, stout, broadened at their bases; pedicel 5–20 mm long, hairy, but these hairs not glandular (usually close to streams; common) . . . .*Rosa californica* (pl. 50)

            California Wild Rose

24b Most leaves either palmately compound or with only 3 leaflets; "fruit" an aggregate of smaller fruit, these attached to one another and to a conical receptacle that is raised above the sepals, as in a raspberry or blackberry; shrubs or vines

    29a Shrubs; aggregate of ripe fruit separating freely from the receptacle (raspberries)

        30a Stem usually covered with a whitish deposit; leaflets 3, occasionally 5; petals white, less than 10 mm long, shorter than the sepals; aggregate of fruit dark purple or red, with a whitish coating . . . . . . *Rubus leucodermis*

        Blackcap Raspberry, Western Raspberry; SCr-n

        30b Stem not covered with a whitish deposit; leaflets 3; petals purple red, 10–15 mm long, longer than the sepals; aggregate of fruit yellow, orange, pink orange, or red (often in moist places) . . . .

        . . . . . . *Rubus spectabilis* [includes *R. spectabilis* var. *franciscanus*] (pl. 49)

        Salmonberry; SCl-Sn

    29b Vines; aggregate of ripe fruit not separating freely from the receptacle (blackberries)

        31a Leaflets deeply lobed (prickles 5–6 mm long, usually curved; widely naturalized) . . . . *Rubus laciniatus*

        Cutleaf Blackberry; eu

        31b Leaflets not deeply lobed

            32a Slender-stemmed vines; flowers in small cymelike clusters; prickles straight, slender, not thickened at their bases (common) . . . .

            . . . .*Rubus ursinus* [includes *R. vitifolius* vars. *eastwoodianus* and *vitifolius*] (fig.)

            California Blackberry

32b Coarse-stemmed vines; flowers in large panicles; prickles, if present, usually curved, stout, thickened at their bases (escape from cultivation)

    33a Almost all stems bearing prickles (widespread) . . . .
. . . . *Rubus discolor [R. procerus]*
Himalayan Blackberry; eua

    33b Stems without prickles . . . . *Rubus ulmifolius* var. *inermis*
Thornless Blackberry; eu

Rosaceae, Subkey: Small trees or shrubs, without prickles or thorns; leaves not compound; bractlets none

1a Leaves narrow and rigid, up to about 1 cm long (petals white, about 3 mm long; ovary superior; shrub; widespread) . . . . *Adenostoma fasciculatum* (pl. 48)
Chamise; Me-s

1b Leaves neither narrow nor rigid, usually much more than 1 cm long

  2a Leaves smooth margined

    3a Leaves deciduous, 5–13 cm long, not woolly on the undersides; staminate and pistillate flowers on separate plants; ovary superior; fruit orange, sometimes becoming blue black; shrub or tree (in shaded areas) . . . .
. . . . *Oemleria cerasiformis [Osmaronia cerasiformis]* (fig.)
Osoberry

    3b Leaves evergreen, up to 4 cm long, woolly on the undersides; all flowers with stamens and a pistil; ovary inferior; fruit red; shrubs

      4a Petals pink; leaves up to 3 cm long . . . . *Cotoneaster franchetii*
Pinkflower Cotoneaster; as

      4b Petals white; leaves up to 4 cm long . . . . *Cotoneaster pannosa*
Silverleaf Cotoneaster; as

  2b Leaves with fine to coarse teeth, or sometimes with 1 or 2 lobes

    5a Teeth limited to upper half or two-thirds of leaf margin (leaf blades less than twice as long as wide)

      6a Some teeth double, the 3–7 large teeth on each side of leaf blade with smaller teeth (flowers less than 5 mm wide, in dense panicles up to 25 cm long; petals white; ovary superior; fruit dry; shrub; widespread) . . . .
. . . . *Holodiscus discolor* [includes *H. discolor* var. *franciscanus*] (pl. 48)
Creambush, Oceanspray

      6b Teeth not double, all more or less equal in size

        7a Evergreen; leaf blade usually wedge shaped at the base and with blunt teeth; petals absent; flowers 1–3 in each cluster; ovary superior; fruit dry; style 1; shrub or tree (style very hairy, persistent, 5–10 cm long; widespread) . . . . *Cercocarpus betuloides* (pl. 48)
Birchleaf Mountain-mahogany

        7b Deciduous; leaf blade usually rounded at the base and with pointed teeth; petals 5, white; flowers several in each cluster; ovary inferior; fruit fleshy, blue black; styles 4 or 5; shrub . . . . . *Amelanchier alnifolia* (pl. 48)
Serviceberry

5b Teeth around all leaf margin, except perhaps at the very base (shrubs or trees)

8a Some leaf blades with 1 or 2 large teeth or lobes below the middle (flower about 2 cm wide; ovary inferior; fruit applelike, yellow or red, up to 15 mm long) . . . . *Malus fusca* (fig.)

Oregon Crab Apple; Sn, Na-n

8b None of the leaf blades with 1 or 2 large teeth or lobes below the middle

9a Leaves stiff and sharply toothed, evergreen

10a Most leaf blades less than 3 times as long as wide; flowers in racemes; ovary superior; fruit 12–25 mm long, red to blue black, juicy . . . .
. . . . *Prunus ilicifolia* (pl. 49)

Hollyleaf Cherry, Islay; Na-s

10b Most leaf blades at least 3 times as long as wide; flowers in panicles; ovary mostly inferior; fruit 5–10 mm long, red, mealy (widespread)
. . . . *Heteromeles arbutifolia* (pl. 48)

Toyon, Christmas-berry

9b Leaves not stiff or sharply toothed, deciduous (ovary superior)

11a Pedicel, if present, less than 3 mm long; flowers 1–3 in each cluster; petals 12–15 mm long, pink, turning white; fruit 25–40 mm long, hairy (tree; leaves 3–10 cm long) . . . . . *Prunus dulcis [P. amygdalus]*

Almond; as?

11b Pedicel 3–15 mm long; flowers 1 to many in each cluster; petals 4–10 mm long, white; fruit 6–25 mm long, usually not hairy

12a Flowers and fruit numerous in each cluster; leaves 6–12 cm long (pedicel 5–8 mm long; fruit 6–14 mm long, red to black; in moist areas) . . . . *Prunus virginiana* var. *demissa* (fig.)

Western Chokecherry

12b Flowers and fruit 1–12 in each cluster; leaves 2.5–8 cm long

13a Pedicel 10–15 mm long; fruit 15–25 mm long, yellow to dark red, 1–7 in each cluster . . . . *Prunus subcordata* (fig.)

Sierra Plum; Mo-n

13b Pedicel 3–12 mm long; fruit 7–14 mm long, dark red to purple, 3–12 in each cluster . . . . *Prunus emarginata* (pl. 49)

Bitter Cherry

RUBIACEAE (MADDER FAMILY)   Coffee, cinchona (the source of quinine), and gardenias are perhaps the best known plants of the large and diverse Madder or Bedstraw Family. Most of our representatives are completely herbaceous, but some are slightly woody at the base. One is a large shrub.

The leaves of Rubiaceae are opposite or whorled, and the flowers of most of our native species have a four-lobed tubular corolla and four stamens. The ovary is inferior, and calyx lobes, above the ovary, may be indistinct, short lived, or absent. In nearly all Bay Area representatives, there are two styles and the fruit is distinctly two lobed; the lobes eventually separate into one-seeded nutlets that may be covered with hooked bristles or hairs. The genus *Galium* is an easy one to recognize, for its stems are four sided, at least when young, and often have stiff hairs that cause them to become interlocked with other stems. In some species, all flowers have

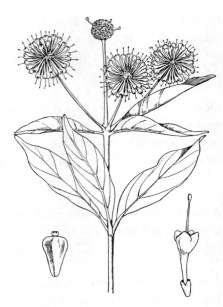

Cephalanthus occidentalis var. *californicus*
California Button-willow

a pistil and stamens, but in other species, the pistillate and staminate flowers are on separate plants. The pistillate flower is usually solitary, whereas staminate flowers are in small clusters.

1a   Leaves opposite, in pairs or in threes; flowers in dense, globular clusters; shrub (corolla white or pale yellow; calyx 4 lobed, the lobes persistent; often more than 5 m tall; in moist areas) . . . . *Cephalanthus occidentalis* var. *californicus* (fig.)
                                California Button-willow; Na-n

1b   Leaves usually in whorls of at least 4; flowers solitary or a few in each inflorescence; herbs, but some woody at the base (stem 4 sided, at least when young)

     2a   Corolla pink or lavender, the 4 lobes much shorter than the tube; calyx 6 lobed (leaves 4–13 mm long, most in whorls of 5 or 6) . . . . *Sherardia arvensis* (pl. 50)
                                  Field Madder; me

     2b   Corolla white, green, yellow, or purple, the 3 or 4 lobes much longer than the tube; calyx lobes not apparent

         3a   Most leaves in whorls of 5–8

             4a   Flowers usually at least 6 in each cluster (corolla white)

                 5a   Leaves less than 7 mm long; annual; stem erect, up to 30 cm tall (fruit not hairy) . . . . *Galium divaricatum*
                                Lamarck Bedstraw; me

                 5b   Leaves 10–30 mm long; perennial; stem sometimes erect, up to 100 cm long

                     6a   Fruit not hairy; leaves usually in whorls of 8; stem erect, or sprawling and with the tip rising; not restricted to wet places . . . .
   . . . . *Galium mollugo*
                                Hedge Bedstraw; eu

6b Fruit somewhat hairy; leaves in whorls of 5–8; stem sprawling; in wet places .... *Galium mexicanum* var. *asperulum*
Rough Bedstraw; SCl-n

4b Flowers usually 1–3, but sometimes 4 or 5 in each cluster

7a Flowers 1 or 2 in each cluster (corolla 4 lobed; fruit sausage shaped, with hooked hairs; annual; up to 7 cm tall; widespread) .... *Galium murale*
Tiny Bedstraw; eu

7b Flowers usually 3, sometimes 4 or 5 in each cluster

8a Corolla usually 3 lobed, white or pink; leaves up to 2 cm long, in whorls of 5 or 6; fruit not hairy; up to 10 cm tall, forming mats (usually perennial; mostly in wet places at elevations above 1,500 ft) ....
.... *Galium trifidum* var. *pusillum*
Trifid Bedstraw

8b Corolla 4 lobed, white or cream colored; leaves up to 4 cm long, in whorls of 6–8; fruit with hooked hairs; stem 20–90 cm long, prostrate, climbing, or erect

9a Plant with hooked bristles; each leaf blade at least 4 times as long as wide, with some bristles on midrib; annual; sprawling or upright (common) .... *Galium aparine* (pl. 50)
Goosegrass; eu?

9b Plant often not hairy; each leaf blade 2–3 times as long as wide, without bristles; perennial; usually sprawling, the stem tips rising .... *Galium triflorum*
Sweet-scented Bedstraw

3b Most leaves in whorls of 4

10a Leaves tough, sharply pointed at the tips (leaves about 1 cm long; corolla yellow; fruit not hairy; perennial; up to 15 cm tall, forming dense mats; on dry slopes) .... *Galium andrewsii* (pl. 50)
Phloxleaf Bedstraw; La-s

10b Leaves not tough or sharply pointed at the tips

11a Plant not woody

12a Leaves up to 3 mm long; annual; not restricted to moist areas (angles of stem not bristly; fruit bristly; corolla green, yellow, or white; up to 7 cm tall; widespread) .... *Galium murale*
Tiny Bedstraw; eu

12b Leaves 4–19 mm long; usually perennial; in moist areas

13a Corolla usually 3 lobed, white or pink; angles of stem bristly; fruit not hairy; forming mats, up to 10 cm tall ....
.... *Galium trifidum* var. *pusillum*
Trifid Bedstraw

13b Corolla 4 lobed, yellow; angles of stem not bristly; fruit hairy; usually climbing, the stem up to 90 cm long ....
.... *Galium californicum* (pl. 50)
California Bedstraw

11b Plant woody at the base

14a Angles of stem bristly; fruit not hairy (corolla yellow or red; leaves up to 18 mm long; perennial; clambering over and through other plants; stem often more than 1 m long)

15a Larger leaves not more than 4 times as long as wide; Coast Ranges .... *Galium porrigens* var. *porrigens* [*G. nuttallii* var. *ovalifolium*] (pl. 50) Climbing Bedstraw

15b Larger leaves more than 5 times as long as wide; inner Coast Ranges .... *Galium porrigens* var. *tenue* [*G. nuttallii* var. *tenue*] Narrowleaf Climbing Bedstraw

14b Angles of stem not bristly, but sometimes hairy; fruit sometimes hairy (plant tufted or climbing)

16a Leaves up to 27 mm long, the surfaces sometimes with hairs only on the veins and at the margins; corolla purple red or yellow; fruit sometimes hairy; stem up to 25 cm long; in dry, open areas .... *Galium bolanderi* [*G. pubens*] Bolander Bedstraw; Sl-n

16b Leaves up to 18 mm long, the surfaces hairy; corolla yellow; fruit hairy; stem up to 90 cm long; in moist, shaded areas .... *Galium californicum* (pl. 50) California Bedstraw

RUTACEAE (RUE FAMILY)   The Rue Family is a large one and includes citrus trees. The only species native to the region is *Ptelea crenulata* (pl. 51) (Hoptree), found mostly in canyons of the inner Coast Ranges from Santa Clara County northward. It is a small tree, up to about 5 m tall, with alternate, compound, deciduous leaves that, when rubbed or bruised, yield a citruslike aroma. Each leaf has three leaflets, 2–7 cm long. The small flowers, clustered in cymes, are greenish white. There are four or five sepals, petals, and stamens. The ovary is superior, and the single pistil develops into a flattened fruit about 1.5 cm wide. Much of the fruit, except for the central portion that is occupied by the two seeds, consists of a thin, nearly membranous wing.

SALICACEAE (WILLOW FAMILY)   There are only two genera in the Willow Family, and both are represented in our area by native trees and shrubs usually found in moist areas. The alternate, deciduous leaves are smooth margined or toothed, and the flowers are concentrated in staminate and pistillate catkins, which are usually on separate plants. There are no petals or sepals, but each flower has a little scalelike bract below it. Furthermore, in *Populus* (cottonwoods), there is a cup-shaped disk under each flower; in *Salix* (willows), one or two rather conspicuous glands are located in approximately the same place. Staminate flowers of *Populus* have eight to many stamens, whereas those of *Salix* have one to eight stamens. The pistillate flowers have a single pistil with two to four stigmas that correspond to the number of seed-producing divisions. The ovary is superior. The numerous seeds that develop in the fruit have long hairs that facilitate dispersal by wind.

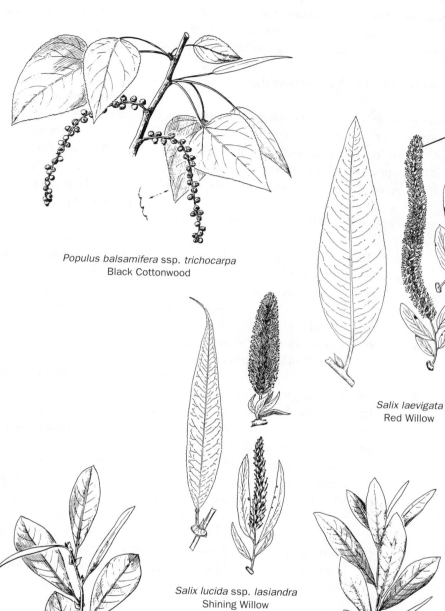

Populus balsamifera ssp. *trichocarpa*
Black Cottonwood

catkin

Salix laevigata
Red Willow

Salix lucida ssp. *lasiandra*
Shining Willow

Salix scouleriana
Scouler Willow

Salix sitchensis
Sitka Willow

1a Leaf blade less than twice as long as wide, broadest near the base (leaf blades 3–7 cm long, usually not hairy; trees; sometimes more than 25 m tall)

  2a Leaves smooth margined .... *Populus balsamifera* ssp. *trichocarpa [P. trichocarpa]* (fig.)
      Black Cottonwood

  2b Leaves toothed (teeth absent from the tip of the leaf, however) ..... *Populus fremontii*
      Fremont Cottonwood, Alamo

1b Leaf blade generally at least twice as long as wide, usually not broadest near the base

  3a Leaves with yellow glands on the teeth of the blade and uppermost portion of the peti-ole (mature leaves toothed, not hairy, up to 17 cm long but not often more than 2 cm wide, the upper surfaces shiny; up to 10 m tall) ....
      .. *Salix lucida* ssp. *lasiandra* [includes *S. lasiandra* vars. *lasiandra* and *lancifolia*] (fig.)
      Shining Willow, Yellow Willow

  3b Leaves without yellow glands on the teeth of the blade and uppermost portion of the petiole

    4a Leaves usually less than 6 mm wide (leaves 5–12 cm long, hairy, about the same color on both surfaces; inflorescence 2–7 cm long; up to 7 m tall) ....
        .... *Salix exigua* [includes *S. hindsiana* vars. *hindsiana* and *leucodendroides*]
        Narrowleaf Willow, Sandbar Willow

    4b Leaves usually at least 6 mm wide

      5a Leaves less than 1.5 cm wide (leaves 3–9 cm long, nearly sessile, sometimes hairy; catkins 2–5 cm long; up to 4 m tall)................. *Salix melanopsis*
          Dusky Willow, Black Willow; Sn-n

      5b Some leaves usually more than 1.5 cm wide

        6a Leaf blades rarely so much as 3 times as long as wide, with margins often obviously inrolled, at least in the lower portion (ovary hairy; catkins ap-pearing before or with the leaves)

          7a Leaves sometimes hairy on the undersides, but not silky hairy; upper surfaces of leaves hairy on the midveins; catkins up to 6.5 cm long; up to 10 m tall; not restricted to wet habitats ....
              .... *Salix scouleriana* (fig.)
              Scouler Willow; Mo-n

          7b Leaves silky hairy on the undersides; upper surfaces of leaves not hairy; up to 7 m tall; catkins up to 9.5 cm long; restricted to wet habi-tats .... *Salix sitchensis* [includes *S. coulteri*] (fig.)
              Sitka Willow; Sn-n

        6b Leaf blades usually at least 4 times as long as wide, the margins inrolled only slightly, if at all

          8a Mature leaves densely hairy on the undersides, sparsely hairy on the upper surfaces; ovary hairy; usually not more than 1 m tall; on ser-pentine (catkins up to 8.5 cm long, appearing before the leaves)
              .... *Salix breweri*
              Brewer Willow; La-SB

8b   Mature leaves not hairy, but young leaves may be hairy; ovary not hairy; often more than 10 m tall; not on serpentine

    9a   Leaf blades 3.5–12.5 cm long, smooth margined or with fine teeth, slightly inrolled; catkins 1.5–7 cm long, appearing before the leaves . . . . *Salix lasiolepis* [includes *S. lasiolepis* var. *bigelovii*]

            Arroyo Willow

    9b   Leaf blades 6.5–15 cm long, with fine teeth, not inrolled; catkins 3.5–11 cm long, appearing with or after the leaves . . . .

      . . . . . *Salix laevigata* [includes *S. laevigata* var. *araquipa*] (fig.)

            Red Willow

SAURURACEAE (LIZARD'S-TAIL FAMILY)   The sole representative of this family in California is *Anemopsis californica* (pl. 51) (Yerba-mansa). It sometimes forms dense stands in somewhat wet, alkaline soils from Santa Clara County southward. *Anemopsis* has creeping underground stems that take root and give rise to new plants. In the past, it was thought to be useful for treating diseases of the skin and blood.

The flowers are concentrated into a thick inflorescence up to about 4 cm long, at the base of which is a ring of five to eight white, often red-tinged bracts. Individual flowers, without sepals or petals, but with a white bract, 5–6 mm long beneath each one, have six or eight stamens and a pistil with three or four styles. The ovary is superior. Most of the leaves are basal, and the blades of the larger ones, up to 20 cm long, typically have a notch on both sides of the petiole.

SAXIFRAGACEAE (SAXIFRAGE FAMILY)   With some exceptions, plants that belong to the Saxifrage Family are perennials that die back each autumn. Many of them grow in moist, shaded areas. The principal leaves are usually concentrated at the base of the plant: the upper leaves are sometimes present but more often are reduced or absent. In each flower, much of the calyx forms a five-lobed cup or floral tube, and there are usually five separate petals attached to this. The corolla is usually regular. Most saxifrages have 5 or 10 stamens, but *Tolmiea* has 3. The ovary is superior, half-inferior, or inferior. There are usually two or three styles. When the fruit is ripe, it is dry and splits open lengthwise.

Two common exotic components of local gardens are in this family. One of them is *Heuchera sanguinea* (coralbells), which grows wild in New Mexico; the other is *Bergenia crassifolia,* from Siberia. A California native sold for indoor cultivation is *Tolmiea menziesii* (Piggy-back-plant). Its common name alludes to its charming habit of sprouting new plants on its leaves). It is perfectly hardy in this region and is a good subject for a shady, reasonably moist outdoor habitat. Some other shade-loving natives that deserve to be grown are *Tiarella trifoliata* var. *unifoliata* (Sugarscoop), and *Tellima grandiflora* (Fringecups). Species of *Heuchera* would be attractive additions to exposed rockeries. Seeds of other plants, such as *Lithophragma*, are also available.

1a   Flower solitary at the top of the stem; petals white, 1–1.5 cm long, not lobed (all leaves basal, smooth margined; stamens with anthers 5, those without 5; stigmas 4; ovary superior; in wet meadows). . . . . *Parnassia californica* [*P. palustris* var. *californica*]

            Grass-of-Parnassus; SB-n

1b   Flowers more than 1 each in inflorescence; petals, if white, either less than 1 cm long or lobed

Heuchera micrantha
Smallflower Alumroot

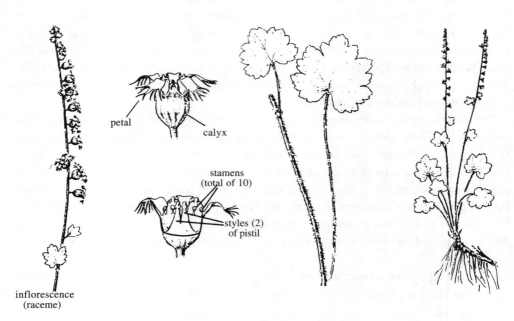

petal

calyx

stamens
(total of 10)

styles (2)
of pistil

inflorescence
(raceme)

Tellima grandiflora
Fringecups

2a Petals 4; calyx irregular; stamens 3 (petals 8–12 mm long, purple brown, almost threadlike; leaves mainly basal, lobed and toothed, some of them sprouting plants that may take root; ovary superior; on moist banks) . . . . *Tolmiea menziesii* (pl. 51)

Piggy-back-plant; SFBR-n

2b Petals 5; calyx regular; stamens 5 or 10

    3a Stamens 5 (leaves lobed and toothed; petals not more than 5 mm long; ovary half-inferior)

        4a Leaves scattered along the stem, as well as basal (leaf blades up to 8 cm wide, lobed and toothed; inflorescence usually more than 30 cm long; petals white; restricted to shady, wet banks and cliffsides) . . . *Boykinia occidentalis [B. elata]*

Brookfoam

        4b Nearly all leaves basal, only 1 or 2 along each stem

            5a Petals yellow green, with 4–7 hairlike lobes, the corolla therefore resembling a snowflake (leaf blades up to 5 cm wide; in wet, shaded areas) . . . . *Mitella ovalis*

Mitrewort; Ma

            5b Petals white or pink, not lobed

                6a Styles 2–4 mm long, protruding out of corolla; inflorescence usually open, the flowers not crowded; hairs on calyx not obvious without a hand lens; in moist, rocky areas (widespread) . . . . *Heuchera micrantha* [includes *H. micrantha* var. *pacifica*] (fig.)

Smallflower Alumroot; SLO-n

                6b Styles less than 2 mm long, not protruding out of corolla; inflorescence usually dense, the flowers crowded; hairs on calyx dense and obvious; coastal . . . . *Heuchera pilosissima*

Seaside Alumroot, Seaside Heuchera; SLO-n

    3b Stamens 10 (leaves mainly basal, the stem leaves much reduced)

        7a Pistil with 2 styles (in moist, shaded areas)

            8a Leaves lobed and toothed, the blades up to 12 cm wide; petals either less than 1.5 mm wide, or broader and lobed

                9a Petals white, up to 4 mm long, less than 1.5 mm wide, not lobed; flowers in a panicle; ovary superior . . . . *Tiarella trifoliata* var. *unifoliata* [*T. unifoliata*]

Sugarscoop; SFBR-n

                9b Petals greenish white to nearly crimson, up to 7 mm long, at least 2 mm wide, with 5–7 slender lobes; flowers in a raceme; ovary half-inferior (petals sometimes falling early) . . . . *Tellima grandiflora* (pl. 51; fig.)

Fringecups; SLO-n

            8b Leaves toothed, the blades less than 10 cm wide; petals at least 2 mm wide, not lobed (petals white, sometimes with darker spots)

                10a Leaf blades longer than wide, not more than 5 cm wide; filaments usually wider below than above; ovary half-inferior (petals 2–5 mm long; flowers often on 1 side of the inflorescence; widespread) . . . . *Saxifraga californica*

California Saxifrage

10b Leaf blades not longer than wide, up to 10 cm wide; filaments usually wider above than below; ovary superior

 11a Corolla irregular, 2 petals about 12 mm long, 3 petals 3–4 mm long; leaves up to 10 cm wide, with white veins on the upper surfaces and red veins on the undersides; new plants produced from prostrate stems that root at the nodes .... *Saxifraga stolonifera [S. sarmentosa]* Strawberry-geranium; as

 11b Corolla regular, all petals 4–5 mm long; leaves up to 7 cm wide, with veins of the same color on both surfaces; new plants not produced from prostrate stems .... *Saxifraga mertensiana* Wood Saxifrage; Sn-n

7b Pistil with 3 styles (petals white)

 12a Basal leaves either palmately compound, or each divided more than halfway to the base of the blade (petals 7–16 mm long, 3 lobed; ovary half-inferior; in open areas) .... *Lithophragma parviflorum* (pl. 51) Prairie Starflower

 12b Basal leaves not compound, each divided less than halfway to the base of the blade

  13a Petals 3 lobed at the tips (petals 5–12 mm long; basal leaves usually lobed and toothed; stem leaves 1–3, alternate; widespread)

   14a Pedicel 3–10 mm long; floral tube conical at the base; ovary half-inferior; seeds smooth; in open areas .... *Lithophragma affine* (pl. 51) Woodland-star

   14b Pedicel not more than 2 mm long; floral tube usually squared off at the base; ovary superior; seeds spiny; in shaded areas (widespread) .... *Lithophragma heterophyllum* Hill Starflower

  13b Petals not 3 lobed but sometimes with a few teeth on the margins (seeds spiny)

   15a Stem with 2 opposite leaves; basal leaves with 3 lobes but not toothed, not more than 2.5 cm wide; pedicel 4–10 mm long (petals 4–8 mm long; ovary half-inferior; in moist, shaded areas) .... *Lithophragma cymbalaria* Mission Starflower, Missionstar; SCl-s

   15b Stem with 1–3 alternate leaves; basal leaves usually lobed and toothed, up to 4 cm wide; pedicel not more than 4 mm long

    16a Inflorescence with 3–12 flowers; petals 5–12 mm long, usually without teeth at their bases; ovary superior; in shaded areas (common) .... *Lithophragma heterophyllum* Hill Starflower

16b Inflorescence with 5–25 flowers; petals 4–7 mm long, usually with fine teeth at their bases; ovary half-inferior; in open areas .... *Lithophragma bolanderi*
Bolander Starflower; SFBR-Me

SCROPHULARIACEAE (SNAPDRAGON FAMILY)
Scrophulariaceae (Snapdragon Family) has numerous representatives here, and some are exceptionally attractive when in flower. Most of our species are herbaceous, but there are a few shrubs. Many members of the family are partial parasites; these including species of *Castilleja, Cordylanthus, Pedicularis,* and *Triphysaria.* They have green leaves and can therefore make part of their own food, but their roots are joined to those of various other plants, which supply them with some of the organic nutrients they require. The results of recent research will probably soon lead to some drastic revisions in classification, to family, of plants presently assigned to Scrophulariaceae and to certain related assemblages.

Bracts are often present below the inflorescence or associated with each flower. There are generally five sepals, sometimes four. The corolla is typically two lipped, the upper lip being divided into two lobes and the lower lip divided into three lobes. The corolla tube or its lower lip may be drawn out into a spur or sac. In certain genera, however, the corolla is nearly regular. In *Verbascum,* for instance, there are five almost equal lobes; in *Synthyris* and *Veronica,* the corolla has four somewhat similar lobes, but the uppermost one, the result of fusion of two lobes, may be slightly larger. There are commonly four or five stamens, but one of these may lack an anther or be absent altogether. The stigma usually has two lobes. The ovary is superior, and the fruit is partitioned lengthwise into two seed-producing halves that separate after it has ripened and dried.

Several California natives are in cultivation. Some of the best perennials are *Galvezia speciosa,* from the Channel Islands, and various species of *Mimulus* and *Penstemon.* One that is especially useful in dry, somewhat shaded situations is *Mimulus aurantiacus* (Bush Monkeyflower). For wet places, including stream banks and edges of pools, *Mimulus cardinalis, M. guttatus,* and *M. moschatus* are excellent subjects. Among the annuals that will provide much color is *Collinsia heterophylla* (Chinesehouses). It will probably do best if partly shaded and not allowed to become too dry.

1a Upper lip of corolla not modified to form a hood or beaklike structure that encloses the stamens and at least much of the pistil
  2a All stem leaves alternate (biennial; sepals separate or nearly so; often more than 1 m tall)
    3a Corolla definitely irregular, 4–6 cm long, white to pink purple, with an obvious tube; anther-bearing stamens 4 (widely established; poisonous if eaten)........
.... *Digitalis purpurea* (fig.)
Foxglove; eu
    3b Corolla nearly regular, less than 1 cm long, yellow, without an obvious tube; anther-bearing stamens 5
      4a Plant woolly, owing to abundant soft, white hairs; pedicel 1–2 mm long (widespread) .... *Verbascum thapsus*
Common Mullein, Woolly Mullein; eua
      4b Plant not woolly but with some glandular hairs, especially above; pedicel 10–15 mm long............................... *Verbascum blattaria* (fig.)
Moth Mullein; eua

Digitalis purpurea
Foxglove

Linaria canadensis
Blue Toadflax

Mimulus tricolor
Tricolor Monkeyflower

Triphysaria pusilla
Dwarf Owl's-clover

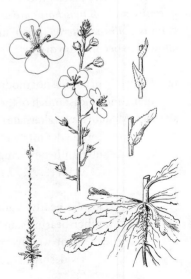

Verbascum blattaria
Moth Mullein

Veronica anagallis-aquatica
Great Water Speedwell

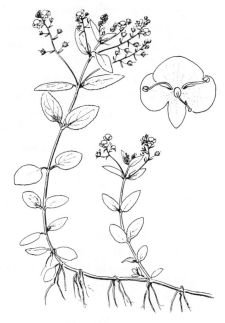

Veronica americana
American Brooklime

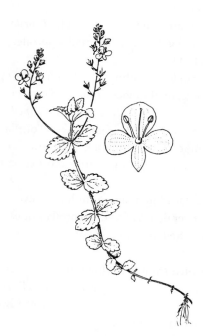

Veronica chamaedrys
Germander Speedwell

Veronica scutellata
Marsh Speedwel

2b Leaves either almost entirely basal, or usually at least the lower stem leaves opposite or whorled

  5a Corolla with a spur or sac (corolla usually irregular; anther-bearing stamens 4)

    6a Corolla tube with a broad sac (at least the lower stem leaves opposite or whorled; annual) .... SCROPHULARIACEAE, SUBKEY 1
      Antirrhinum, Collinsia

    6b Corolla tube with a slender spur (spur on the lower side of corolla tube near the base; lower leaves usually opposite or whorled)

      7a Corolla blue or violet, less than 1 cm long (excluding spur); annual (mostly in sandy habitats)..
        .... *Linaria canadensis* [includes *L. canadensis* var. *texana*] (fig.)
        Blue Toadflax; SCl-n

      7b Corolla primarily yellow, with orange inside, 1.5–2 cm long (excluding spur); perennial.................................... *Linaria vulgaris*
        Butter-and-eggs; me

  5b Corolla without a spur or sac

    8a With 4 anther-bearing stamens and 1 antherless stamen, this reduced to a scale in *Scrophularia californica* (leaves opposite or whorled) ....
      .... SCROPHULARIACEAE, SUBKEY 2

    8b With 2 or 4 anther-bearing stamens and no antherless stamens

      9a With 2 anther-bearing stamens (corolla often nearly regular, usually 4 lobed, the uppermost lobe usually the largest).......................
        .... SCROPHULARIACEAE, SUBKEY 3

      9b With 4 anther-bearing stamens

        10a Corolla irregular, more than 7 mm long, except in *Tonella tenella;* most leaves distributed along the stem, the leaves opposite; petioles usually not longer than leaf blades

          11a Corolla about 2.5 mm long, without an obvious tube (corolla white, turning to violet on the lobes; annual; up to 30 cm tall; in shaded areas) ....*Tonella tenella*
            Smallflower Tonella; SCl-n

          11b Corolla generally more than 7 mm long, and with an obvious tube .... SCROPHULARIACEAE, SUBKEY 4
            Mimulus

        10b Corolla nearly regular, not more than 3 mm long; leaves entirely basal; petioles, if any, longer than leaf blades (corolla white, pink, or lavender; annual; often tufted or forming mats; less than 10 cm tall; on muddy shores)

          12a Leaf blades not distinct from the petioles (leaves 1–3 cm long)
            .... *Limosella subulata*
            Delta Mudwort; na

          12b Leaf blades distinct from the petioles

            13a Lobes of corolla rounded, white or violet tinged; leaves 1–6 cm long, the blades rarely more than 2 mm wide; style 0.5–1 mm long .... *Limosella acaulis*
              Southern Mudwort; Ma-s

13b Lobes of corolla pointed, white or pale pink; leaves 3–12 cm long, sometimes more, the blades up to 8 mm wide; style less than 0.5 mm long . . . . *Limosella aquatica* Northern Mudwort; eua

1b Upper lip of corolla modified to form a narrow hood or beaklike structure that usually encloses the stamens and at least much of the pistil (herbs)

14a Leaves opposite, except for the leaflike bracts in the inflorescence (annual)

15a Corolla yellow; inflorescence somewhat loose; calyx lobes equal (coastal) . . . . *Parentucellia viscosa* Yellow Parentucellia; eu

15b Corolla pink and white; inflorescence dense; calyx lobes unequal . . . . *Bellardia trixago* (pl. 52); me

14b Leaves alternate

16a Each stamen with 2 pollen-producing sacs, these attached at the same level to the filament; most leaves at least 10 cm long, pinnately lobed, the lobes toothed (corolla purple red; perennial)

17a Corolla about 25 mm long, its lower lip only about one-eighth as long as the upper lip; leaves shorter than the flowering stem (common) . . . . *Pedicularis densiflora* (pl. 54) Indian-warrior

17b Corolla about 18 mm long, its lower lip about half as long as the upper lip; leaves often as long or longer than the flowering stem . . . . *Pedicularis dudleyi* Dudley Lousewort; SM; 1b

16b Each stamen either with 1 pollen-producing sac or with 2 sacs, one of these attached below the other; leaves rarely more than 6 cm long, and if pinnately lobed, the lobes smooth margined (bracts and calyx usually not green, at least at the tips)

18a Calyx not boat shaped; inflorescence usually with more than 10 flowers . . . . SCROPHULARIACEAE, SUBKEY 5 *Castilleja, Triphysaria*

18b Calyx a boat-shaped structure, sometimes notched at the tips, extending slightly above the corolla; inflorescence usually with fewer than 10 flowers, rarely 15 (lower lip of corolla forming a pouch; bract closest to the calyx [inner bract] resembling the calyx; outer bract below 1 or several flowers, leaflike)

19a Leaves mostly less than 8 times as long as wide (inflorescence 2–15 cm long; in salt marshes)

20a None of the leaves lobed; inner bract with 2 teeth separated by a notch . . . . *Cordylanthus maritimus* Salt Marsh Bird's-beak; 1b

20b Some of the upper leaves with 1 or 2 pairs of short side lobes; inner bract with 3–7 lobes

21a Plant bristly, and usually branching near the base; corolla pouch and tube only slightly hairy (in inland saline habitats) . . . . *Cordylanthus mollis* ssp. *hispidus* [*C. hispidus*] Hispid Bird's-beak; Sl; 1b

21b Plant usually soft hairy, usually branching near the middle; corolla pouch and tube densely hairy (in coastal salt marshes) .... *Cordylanthus mollis* ssp. *mollis*

Soft Bird's-beak; SF; 1b

19b Lower leaves mostly more than 8 times as long as wide, the upper ones wider and sometimes lobed

22a Plant bristly, not glandular; inflorescence usually with 5–15 flowers, but sometimes 3 or 4 (corolla yellow with a maroon mark, the pouch white) .... *Cordylanthus rigidus*

Stiff Bird's-beak; SCl-s

22b Plant not bristly, but usually with some gland-tipped hairs; inflorescence with 1–4 flowers

23a Corolla white, the inside with purple streaks; outer bracts 3 lobed nearly to their bases (on serpentine) .... *Cordylanthus nidularius*

Mount Diablo Bird's-beak; CC (MD); 1b

23b Corolla white, but usually with yellow tips and some maroon markings; outer bracts either not lobed or lobed for not more than two-thirds their length (these species difficult to separate)

24a Corolla only slightly marked with maroon (tips of outer bracts not lobed, but sometimes with 3 rounded teeth) .... *Cordylanthus pilosus*

Hairy Bird's-beak; SFBR

24b Corolla conspicuously marked with maroon (on serpentine)

25a Outer bracts divided, for about half their length, into 3 slender lobes .... *Cordylanthus tenuis* ssp. *capillaris*

Pennell Bird's-beak; Sn; 1b

25b Outer bracts not lobed .... *Cordylanthus tenuis* ssp. *brunneus*

Serpentine Bird's-beak; Na, Sn; 4

Scrophulariaceae, Subkey 1: Upper lip of corolla not modified to form a hood or beaklike structure; at least the lower stem leaves opposite; corolla irregular, the tube with a broad sac; anther-bearing stamens 4; annual .... *Antirrhinum, Collinsia*

1a Corolla tube with a sac on the lower side near the base; leaves opposite below and alternate above

2a Corolla 3–5 cm long; plant robust .... *Antirrhinum majus*

Common Snapdragon; me

2b Corolla rarely more than 1.5 cm long, except in *Antirrhinum multiflorum*; plant mostly delicate and often clinging to other vegetation by slender branches or pedicels

3a Stem twining and clinging by slender pedicels; pedicel 3–9 cm long (corolla blue, about 1.5 cm long) .... *Antirrhinum kelloggii* (pl. 52)

Lax Snapdragon, Kellogg Snapdragon; Ma-s

3b Stem sometimes clinging by slender branches, but not by pedicels; pedicel not more than 1 cm long

    4a Plant with slender, somewhat twisted branches that usually cling to other vegetation (corolla lavender violet or light purple; pedicel 1–4 mm long)

        5a Corolla 11–17 mm long, without dark veins; upper calyx lobes 8–12 mm long, the others 7–9 mm long, longer than the fruit (coastal) . . . . *Antirrhinum vexillo-calyculatum* ssp. *vexillo-calyculatum*

            Wiry Snapdragon; Sn-SB

        5b Corolla 8–12 mm long, with some dark veins; upper calyx lobes 4–7 mm long, the others 3–5 mm long, shorter than the fruit. . . . . *Antirrhinum vexillo-calyculatum* ssp. *breweri* [*A. breweri*] (pl. 52)

            Brewer Snapdragon; Sn-n

    4b Plant supporting itself without the help of twisted branches that cling to other vegetation

        6a Flowers solitary in the leaf axils; corolla 9–11 mm long, white with violet veins, the sac on lower side deeper than wide; pedicel 1–2 mm long. . . . . *Antirrhinum cornutum*

            Spurred Snapdragon; Na-n

        6b Flowers in racemes at stem ends; corolla 13–18 mm long, pink to red, the sac on lower side not deeper than wide; pedicel 2–10 mm long

            7a Plant hairy; calyx lobes unequal; sac on lower side of corolla wider than deep; annual; usually less than 1 m tall (often on fire burns) . . . . *Antirrhinum multiflorum*

                Manyflower Snapdragon, Sticky Snapdragon; SCl-s

            7b Plant not hairy; calyx lobes equal; sac on lower side of corolla about as wide as deep; perennial; often more than 1 m tall (sometimes on serpentine) . . . . *Antirrhinum virga*

                Tall Snapdragon; Sn; 4

1b Corolla tube with one sac on upper side and another on the middle lobe of the lower lip, this one enclosing the stamens; leaves all opposite

    8a Flowers, even the upper ones, mostly in pairs in the leaf axils; pedicel usually more than 1.5 cm long (corolla purple, lavender, or white, the upper lip usually lighter; middle lobe of lower lip hairy)

        9a Corolla 7–13 mm long . . . . *Collinsia sparsiflora* var. *sparsiflora* (pl. 53)

            Fewflower Blue-eyed-Mary; CC-n

        9b Corolla 12–20 mm long . . . . *Collinsia sparsiflora* var. *arvensis*

            Largeflower Blue-eyed-Mary; Na, Sn, CC

    8b Flowers, at least the upper ones, in whorls of at least 3; pedicel usually less than 1.5 cm long, but up to 2 cm long in *Collinsia multicolor*

        10a Side lobes of lower lip of corolla hairy on the inner surfaces (plant imparting a brown stain to the skin when handled; corolla yellow to pale green, with some spots) . . . . *Collinsia tinctoria*

            Sticky Chinesehouses; Sn-n

        10b Side lobes of lower lip of corolla not hairy on the inner surfaces

11a Upper lip of corolla less than half as long as the lower lip; flowers in a single cluster (upper lip of corolla blue, lower lip white; coastal, usually on sandy backshores)... *Collinsia corymbosa*

Round-headed Chinesehouses; SF-n; 1b

11b Upper lip of corolla more than half as long as the lower lip; flowers not in a single cluster

12a Pedicel longer than the calyx (lower flowers single or in pairs, the upper ones in whorls; upper lip of corolla white with spots, lower lip blue, the middle lobe purple; calyx lobes pointed at the tips; in moist, shaded areas) .... *Collinsia multicolor*

San Francisco Collinsia; SF-Mo; 1b

12b Pedicel equal to or shorter than the calyx

13a Corolla dark purple; pedicel about equal to the calyx; filaments of upper 2 stamens hairy only near their bases (calyx lobes blunt at the tips; sometimes on serpentine) ....*Collinsia greenei*

Green Collinsia; Sn-n

13b Corolla not dark purple; pedicel shorter than the calyx; filaments of upper 2 stamens hairy for much of their length

14a Upper lip of corolla white, the lower lip violet; calyx lobes pointed at the tips; filaments of upper 2 stamens with slender outgrowths that project upward into the sac of the corolla tube; common at the margins of woods ....
.... *Collinsia heterophylla* (pl. 53)

Chinesehouses

14b Corolla uniformly rose to white; calyx lobes blunt at the tips; filaments of upper 2 stamens without outgrowths; in open, sandy areas .... *Collinsia bartsiifolia* [*C. bartsiaefolia*]

White Chinesehouses; La-s

Scrophulariaceae, Subkey 2: Upper lip of corolla not modified to form a hood or beaklike structure; leaves opposite or whorled; corolla usually irregular; anther-bearing stamens 4; antherless stamen 1, this reduced to a scale in *Scrophularia californica*

1a Corolla 8–12 mm long, the 2 upper lobes larger than the other 3; stem 4 sided (corolla red brown, the lower lobe turned downward; leaves coarsely toothed; herb; often more than 1 m tall)... *Scrophularia californica* (pl. 54)

Beeplant, California Figwort

1b Corolla usually at least 15 mm long, the 2 upper lobes equal to or smaller than the other 3; stem usually not 4 sided

2a Corolla scarlet, with a slender tube and 5 nearly equal lobes (plant not glandular hairy; herb)... *Penstemon centranthifolius*

Scarlet-bugler; La-s

2b Corolla not scarlet, but occasionally some other shade of red, distinctly 2 lipped

3a Corolla mostly yellow, but with a brown upper lip and with some purple lines, less than 1.5 cm long (calyx glandular hairy; shrub) .... *Keckiella lemmonii*

Lemmon Keckiella; Sl-n

3b   Corolla not yellow, usually at least 1.5 cm long

    4a   Corolla white, tinged with pink or lavender, and with darker lines

        5a   Corolla usually 2–3 cm long, tinged with violet or blue violet; herb (calyx glandular hairy) . . . . *Penstemon grinnellii* var. *scrophularioides*
Grinnell Penstemon; SCl (MH)

        5b   Corolla usually 1.5–2 cm long, tinged with pink; shrubs

            6a   Calyx glandular hairy . . . . *Keckiella breviflora* var. *breviflora*
Gaping Keckiella, Yawning Penstemon; Al-s

            6b   Calyx not hairy . . . . *Keckiella breviflora* var. *glabrisepala;* Me-Na

    4b   Corolla pink, rose, red, magenta, blue, lavender, violet, or purple

        7a   Leaves rarely more than 5 mm wide, and usually at least 8 times as long as wide; calyx not glandular, but sometimes with hairs (corolla magenta to blue; woody, at least at the base)

            8a   Plant not obviously hairy, except perhaps in the lower portion (widespread) . . . *Penstemon heterophyllus* var. *heterophyllus* (pl. 54)
Foothill Penstemon, Chaparral Penstemon

            8b   Plant with short hairs almost throughout . . . .
. . . .*Penstemon heterophyllus* var. *purdyi*
Purdy Penstemon; SB-n

        7b   Wider leaves mostly at least 10 mm, and usually not more than 5 times as long as wide; calyx glandular hairy

            9a   Larger leaves commonly 10 cm long and 2 cm wide; herb; up to about 100 cm tall (corolla lavender, violet purple, or red purple) . . . . *Penstemon rattanii* var. *kleei* [*P. rattanii* ssp. *kleei*]
Santa Cruz Mountains Beardtongue; SCl; 1b

            9b   Larger leaves up to 4 cm long and less than 2 cm wide; shrubs; mostly less than 50 cm tall

                10a   Inflorescence a raceme; corolla dark rose purple; leaves with more than 5 teeth . . . .*Penstemon newberryi* var. *sonomensis*
Sonoma Beardtongue; Na, Sn; 1b

                10b   Inflorescence a nearly flat-topped corymb; corolla pink to red; leaves either smooth margined or with 3–5 teeth . . . .
. . . . *Keckiella corymbosa*
Redwood Keckiella, Cliff Penstemon; Mo-n

Scrophulariaceae, Subkey 3:   Upper lip of corolla not modified to form a hood or beaklike structure; leaves either almost entirely basal, or opposite or whorled; corolla often nearly regular, usually 4 lobed, the uppermost lobe usually the largest; anther-bearing stamens 2; antherless stamens none

1a   Leaves whorled, tough and leathery, smooth margined; shrubs (leaves sessile; often more than 1 m tall; on bluffs and other coastal locations)

    2a   Corolla purple blue, about 8 mm wide; leaves up to 10 cm long . . . . *Hebe speciosa*
Showy Hebe; au

    2b   Corolla lilac, about 10 mm wide; leaves about 3 cm long (believed to be a hybrid) . . . .
. . . . *Hebe* × *franciscana*

1b   Leaves opposite or almost entirely basal, not tough and leathery, often toothed; herbs

    3a   Leaves almost entirely basal, the stem leaves, if present, very reduced and alternate; petioles at least 5 cm long; leaf blades usually more than 2 cm wide, heart shaped (leaves coarsely toothed; corolla purple blue; in moist, shaded areas)................ .... *Synthyris reniformis* [includes *S. reniformis* var. *cordata*] Snowqueen; Ma-n

    3b   Leaves opposite, scattered along the stem, but each flower has a leaflike bract below it; petioles absent or less than 0.5 cm long; leaf blades rarely so much as 2 cm wide, not heart shaped, except sometimes in *Veronica filiformis*

        4a   Flowers in racemes in the leaf axils, typically 2 or more racemes on each main stem (perennial)

            5a   Leaves toothed, with short but distinct petioles

                6a   Leaves with very small teeth; corolla 7–10 mm wide, violet blue to lilac (in wet habitats) .... *Veronica americana* (fig.) American Brooklime

                6b   Leaves with coarse teeth; corolla 10–12 mm wide, bright blue with white dots .... *Veronica chamaedrys* (fig.) Germander Speedwell; eu

            5b   Leaves sometimes toothed, sessile (corolla lilac to lavender blue; in wet places)

                7a   Leaves commonly at least 8 times as long as wide; corolla 5–7 mm wide .... *Veronica scutellata* (fig.) Marsh Speedwell; Ma-n

                7b   Leaves not often so much as 4 times as long as wide; corolla 4–5 mm wide .... *Veronica anagallis-aquatica* (fig.) Great Water Speedwell; eu

        4b   Flowers either solitary in the leaf axils or in racemes at the tip of the stem (leaves with short petioles)

            8a   Leaves sometimes toothed, the lower one several times as long as wide; corolla white (corolla 2–3 mm wide; flowers in racemes; annual; usually in moist areas)........................... *Veronica peregrina* ssp. *xalapensis* Purslane Speedwell

            8b   Leaves usually toothed, the lower ones not so much as twice as long as wide; corolla blue

                9a   Corolla 2–3 mm wide; pedicel 1–2 mm long (flowers in racemes; annual) .... *Veronica arvensis* Common Speedwell, Corn Speedwell; eua

                9b   Corolla usually at least 7 mm wide; pedicel much more than 2 mm long

                    10a   Flowers in racemes; corolla 7–11 mm wide; longer leaf blades more than 1 cm, longer than wide, not heart shaped; annual .... *Veronica persica* Persian Speedwell; as

                    10b Flower solitary; corolla 10–15 mm wide; leaf blades rarely more than 1 cm long, about as long as wide, often somewhat heart shaped; perennial .... *Veronica filiformis* Asian Speedwell; as

Scrophulariaceae, Subkey 4: Upper lip of corolla not modified to form a hood or beaklike structure; leaves opposite; corolla irregular; anther-bearing stamens 4, antherless stamens none . . . . *Mimulus*

1a   Corolla scarlet (herb; up to 80 cm tall; along streams) . . . . *Mimulus cardinalis* (pl. 53)
                                                                     Scarlet Monkeyflower

1b   Corolla not scarlet, but sometimes purple red

    2a   Corolla usually dull orange; shrub (often more than 1 m tall; widespread on dry hillsides) . . . . *Mimulus aurantiacus* (pl. 53)
                                         Bush Monkeyflower, Sticky Monkeyflower

    2b   Corolla yellow, purple red, purple, or pink; herbs

        3a   Corolla mostly yellow, often with red spots or a red blotch (growing in wet, often partly shaded places)

            4a   Stem and leaves slimy; corolla almost entirely yellow, without prominent red spots, but sometime with very small brown dots (leaves with short hairs; corolla about 2 cm long; up to 30 cm tall) . . . . *Mimulus moschatus*
                                   Muskflower, Musk Monkeyflower; SCr-n

            4b   Stem and leaves not slimy, except sometimes in *Mimulus floribundus;* corolla usually with prominent red or red brown spots on the lips or inside the tube

                5a   Calyx of mature flower flattened from side to side, inflated in fruit; corolla usually at least 2 cm long (up to 1 m tall, but usually much shorter; widespread) . . . . . . *Mimulus guttatus* [include *M. guttatus* sspp. *arvensis, micranthus,* and *M. nasutus*] (pl. 53)
                                   Common Monkeyflower

                5b   Calyx of mature flower cylindrical, not obviously inflated in fruit; corolla rarely so much as 1.5 cm long

                    6a   Midribs of calyx lobes raised into ridges; leaves usually not sessile, the blades usually less than twice as long as wide, toothed; up to 50 cm tall . . . . *Mimulus floribundus*
                        Longflowering Monkeyflower, Floriferous Monkeyflower

                  6b   Midribs of calyx lobes not raised into ridges; leaves sessile, the blades usually more than twice as long as wide, smooth margined; up to 35 cm tall . . . . *Mimulus pilosus*
                          Downy Monkeyflower, Downy Mimetanthe

        3b   Corolla mostly purple red, purple, or pink, but often with yellow or white inside, and in *Mimulus tricolor* and *M. angustatus,* tube is yellow

            7a   Tube of corolla yellow (corolla 3–4.5 cm long, the lips purple and spotted; not more than 12 cm tall; mostly around drying pools)

                8a   Tube of corolla twice as long as the calyx; throat of corolla white with a purple-spotted yellow blotch; up to 12 cm tall . . . . *Mimulus tricolor* (fig.)
                                Tricolor Monkeyflower; Sn-n

                8b   Tube of corolla 3–6 times as long as the calyx; throat of corolla purple with darker spots; not more than 2 cm tall . . . . *Mimulus angustatus*
                            Narrowleaf Monkeyflower; Me-Na

7b Tube of corolla not yellow

    9a Pedicel much longer than the calyx; corolla usually falling away after withering (corolla red purple; pedicel 15–25 mm long; up to 8 cm tall) .... *Mimulus androsaceus*
Androsace Monkeyflower; SCl

    9b Pedicel usually shorter than the calyx; corolla usually adhering to the pistil after withering

        10a Corolla 3–4.5 cm long (fruit lopsided, thick walled, breaking open late or not at all)

            11a Corolla mostly purple, the throat often streaked with gold, sometimes closed, the lower lip less than one-third as long as the upper lip; up to 6 cm tall (on serpentine) .... *Mimulus douglasii*
Purple Monkeyflower, Chinless Monkeyflower; SB-n

            11b Corolla mostly red purple, the throat yellow or gold, not closed, the lower lip more than one-third as long as the upper lip; up to 30 cm tall (in moist areas) ....*Mimulus kelloggii*
Kellogg Monkeyflower; Na-n

        10b Corolla not more than 2.5 cm long

            12a Corolla lobes essentially equal; stigma lobes nearly equal in size and shape (corolla 13–20 mm long, mostly red purple, the throat white with dark spots; up to 20 cm tall; in sandy areas) .... *Mimulus layneae*
Layne Monkeyflower; Na-n

            12b Corolla lobes obviously unequal; stigma lobes unequal in size, shape, or both

                13a Corolla 8–10 mm long; calyx less than 10 mm long, usually longer than the longest bracts (corolla red purple or rose; fruit symmetrical, thin walled, usually breaking open soon after ripening; up to 15 cm tall) .... *Mimulus rattanii*
Rattan Monkeyflower; Ma

                13b Corolla 15–25 mm long; calyx mostly more than 10 mm long, shorter than the longest bracts

                    14a Plant up to 60 cm tall, with a tobaccolike odor; corolla mostly red purple, the throat with some white on the lower side; fruit symmetrical, thin walled, usually breaking open soon after ripening; in dry, open places .... *Mimulus bolanderi*
Bolander Monkeyflower; Me-s

                    14b Plant up to 8 cm tall, without a tobaccolike odor; corolla red purple; fruit lopsided, thick walled, breaking open late or not at all; in damp places .... *Mimulus congdonii*
Congdon Monkeyflower; Me-s

Scrophulariaceae, Subkey 5: Upper lip of corolla modified to form a narrow hood or beaklike structure; bracts and calyx usually not green, at least at the tips; leaves alternate; pollen-producing sacs either 1 or 2, one of the sacs attached below the other .... *Castilleja, Triphysaria*

1a Corolla mostly green, the upper lip at least 3 times longer than the lower lip (pollen-producing sacs 2 on each stamen)

    2a Stem and leaves white woolly (bract tips usually orange red; corolla 18–25 mm long; perennial, sometimes woody; widespread)............... *Castilleja foliolosa* (pl. 52)

                    Woolly Indian Paintbrush

    2b Stem and leaves not white woolly, but sometimes hairy

        3a Bracts and leaves not lobed; annual (bract tips red; corolla 15–35 mm long, the lower lip yellow or purple, upper lip yellow)................................

               Castilleja minor ssp. *spiralis* [*C. stenantha* sspp. *spiralis* and *stenantha*]

                 Largeflower Paintbrush; La-s

        3b Most or all bracts lobed; perennial (leaves often lobed)

            4a Calyx divided more deeply in front compared with back (bract tips usually red or orange red, occasionally yellow)

                5a Bracts 2–6 cm long, not lobed; corolla 3–5 cm long, the upper lip about 1.5 cm long; leaves up to 8 cm long, sometimes lobed; often more than 100 cm tall........ *Castilleja subinclusa* ssp. *franciscana* [*C. franciscana*]

                      Franciscan Paintbrush; Sn-SM

                5b Bracts 1.5–3.5 cm long, with 1–5 lobes or none; corolla 2–4 cm long, the upper lip about 1 cm long; leaves up to 6 cm long, not lobed; not more than 50 cm tall.............. *Castilleja miniata* [includes *C. uliginosa*]

                      Red Paintbrush; Sn

            4b Calyx divided about as deeply in front as in back

                6a Stem and leaves obviously glandular hairy below the inflorescence (bract tips red or yellow; corolla 2.5–4 cm long, the upper lip about 1.5 cm long; calyx lobes 2–4 mm long) ....

                    .... *Castilleja applegatei* ssp. *martinii* [*C. roseana*]

                    Wavyleaf Paintbrush, Martin Paintbrush; La-s

                6b Stem and leaves not obviously glandular hairy below the inflorescence (*Castilleja wightii*, usually with some gland-tipped hairs lower on the stem)

                    7a Calyx lobes 2–3 mm long, blunt at the tips (bract tips red to yellow; corolla 21–25 mm long, the upper lip 13–15 mm long; coastal) .... *Castilleja wightii* (pl. 53)

                      Seaside Paintbrush; Me-SM

                  7b Calyx lobes 3–6 mm long, pointed at the tips

                      8a Corolla 18–22 mm long, the upper lip 9–10 mm long; bract tips usually yellow; on serpentine................................

                      .... *Castilleja affinis* ssp. *neglecta* [*C. neglecta*]

                      Tiburon Indian Paintbrush; Ma; 1b

                    8b Corolla 25–35 mm long, the upper lip more than 16 mm long; bract tips red, purple red, orange red or yellow; not on serpentine

9a   Leaves 3–9 cm long, often with 2–6 lobes, these not more than 5 mm wide; hairs, if present, stiff; bract tips red, orange red, or yellow; up to 60 cm tall; in dry areas; widespread …
.....*Castilleja affinis* ssp. *affinis* [includes *C. inflata*] (pl. 52)
Common Indian Paintbrush, Coast Paintbrush; SFBR-s

9b   Leaves mostly 2–5 cm long, not lobed; hairs not stiff; bract tips purple red; up to 100 cm tall; in swampy areas; rare
.... *Castilleja chrymactis*
Alaska Paintbrush; na

1b   Corolla not mostly green, the upper lip not more than 2 times longer than the lower lip (annual)

10a   Each stamen with only 1 pollen-producing sac

11a   Corolla 4–6 mm long, purple brown; plant branching from the base, without a distinct main stem .... *Triphysaria pusilla [Orthocarpus pusillus]* (fig.)
Dwarf Owl's-clover; SLO-n

11b   Corolla 10–25 mm long, not purple brown; plant branching well above the base, with a distinct main stem

12a   Stamens protruding beyond upper lip of corolla; corolla white or cream colored (corolla 10–14 mm long) ....
.... *Triphysaria floribunda [Orthocarpus floribundus]*
San Francisco Owl's-clover; Ma-SM; 1b

12b   Stamens not protruding beyond upper lip of corolla; corolla not white or cream colored

13a   Upper lip of corolla yellow; plant scarcely hairy, except perhaps on the inflorescence (lower lip of corolla 2–4 mm deep, pale yellow with purple dots on the margins)..
..... *Triphysaria versicolor* ssp. *faucibarbata [Orthocarpus faucibarbatus]*
Smooth Owl's-clover; Me-Ma

13b   Upper lip of corolla purple; plant very hairy

14a   Lower lip of corolla 3–4 mm deep, yellow, purple at the base (common) .... *Triphysaria eriantha* ssp. *eriantha*
[Orthocarpus erianthus] (pl. 54)
Yellow Johnnytuck, Butter-and-eggs; SLO-n

14b   Lower lip of corolla 4–5 mm deep, rose or white ....
.... *Triphysaria eriantha* ssp. *rosea*
[Orthocarpus erianthus var. *roseus]*
Rose Johnnytuck; Me-SLO

10b   Each stamen with 2 pollen-producing sacs, 1 at the tip, the other lower

15a   Bracts entirely green; upper lip of corolla not protruding much beyond the lower lip (corolla yellow or white, usually with 2 purple spots)

16a   Bracts not lobed; corolla 15–25 mm long, the tube 1–2 mm wide (up to 25 cm tall; in moist areas, including vernal pools)..............................
.... *Castilleja campestris [Orthocarpus campestris]*
Field Owl's-clover; Sn-n

16b Bracts lobed; corolla 20–28 mm long, the tube more than 2 mm wide just
below the level where the upper and lower lip diverge (in grassland)
    17a Corolla yellow (widespread)......................................
.... *Castilleja rubicundula* ssp. *lithospermoides*
[Orthocarpus lithospermoides var. *lithospermoides]*
Yellow Creamsacs; SCl-n
    17b Corolla white, turning pink with age ....
.... *Castilleja rubicundula* ssp. *rubicundula*
[Orthocarpus lithospermoides var. *bicolor]*
White Creamsacs; Na; 1b

15b Bracts with purple or yellow tips; upper lip of corolla definitely protruding beyond the lower lip
    18a Upper lip of corolla densely hairy, with a hooklike tip; stem mainly purple or
red (corolla purple, the lower lip often white or yellow at the tip and with
purple dots)
        19a Inflorescence purple at the tip but green for more than half its length,
and without a banded appearance; most bract tips 1 mm wide, usually
purple; on grassy hills (common) .... *Castilleja exserta* ssp. *exserta*
[Orthocarpus purpurascens var. *purpurascens]* (pl. 52)
Purple Owl's-clover, Escobita; Me-s
        19b Inflorescence purple for more than half its length and with a banded ap-
pearance because of bract tips; most bract tips 2–3 mm wide, white,
lavender, or purple; on coastal dunes ..... *Castilleja exserta* ssp. *latifolia*
[Orthocarpus purpurascens var. *latifolius]*
Banded Owl's-clover; SLO-n
    18b Upper lip of corolla only slightly hairy, nearly straight; stem not mainly pur-
ple or red
        20a Corolla slender, not noticeably broader at the tip; stigma not protruding
beyond corolla (corolla white or purple, with purple dots) ....
.... *Castilleja attenuata [Orthocarpus attenuatus]* (pl. 52)
Valley-tassels
        20b Corolla becoming broader toward the tip; stigma protruding beyond
corolla
            21a Corolla mostly purple, but the lower lip often yellow; leaves usually
not more than 3 mm wide, and usually with lobes that are as long as
the basal portions; upright; in grassy fields (widespread) ....
.... *Castilleja densiflora [Orthocarpus densiflorus]*
Common Owl's-clover; Me-s
            21b Corolla mostly pale yellow, but with some purple dots; leaves often
more than 3 mm wide and either smooth margined or with lobes
that are shorter than the basal portions; somewhat sprawling; in salt
marshes .... *Castilleja ambigua [Orthocarpus castillejoides]*
Johnny-nip; Mo-n

SIMAROUBACEAE (QUASSIA FAMILY)   Our only representative of the Quassia or Simarouba Family is *Ailanthus altissima* (Tree-of-heaven). Imported from China, it has become so successfully established in some portions of California that it may seem to be native. In many areas, it is considered a noxious weed because of its invasive roots and rapid growth.

This tree, which grows to a height of about 20 m, has alternate, deciduous leaves that are pinnately compound and may be more than 75 cm long. The 13–25 leaflets resemble those of *Juglans,* our native walnut; the leaflets of a walnut, however, are toothed and sessile, whereas those of *Ailanthus* have only a couple of teeth at the base and usually are stalked. The small flowers, in panicles at stem ends, have five or six separate greenish yellow petals and five or six sepals united at their bases. The ovary is superior. The pistillate flowers, with two to five pistils, and the staminate flowers, with 10–12 stamens, are generally on separate plants, mixed with some flowers that have both pistils and stamens. The 2–5 styles are twisted around each other. Each pistil consists of a few divisions that eventually separate, forming one-seeded, winged structures.

SOLANACEAE (NIGHTSHADE FAMILY)   From one's seat at the dinner table, the Nightshade Family appears to be very important, for it includes the potato, tomato, eggplant, and many varieties of sweet pepper and chili pepper (but not black pepper). It also includes tobacco and plants of medicinal value, as well as some whose fruit or leaves are poisonous. *Petunia* and *Salpiglossis* are well-known genera of garden plants. Several of the species common in our area are well-established weeds.

The leaves are usually alternate and often lobed. The flower generally has a five-lobed calyx and corolla and five stamens attached to the corolla tube. The ovary is superior. The fruit, partitioned into two to five seed-producing divisions, is sometimes dry when mature, or sometimes a fleshy berry.

1a   Corolla with a definite and often long tube
  2a   Shrub or small tree; often more than 5 m tall (leaves bluish, smooth margined; flowers in panicles; corolla greenish yellow, 3–4 cm long, mostly tubular, but with short lobes; fruit dry) . . . . *Nicotiana glauca* (pl. 54)
Tree-tobacco; sa
  2b   Herbs; not more than 2 m tall
    3a   Corolla tube becoming slightly narrower toward the opening; fruit a fleshy berry (corolla white; fruit 10–15 mm wide, white or yellow; leaves smooth margined; perennial; stem usually trailing or climbing, up to 1.5 m long) . . . .
. . . . *Salpichroa origanifolia* [*S. rhomboidea*]
Lily-of-the-valley Vine; sa
    3b   Corolla tube becoming broader toward the opening; fruit dry, cracking apart into 2 or 4 divisions
      4a   Corolla less than 1 cm long, the tube white, the lobes pale blue or purple and slightly unequal; fruit cracking apart into 2 divisions (leaves smooth margined, the blades about 1 cm long, sometimes opposite; stem generally falling down) . . . . *Petunia parviflora*
Wild Petunia

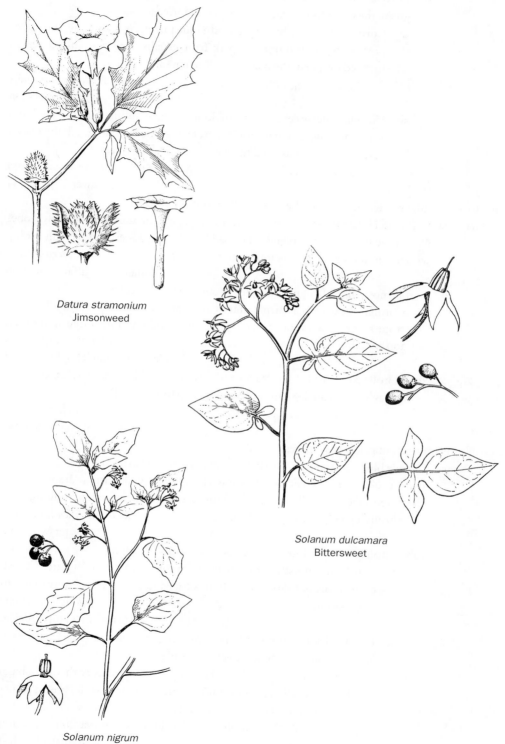

Datura stramonium
Jimsonweed

Solanum dulcamara
Bittersweet

Solanum nigrum
Black Nightshade

4b Corolla more than 2 cm long, white, white tinged with blue purple, or pale green, the lobes equal; fruit cracking apart into 4 divisions

    5a Corolla white or white tinged with blue purple, 6–8 cm long, the upper portion broadly funnel shaped; fruit usually with soft spines; leaves toothed or lobed, the blades up to 20 cm long................................... *Datura stramonium* (fig.)
Jimsonweed; sa

    5b Corolla pale green, 2.5–4 cm long, mostly tubular, but with lobes 1–1.5 cm long; fruit without spines; leaves smooth margined, the blades rarely more than 10 cm long (calyx with 5 dark stripes)............... *Nicotiana acuminata* var. *multiflora*
Manyflower Tobacco; sa

1b Corolla with an extremely short tube (leaves often toothed or lobed)

  6a Plant with prickles on the stem and sometimes with spines on the leaves and calyx lobes

    7a Corolla yellow; fruit completely enclosed by the calyx lobes; annual (calyx lobes with numerous spines; leaves lobed; up to 70 cm tall) .... *Solanum rostratum*
Buffalobur, Buffalo-berry; na

    7b Corolla not yellow; fruit not completely enclosed by the calyx lobes; shrubs

      8a Corolla white, with a purple, star-shaped mark; leaves and calyx lobes with stout spines; leaves lobed; fruit 30–40 mm wide, yellow; often more than 100 cm tall .... *Solanum marginatum*
White-margined Nightshade; af

      8b Corolla pale purple blue; leaves and calyx lobes without spines; leaves sometimes lobed; fruit 6–8 mm wide, orange; up to 50 cm tall.................. *Solanum lanceolatum*
Lanceleaf Nightshade; sa

  6b Plant without prickles or spines

    9a Corolla blue, violet, or purple (pale purple in *Solanum aviculare*)

      10a Corolla deeply lobed, the lobes much longer than wide (fruit red; some leaves three lobed; vinelike plant, sometimes woody at the base, climbing over shrubs; usually in wet places) .... *Solanum dulcamara* (fig.)
Bittersweet; eua

      10b Corolla shallowly lobed, the lobes much wider than long

        11a Leaf blades mostly more than 10 cm long and often pinnately lobed; corolla pale purple (corolla about 3 cm wide; shrub; often more than 2 m tall; coastal) .... *Solanum aviculare*
Australian Nightshade; au

        11b Leaf blades rarely more than 8 cm long, sometimes with 1 or 2 basal lobes; corolla deep blue, violet, or purple

          12a Hairs on stem not branched; leaves sometimes with 1 or 2 basal lobes (corolla 15–30 mm wide; fruit 10–15 mm wide, green; sometimes woody at the base) ....
.... *Solanum xanti* [includes *S. xanti* var. *intermedium*]
Purple Nightshade; Me-s

12b Some hairs on upper stem branched; leaves toothed or smooth margined

13a Corolla 16–25 mm wide, deep blue or violet, the back of the lobes without star-shaped hairs; fruit 12–14 mm wide, white with a green base; upright, sometimes woody at the base (widespread) .... *Solanum umbelliferum*
[includes *S. umbelliferum* var. *incanum*] (pl. 54)
Blue Nightshade, Blue Witch; Me-s

13b Corolla 10–18 mm wide, purple, the back of the lobes with star-shaped hairs; fruit 5–6 mm wide, yellow; climbing, not woody
.... *Solanum furcatum*
Forked Nightshade; sa

9b Corolla yellow, white, or white tinged with purple or lavender

14a Corolla yellow, with a purple center, shallowly lobed; mature calyx enlarged, forming a loose bladderlike covering around the entire fruit (leaves sometimes toothed, sometimes opposite; annual) ....
.... *Physalis philadelphica [P. ixocarpa]*
Tomatillo; mx

14b Corolla white or white tinged with purple or lavender, deeply lobed; mature calyx, if slightly enlarged, not completely covering the fruit

15a Stem and calyx very hairy; mature calyx slightly enlarged and partly covering the fruit; fruit yellow; leaves sometimes lobed (annual; most if not all Bay Area specimens misidentified as *Solanum sarrachoides*) ....
.... *Solanum physalifolium* (pl. 54)
Hairy Nightshade; sa

15b Stem and calyx only slightly if at all hairy; mature calyx not enlarged; fruit black or dark purple; leaves sometimes toothed, but not lobed

16a Inflorescence usually branched into 2 umbel-like flower clusters (corolla 10–18 mm wide; perennial) .... *Solanum furcatum*
Forked Nightshade; sa

16b Inflorescence usually consisting of a single umbel or raceme

17a Corolla 3–6 mm wide; calyx lobes turned back when the fruit is ripe (fruit glossy; annual or perennial; up to 80 cm tall)
.... *Solanum americanum [S. nodiflorum]*
Smallflower Nightshade; sa

17b Corolla 10 mm wide; calyx lobes not turned back

18a Corolla with green spots below the bases of the lobes; anthers 2.5–4 mm long; perennial; up to 2 m tall ....
.... *Solanum douglasii*
Douglas Nightshade; SM-s

18b Corolla with yellow spots below the bases of the lobes; anthers less than 2.5 mm long; annual; about 0.5 m tall
.... *Solanum nigrum* (fig.)
Black Nightshade; eu

STERCULIACEAE (CACAO FAMILY)   Nearly all members of Sterculiaceae are found in the tropics. Most famous of them is the cacao tree, whose seeds yield chocolate; it is a native of South America. Our only indigenous species, *Fremontodendron californicum* [includes *F. californicum* sspp. *crassifolium* and *napense*] (pl. 55) (Flannelbush, Fremontia) is interesting botanically and a valuable shrub for gardens, for it is easily grown, drought resistant, and deer resistant, and it produces large flowers through much of the year.

The family as a whole is characterized by a flower that usually has a five-lobed calyx, five stamens whose filaments are attached to the calyx and also partly united to each other, and either five petals or none. There may also be five additional stamens that lack anthers. The ovary is superior, and the fruit is dry at maturity.

In *Fremontodendron*, petals are lacking, but the calyx is up to 2.5 cm wide and bright yellow when fresh. The fruit is partitioned into five few-seeded divisions that separate from one another at maturity. The lobed leaves are covered with star-shaped hairs that readily detach and may irritate the skin.

TAMARICACEAE (TAMARISK FAMILY)   Several species of *Tamarix*, including *T. ramosissima [T. pentandra]* (Salt-cedar or Tamarisk), native to Asia, are now well established in various parts of California, generally along streams. So aggressive are they in many areas of the southwestern United States that the competing native flora has been eliminated. Their very deep roots and their association with watercourses make them difficult to eradicate without poisoning the water.

Tamarisks are shrubs or small trees with long, slender branches. The leaves, rarely more than 4 mm long, are needlelike or scalelike and excrete salt. It is the habit of these plants to shed, once a year, many of their young twigs, as well as their leaves. The flowers, concentrated in slender racemes, are usually only 1–2 mm wide, usually with five white or pink petals and an equal number of sepals and stamens. The ovary is superior. The single pistil has three styles and matures into a small dry fruit. Each of the several seeds within it has a tuft of long hairs.

THYMELAEACEAE (DAPHNE FAMILY)   Thymelaeaceae should be appreciated by gardeners as the family that has given them a few European species of *Daphne*. The group is best represented, however, in the tropics. *Dirca occidentalis* (pl. 55) (Western Leatherwood), our only native species, grows only in the San Francisco Bay region and is not common. It is a deciduous shrub, up to about 2 m tall, with branches that are exceptionally pliable when they are young. The smooth-margined leaves, usually appearing after the flowers, are more or less oval, 3–7 cm long, bright green on the upper surfaces, and paler and slightly hairy on the undersides. The sessile flowers are usually turned downward and have a bright yellow, four-lobed calyx about 1 cm long that resembles a corolla. Real petals are absent. The calyx widens toward the top and has 8 stamens protruding from it. The ovary is superior. The single pistil, if it develops, ripens into an egg-shaped, yellowish green fruit 8–10 mm long.

TROPAEOLACEAE (NASTURTIUM FAMILY)   The Nasturtium Family is native in Central and South America. One species, *Tropaeolum majus* (Garden Nasturtium), has become naturalized in our area. It may be annual or perennial, and it sometimes becomes almost vinelike, with stems more than 1 m long. The leaf petioles are inserted close to the center of nearly circular blades. The irregular flower, up to 6 cm wide, usually has orange petals, but the color varies from nearly white to deep red. Two of the five petals are slightly different from the other 3, and one of the five sepals has a long spur. There are eight stamens in two whorls, and they are not all the same size. The ovary is superior. The fruit-forming part of the pistil is three lobed, and there are three stigmas. At maturity, each division encloses a single large seed.

URTICACEAE (NETTLE FAMILY)   All nettles in our region are annual or perennial herbs, and most of them have stinging hairs on the stems and leaves. Contact with them may result in a moderately painful experience not quickly forgotten!

The flowers, up to 4 mm long, are concentrated in inflorescences that arise from the leaf axils. Generally each plant, and sometimes each inflorescence, has flowers with 4 stamens as well as flowers with a pistil that produces a single seed. Sometimes flowers with a pistil and stamens are also present. Each flower has 2–4 sepals that are either separate or united. Petals are absent. The ovary is superior.

1a   Leaves without stinging hairs, alternate, smooth margined (leaves 1–9 cm long; sepals united at their bases; flowers often with both pistil and stamens present; up to 80 cm tall) .... *Parietaria judaica*
Western Pellitory; eua

1b   Leaves with stinging hairs, opposite, toothed

   2a   Flowers in nearly globular clusters; sepals of pistillate flowers united at the tips, sepals of staminate flowers usually separate (leaves up to 4 cm long; weak stemmed, up to 50 cm tall; on chaparral and oak woodland) .... *Hesperocnide tenella*
Western Nettle; Na-s

   2b   Flowers in racemes or panicles; sepals of all flowers usually separate

      3a   Leaf blades usually less than 4 cm long; each inflorescence up to 2.5 cm long, with both pistillate and staminate flowers; annual; up to 60 cm tall .... *Urtica urens*
Dwarf Nettle; eu

      3b   Leaf blades usually more than 4 cm long; each inflorescence up to 7 cm long, with either pistillate or staminate flowers; perennial; often more than 100 cm tall (in moist areas)

         4a   Stem with many whitish, nonstinging hairs; most leaves at least twice as long as wide (widespread) .... *Urtica dioica* ssp. *holosericea [U. holosericea]* (pl. 55)
Hoary Nettle

         4b   Stem without whitish, nonstinging hairs; most leaves less than twice as long as wide (coastal) .... *Urtica dioica* ssp. *gracilis [U. californica]*
American Stinging Nettle; Sn-SM

VALERIANACEAE (VALERIAN FAMILY)   All of our native species of the Valerian Family are small herbaceous annuals of the genus *Plectritis*. These usually form carpets of many plants, but only some of the more richly colored forms of *P. congesta* are likely to become popular in gardens. Species of *Plectritis* have opposite leaves, four-sided stems, and crowded inflorescences. The irregular corolla, with five lobes, is slightly to markedly two lipped, and there is usually a saclike spur originating from the lower side of the long tube. The three stamens are attached to the inside of the corolla tube. The ovary is inferior, and the calyx, united with it, does not have noticeable lobes above the ovary. The fruit contains a single seed and is dry when mature.

Some members of the family, including *Centranthus ruber* (Red Valerian), which is widely cultivated, have a nearly regular corolla, although the tube does have a spur. The stem of this plant is not four sided. Red Valerian often escapes and becomes so well established, especially along the coast, that it may seem to be native.

1a   Calyx distinct, the lobes curving inward; corolla 14–18 mm long, white, lavender, or purple red; stem not 4 sided . . . . *Centranthus ruber*
                                                            Red Valerian; me

1b   Calyx not apparent; corolla 2–10 mm long, white or pink; stem often 4 sided

     2a   Corolla 4–10 mm long (corolla pink, definitely 2 lipped)

         3a   Corolla usually dark pink, with 1 or 2 darker spots on the lower lip; spur more than half as long as the entire corolla (widespread) . . . . *Plectritis ciliosa* ssp. *ciliosa*
                                                   Long-spurred Plectritis; Me-Mo

         3b   Corolla light to dark pink, not spotted; spur about a third as long as the entire corolla (on coastal bluffs) . . . . *Plectritis congesta* (pl. 55)
                                          Seablush, Pink Plectritis; SLO-n

     2b   Corolla 1–4 mm long

         4a   Corolla definitely 2 lipped, usually dark pink, with 1 or 2 darker spots on the lower lip (spur nearly half as long as the entire corolla; widespread) . . . . *Plectritis ciliosa* ssp. *insignis*
                                                 Pink Plectritis

         4b   Corolla distinctly or indistinctly 2 lipped; white to pale pink, not spotted

             5a   Spur stout, about a third as long as the entire corolla . . . . *Plectritis macrocera* [includes *P. macrocera* var. *grayii*] (pl. 55)
                                     Long-horned Plectritis, White Plectritis

             5b   Spur slender, often absent, less than a third as long as the entire corolla (on coastal bluffs) . . . . *Plectritis brachystemon* [includes *P. congesta* var. *major* and *P. magna* var. *nitida*]
                                                 Pale Plectritis; Mo-n

VERBENACEAE (VERBENA FAMILY)   The leaves of verbenas and their relatives are opposite or whorled and often toothed or lobed. The flowers, each usually with a bract beneath it, are in several elongated inflorescences or dense heads on each plant. The stems are often four sided. The calyx is tubular, with five lobes in *Verbena* and two to four lobes in *Phyla*. The corolla has four or five lobes and sometimes is decidedly two lipped. There are usually four stamens in two pairs and a single pistil. The ovary is superior. The fruit breaks apart into two or four nutlets, each containing a single seed.

Some members of the family are attractive and widely cultivated, but species of *Lantana* may become pests in warm-temperate and tropical areas. *Phyla nodiflora* has been used in erosion control and in place of lawn grasses. In Europe, *Verbena officinalis* is used for making tea and as an herbal remedy.

Verbena bracteata
Bracted Verbena

1a Inflorescence up to 1 cm long, on a peduncle 1.5–9 cm long; corolla pink or white; calyx with 2–4 lobes; leaves usually toothed in the upper half; prostrate, forming tight mats (leaf blades 5–30 mm long) . . . . *Phyla nodiflora* [includes *Lippia nodiflora* var. *rosea*]
Lemon Verbena, Garden Lippia; sa

1b Inflorescence often much more than 1 cm long, the peduncle usually shorter than its inflorescence; corolla usually blue or purple; calyx 5 lobed; leaves coarsely toothed or lobed; upright

2a Leaves usually 1–3 cm long, sometimes up to 6 cm; bract beneath each flower 4–8 mm long, longer than the flower (inflorescence 2–10 cm long) . . . . *Verbena bracteata* (fig.)
Bracted Verbena

2b Leaves 4–15 cm long; bract beneath each flower 3–5 mm long, shorter than the flower

3a Leaves sessile, 7–15 cm long, those on the upper portions of stem usually not lobed; inflorescences dense, each usually 5–10 cm long, usually 8–17 on each stem
. . . . *Verbena bonariensis*
Clusterflower Verbena; sa

3b Leaves not sessile, 4–10 cm long, those on the upper portions of stem usually lobed; inflorescences interrupted below, each usually 10–20 cm long, 1–3 on each stem

4a Inflorescence 10–25 cm long when fruit are maturing; upper surfaces of leaves grayish green, not rough to the touch . . . .
. . . . *Verbena lasiostachys* var. *lasiostachys*
Western Verbena, Western Vervain

4b   Inflorescence 3–10 cm long when fruit are maturing; upper surfaces of leaves green, slightly rough to the touch ....
.... *Verbena lasiostachys* var. *scabrida [V. robusta]* (pl. 55)
Robust Verbena; Ma-s

VIOLACEAE (VIOLET FAMILY)   The Violet Family includes our familiar garden pansies, derived from the European *Viola tricolor* (Wild Pansy). Violets have a distinctly irregular flower composed of five separate petals, the two upper ones, the two at the side, and the lower one being different. The lower petal, furthermore, has a saclike spur near its base. The five separate sepals are also unequal. There are five stamens, the two lower ones having winglike lobes that go down into the spur of the lower petal. The ovary is superior. The fruit is not partitioned lengthwise into chambers, but its wall consists of three valves, and these split apart when the fruit is ripe. The leaves, with palmately arranged veins, are often mainly basal.

Gardeners find two of our native species fairly easy to grow. These are *Viola glabella* (Stream Violet), and *V. sempervirens* (Evergreen Violet). Both have yellow flowers and do well in shaded situations that remain at least slightly moist during summer.

1a Petals not orange yellow or mostly yellow (leaves toothed)
    2a   Petals 15–25 mm long (petals blue violet or a combination of yellow, white, and blue; leaf blades up to 2 cm wide; basal leaves absent) .... *Viola tricolor*
Wild Pansy; eu

    2b   Petals 7–13 mm long
        3a   All 5 petals similar in coloration, violet or pale violet; leaf blades nearly round, up to 4 cm wide (leaves basal and on the stem; widespread) .... *Viola adunca* (pl. 56)
Western Dog Violet; Mo-n

Viola sempervirens
Evergreen Violet

3b   Petals either not all alike in their coloration or primarily cream colored; leaf blades usually longer than wide, up to 2 cm wide

    4a   Petals mostly cream colored, often tinged with violet; basal leaves absent . . . . *Viola arvensis*
Field Violet; eu

    4b   Each petal white with yellow at its base, the 2 side petals also with a purple spot near their bases, the backs of the 2 upper petals red violet; leaves basal and on the stem (common in redwood forests, but sometimes found elsewhere) . . . . *Viola ocellata* (pl. 56)
Western Heart's-ease; Mo-n

1b   Petals orange yellow or mostly yellow, sometimes with dark lines, and sometimes darker on the back

  5a   Leaves compound or deeply lobed (leaves basal and also on the stem)

    6a   Most leaves pinnately compound (leaflets 3–5, deeply lobed; in moist areas, often on serpentine) . . . . *Viola douglasii*
Douglas Violet; SLO-n

    6b   Leaves palmately compound or lobed

      7a   Leaves compound, the 3 leaflets deeply lobed (in shaded areas) . . . . *Viola sheltonii*
Shelton Violet; SCl-n

      7b   Leaves lobed, some lobes shallowly lobed again . . . . . *Viola lobata* ssp. *lobata*
Lobed Pine Violet; Na-n

  5b   Leaves not compound or deeply lobed

    8a   All or most stem leaves located just below the flowers (basal leaves 1–4 or none)

      9a   Backs of 2 upper petals brown; stipules with prominent teeth . . . . *Viola lobata* ssp. *integrifolia*
Pine Violet; Na-n

      9b   Backs of 2 upper petals not brown; stipules without prominent teeth (in moist, shaded areas) . . . . *Viola glabella* (pl. 56)
Stream Violet; Mo-n

    8b   Leaves scattered along the stem

      10a   Stems creeping, rooting at the nodes, forming mats (leaves evergreen, often with purple spots; basal leaves present; in shaded, moist coastal woods) . . . . *Viola sempervirens* (fig.)
Evergreen Violet; Mo-n

      10b   Stems upright or with the tips rising, not rooting at the nodes; leaves deciduous, not spotted

        11a   Leaves without a purplish tinge, either not hairy or hairy on both surfaces; basal leaves lacking; petals orange yellow (widespread) . . . . *Viola pedunculata* (pl. 56)
Johnny-jump-up; Sn-s

        11b   Leaves often with a purplish tinge, usually hairy only on the undersides; basal leaves 1–5; petals yellow . . . . *Viola purpurea*
Mountain Violet

VISCACEAE (MISTLETOE FAMILY)   Viscaceae, sometimes included in the Loranthaceae, includes our familiar Christmas mistletoes, which belong to *Phoradendron*, and *Viscum album*, believed to have been introduced to California by Luther Burbank. The species in our region are perennial and parasitize trees and shrubs, sometimes causing abnormal branching of the host, called a witches' broom. Generally, a particular species of mistletoe grows only on a certain host species. The leaves, which are opposite and smooth margined, may be well-developed, leathery structures, or they may be reduced to small scales. The pistillate and staminate flowers are on separate plants and lack petals. The staminate flowers have three or four stamens and three or four sepals. The pistillate flowers have one pistil and two to four sepals. The ovary is inferior. The fruit is a sticky or mucilaginous berry containing a single seed. In *Arceuthobium*, the seed is explosively expelled from the fruit, sometimes for a distance of up to 15 m. Seeds of *Phoradendron* and *Viscum* are dispersed by birds.

1a   Plant yellow green, yellow, orange yellow, or brownish; stem angled at least when young; leaves scalelike; fruit slightly flattened, purplish, with a whitish coating, explosive; parasitic on coniferous trees

    2a   On *Pseudotsuga menziesii*; flowering April to June .... *Arceuthobium douglasii*
        Douglas-fir Dwarf-mistletoe

    2b   On conifers other than *Pseudotsuga menziesii*, including species of *Pinus, Abies*, and *Tsuga;* flowering Augustand September (the following are so similar to each other that some taxonomists refer all of them to one species, *Arceuthobium campylopodum*)
    .... *Arceuthobium abietinum [A. campylopodum* forma *abietinum]*
        Fir Dwarf-mistletoe; on *Abies grandis, A. concolor*
    .... *Arceuthobium californicum*
        Sugar Pine Dwarf-mistletoe; on *Pinus lambertiana*
    .... *Arceuthobium campylopodum* (pl. 56)
        Western Dwarf-mistletoe; on *Pinus sabiniana, P. coulteri*
    .... *Arceuthobium tsugense [A. campylopodum* forma *tsugensis]*
        Hemlock Dwarf-mistletoe; on *Tsuga heterophylla*

1b   Plant yellow green or green; stem cylindrical; leaves well developed or scalelike; fruit globular, white, straw colored, or pink, not explosive; some species on coniferous trees, others on nonconiferous woody trees and shrubs

    3a   Leaves reduced to scales

        4a   Parasitic on *Calocedrus decurrens;* fruit pinkish white or straw colored..........
        .... *Phoradendron libocedri [P. juniperinum* var. *libocedri]*
        Incense-cedar Mistletoe

        4b   Parasitic on *Juniperus californica;* fruit pinkish white ....
        .... *Phoradendron juniperinum*
        Juniper Mistletoe

    3b Leaves well developed

        5a   Leaves 5–8 cm long, greenish yellow (fruit white; parasitic on apple trees and maples in a few localities in Sonoma County) .... *Viscum album*
        European Mistletoe; eua

5b   Leaves mostly less than 4 cm long, green

      6a   Clusters of pistillate flowers with 2 flowers at each node (fruit white, straw colored, or pinkish)

          7a   Leaves mostly 1–1.5 cm long, sessile; parasitic on *Juniperus californica, Cupressus sargentii,* and perhaps other species of *Cupressus* . . . . *Phoradendron densum* [*P. bolleanum* var. *densum*]
               Dense Mistletoe

          7b   Leaves mostly 1.5–3 cm long, with short petioles; parasitic on *Abies concolor* . . . . *Phoradendron pauciflorum* [*P. bolleanum* var. *pauciflorum*]
               Fir Mistletoe

      6b   Clusters of pistillate flowers with more than 5 flowers at each node

          8a   Leaves densely hairy; fruit hairy near the tip, pinkish white; parasitic on various species of *Quercus,* sometimes on other trees and shrubs, including *Umbellularia californica, Adenostoma fasciculatum,* and species of *Arctostaphylos* (widespread). . . . . *Phoradendron villosum* (pl. 56)
               Oak Mistletoe

          8b   Leaves not hairy or hairy only when young; fruit not hairy, white, sometimes tinged pink; parasitic on *Fraxinus latifolia, Platanus racemosa, Populus fremontii,* species of *Salix,* and cultivated fruit trees . . . . *Phoradendron macrophyllum* [*P. tomentosum* ssp. *macrophyllum*]
               Bigleaf Mistletoe

VITACEAE (GRAPE FAMILY)   The Grape Family consists of woody climbers with branching tendrils and includes, besides grapes, *Parthenocissus quinquefolia* (Virginia Creeper) and *P. tricuspidata* (Japanese Ivy). *Vitis californica* (pl. 56) (California Wild Grape) is common in open canyons from San Luis Obispo County northward. It resembles rampant cultivated grapes in the way it grows. In one season, the vine may grow 5 m. This species is especially attractive in autumn, when its deciduous leaves turn reddish. It can be propagated by cuttings and is easily grown in sunny or partly shaded locations. Cutting it back to the base annually keeps it under control.

Vitis californica has palmately veined leaves, 7–14 cm wide, that are toothed and usually lobed. The flowers are in clusters whose stalks arise opposite the leaf petioles. Sometimes pistillate and staminate flowers are mixed with flowers that have both a pistil and stamens. Each flower has a shallow calyx tube with scarcely any lobes, five yellowish petals that are united at the tips and that fall away early, and three to nine stamens. The ovary is superior. The fruit, usually 6–10 mm wide and purplish when ripe, is too sour to be popular.

ZYGOPHYLLACEAE (CALTROP FAMILY)   Zygophyllaceae is a mostly tropical family. The only species in our area is *Tribulus terrestris* (pl. 56) (Puncture-vine or Caltrop). It forms extensive mats that hug the ground in disturbed areas, such as vacant lots and along railroad tracks. It was introduced from the Mediterranean region of Europe and is now widely established in California.

The stems of *Tribulus* are usually bristly, and the opposite leaves are pinnately compound. The flower has five separate sepals, five separate yellow petals, and 10 stamens. The ovary is superior. The fruit eventually splits apart into five hard nutlets, each with two to four formidable spines. The name caltrop, in fact, is derived from an old English word for an ancient four-pointed weapon thrown on the ground to disable cavalry horses. The fruit and its component nutlets are a menace to animals, and also to humans who enjoy walking barefooted.

Monocotyledonous Families

ALISMATACEAE (WATER-PLANTAIN FAMILY) Members of Alismataceae grow mostly in shallow water at the edges of ponds, lakes, and sluggish streams. The leaves are entirely basal, and in our species they are large and broad bladed. The flowers are numerous, in a whorled and sometimes compound inflorescence, which often rises above the leaves. There may be some leaflike bracts just below it. Typically, the flowers have three sepals, three delicate petals, six to many stamens, and six to many pistils; in species of *Sagittaria*, however, some or all flowers of the inflorescence may be pistillate or staminate. The fruiting part of the pistil, dry at maturity, is joined tightly to the single seed within.

1a　Leaf blades usually arrowhead shaped, with pointed basal lobes; stamens many; pistils many (petals white)

    2a　Petals 10–20 mm long; leaf blades 6–30 cm long; bracts below flower whorls often somewhat boat shaped, blunt at the tips, rarely so much as 1.5 cm long . . . . *Sagittaria latifolia* (fig.)

        Wapato

    2b　Petals 6–10 mm long; leaf blades 5–15 cm long; bracts below flower whorls flat, pointed at the tips, often more than 1.5 cm long . . . . *Sagittaria cuneata*

        Arrowhead; Ma

1b　Leaf blades oval or somewhat elongated; stamens 6; pistils 6 to many

    3a　Petals 8–10 mm long, yellow or white, the margins fringed; leaf blades 3–9 cm long; pistils 6–15; fruit with a sharp beak. . . . . *Damasonium californicum [Machaerocarpus californicus]*

        Fringed Water-plantain; Sn

    3b　Petals 3–6 mm long, rose, pink, or white, the margins not fringed; leaf blades 5–15 cm long; pistils many; fruit without a beak

        4a　Leaf blades usually not more than twice as long as wide; petals rounded at the tips, pink or white . . . . *Alisma plantago-aquatica [A. triviale]* (fig.)

        Common Water-plantain

        4b　Leaf blades 3–4 times as long as wide; petals pointed at the tips, rose . . . . *Alisma lanceolatum*

        Lanceleaf Water-plantain; eua

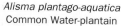
Alisma plantago-aquatica
Common Water-plantain

Sagittaria latifolia
Wapato

APONOGETONACEAE (CAPE-PONDWEED FAMILY) The Cape-pondweed Family consists of a single genus and about 40 species of freshwater plants. The basal leaves have a characteristic pattern of venation: the several parallel primary veins are linked by many cross veins. The simple or branched inflorescence, held above the water, has a bract beneath it. The flower has one to six perianth segments and six or more stamens arranged in two circles around the three to six pistils, whose lower portions are united. The ovary is superior.

Aponogeton distachyon [A. distachyus] (Cape-pondweed), native to South Africa, is cultivated in garden pools. It has become naturalized in a few places in California. The leaf blades float, and the flowers, crowded onto a two-branched inflorescence, have one elliptical, white perianth segment about 1.5 cm long, many stamens with purple anthers, and three pistils.

ARACEAE (ARUM FAMILY) The calla-lilies, from Africa, and species of *Anthurium,* from tropical America, are familiar examples of Araceae. The flowers are small and crowded onto a clublike inflorescence called the spadix. Originating below the spadix, and sometimes surrounding it almost completely, is a sort of inflorescence bract about 20 cm long called a spathe. This is often richly colored. All flowers may have a few stamens and a pistil, or the staminate flowers may be concentrated in the upper part of the spadix and the pistillate ones in the lower portion. The ovary is superior or half-inferior. Small scalelike structures, if present, are the only perianth parts. In many members of this large family, the inflorescences produce unpleasant odors. These attract insects that normally lay their eggs on dung or dead animals. The insects thus inadvertently pollinate the flowers.

1a   Inflorescence bract cream colored or white; leaf blades arrowhead shaped, 15–45 cm long
. . . . *Zantedeschia aethiopica*
Calla-lily; af

1b   Inflorescence bract yellow; leaf blades elongate oval, often more than 100 cm long (in
swampy habitats near the coast) . . . . *Lysichiton americanum* (pl. 57)
Yellow Skunk-cabbage; SM-n

COMMELINACEAE (SPIDERWORT FAMILY)   The only representative of Commelinaceae in our region is *Tradescantia fluminensis* (Spiderwort), which has become established in some damp habitats. It is a native of South America, much cultivated in gardens and as a pot plant. Its creeping stems, rooting freely at the nodes, bear oval leaves about 4 cm long. The leaves are often tinged with violet on the undersides, and their bases form sheaths around the stem. The flowers, about 1 cm wide, are in clusters, each of which has a pair of elongated bracts beneath it. The three petals are white, and the three sepals are green. There are six stamens and one pistil, which has a superior ovary and matures as a dry capsule containing several seeds.

CYPERACEAE (SEDGE FAMILY)   Most members of Cyperaceae grow in aquatic habitats, or at least in places that are wet for part of the year. They are somewhat grasslike, but the stems are usually solid and three sided. The leaves usually do not have ligules. They are slender, but the basal portion of each one generally has a substantial sheath that clasps the stem; the edges of the sheath, as is the case in certain grasses, are fused. In the inflorescence, which may have one or more bracts below it, the flowers, called florets, are clustered in spikelets. A spikelet may also have a bract, and still another kind of bract, termed a scale, is associated with each floret. The scales persist to the fruiting stage, and they sometimes have an awn at the tip; furthermore, the lowest scale of a spikelet may be enlarged. A floret ordinarily has three stamens (rarely two) and a pistil with two to four feathery stigmas. Most sedges have no perianth parts, but in some genera there are bristles that may correspond to them. In *Carex*, which includes about a third of the species of Cyperaceae in the world, each pistillate floret and the fruit that develops from it lies within a sac that is open only at the tip. The stigmas usually protrude from the opening.

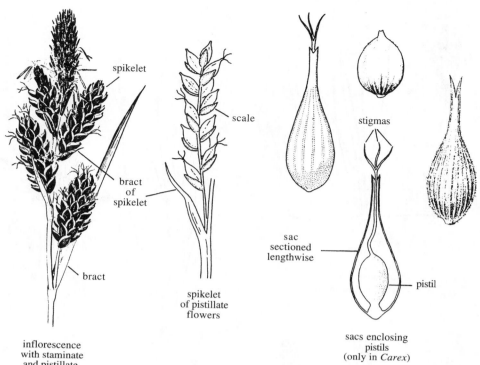

spikelet

scale

bract
of
spikelet

bract

stigmas

sac
sectioned
lengthwise

pistil

inflorescence
with staminate
and pistillate
flowers

spikelet
of pistillate
flowers

sacs enclosing
pistils
(only in *Carex*)

Cyperaceae
(Sedge Family)

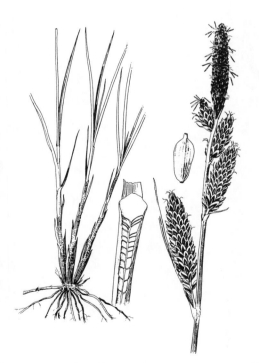

Carex nudata
Torrent Sedge

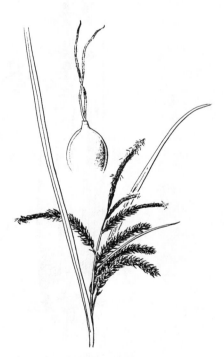

Carex obnupta
Slough Sedge

Eleocharis pauciflora
Fewflower Spikerush

Rhynchospora alba
White Beakrush

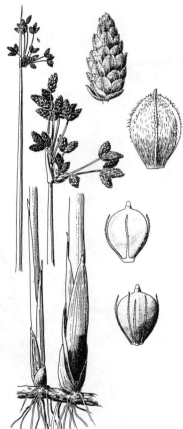

Scirpus acutus var. *occidentalis*
Hardstem Bulrush

1a   Each floret or fruit not enclosed within a sac; all florets with a pistil and stamens

    2a   Stem cylindrical or triangular; spikelet not flattened, the florets and scales in several rows and arranged spirally

        3a   Plant not more than 4 cm tall; floret without bristles (bracts below the inflorescence 1–3, each not more than 1 cm long; scale tip elongated, narrow; annual)
            . . . . *Lipocarpha occidentalis [Hemicarpha occidentalis]*
            Western Lipocarpha

        3b   Plant at least 7 cm tall; floret usually with bristles (lift the scale to see the bristles)

            4a   Floret with numerous white bristles, these extending far beyond the scale (bract below the inflorescence single, 1–2 cm long; up to 60 cm tall; mostly in sphagnum bogs) . . . . *Eriophorum gracile*
                Slender Cottongrass; SF, Sn-n

            4b   Floret with 1–12 bristles, these not extending far, if at all, beyond the scale and absent in some species

                5a   Spikelet solitary at stem tip; inflorescence without a bract below it; scale without an awn (stem cylindrical; leaves reduced to sheaths; plant forming clumps) . . . . CYPERACEAE, SUBKEY 2
                    Eleocharis

                5b   Spikelets in clusters at stem tip or in leaf axils; inflorescence with at least 1 bract below it; scale usually with a short awn . . . CYPERACEAE, SUBKEY 3
                    Rhynchospora, Scirpus

    2b   Stem triangular; spikelet flattened, the florets and scales in 2 rows, but these may not be obvious in *Kyllinga brevifolia*, because the inflorescence is globular

        6a   Inflorescence a singular globular head of sessile spikelets, the head 4–8 mm long; spikelet 3 mm long, with 1 or 2 florets (bracts 2–4, each 1–4 cm long) . . . .
            . . . . *Kyllinga brevifolia;* sa

        6b   Inflorescence either consisting of several clusters of spikelets on stalks or, if a single globular head of sessile spikelets, the head at least 10 mm long; spikelet 3–10 mm long, with several florets

            7a   Stalk of inflorescence with ridges (separate the spikelets to see the stalk)

                8a   Bracts below the inflorescence 1–4, each 2–4 cm long; leaves 2 or 3 on each stem; not more than 20 cm tall (spikelet 4–10 mm long; annual). . .
                    . . . . *Cyperus squarrosus [C. aristatus]*
                    Awned Cyperus

                8b   Bracts 3–11, each at least 8 cm long; leaves at least 5 on each stem; often more than 50 cm tall

                    9a   Bracts 3–7, each 8–12 cm long; spikelet 6–30 mm long; up to 50 cm tall; perennial . . . . *Cyperus esculentus*
                      Yellow Nutgrass

                  9b   Bracts 3–11, each often more than 20 cm long; spikelet 3–10 mm long; up to 100 cm tall; annual . . . . *Cyperus erythrorhizos*
                    Red-rooted Cyperus

            7b   Stalk of inflorescence without ridges

10a Bracts below the inflorescence 4–22, each often over 20 cm long; leaves
    sometimes lacking blades, more than 5 on each stem; up to 90 cm tall
    (perennial)

    11a Bracts 4–8; spikelet 10–20 mm long; leaf blades present . . . .
    . . . . *Cyperus eragrostis* (pl. 57)
             Tall Cyperus

    11b Bracts 10–22; spikelet 5–10 mm long; leaf blades absent . . . .
    . . . . *Cyperus involucratus*
        African Cyperus; af

10b Bracts 2–6, each not more than 18 cm long; leaves with blades, 2–4 on
    each stem; up to 50 cm tall (spikelet 4–10 mm long)

    12a Leaves 2 on each stem, shorter than the stem; stigmas 2; perennial
    (bracts 2 or 3) . . . . . . *Cyperus niger* [includes *C. niger* var. *capitatus*]
        Brown Umbrella Sedge; Sn-s

    12b Leaves 2–4 on each stem, as long as the stem; stigmas 3; annual

        13a Scales with pointed tips curving outward; bracts 3–6 . . . .
        . . . . *Cyperus acuminatus*
            Short-pointed Cyperus

        13b Scales with rounded, straight tips; bracts 2 or 3 . . . .
        . . . . *Cyperus difformis*
            Flat Sedge; eu

1b Each pistillate floret or fruit enclosed within a sac, which often has a characteristic beak;
  florets either pistillate or staminate (spikelet typically either staminate or pistillate, and
  both types usually present in the inflorescence; often both the scale of the lowest floret of a
  spikelet and the bract of the lowest spikelet in the inflorescence enlarged) . . . . *Carex*

  14a At least some spikelets on stalks

    15a Sac around pistillate floret or fruit not hairy

        16a Lowest bract of inflorescence leaflike, longer than the lowest spikelet and
        often longer than rest of the inflorescence. . . . . CYPERACEAE, SUBKEY 1
            Carex (in part)

        16b Lowest bract, if present, hairlike or bristlelike, not leaflike, seldom much
        longer than the lowest spikelet, except in *Carex stipata* (stigmas 2; in moist or
        wet habitats)

            17a Scale of floret with a conspicuous awn (inflorescence not interrupted,
            crowded; leaves 4–7 mm wide; up to 70 cm tall). . . . . *Carex dudleyi*
                Dudley Sedge; Mo-n

            17b Scale with a pointed tip, but without an awn

                18a Inflorescence interrupted (scale of floret sometimes longer than
                the sac)

                    19a Leaves 3–6 mm wide; inflorescence up to 8 cm long; sac brown-
                    ish black; up to 120 cm tall . . . . *Carex cusickii*
                      Cusick Sedge; SLO-n

                    19b Leaves 1–3 mm wide; inflorescence up to 5 cm long; sac brown;
                    up to 70 cm tall . . . . *Carex diandra*
                      Panicled Sedge

18b Inflorescence not interrupted (leaves 3–8 mm wide)

20a Inflorescence up to 5 cm long; ligules conspicuous; sac widest near the middle (inflorescence very compact) .... *Carex densa* [includes *C. breviligulata* and *C. vicaria*] Dense Sedge

20b Inflorescence up to 10 cm long; ligules not conspicuous; sac widest at the base .... *Carex stipata* Awlfruit Sedge; Sn-s

15b Sac around pistillate floret or fruit covered with fine hairs (use a hand lens; lowest bract of the inflorescence leaflike; stigmas 3)

21a Leaves hairy, especially along the edges, up to 12 mm wide (in moist habitats) .... *Carex gynodynama* Olney Hairy Sedge; Mo-n

21b Leaves not hairy, less than 5 mm wide (stem may be red brown at the base; pistillate spikelets often hidden among the basal leaves)

22a Lowest spikelet of inflorescence usually more than 2 cm long, compact, with more than 15 florets; lowest bract often more than 20 cm long; sac densely hairy; up to 100 cm tall; in wet or dry habitats, sometimes a garden weed (scale purple brown, with a paler center, pointed, narrower than the sac; beak of floret sac with teeth about 1 mm long)............ .... *Carex lanuginosa* Woolly Sedge

22b Lowest spikelet usually less than 1.5 cm long, not compact, few flowered; lowest bract not more than 10 cm long; sac not densely hairy; up to 35 cm tall; in dry habitats

23a Uppermost spikelet of inflorescence often less than 1 cm long, with fewer than 10 florets; edges of lowest bract red brown where it clasps the stem; beak of floret sac with very small teeth; up to 15 cm tall ... . *Carex brevicaulis* Shortstem Sedge; SlO-n

23b Uppermost spikelet 1–2 cm long, with more than 10 florets; edges of lowest bract not red brown; beak with obvious teeth; usually more than 15 cm tall .... *Carex globosa* Roundfruit Sedge

14b Spikelets sessile

24a Inflorescence consisting of 1 or 2 spikelets, with not more than 10 florets on each stem; leaves less than 2 mm wide; stigmas 3

25a Sac of floret 5–7 mm long; awn of scale of lowest floret of each spikelet 10–40 mm long, the awns of the other scales usually shorter; inflorescence up to 7 mm wide; in dry forests (stem cylindrical) .... *Carex multicaulis* Manystem Sedge

25b Sacs 3–5 mm long; awn not more than 5 mm long, the awns of the other scales, if present, shorter; inflorescence less than 4 mm wide; in wet areas ... . *Carex leptalea* Flaccid Sedge; Ma-n; 2

24b Inflorescence consisting of 2 or more spikelets, each with more than 10 florets; leaves 2–5 mm wide; stigmas 2 (inflorescence usually at least 5 mm wide)

    26a Lowest bract of inflorescence at least as long as rest of the inflorescence

        27a Plant of moist or wet areas; inflorescence usually more than 10 mm wide; lowest bract leaflike (inflorescence 1–2 cm long; leaves sometimes longer than the stem; up to 60 cm tall)................... *Carex athrostachya*
            Slenderbeak Sedge; Ma-n

        27b Plant of dry, open areas; inflorescence less than 10 mm wide; bract not leaflike

            28a Scale of floret longer than the sac; scale with an awn; beak of floret sac with conspicuous teeth; inflorescence 2–5 cm long; leaves up to 3 mm wide; sometimes not forming clumps; up to 80 cm tall ....*Carex tumulicola*
                Foothill Sedge; Mo-n

            28b Scale slightly shorter than the sac; scale mostly without an awn; beak with very small teeth, if any; inflorescence up to 2.5 cm long; leaves up to 4 mm wide; forming clumps; up to 45 cm tall .... *Carex subfusca [C. teneraeformis]*
                Rusty Sedge

    26b Lowest bract shorter than rest of the inflorescence

        29a Lowest bract at least as long as the lowest spikelet, sometimes nearly as long as rest of the inflorescence (often more than 65 cm tall)

            30a Weed in lawns and other disturbed areas (inflorescence 1–2 cm long; scale of floret shorter than the sac) ...... *Carex leavenworthii*
                Leavenworth Sedge; na

            30b Native plants in bogs, marshes, or other wet habitats

                31a Beak of floret sac with very small teeth, if any; scale nearly as long as or longer than the sac

                    32a Beak about one-third the length of the floret body; scale with a paler midrib; inflorescence not dense, 1.5–4 cm long; leaves 2–4 mm wide (beak often red) .... *Carex ovalis [C. tracyi]*
                      Tracy Sedge; Ma-n

                    32b Beak much less than one-third the length of the floret body; scale often with white margins; inflorescence dense, 1–5 cm long; leaves 1–3 mm wide ....*Carex praegracilis*
                      Clustered Field Sedge

                31b Beak with conspicuous teeth; scale definitely shorter than the sac

                  33a Scale with a prominent awn, the awn about one-fourth as long as rest of scale (inflorescence usually less than 1 cm wide and not more than 5 cm long; leaves 3–5 mm wide) .... *Carex bolanderi*
                    Bolander Sedge; Mo-n

33b Scale without a prominent awn, the awn, if present, much less than one-fourth as long as rest of scale (leaves sometimes longer than the stem)

    34a Scale brown with white margins; leaves 2–3 mm wide; inflorescence often more than 7 mm wide; in bogs and brackish areas . . . . *Carex echinata* ssp. *phyllomanica [C. phyllomanica]*
    Coastal Sedge; SCr-n

    34b Scale white; leaves 3–5 mm wide; inflorescence not more than 6 mm wide; in moist woods . . . . *Carex deweyana* ssp. *leptopoda [C. leptopoda]*
    Short-scaled Sedge; Scr-n

29b Lowest bract, if present, shorter than the lowest spikelet (beak of floret sac without conspicuous teeth; in bogs, marshes, or other habitats that are at least seasonally wet)

  35a Scale of floret longer than the sac; scale with an awn (inflorescence 1–2.5 cm long and usually less than 1 cm wide; leaves sometimes longer than the stem; up to 50 cm tall) . . . . *Carex simulata*
  Shortbeak Sedge

  35b Scale not longer than the sac; scale without an awn

    36a Inflorescence usually over 1 cm wide, so dense that individual spikelets are often obscured; often more than 80 cm tall

      37a Beak of floret sac about half as long as the scale; moist areas at elevations of up to 5,200 ft . . . . *Carex subbracteata*
      Small-bracted Sedge

      37b Beak less than one-fourth as long as the scale; in coastal marshes or boggy areas at elevations of up to 1,000 ft . . . . *Carex harfordii* [includes *C. montereyensis*]
      Harford Sedge, Monterey Sedge; SLO-n

    36b Inflorescence usually not more than 1 cm wide, not so dense as to obscure individual spikelets

      38a Inflorescence up to 8 cm long and 1 cm wide, usually with 10 or more spikelets; each spikelet with about 20 florets (beak of floret sac less than one-fourth as long as the scale; often more than 1 m tall) . . . . *Carex feta*
      Green-sheathed Sedge; SCl-n

      38b Inflorescence not more than 3.5 cm long and less than 1 cm wide, with fewer than 10 spikelets; each spikelet with about 10 florets

        39a Beak of floret sac about as long as the scale; leaves sometimes longer than the stem; up to 60 cm tall . . . . *Carex echinata* ssp. *phyllomanica [C. phyllomanica]*
        Coastal Sedge; SCr-n

39b Beak half as long as the scale; leaves shorter than the stem; up to 120 cm tall (in vernal pools) . . . . *Carex gracilior*
Slender Sedge

Cyperaceae, Subkey 1: Each pistillate floret or fruit enclosed within a sac, this not hairy; at least some spikelets on stalks; lowest bract of inflorescence leaflike, longer than the lowest spikelet and often longer than rest of the inflorescence . . . . *Carex* (in part)

1a Leaves often more than 1 cm wide, with very small bumps on the upper surfaces, and sometimes with a rectangular pattern on the undersides (stigmas 3; in moist or wet habitats)

2a Lowest spikelet of inflorescence up to 7 mm wide and 5–16 cm long; sac of floret 3 mm long, the beak 1 mm long (beak curved, and with very small teeth; leaves 8–20 mm wide) . . . . *Carex amplifolia*
Ampleleaf Sedge; SM-n

2b Lowest spikelet of inflorescence up to 14 mm wide and 7–10 cm long; sac 5–6 mm long, the beak 1–4 mm long

3a Sac of floret 1 mm wide, the tapered beak 2–4 mm long, with teeth 1–2 mm long; lowest spikelet of inflorescence up to 7 cm long; leaves 6–16 mm wide (inflorescence bristly) . . . . *Carex comosa*
Bristly Sedge; La-SCr; 2

3b Sac 2–3 mm wide, the beak less than 2 mm long, with very small teeth; lowest spikelet up to 10 cm long; leaves 2–12 mm wide . . . . . *Carex utriculata [C. rostrata]*
Beaked Sedge; SCr-n

1b Leaves often less than 1 cm wide, without bumps or a rectangular pattern

4a Leaves not more than 4 mm wide (lowest spikelet of inflorescence 2–5 cm long; in moist or wet habitats; often more than 80 cm tall)

5a Scale of floret longer than the sac; uppermost spikelet of inflorescence pistillate at the tip and staminate below (lowest bract shorter than or as long as the inflorescence; stigmas 3) . . . . *Carex buxbaumii*
Buxbaum Sedge; Ma-n; 4

5b Scale shorter than the sac; uppermost spikelet entirely staminate or at least staminate at the tip

6a Lowest bract of inflorescence longer than rest of inflorescence (uppermost spikelet of inflorescence staminate at the tip and pistillate below; stigmas 3; sometimes found on serpentine) . . . . *Carex serratodens*
Bifid Sedge

6b Lowest bract shorter than rest of inflorescence

7a Scale, with white margins, about as wide as the sac; beak of sac one-fourth as long as the scale and with teeth; stigmas 3; sometimes on serpentine . . . . *Carex mendocinensis*
Mendocino Sedge; Ma-n

7b Scale, with dark brown margins, much narrower than the sac; beak less than one-fourth as long as the scale and without teeth; stigmas 2; not on serpentine . . . . *Carex nudata* (fig.)
Torrent Sedge

4b Some leaves more than 4 mm wide

    8a Leaves hairy (lowest bract of inflorescence usually longer than rest of the inflorescence; leaves 3–12 mm wide; scale of floret shorter than the sac; beak of sac toothed; stigmas 3; up to 90 cm tall; in moist or wet habitats) . . . . *Carex gynodynama*

                      Olney Hairy Sedge; Mo-n

    8b Leaves not hairy

        9a Leaves longer than the stem (lowest bract of inflorescence longer than rest of the inflorescence; leaves 3–7 mm wide; scale of floret about half as long as the sac; beak of sac 2–3 mm long, with obvious teeth; stigmas 3; up to 1 m tall; in moist or wet habitats). . . . . *Carex vesicaria* var. *major* [*C. exsiccata*]

                          Inflated Sedge; SCl-n

        9b Leaves shorter than the stem, usually much shorter

            10a Plant not more than 15 cm tall (lowest bract of inflorescence longer than rest of the inflorescence; leaves 2–5 mm wide; sac of floret without a beak; stigmas 2; in moist or wet habitats) . . . .

            . . . . *Carex saliniformis* [*C. salinaeformis*]

                        Deceiving Sedge; SCr-n; 1b

            10b Plant more than 50 cm tall

                11a Lowest bract of inflorescence equal to or longer than rest of the inflorescence (beak of floret sac less than one-third as long as the scale; up to 1 m tall)

                    12a Lowest bract less than 10 cm long (leaves up to 9 mm wide)

                        13a Awn of floret scale about as long as rest of the scale; lowest spikelet of inflorescence up to 8 cm long, the uppermost one up to 6 cm long; in vernal pools . . . . . . *Carex barbarae*

                                  Santa Barbara Sedge

                      13b Awn, if present, much shorter than rest of scale; lowest spikelet up to 6 cm long, the uppermost one up to 4 cm long; in dry or wet habitats, sometimes a garden weed (scale dark, narrower than the sac) . . . . . *Carex nebrascensis*

                                  Nebraska Sedge

                  12b Lowest bract at least 10 cm long, often more than 19 cm (sac of floret not more than 4 mm long, the scale often longer; in moist or wet habitats)

                      14a Scale of floret without an awn; uppermost spikelet of inflorescence 8–14 cm long, the lowest one up to 20 cm long (lowest bract up to 20 cm long; leaves 6–12 mm wide) . . . .

                        . . . . *Carex schottii*

                              Schott Sedge; SCl-s

                    14b Scale with an awn; uppermost spikelet not more than 8 cm long, the lowest one not more than 14 cm long

                      15a Leaves not more than 5 mm wide; lowest spikelet up to 14 cm long (uppermost spikelet up to 6 cm long; lowest bract up to 50 cm long) . . *Carex obnupta* (fig.)

                                Slough Sedge; SlO-n

15b Widest leaves at least 9 mm; lowest spikelet 8 cm long

16a Uppermost spikelet not more than 4 cm long; lowest bract up to 25 cm long (awn of scale sometimes longer than rest of the scale) . . . .
. . . . *Carex lyngbyei*
Lyngbye Sedge; Ma-n; 2

16b Uppermost spikelet up to 8 cm long; lowest bract up to 50 cm long . . . .
. . . . *Carex aquatilis* var. *dives* [*C. sitchensis*]
Sitka Sedge; SCr-n

11b Lowest bract shorter than rest of the inflorescence (in moist or wet habitats)

17a Leaves not more than 15 cm long, strictly at the base of the plant; lowest bract of inflorescence 2–3 cm long (leaves about 5 mm wide; spikelet 1–2 cm long; stigmas 3; up to 90 cm tall) . . .
. . . . *Carex luzulina*
Luzula-like Sedge; Ma-n

17b Leaves more than 15 cm long, or if shorter not confined to the base of the plant; lowest bract more than 3 cm long

18a Leaves not more than 5 mm wide (lowest bract not more than 10 cm long; sac of floret less than 5 mm long)

19a Scale of floret green, this about as wide as the sac, and usually with an awn; spikelet not more than 2 cm long; stigmas 3; up to 60 cm tall . . . . *Carex albida*
White Sedge; Sn; 1b

19b Scale dark brown, much narrower than the sac, and without an awn; spikelet up to 5 cm long; stigmas 2; up to 100 cm tall . . . . *Carex senta*
Rough Sedge; Sl-s

18b Widest leaves more than 5 mm (scale of floret usually with a short awn)

20a Lowest bract of inflorescence 10–15 cm long; lowest spikelet of inflorescence not more than 4 cm long; sac of floret 5–6 mm long; stigmas 3 (uppermost spikelet not more than 3 cm long; in moist habitats) . . . .
. . . . *Carex hendersonii*
Henderson Sedge; Sn-n

20b Lowest bract less than 10 cm long; lowest spikelet up to 8 cm long; sac less than 5 mm long; stigmas 2

21a Awn of floret scale about as long as the rest of the scale; lowest spikelet up to 8 cm long, the uppermost one up to 6 cm long; in vernal pools . . . .
. . . . *Carex barbarae*
Santa Barbara Sedge

21b Awn, if present, much shorter than rest of the scale; lowest spikelet up to 6 cm long, the uppermost one up to 4 cm long; in dry or wet habitats, sometimes a garden weed (scale dark, narrower than the sac) . . . . *Carex nebrascensis*
Nebraska Sedge

Cyperaceae, Subkey 2: Inflorescence without a bract below it; spikelet solitary at stem tip; florets and scales in several rows and arranged spirally; floret with a pistil and stamens, but without a sac; floret with 1–6 bristles, but these absent in *Eleocharis acicularis;* leaves reduced to sheaths . . . . *Eleocharis*

1a Spikelet usually at least 8 mm long, with at least 10 florets (floret bristles 1–6)

   2a Leaf sheaths less than 2 cm long; annual (scale of floret brown; stigmas 2; up to 50 cm tall; in marshes and ponds) . . . . *Eleocharis obtusa*
Blunt Spikerush; Ma-n

   2b Leaf sheaths at least 2 cm long; perennial

      3a Spikelet up to 25 mm long; scale of floret with a green midrib; stigmas 2 (up to 1 m tall; in marshes) . . . . *Eleocharis macrostachya*
Common Spikerush, Pale Spikerush

      3b Spikelet up to 15 mm long; scale without a green midrib; stigmas 3

         4a Scale of floret light colored; up to 150 cm tall (in saline or alkaline habitats) . . . . *Eleocharis rostellata*
Beaked Spikerush, Walking Sedge; Ma, Sn

         4b Scale dark colored; up to 40 cm tall (in moist areas) . . . . *Eleocharis montevidensis* [includes *E. montevidensis* var. *parishii*]
Montevideo Spikerush, Dombey Spikerush

1b Spikelet less than 8 mm long, usually with fewer than 10 florets (stigmas 3)

   5a Leaf sheaths 2–3 cm long; stem not delicate, up to 40 cm tall; in moist meadows (spikelet 4–7 mm long; floret bristles 1–6) . . . . *Eleocharis pauciflora* [includes *E. pauciflora* var. *suksdorfiana*] (fig.)
Fewflower Spikerush; Ma-n

   5b Leaf sheaths not more than 1 cm long; stem delicate, not more than 20 cm tall; found in wet habitats, including salt marshes

      6a Spikelet up to 7 mm long; scale of floret dark brown; floret bristles none; up to 20 cm tall . . . . *Eleocharis acicularis*
Needle Spikerush; Sn, Ma, Me

      6b Spikelet not more than 4 mm long; scale straw colored, green, or yellow, sometimes red brown; floret bristles 1–6; up to 8 cm tall

         7a Scale of floret straw colored . . . . *Eleocharis radicans*
Spongystem Spikerush

         7b Scale green or yellow, sometimes red brown . . . . *Eleocharis parvula* [includes *E. parvula* var. *coloradoensis*]
Small Spikerush; Na, Ma; 4

Cyperaceae, Subkey 3:   Inflorescence with at least one bract below it; spikelets in clusters at stem tip or in leaf axils; florets and scales in several rows and arranged spirally; floret without a sac, but with a pistil and stamens and sometimes with 2–12 bristles, but these absent in *Scirpus cernuus* and *S. koilolepis;* scale usually with a short awn . . . .*Rhynchospora, Scirpus*

1a   Clusters of spikelets 1–5, in leaf axils; leaves usually less than 3 mm wide (spikelet 4–6 mm long; in bogs)

  2a   Scale of floret tan to white

    3a   Bristles of floret 10–12, with barbs turned down . . . . *Rhynchospora alba* (fig.)
         White Beakrush; Sn; 2

    3b   Bristles 5 or 6, with barbs turned up . . . . *Rhynchospora globularis*
         Roundhead Beakrush; Sn; 2

  2b   Scale red to dark brown

    4a   Scale of floret red to dark brown; bristles of floret 6 or 7, the barbs turned up . . . .
         . . . . *Rhynchospora californica*
         California Beakrush, Brown Beakrush; Ma (PR); Sn; 1b

    4b   Scale dark brown; bristles 5–7, the barbs turned down . . . .
         . . . . *Rhynchospora capitellata [R. glomerata* var. *capitellata]*
         Brown Beakrush; Sn; 2

1b   Clusters of spikelets usually more than 5, at stem tip; leaves usually more than 3 mm wide, except in *Scirpus cernuus* and *S. koilolepis*

  5a   Bracts of the inflorescence 2–5, each up to 30 cm long; stem leafy (up to 1.5 m tall)

    6a   Stem cylindrical; spikelet 3–6 mm long; scale of floret without an awn; floret bristles usually 4 (in freshwater habitats) . . . . *Scirpus microcarpus*
         Smallfruit Bulrush, Panicled Bulrush

    6b   Stem sharply triangular; spikelet 10–25 mm long; scale with a stout awn; bristles usually 6

      7a   Inflorescence open, spreading; most of the clusters of spikelets on stalks more than 2.5 cm long; leaves 8–16 mm wide; stigmas 3; in freshwater habitats
           . . . . *Scirpus fluviatilis*
           River Bulrush; Na, SM

      7b   Inflorescence usually dense, compact; clusters of spikelets usually sessile just above the bracts; leaves usually 4–6 mm wide, occasionally up to 15 mm wide; stigmas 2; in salt marshes and freshwater habitats. . . . . *Scirpus robustus*
           Robust Bulrush, Prairie Bulrush

  5b   Bract single, not more than 10 cm long; leaves mainly near the base

    8a   Inflorescence less than 1 cm long; spikelet not more than 5 mm long; scale folded around the floret, usually with a green midrib; annual; not more than 20 cm tall (floret bristles none; spikelets 1–3; awn of scale, if present, less than 1 mm long; in saltwater and freshwater marshes)

      9a   Inflorescence with 1–3 spikelets, its bract 10–25 mm long, much longer than rest of the inflorescence (scale of floret with a keel on the back) . . . .
           . . . . *Scirpus koilolepis*
           Keeled Bulrush, Dwarf Club Rush; Me-s

9b Inflorescence with 1 spikelet, its bract 2–5 mm long, scarcely longer than the spikelet. . . . . *Scirpus cernuus* [includes *S. cernuus* var. *californicus*]

Low Bulrush, Low Club Rush

8b Inflorescence at least 1 cm long; spikelet 5–18 mm long; scale flattened and keeled, not folded around the floret, usually flecked with brown or red; perennial; at least 50 cm tall

10a Inflorescence usually spreading, at least 3 cm wide, its bract shorter than rest of the inflorescence; scale of floret usually flecked with red or orange; up to 4 m tall; in freshwater habitats and in ponds or seeps at the edges of salt marshes

11a Stem sharply triangular; bract 3–8 cm long; spikelets 20 to many; floret bristles 2–4; leaf blades absent . . . . *Scirpus californicus*

California Bulrush; Ma-s

11b Stem cylindrical; bract 1–4 cm long; spikelets 3 to many; bristles usually 6; leaf blades, if present, up to 8 cm long . . . . *Scirpus acutus* var. *occidentalis* (fig.)

Hardstem Bulrush, Common Tule

10b Inflorescence not spreading, not more than 2 cm wide, its bract longer than rest of the inflorescence; scale brown; up to 2 m tall; in saltwater and freshwater habitats (stem triangular; floret bristles 2–7)

12a Inflorescence with 1–7 spikelets, its bract 3–10 cm long; lower scales of spikelets with awns more than 4 mm long; stem usually not more than 4 mm wide; up to 1 m tall. . . . . *Scirpus pungens* [*S. americanus* vars. *longispicatus* and *monophyllus*]

Common Threesquare

12b Inflorescence with 5–12 spikelets, its bract not more than 3 cm long; lower scales of spikelets with awns less than 1 mm long; stem up to 10 mm wide; up to 2 m tall . . . . *Scirpus americanus*

American Bulrush, Threesquare

HYDROCHARITACEAE (WATERWEED FAMILY) Our representatives of Hydrocharitaceae are submerged aquatics. Their leaves, opposite or arranged in whorls, are at least several times as long as wide and sometimes very slender. Individual flowers, borne in the leaf axils, usually originate above a bract or within a tubular structure. The flowers are either staminate or pistillate, the two types commonly on separate plants. The extent to which the perianth parts and stamens develop varies greatly, but there is a single pistil. The fruit is either an achene or somewhat berrylike.

The genus *Najas* has, until recently, been placed in a separate family, Najadaceae. Two of its species have extremely flexible stems and leaves that closely resemble those of *Zannichellia palustris* (Zannichelliaceae). In *Zannichellia*, however, the pairs of opposite leaves are clearly spaced; in *Najas* they often appear to be in whorls because successive pairs are close together.

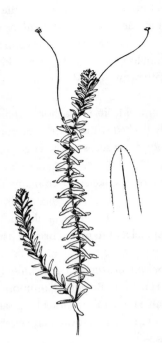

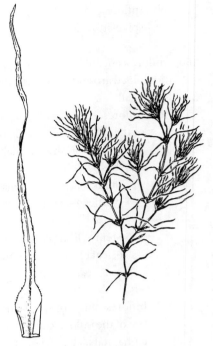

Elodea canadensis
Common Waterweed

Najas flexilis
Common Waterweed

1a Leaves and stem very flexible, except in *Najas marina;* leaves opposite, sometimes appearing to be in whorls because successive pairs are close together; flowers without perianth segments; fruit an achene with a networklike pattern (visible with a dissecting microscope)

2a Leaves stiff, up to 3 mm wide and 40 mm long, with spiny teeth about 1 mm long; stem often with spinelike outgrowths; staminate and pistillate flowers on separate plants . . . . *Najas marina*
Hollyleaf Waternymph, Large Najas; La-s

2b Leaves very flexible, not more than 2 mm wide and 25 mm long, with extremely fine teeth; stem without spinelike outgrowths; staminate and pistillate flowers on same plant

3a Surface of fruit dull, the networklike pattern of the surface visible with a hand lens; anthers usually with only 1 chamber; leaves rarely so much as 1 mm wide (often in brackish water) . . . . *Najas guadalupensis*
Common Waternymph, Guadaloupe Najas

3b Surface of fruit shiny, the networklike pattern of the surface usually not visible with a hand lens; anthers usually with 4 chambers; leaves usually slightly more than 1 mm wide . . . . *Najas flexilis* (fig.)
Slender Waternymph

1b Leaves and stem not especially flexible, rather firm to the touch; at least some leaves in whorls of at least 3; flowers, if formed, reaching the surface when fully developed, usually with at least some perianth segments, and, unless free floating, on a long tube of the perianth that resembles a peduncle; fruit somewhat berrylike, elongated

    4a Undersides of leaves, along the midribs, roughened by conical bumps; horizontal stems thickened to form tuberous structures (leaves 1–2 cm long, toothed, in whorls of 4–8; staminate flowers breaking off and floating free; pistillate flowers remaining attached by the tube of the perianth; a serious pest) . . . . *Hydrilla verticillata;* eua

    4b Undersides of leaves, along the midribs, smooth; horizontal stems not thickened to form tuberous structures (some populations not forming flowers)

        5a Leaves 2–4 cm long, most of them in whorls of 3, 4, or 6, but the ones lower on the stem opposite; petals 6–10 mm long, about twice as long as the sepals (escape from aquaria and garden pools; all plants in California seem to be staminate). . . . .
. . . . *Egeria densa [Elodea densa]*
Brazilian Waterweed; sa

        5b Leaves up to 1.5 cm long, mostly in whorls of 3, at least on upper portion of the stem; petals, if present, less than 5 mm long

            6a Leaves up to 4 mm wide, blunt or abruptly pointed at tips; petals of pistillate flower about 2.5 mm long, those of staminate flower about 5 mm long . . . . . .
. . . . *Elodea canadensis* (fig.)
Common Waterweed, Rocky Mountain Waterweed

            6b Leaves up to 2 mm wide, tapering gradually to pointed tips; all petals, when present, less than 1 mm long (staminate flowers breaking off, free floating)
. . . . *Elodea nuttallii*
Nuttall Waterweed

IRIDACEAE (IRIS FAMILY) Members of Iridaceae are perennials with rather narrow alternate or basal leaves. Flowers arise from between a pair of substantial, leaflike bracts. The three sepals and three petals of each flower are sometimes similar and fused together at their bases to form a tube, which may be prolonged for a considerable distance above the floral tube. The floral tube is fused to the fruit-forming part of the pistil, so the ovary is inferior. The three stamens are attached to the sepals. The style of the pistil is usually three lobed, each lobe bearing a stigma on its underside. In *Iris,* the style lobes are broad and each divides into two nearly petal-like branches. The flowers of *Iris* are admirably fitted for cross-pollination by insects that search for nectar at the bases of the sepals; as they do this, they are brought into contact with the anthers and stigmas. The fruit of all members of the family are dry capsules.

This family gives us not only cultivated irises, but also crocuses, tigridias, and gladioli. Some of them, including *Freesia alba* of South Africa, sometimes escape and become established in the wild. Our native *Iris douglasiana* (Douglas Iris) is an excellent subject for sunny or somewhat shaded garden situations, such as under oaks. *Sisyrinchium bellum* (Blue-eyed-grass) readily reseeds itself in a variety of sunny habitats. *Sisyrinchium californicum* (Golden-eyed-grass) requires moisture; wet soil just beyond the edges of a pool should suit it nicely.

1a Flower orange red; inflorescence usually with more than 15 flowers (a garden hybrid). . . . .
. . . . *Crocosmia* × *crocosmiiflora* [*C. crocosmiflora*]
Montbretia; af

1b Flower not orange red; inflorescence with 1–6 flowers, except in *Iris pseudacorus*

　2a Petals and sepals similar and not more than 1.5 cm long; stem with 2 winglike margins (inflorescence slightly umbel-like)

　　3a Petals and sepals blue purple; filaments nearly completely united; widespread in grassy places . . . . *Sisyrinchium bellum* (pl. 57)
Blue-eyed-grass

　　3b Petals and sepals yellow; filaments nearly completely separate; mostly restricted to wet situations (coastal). *Sisyrinchium californicum* (pl. 57)
Golden-eyed-grass; Mo-n

　2b Petals and sepals decidedly different and usually more than 2.5 cm long; stem cylindrical (petals and sepals usually with darker veins)

　　4a Inflorescence with many flowers; often more than 100 cm tall; rooted in mud at the edges of lakes, ponds, and streams (flower bright yellow or cream colored; tube of perianth 12 mm long) . . . . *Iris pseudacorus*
Yellowflag, Sword Iris; eu

　　4b Inflorescence with 1–6 flowers; rarely more than 40 cm tall; strictly terrestrial

　　　5a Tube of perianth 5–28 mm long, about the same length as or shorter than the floral tube; inflorescence with 2–6 flowers (sepals 5–10 cm long)

　　　　6a Tube of perianth 10–28 mm long, about as long as the floral tube; tip of floral tube with several small lobes; flowers 2 or 3, usually dark or pale lilac or purple but sometimes cream colored or red purple; leaves usually at least 1 cm wide (on grassy hillsides near the coast) . . . .
. . . . *Iris douglasiana* (pl. 57)
Douglas Iris

　　　　6b Tube of perianth 5–13 mm long, much shorter than the floral tube; tip of floral tube without lobes; flowers 3–6, mostly lavender blue; leaves usually less than 1 cm wide . . . . *Iris longipetala*
Coast Iris, Long-petaled Iris; Me-Mo

　　　5b Tube of perianth 28–86 mm long, much longer than the floral tube; inflorescence with 1 or 2 flowers

　　　　7a Stem usually with more than 3 leaves, these overlapping and clasping the stem for most of their length; tube of perianth 3–5 cm long, not widening at the top; sepals up to 8.5 cm long (flowers 2, cream colored, pale yellow, or white with a lavender tinge; leaves often tinged purple red) . . . .
. . . . *Iris purdyi*
Purdy Iris; Sn-n

　　　　7b Stem usually with not more than 3 leaves, these not overlapping or clasping the stem for most of their length; tube of perianth 3–8.5 cm long, widening at the top; sepals up to 7 cm long

8a   Tube of perianth 3–6 cm long, widening gradually at the top; flowers 2, cream colored, tinged with yellow (in slightly shaded situations, including oak woods) .... *Iris fernaldii* (pl. 57)
                                            Fernald Iris; Sn, Sl-SCr

8b   Tube of perianth 3.5–8.5 cm long, widening abruptly at the top; flowers 1 or 2, golden yellow, cream colored, or purple blue (on grassy slopes and in oak woods) .... *Iris macrosiphon*
                                            Ground Iris; SCl-n

JUNCACEAE (RUSH FAMILY)   In their general appearance, and in the way their leaf sheaths envelop the stems, members of Juncaceae resemble grasses. The leaves are often reduced, however, to the point that only the sheaths remain. Furthermore, when leaf blades are present, these are more often cylindrical rather than flattened.

The flowers, usually concentrated in inflorescences at stem ends, differ from those of grasses in that they have definite perianth segments. Three of these correspond to sepals, and three others correspond to petals. There are two, three, or six stamens and a pistil with two or three stigmas. The ovary is superior. The fruit, often with a little beak at the tip, either contains numerous seeds in three chambers or three seeds in a single chamber.

The inflorescence typically has two bracts below it. The lower one is longer than the upper one and may appear to be a continuation of the stem. There are usually one or two bractlets associated with each flower and sometimes with clusters of flowers. In identification of rushes, the appearance and disposition of bracts, leaf blades, and sheaths are important. It may also be necessary to have mature fruit.

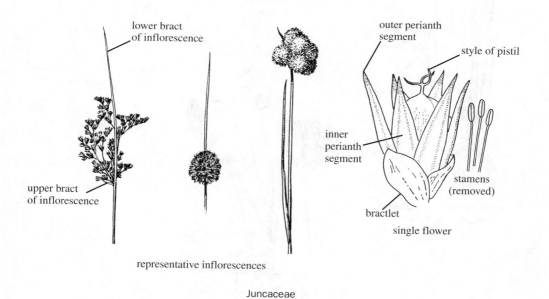

representative inflorescences

Juncaceae
(Rush Family)

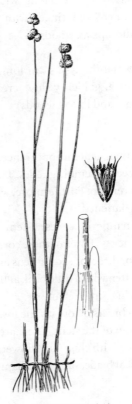

Juncus bolanderi
Bolander Rush

Juncus bufonius var. *bufonius*
Common Toad Rush

Juncus effusus var. *pacificus*
Pacific Bog Rush

Juncus falcatus
Sickleleaf Rush

Juncus kelloggii
Kellogg Rush

Juncus hemiendytus
Self-pollinating Rush

Juncus lesueurii
Salt Rush

Luzula comosa
Common Wood Rush

1a   Leaf blades very flexible and flattened, with long hairs, and often shredding at the bases when young; edges of leaf sheaths fused together; stem hollow; mostly in woodland habitats . . . . *Luzula comosa* [includes *L. subsessilis*] (pl. 57; fig.)

                                                           Common Wood Rush

1b   Leaf blades, if present, generally stiff and usually not flattened, without hairs, and not shredding at the bases; edges of leaf sheaths not fused; stem not hollow; mostly in wet places, but sometimes on backshores of sandy beaches

     2a   Stem branching below the inflorescence (flowers usually solitary in leaf axils; annual; rarely so much as 20 cm tall; bractlets beneath single flowers, or clusters of flowers, membranous, up to 2 mm long)

         3a   Inner 3 perianth segments rounded at the tips, the outer 3 sharply pointed; tip of fruit bluntly rounded (flowers often concentrated near branch tips; mostly in coastal saline habitats) . . . . *Juncus ambiguus* [*J. bufonius* var. *halophilus*]

                                                                    Coastal Rush

         3b   Inner 3 perianth segments, like the outer ones, sharply pointed at the tips; tip of fruit conical

            4a   Outer perianth segments 3–4 mm long (flowers scattered along the branches; in mud of drying pools and on stream banks) . . . . *Juncus bufonius* var. *occidentalis*

                                      Western Toad Rush, Roundfruit Toad Rush

            4b   Outer perianth segments 4–7 mm long

                5a   Flowers scattered along the branches (sometimes a weed in gardens, but also in many other moist habitats) . . . . *Juncus bufonius* var. *bufonius* (fig.)

                                                 Common Toad Rush

                5b   Most flowers concentrated near branch tips, which are often slightly coiled (in saline habitats) . . . . *Juncus bufonius* var. *congestus*

                                                Congested Toad Rush

     2b   Stem not branching below the inflorescence, originating from the base of the plant

         6a   Flowers usually not more than 3 at stem tip; stamens 2 or 3; not often more than 5 cm tall (annual)

            7a   Flowers commonly 3 at stem tip; usually 3–5 cm tall (flower red) . . . . *Juncus kelloggii* (fig.)

                                                   Kellogg Rush

            7b   Flower usually solitary at stem tip; usually less than 3 cm tall (plant with few leaves)

                8a   Fruit protruding 1 mm beyond the flower; perianth segments with a red midrib; at elevations above 2,000 ft (plant with few stems) . . . . *Juncus hemiendytus* (fig.)

                                        Self-pollinating Rush; Na-n

                8b   Fruit not protruding so much as 1 mm beyond the flower; perianth segments with a green midrib; from lowland habitats to elevations of 4,500 ft (plant forming clumps) . . . . *Juncus uncialis*

                                        Inch-high Dwarf Rush; SLO-n

         6b   Flowers more than 5, usually many in each inflorescence; stamens usually 6, but sometimes 3; more than 10 cm tall

9a   Lower bract below the inflorescence usually flattened and either shorter than the inflorescence or directed away from the axis of the stem so that the inflorescence appears to be at the top; leaf blades usually present . . . .
. . . . JUNCACEAE, SUBKEY

9b   Lower bract below the inflorescence cylindrical (use hand lens), appearing to be a continuation of the stem, and longer than the inflorescence, so that the inflorescence appears not to be at the top; leaves often reduced to only the sheaths (lower bracts of *Juncus occidentalis* and *J. tenuis,* in the subkey, may appear to be continuations of the stem, but both of these rushes have flattened bracts and some distinct leaf blades)

    10a   Flower 4–8 mm long, sometimes only 3.5 mm long in *Juncus mexicanus* and *J. balticus;* fruit with a sharp beak (stamens 6)

        11a   Flower usually 6–7 mm long; perianth segments with red brown margins (lower bract 10–25 cm long; stem cylindrical; leaf blades absent, the sheaths dark to light brown; stems often more than 90 cm tall, often in rows arising from horizontal underground stems; at the edges of salt marshes or on backshores of sandy beaches) . . . . . .
. . . . *Juncus lesueurii* (fig.)
Salt Rush; SLO-n

        11b   Flower 4–5 mm long, rarely 6 mm long in *Juncus mexicanus;* perianth segments with transparent, white margins

            12a   Stem flattened and twisted; lower bract up to 15 cm long; perianth segments green, striped on both sides with red brown; up to 60 cm tall (leaf sheaths yellow to brown, the blades rarely present) . . . . *Juncus mexicanus*
Mexican Rush; La-s

            12b   Stem cylindrical or somewhat flattened, not twisted; lower bract up to 20 cm long; perianth segments brown or green, with a green midrib; often more than 100 cm tall (leaf sheaths sometimes with a bristle up to 1 cm long; stems often in rows, arising from horizontal underground stems) . . . *Juncus balticus*
Baltic Rush, Wire Rush

    10b   Flower not more than 3.5 mm long; fruit without a sharp beak, but it may be pointed (leaves without blades, the sheaths usually dull to dark brown; often more than 90 cm tall)

        13a   Inflorescence usually at least 4 cm long; stem 2–5 mm wide (lower bract rarely as long as one-third the length of the stem; leaf sheaths 5–15 cm long, brown to black brown; perianth segments stiff, pale brown to straw colored, the edges light tan; stamens 3; plant forming clumps; in boggy areas inland) . . . .
. . . . *Juncus effusus* var. *pacificus* (fig.)
Pacific Bog Rush

        13b   Inflorescence usually less than 4 cm long; stem usually less than 2 mm wide (perianth segments usually with a green midrib; plant forming clumps)

14a Perianth segments upright at maturity, the margins dark brown; fruit somewhat 3 sided; inflorescence up to 4 cm wide; lower bract usually less than one-third the length of the stem; leaf sheaths 4–15 cm long, red brown at their bases; stamens 3; in coastal boggy areas .... *Juncus effusus* var. *brunneus*
Brown Bog Rush

14b Perianth segments spreading apart at maturity, the margins mostly red brown; fruit globular; inflorescence up to 6 cm wide; lower bract usually more than one-third the length of the stem; leaf sheaths 2–10 cm long, dark brown at their bases; stamens 6; in moist but not boggy areas (stem sometimes blue green) .... *Juncus patens*
Spreading Rush

Juncaceae, Subkey: Flowers more than 5, usually many, in each inflorescence; lower bract below the inflorescence usually flattened and either shorter than the inflorescence or directed away from the axis of the stem so that the inflorescence appears to be at the top; leaf blades usually present

1a Inflorescence loose, with single flowers scattered along its branches and with bractlets below each flower (leaves basal, 5–20 cm long, the blades more than half as long as the stem; flower green to reddish, about 4 mm long; stamens 6; up to 60 cm tall) ...... *Juncus tenuis*
Slender Rush

1b Inflorescence consisting of 1 or more dense clusters of flowers, with bractlets below each cluster and sometimes also below each flower

2a Lower bract below inflorescence protruding beyond the inflorescence at a slight angle, superficially stemlike (inflorescence less than 1 cm wide; flower more than 4 mm long; stamens 6; leaves less than 1 mm wide, mainly basal, less than half as long as the stem; up to 60 cm tall, forming clumps) .... *Juncus occidentalis* [*J. tenuis* var. *congestus*]
Western Rush

2b Lower bract either not protruding beyond the inflorescence or, if longer than the inflorescence, directed to the side at a wide angle and not stemlike

3a Leaf blades either cylindrical or flattened but not folded and not clasping the stem, sometimes with ridges that make them appear jointed; stem cylindrical or partly flattened

4a Inflorescence 5–15 cm wide, often with 2 or more whorls of spreading branches; flower clusters more than 10 in each inflorescence, each cluster 4–6 mm wide; leaves shorter than the stem (flower cluster brown, with fewer than 15 flowers; each leaf blade 1–3 mm wide, appearing to be jointed, with a white membrane 4–6 mm long where it joins the sheath; fruit protruding beyond the perianth; stamens 6; up to 1 m tall) .... *Juncus dubius*
Mariposa Rush

4b Inflorescence less than 5 cm wide, without whorls of spreading branches; flower clusters fewer than 10 in each inflorescence, each cluster more than 6 mm wide, but sometimes only 5 mm in *Juncus covillei*, leaves sometimes as long as the stem

5a Flower cluster with at least 20 flowers; each leaf blade cylindrical, appearing to be jointed, and with a white membrane 4–6 mm long where it joins the sheath; stamens 3; up to 100 cm tall (flower dark brown; leaf blades 1–3 mm wide) . . . . *Juncus bolanderi* (fig.)
Bolander Rush; SCl-n

5b Flower cluster with not more than 15 flowers; leaf blades flattened, 2–4 mm wide, not appearing to be jointed, and without white membranes; stamens 6; up to 30 cm tall

6a Each inflorescence with 1–6 flower clusters, these 5–10 mm wide; flower light brown, 2–5 mm long; fruit protruding up to 1 mm beyond the perianth; leaf blades not sickle shaped (plant forming clumps) . . . .*Juncus covillei*
Coville Rush; Ma-n

6b Each inflorescence with 1–3 flower clusters, these 10–15 mm wide; flower dark brown, 4–6 mm long; fruit either not protruding beyond the perianth, or just barely; leaf blades slightly sickle shaped (stem sometimes growing horizontally and rooting) . . . .
. . . . *Juncus falcatus* (fig.)
Sickleleaf Rush

3b Leaf blades flattened, folded, and clasping the stem, similar to the leaf blades of an iris, the ridges, if present, usually not extending completely across them, except sometimes in *Juncus phaeocephalus;* stem flattened (following species difficult to separate)

7a Fruit not protruding beyond the perianth; flower 4–6 mm long (leaf blades usually not more than 5 mm wide; flower dark brown or green brown; beak less than one-fourth as long as the body of the fruit; stamens 6)

8a Inflorescence 2–5 cm wide, its few branches ascending and bearing 1–5 flower clusters, each 10–15 mm wide and often with more than 50 flowers; up to 50 cm tall . . . . *Juncus phaeocephalus* var. *phaeocephalus*
Brown-headed Rush

8b Inflorescence up to 12 cm wide, its many branches spreading and bearing up to 25 flower clusters, each 5–10 mm wide and with fewer than 16 flowers; up to 90 cm tall . . . .
. . . . *Juncus phaeocephalus* var. *paniculatus*
Panicled Brown-headed Rush

7b Fruit protruding beyond the perianth, sometimes just barely; flower 3–4 mm long

9a Beak about one-fourth as long as the body of the fruit, the tip tapering gradually (inflorescence up to 12 cm wide, its branches many and spreading; each flower cluster 5–10 mm wide, with not more than 15 flowers; each inflorescence with not more than 15 flowers; leaf blades usually not more than 5 mm wide; flower light to dark brown; stamens 6; up to 60 cm tall) . . . . *Juncus oxymeris*
Pointed Rush

9b Beak less than one-fourth as long as the body of the fruit, the tip not tapering gradually

　　10a Leaf blades 2–5 mm wide; flower green brown or dark brown; stamens 3; not more than 50 cm tall .... *Juncus ensifolius*
　　　　　　Threestem Rush; Na-n

　　10b Leaf blades 3–12 mm wide; flower straw colored to dark brown; stamens 6; up to 90 cm tall .... *Juncus xiphioides*
　　　　　　Irisleaf Rush

JUNCAGINACEAE (ARROW-GRASS FAMILY)　Members of the Arrow-grass Family are semiaquatic, or at least grow in wet places. In our region, one or more species of *Triglochin* are found in almost any salt marsh. *Lilaea scilloides* (Flowering Quillwort), sometimes accorded its own family (Lilaeaceae), is occasionally encountered in freshwater habitats, particularly vernal pools. The leaves are all basal and somewhat grasslike in general appearance because they have sheaths at their bases. In species of *Triglochin*, the flowers are in crowded, elongated racemes on unbranched stems that usually rise above the leaves. They have three to six green perianth segments, three to six stamens, and one pistil, the fruiting part of which is partitioned lengthwise into three to six divisions. In *Lilaea*, the flower has one perianth segment or none. Those in inflorescences that are out of the water may be either staminate or have both a stamen and a pistil. Pistillate flowers, located at the base of the plant, are peculiar in that the style of the pistil is often more than 6 cm long.

1a Inflorescence up to 2 cm long; flowers with 1 perianth segment or none, and either with 1 stamen or with both a stamen and a pistil; flowers at the base of the plant with only a pistil, its style 6–20 cm long; in or near freshwater (stem 6–20 cm long, equal to or shorter than the leaves, these 1–5 mm wide, cylindrical) .... *Lilaea scilloides*
　　　　Flowering Quillwort

1b Inflorescence more than 2 cm long, usually at least 10 cm; flowers with 3–6 perianth segments, all with 1–6 stamens and a pistil; flowers absent at the base of the plant; in saline or alkaline habitats

　　2a Stem usually 1 or 2, equal to or shorter than the leaves; sepals 3–5; stamens 1–3; fruiting portion of the pistil consisting of 3 divisions; fruit about as wide as long, each division 3 ribbed (leaves flattened, 1–2 mm wide; racemes up to 12 cm long; stem up to 20 cm tall) .... *Triglochin striata*
　　　　　　Three-ribbed Arrow-grass

　　2b Stem usually 2 or more, longer than the leaves; sepals 6; stamens 6; fruiting portion of the pistil consisting of 6 divisions; fruit longer than wide, the divisions not obviously ribbed

　　　　3a Stem up to 70 cm tall; racemes up to 40 cm long; leaves flattened, 1–5 mm wide, the ligules usually pointed, rarely slightly notched..... *Triglochin maritima* (fig.)
　　　　　　　　Seaside Arrow-grass; SFBR

　　　　3b Stem usually not more than 40 cm tall; racemes up to 15 cm long; leaves cylindrical, about 1 mm wide, the ligules 2 lobed .... *Triglochin concinna* (fig.)
　　　　　　　　Salt Marsh Arrow-grass

Triglochin concinna
Salt Marsh Arrow-grass

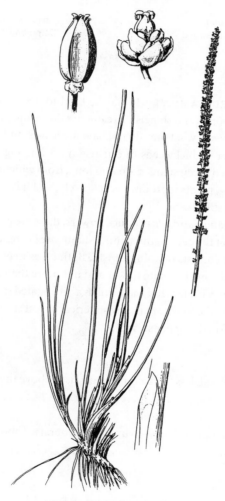

Triglochin maritima
Seaside Arrow-grass

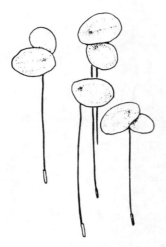

Lemna minor
Common Duckweed

Spirodela polyrrhiza
Common Duckmeat

LEMNACEAE (DUCKWEED FAMILY)   It may be hard to believe that the little duckweeds that float at the surface of ponds and sluggish streams are flowering plants. They are indeed capable of producing tiny flowers; what they do not have are leaves, and some also lack roots. The more commonly encountered duckweeds of our region, belonging to the genus *Lemna* and *Spirodela*, consist of flattened green stems, 2–10 mm long, from which one or more unbranched roots originate. By budding, plants reproduce vegetatively, and it is usual to see the parent and one or more of its progeny still attached.

When species of *Lemna* and *Spirodela* produce flowers, these are practically microscopic and partly hidden in a little pit at the edge of the stem. A staminate flower consists of just one or two stamens; a pistillate flower consists of a single pistil. Both types may be found in the same pit, but in that case there is only one pistillate flower and not more than two staminate flowers. In some references, the complex of stamens and pistil in a pit is stated to be a single flower.

In *Wolffia* and *Wolffiella*, the stems are nearly spherical or just slightly flattened, there are no roots, and the flower pit is on the upper surface.

1a   Plant without roots
    2a   Plant nearly spherical, less than 2 mm long, raised above the surface of the water . . . . .
. . . . *Wolffia columbiana*
Water-meal

    2b   Plant flattened on upper side, 4–10 mm long, not raised above the surface of the water
. . . . *Wolffiella lingulata*
Mudmidget; SM-n

1b   Plant with at least 1 root
    3a   Plant with at least 2 roots, sometimes only 1 in *Spirodela punctata*
        4a   Plant 5–10 mm long, with 7–12 veins and 5–16 roots . . . . *Spirodela polyrrhiza* (fig.)
Common Duckmeat, Greater Duckweed

4b   Plant 3–5 mm long, with 3–5 veins and usually 2–6 roots, but sometimes only 1 . . . . *Spirodela punctata [S. oligorrhiza]*

Small Duckmeat; Al

3b   Plant with only 1 root

5a   Upper surface of the stem smooth, with only 1 vein or none; plant pale green or nearly transparent

6a   Plants in groups of 4–8, each 2–4 mm long, narrowly elliptical, usually about 3 times as long as wide . . . . *Lemna valdiviana*

Uncommon Duckweed; La-s

6b   Plant solitary or in pairs, each 1–3 mm long, oval, less than twice as long as wide . . . . *Lemna minuta [L. minima* and *L. minuscula]*

Least Duckweed

5b   Upper surface of the stem bumpy, with 3–5 veins, but these sometimes indistinct; plant light or dark green, sometimes mottled with red in *Lemna gibba*

7a   Plant light green, in pairs, each 2–4 mm long, the veins not distinct; upper part of root with winglike membrane . . . . *Lemna aequinoctialis*

Valley Duckweed

7b   Plant dark green or light green and mottled with red, in groups of 2 or 3, each 2–6 mm long, the veins rather distinct; upper part of root without winglike membranes

8a   Plant usually symmetrical, the lower surface nearly flat; plant dark green, the length usually almost twice the width . . . . *Lemna minor* (fig.)

Common Duckweed, Lesser Duckweed

8b   Plant usually lopsided, the lower surface usually inflated; plant light green and mottled with red, the length not more than 1.5 times the width . . . . *Lemna gibba*

Swollen Duckweed, Gibbous Duckweed

LILIACEAE (LILY FAMILY)   The Lily Family is well represented in the region. A few of our genera have sometimes been placed in Amaryllidaceae (Amaryllis Family). All species are perennial, but most are herbaceous and die back, after flowering and fruiting, to underground bulbs, corms, or rhizomes. There are usually three sepals and three petals; when these components of the flower are much alike, they are all referred to as perianth segments. There are typically six stamens. The ovary is superior in all our representatives, and the fruit, usually dry and cracking apart at maturity, but fleshy in certain types, is partitioned into three divisions. There are a few exceptions to the formula stated above. For instance, in *Maianthemum*, there are only four perianth segments and four stamens, and the fruit consists of two divisions; in *Smilax*, which is a woody vine, some flowers are pistillate, and others are staminate.

Hundreds of exotic species of Liliaceae are in cultivation. They include hyacinths, tulips, and onions, as well as true lilies. Many natives of our western states have been grown not only in North America, but also in other parts of the world. Some of them are easily started from seed, but seedlings cannot be expected to bloom until about the fourth year. Both seeds and flowering-size bulbs raised from seed are often available commercially. It is important to

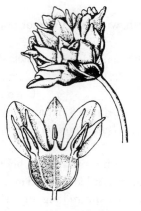

Dichelostemma capitatum
Bluedicks

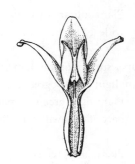

Dichelostemma congestum
Ookow

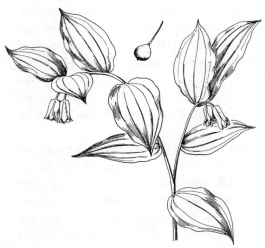

Disporum smithii
Largeflower Fairybells

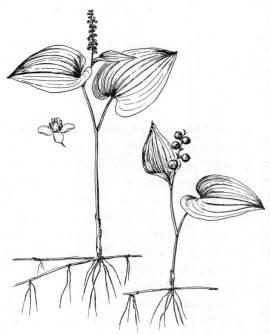

Maianthemum dilatatum
False Lily-of-the-valley

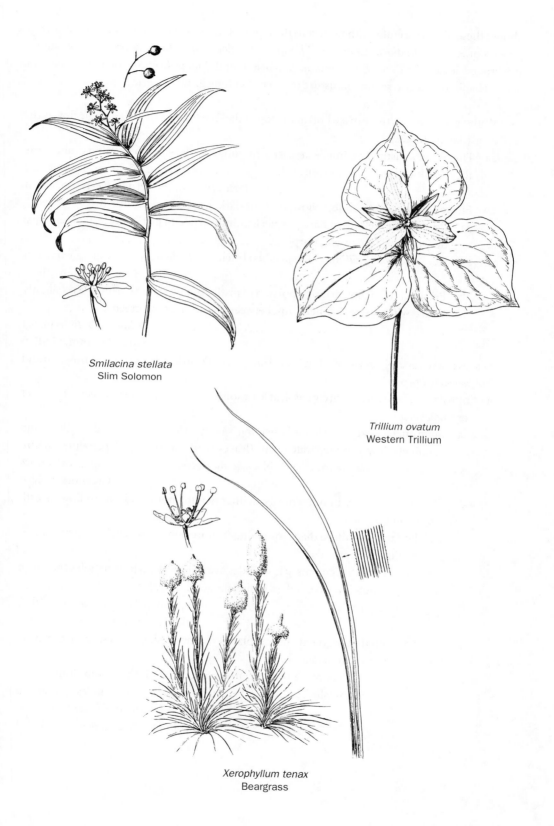

Smilacina stellata
Slim Solomon

Trillium ovatum
Western Trillium

Xerophyllum tenax
Beargrass

choose these plants carefully with respect to the type of situation in which they are to be grown. Species of *Allium, Brodiaea, Camassia, Lilium,* and *Calochortus* generally prefer open, sunny places; species of *Trillium, Erythronium, Smilacina,* and *Maianthemum* usually do best in shaded habitats and also require moisture over a rather long growing period.

1a   With 3 broad leaves, these in a whorl at the top of the stem, where there is a single flower (in moist, shaded areas)

     2a   Flower on a distinct peduncle at least 2 cm long; petals white, becoming dull rose as they age; leaves uniformly green............................ *Trillium ovatum* (fig.)

           Western Trillium, Western Wake-robin; Mo-n

     2b   Flower sessile; petals white, yellow, pink, or dark purple; leaves mottled

         3a   Petals white, yellow, or dark purple; fruiting portion of pistil purple; filament tips, between the anther sacs, purple ....

              .... *Trillium chloropetalum* [includes *T. chloropetalum* var. *giganteum*]

               Giant Trillium; Mo-n

         3b   Petals white to pink; fruiting portion of pistil usually green but sometimes slightly tinged with purple; filament tips, between the anther sacs, green ....

          .... *Trillium albidum* (pl. 61)

               Sweet Trillium; SFBR-n

1b   Leaves not necessarily in whorls, but if so arranged, not restricted to the top of the stem and not especially broad

     4a   Substantial leaves not entirely basal, at least some on the stem, except in certain species of *Calochortus*

         5a   Stem often prickly, with tendrils, twining, woody, at least in old growth; pistillate and staminate flowers on separate plants (leaves nearly heart shaped; perianth 3–6 mm long, green or yellow; fruit fleshy, black; in moist areas) .... *Smilax californica*

               Greenbrier; Na-n

         5b   Stem not prickly, without tendrils, neither twining nor woody; most flowers with a pistil and stamens

             6a   Perianth not more than 7 mm long (inflorescence 2–12 cm long, a raceme or panicle at stem ends)

                 7a   Perianth segments 4; leaf blades heart shaped, on long petioles (perianth 2.5 mm long; coastal, in moist, shaded areas).........................

                    .... *Maianthemum dilatatum* (fig.)

                     False Lily-of-the-valley; SM, Ma-n

                 7b   Perianth segments 6; leaf blades not heart shaped, sessile or nearly so (mostly in shaded habitats)

                     8a   Flowers more than 20, in a panicle; perianth 1–2 mm long; up to 90 cm tall .... *Smilacina racemosa* [includes *S. racemosa* var. *amplexicaulis*] (pl. 60)

                       Fat Solomon, Western Solomon-seal

8b   Flowers 5–15, in a loose raceme; perianth 4–6 mm long; rarely more than 40 cm tall . . . .

. . . . *Smilacina stellata* [includes *S. stellata* var. *sessilifolia*] (fig.)

Slim Solomon, Nuttall Solomon-seal

6b   Perianth more than 10 mm long

9a   Plant branching sparingly if at all

10a Perianth segments all similar . . . . LILIACEAE, SUBKEY 1

Fritillaria, Lilium

10b Petals and sepals distinctly different . . . . LILIACEAE, SUBKEY 2

Calochortus

9b   Plant branching repeatedly (leaves 3–15 cm long, sessile; flowers 1–7 at stem tips; perianth usually white or cream colored, sometimes pale green; in moist coniferous woods)

11a Stamens about as long as or longer than the perianth; stigma not lobed . . . . *Disporum hookeri*

Fairybells; Mo-n

11b Stamens shorter than the perianth; stigma 3 lobed . . . .

. . . . *Disporum smithii* (fig.)

Largeflower Fairybells; SCr-n

4b Substantial leaves entirely basal, stem leaves, if present, much reduced (in some species leaves wither before flowering time; perianth segments similar, at least in color)

12a Flowers in an umbel or umbel-like racemes, this sometimes with smaller umbels or individual flowers below it (leaves entirely basal)

13a Perianth segments united for some distance above their bases, forming a definite tube

14a Umbel with 2 bracts below it (flower fragrant; each perianth segment dull white, green at the base, and with a red or brown midrib on its outer surface) . . . . *Nothoscordum inodorum*

False Garlic; sa

14b Umbel with at least 3 bracts below it . . . . LILIACEAE, SUBKEY 3

13b Perianth segments separate to their bases or nearly so, not forming a tube

15a Leaves 2, 5–10 cm wide, often with dark blotches; umbel nearly sessile (leaves 10–20 cm long; perianth 1.5 cm long, mottled green and purple; pedicel 10–20 cm long; in moist, shaded areas; Coast Ranges) . . . .

. . . . *Scoliopus bigelovii*

Fetid Adder's-tongue; SCr-n

15b Leaves more than 2 or if only 2, not more than 1.5 cm wide, usually without blotches; umbel usually not sessile

16a Leaves 5–12 cm wide; flowering stem often more than 50 cm tall, sometimes with smaller umbels or single flowers below the main umbel; in redwood forests (perianth rose purple, 10–18 mm long) . . .

. . . . *Clintonia andrewsiana* (pl. 60)

Red Bead Lily, Red Clintonia; Mo-n

16b Leaves not more than 1.5 cm wide, sometimes withered by flowering time; flowering stem usually less than 50 cm tall, rarely with flowers below the main umbel; mostly in exposed sunny habitats

17a Umbel, before flowering, enclosed by 3 or more separate bracts; each flower with very small bract; leaves, if bruised, without the odor of onion (perianth segments 3–6 mm long, pale green, with brown midveins; may be on serpentine) . . . . *Muilla maritima* (pl. 60)

Common Muilla

17b Umbel, before flowering, enclosed by a sheath that later splits apart into 2–4 bracts; each flower without a bract; leaves, if bruised, with the odor of onion . . . . LILIACEAE, SUBKEY 4

Allium

12b Flowers in racemes or panicles

18a Leaves 2, mottled, commonly more than 2 cm wide (flowers 1 to several in a raceme)

19a Perianth 3.5–4 cm long, white with a yellow base, the base without a band of contrasting color; anthers yellow; in damp woods above 1,500 ft . . . . *Erythronium helenae*

Napa Fawn Lily, Saint Helena Fawn Lily; Na, Sn; 4

19b Perianth 2.5–3.5 cm long, white to cream colored with a yellow green base, the base with a band of yellow orange or brown; anthers white; in dry areas, from the lowlands to elevations of often more than 1,500 ft . . . . *Erythronium californicum* (pl. 60)

California Fawn Lily; Sn-n

18b Leaves several, not mottled and not often more than 2 cm wide

20a Flowering stem much branched (flowers scattered, up to about 3 cm wide, opening abruptly in late afternoon; perianth segments white, with green or purple midveins; leaves 20–70 cm long and 6–25 mm wide, with wavy margins)

21a Flowering stem often more than 100 cm tall, branching above but not from near the base (widespread) . . . . *Chlorogalum pomeridianum* var. *pomeridianum*

Common Soap-plant, Amole

21b Flowering stem prostrate or up to 40 cm tall, branching from near the base (coastal) . . . . *Chlorogalum pomeridianum* var. *divaricatum* (pl. 59)

Soap-plant; Sn-Mo

20b Flowering stem rarely if ever branched, but there may be several to many pedicels

22a Leaves at least 50, these persisting from year to year; flowering stem with hundreds of flowers (flowering stem often more than 150 cm tall; perianth cream colored; leaves 2–6 mm wide, 30–100 cm long) . . . . *Xerophyllum tenax* (fig.)

Beargrass; Mo-n

22b Leaves fewer than 15, not persisting from year to year; flowering stem usually with fewer than 50 flowers

    23a Leaves entirely basal (leaves less than 1 cm wide)

        24a Perianth 3–6 mm long, white, yellow, or green; leaves 5–20 cm long . . . . *Tofieldia occidentalis* [*T. glutinosa* ssp. *occidentalis*]

            Western Tofieldia; Sn-n

        24b Perianth 20–35 mm long, blue or purple blue, occasionally white; leaves 15–60 cm long (usually in habitats that are moist in winter) (*Camassia quamash* and *C. leichtlinii* ssp. *suksdorfii* often considered distinct: in the former, the perianth segments, one spaced farther apart than the others, turn back as they wither, whereas in the latter, the segments are evenly spaced and twist around developing fruit) . . . . . . *Camassia quamash* [includes *C. quamash* ssp. *linearis* and *C. leichtlinii* ssp. *suksdorfii*] (pl. 59)

            Common Camas; Ma, Na-n

    23b Leaves mainly basal, but some stem leaves present, these reduced (perianth yellow green, cream colored, or white)

        25a Style 1; perianth segments united at their bases, forming a slender tube about as long as the free lobes; flowering stem up to 50 cm tall (leaves 10–30 cm long, less than 1 cm wide; Coast Ranges) . . . . *Odontostomum hartwegii*

            Hartweg Odontostomum; Na

        25b Styles 3; perianth segments more or less free, not forming a tube; flowering stem often more than 50 cm tall

            26a Leaves 4–10 mm wide; stamens in mature flower as long as or longer than the perianth (perianth 4–6 mm long; leaves 10–40 cm long; in moist or dry, grassy places) . . . . *Zigadenus venenosus* (pl. 61)

                Death Camas

            26b Leaves usually more than 10 mm wide; stamens in mature flower shorter than the perianth

                27a Perianth 5–7 mm long; leaves often longer than the stem; on serpentine soils and usually where wet . . *Zigadenus micranthus* var. *fontanus* [*Z. fontanus*]

                    Serpentine Star Lily; Ma; 4

                27b Perianth 8–15 mm long; leaves shorter than the stem; widespread on dry slopes . . . . *Zigadenus fremontii* (pl. 61)

                  Common Star Lily, Fremont Star Lily

Liliaceae, Subkey 1:   Usually at least some substantial leaves scattered along the stem or in whorls; perianth more than 1 cm long, the segments all similar . . . . *Fritillaria, Lilium*

1a   Style not lobed, but the stigma 3 lobed; perianth 4–11 cm long

    2a   Flower upright; perianth white with purple spots when first open, then becoming red (perianth 4–7 cm long; in chaparral or forests) . . . . *Lilium rubescens*

                            Redwood Lily; SCr-n; 4

    2b   Flower nodding; perianth primarily orange or yellow when first open (in moist areas)

        3a   Perianth 5–11 cm long, mostly yellow, with red tinges and maroon spots . . . .
. . . . *Lilium pardalinum* ssp. *pardalinum* (pl. 60)

                            Leopard Lily

        3b   Perianth 5–7 cm long, mostly deep orange, but green at the center, crimson on the outside, and usually maroon dotted . . . .
. . . . *Lilium pardalinum* ssp. *pitkinense* [*L. pitkinense*]

                            Pitkin Marsh Lily; Sn; 1b

1b   Style with 3 slender lobes; perianth not more than 4 cm long

    4a   Leaves present on the upper part of the stem, but not on the lower part (flower nodding)

        5a   Perianth segments orange red or scarlet, with some yellow markings on the inner surfaces (perianth segments 1.5–4 cm long, the tips very curled; in shaded areas)
. . . . *Fritillaria recurva* [includes *F. recurva* var. *coccinea*] (pl. 60)

                            Scarlet Fritillary; Na, Sl-n

        5b   Perianth segments not orange red

            6a   Perianth segments 1–4 cm long, mostly purplish brown, with yellow green markings, but sometimes yellow green with purple mottling almost through-out, the tips straight (in shaded areas) . . . .
. . . . *Fritillaria affinis* [*F. lanceolata*] (pl. 60)

                            Checker Lily, Mission-bells

            6b   Perianth segments 1–2 cm long, mostly yellow green, but often marked with red purple, the tips slightly bent back . . . .
. . . . *Fritillaria eastwoodiae* [*F. phaeanthera*]

                            Butte County Fritillary; Na

    4b   Leaves present on the lower part of the stem

        7a   Perianth segments mottled, at least on 1 surface (usually on serpentine)

            8a   Flower nodding; each perianth segment up to 3 cm long, white, with purple spots and lines, the tip sometimes bent back; leaves not sickle shaped . . . . . .
. . . . *Fritillaria purdyi*

                            Purdy Fritillary; Na-n; 4

             8b   Flower upright; each perianth segment up to 2.5 cm long, green on the outer surface, yellow and rust brown inside, the tip straight; leaves sickle shaped. .
. . . . *Fritillaria falcata*

                            Talus Fritillary; SCl; 1b

        7b   Perianth segments not mottled, but they may have green lines (tips of segments mostly straight; flower nodding)

            9a   Perianth segments dark brown, purple, or purple green on both surfaces (perianth up to 4 cm long) . . . . *Fritillaria biflora*

                        Chocolate Fritillary, Chocolate Lily; Na, SM-s

9b   Perianth segments not dark brown, purple, or purple green on both surfaces, at least the outer surfaces pale
    10a   Perianth segments up to 16 mm long, white, with green lines; flower slightly fragrant . . . . *Fritillaria liliacea*
        Fragrant Fritillary; Sn-Mo; 1b
    10b   Perianth segments 18–35 mm long, pale green on the outer surfaces, purple brown inside; flower with an unpleasant odor . . . . *Fritillaria agrestis*
        Stinkbells; Me-SLO; 4

Liliaceae, Subkey 2:   Usually at least some substantial leaves scattered along the stem; perianth more than 1 cm long, the petals and sepals distinctly different . . . . *Calochortus*
1a   Flower bell shaped, generally nodding
    2a   Petals white (petals 2–2.5 cm long) . . . . *Calochortus albus*
        White Globe Lily, Fairy-lantern; SF-s
    2b   Petals yellow
        3a   Inner surfaces of petals without hairs, or with only a few hairs near the basal glands (petals 16–20 mm long, deep yellow; north Coast Ranges) . . . .
        . . . . *Calochortus amabilis*
            Golden Globe Lily, Diogenes' Lantern; Ma, Na, Sn-n
        3b   Much of the inner surfaces of petals hairy (petals pale yellow)
            4a   Petals 25–33 mm long, about as long as the sepals, sparsely hairy on the inner surfaces; stem usually branched; in woodlands or chaparral . . . .
            . . . . *Calochortus pulchellus* (pl. 59)
                Mount Diablo Fairy-lantern; CC (MD); 1b
            4b   Petals 35–45 mm long, longer than the sepals, with long hairs on the inner surfaces; stem mostly not branching; in open areas, on serpentine. . . . .
            . . . . *Calochortus raichei*
                Cedar's Fairy-lantern; Sn; 1b
1b   Flower broadly bowl shaped, generally facing up (in open areas)
    5a   Petals mainly yellow or yellow green, sometimes with red brown or purple on the inner surfaces
        6a   Each petal fan shaped at the tip, deep yellow, sometimes with lines or a red brown mark inside, scarcely hairy, except for short hairs at the base of the inner surface.
        . . . . *Calochortus luteus* (pl. 58)
            Yellow Mariposa Lily; Me-s
        6b   Each petal pointed at the tip, yellow green with purple markings inside, with long hairs over much of the inner surface or at least along the margins and at the base (on serpentine) . . . . *Calochortus tiburonensis* (pl. 59)
            Tiburon Mariposa Lily; Ma; 1b
    5b   Petals white, cream colored, lavender, lilac, or occasionally yellow, often with red, yellow, or purple on the inner surfaces
        7a   Each petal with long hairs over much of its inner surface (petals 12–25 mm long, white, pink, or purple, without markings at their bases) . . . . . . *Calochortus tolmiei*
            Pussy-ears; SCr-n

7b   Each petal with hairs only at the base of its inner surface

      8a   Petals 12–18 mm long (each petal white or pink, the inside often with a purple mark at the base; stem usually branched, up to 25 cm tall; often on serpentine) . . . . *Calochortus umbellatus* (pl. 59)

                             Oakland Star-tulip; La-SCl; 4

      8b   Petals usually at least 20 mm long, but smaller in *Calochortus uniflorus*

          9a   Gland at the base of the inner surface of each petal bordered by a fringed membrane (each petal 2–4 cm long, white with green stripes, the inside often with purple spots at the base; basal leaves 2–4 mm wide; stem rarely branched; up to 50 cm tall; in dry areas; at elevations above 4,000 ft) . . . . *Calochortus invenustus*

                          Plain Mariposa Lily; SCl (MH)-s

          9b   Gland at base of the inner surface of each petal not bordered by a membrane

              10a   Petals 15–28 mm long; basal leaves 5–20 mm wide; stem rarely branched; less than 5 cm tall; in moist areas (each petal lilac, the inside sometimes with a purple spot near the base) . . . . *Calochortus uniflorus*

                          Largeflower Star-tulip; Me-Mo

              10b   Petals usually more than 25 mm long; basal leaves not more than 6 mm wide; stem usually branched; up to 60 cm tall; in dry areas

                  11a   Each petal deep lilac, the inner surface sometimes with a purple spot; inner petal surface with long hairs scattered on the lower half and hairs at the base that collectively resemble a growth of fungus (Coast Ranges) . . . . *Calochortus splendens*

                          Splendid Mariposa Lily; La-s

                  11b   Each petal white to lavender or pale lilac, occasionally yellow, the inner surface with a red, purple, or brown mark, this often surrounded by yellow; inner petal surface without long hairs, but with short, sparse hairs about 5 mm from the base, these not funguslike

                      12a   Concentration of hairs on the inner surface of each petal rectangular (inner surface with a dark red mark in the center and usually with a paler red blotch above) . . . . *Calochortus venustus* (pl. 59)

                          Butterfly Mariposa Lily; SF-s

                  12b   Concentration of hairs on the inner surface of each petal not retangular

                      13a   Concentration of hairs on the inner surface of each petal shaped like a crescent or an inverted V; inner surface with a brown or purple central mark, this surrounded by bright yellow . . . . . . *Calochortus superbus*

                          Superb Mariposa Lily

13b Concentration of hairs on the inner surface of each petal shaped like 2 crescents or inverted Vs; inner surface with a red brown mark, this surrounded by pale yellow (at elevations above 1,500 ft) .... *Calochortus vestae* Clay Mariposa Lily; Na, Sn-n

Liliaceae, Subkey 3: Leaves usually basal; flowers in an umbel or umbel-like raceme, this usually with at least 3 bracts below it; perianth segments similar, united at their bases, forming at least a short tube

1a Flowering stem often more than 100 cm long, often twining through shrubbery or lying on the ground, but the stem contorted even if upright (perianth rose, pink, or purple, narrowed in the middle, the tube 5–7 mm long; anther-bearing stamens 3; inland, in dry areas) .... *Dichelostemma volubile [Brodiaea volubilis]* Snake Lily, Twining Brodiaea; Sl-n

1b Flowering stem rarely more than 50 cm long, upright, not contorted

2a Perianth some shade of yellow, each segment with a dark midvein on the outer surface (anther-bearing stamens 6, 3 longer than the others; in dry areas)

3a Perianth deep yellow, the tube 7–10 mm long; filaments deeply 2 lobed, the anthers in the clefts (coastal) .... *Triteleia ixioides [Brodiaea lutea]* Golden Triteleia, Pretty-face; SM-SLO

3b Perianth deep to pale yellow, the tube 4–5 mm long; filaments not lobed .... *Triteleia lugens [Brodiaea lugens]* (pl. 61) Uncommon Triteleia; Sn-Sl; 4

2b Perianth purple, violet, lavender, blue, or white

4a Umbel or umbel-like raceme dense, congested; pedicel not more than 1.5 cm long

5a Anther-bearing stamens 6; tube of perianth not narrowed just below the lobes; peduncle usually smooth (stamens of 2 different sizes; perianth blue, blue purple, or pink purple, but sometimes white; widespread) .... *Dichelostemma capitatum [Brodiaea pulchella]* (fig.) Bluedicks

5b Anther-bearing stamens 3; tube of perianth narrowed just below the lobes; peduncle often slightly rough

6a Inflorescence dense, with 6–15 flowers; pedicel 1–6 mm long; antherless stamens 3 ..... *Dichelostemma congestum [Brodiaea congesta]* (pl. 59; fig.) Ookow; SCl-n

6b Inflorescence open, with 10–35 flowers; pedicel 10–35 mm long; antherless stamens absent (perianth pink purple or violet) .... *Dichelostemma multiflorum [Brodiaea multiflora]* Manyflower Bluedicks, Wild Hyacinth; SM, SCl-n

4b Umbel not dense; most pedicels at least 2 cm long

7a Perianth white, very pale blue, or lilac

8a Anther-bearing stamens 3; antherless stamens 3; perianth segments without green midveins (perianth 3–4 cm long; often on serpentine) .. .. *Brodiaea californica* var. *leptandra* California Brodiaea; Sn, Na; 1b

8b   Anther-bearing stamens 6; antherless stamens none; perianth segments with green midveins

    9a   Perianth 1–1.5 cm long; pedicel up to 5 cm long; all stamens the same length and attached at the same level . . . .
. . . .*Triteleia hyacinthina [Brodiaea hyacinthina]*
White Triteleia, Wild Hyacinth; Mo-n

    9b   Perianth 1.5–2.5 cm long; pedicel often more than 10 cm long; stamens of different lengths and attached at different levels (often on serpentine) . . . . *Triteleia peduncularis [Brodiaea peduncularis]*
Long-rayed Triteleia; Mo-n

7b   Perianth bright to dark blue, violet, or purple, occasionally white in *Triteleia laxa*

    10a   Antherless stamens none; anther-bearing stamens 6, these attached at 2 levels (perianth blue, blue purple, or white, 2–3.5 cm long; pedicel up to 9 cm long; widespread) . . . . *Triteleia laxa [Brodiaea laxa]* (pl. 61)
Ithuriel's-spear, Grass-nut

    10b   Antherless stamens 3; anther-bearing stamens 3, all attached at the same level

        11a   Perianth 15–25 mm long; flowering stem not more than 7 cm tall (antherless stamens 4–8 mm long, about equal to the anther-bearing stamens)

            12a   Pedicel 1–5 cm long; anthers hooked at the tips, the filaments with 2 slender outgrowths (Coast Ranges) . . . *Brodiaea stellaris*
Starflower Brodiaea; Sn-n

            12b   Pedicel 3–15 cm long; anthers not hooked, and filaments lacking outgrowths . . . .
. . . . . .*Brodiaea terrestris [B. coronaria* var. *macropoda]* (pl. 58)
Dwarf Brodiaea; SLO-n

        11b   Perianth usually 25–35 mm long; flowering stem more than 10 cm tall

            13a   Antherless stamens flat, 6–9 mm long, usually equal to the anther-bearing stamens (pedicel 5–10 cm long; anthers not hooked and filaments lacking outgrowths) . . . .
. . . .*Brodiaea elegans* (pl. 58)
Elegant Brodiaea; Mo-n

            13b   Antherless stamens slightly inrolled, 8–15 mm long, longer than the anther-bearing stamens

                14a   Pedicel 4–10 cm long; anthers hooked at the tips, the filaments with 2 slender outgrowths . . . .
. . . .*Brodiaea appendiculata*
Grassland Brodiaea; Na-SCl

                14b   Pedicel 1–5 cm long; anthers not hooked and filaments lacking outgrowths (mostly inland) . . . *Brodiaea coronaria*
Harvest Brodiaea

Liliaceae, Subkey 4: Leaves basal, not more than 1.5 cm wide, sometimes withered by flowering time; flowers in umbels, these with 2–4 united bracts beneath them; perianth segments similar, not forming a tube .... *Allium*

1a Leaves often more than 5 mm wide, sometimes sickle shaped; flowering stem flattened or 3 sided, sometimes cylindrical in *Allium cratericola* (leaves flattened; perianth 7–15 mm long; in dry areas)

    2a Perianth white; pedicel 15–35 mm long; leaves 2 or 3, not sickle shaped; flowering stem 3 sided, 2 of the edges somewhat winged .... *Allium neapolitanum*

                  Naples Onion; me

    2b Perianth rose to purple; pedicel 5–18 mm long; leaves 1 or 2, generally sickle shaped; flowering stem usually flattened (often on serpentine)

        3a Stem winged; leaves 2, sickle shaped; umbel with 10–30 flowers; perianth rose to purple; mostly at elevations above 3,000 ft .... *Allium falcifolium* (pl. 58)

               Sickleleaf Onion, Scytheleaf Onion; SCr-n

        3b Stem not winged; leaf usually 1, sometimes straight; umbel with 20–35 flowers; perianth purple, but sometimes pale; usually at elevations below 2,500 ft .... *Allium cratericola*

               Crater Onion; Na-n

1b Leaves generally less than 5 mm wide, not sickle shaped; flowering stem cylindrical

    4a Plants generally in dense colonies and producing bulbs along underground stems; in moist areas (leaves 2 or 3; umbel with 15–35 flowers; pedicel 15–40 mm long; perianth 11–15 mm long, pink or lilac, occasionally white; fruiting portion of pistil with 6 ridges at the tip) .... *Allium unifolium* (pl. 58)

               Clay Onion, Oneleaf Onion; Mo-n

    4b Plants not both in dense colonies and producing bulbs from underground stems; in dry areas

        5a Fruiting portion of pistil with 6 prominent outgrowths at the tip

            6a Leaves 2–4 (perianth 5–9 mm long)

                7a Umbel with 10–15 flowers; each perianth segment white or pink tinged, nearly 3 times as long as wide; pedicel 4–16 mm long; leaves 2–4, cylindrical (sometimes on serpentine) .... *Allium amplectens*

                   Paper Onion, Narrowleaf Onion; Al, SCl

                7b Umbel with 10–50 flowers; each perianth segment rose to purple, rarely white, with a darker base and not more than twice as long as wide; pedicel 10–20 mm long; leaves 2, more or less flattened .... *Allium campanulatum*

                   Sierra Onion; Mo-n

            6b Leaf 1 (leaf cylindrical)

                8a Umbel with 6–35 flowers; perianth 6–12 mm long, dark red purple, the stamens shorter than it; pedicel up to 20 mm long (often on serpentine) .... *Allium fimbriatum*

                 Fringed Onion; Na-s

8b   Umbel with 15–40 flowers; perianth 5–8 mm long, lavender or white, the stamens equal to or slightly longer than it; pedicel up to 15 mm long. . . . . *Allium howellii*
Howell Onion; SCl

5b   Fruiting portion of pistil with 3 very small outgrowths at the tip (stamens shorter than the perianth)

9a   Perianth segments 4–9 mm long, with prominent green or red midveins; leaves 2 (perianth white to pale pink; pedicel 5–12 mm long; sometimes on serpentine) . . . . *Allium lacunosum* (pl. 58)
Wild Onion, Pitted Onion; Ma-s

9b   Perianth segments 8–15 mm long, without prominent green or red midveins; leaves often more than 2

10a   Each inner perianth segment roughened by very small, nearly hemispherical outgrowths on margins of the upper portion (use hand lens) (leaves 2–3)

11a   Outer perianth segment tips curving backward (umbel with 10–40 flowers; perianth white to purple rose; pedicel 6–25 mm long) . . . . *Allium acuminatum* (pl. 58)
Hooker Onion; CC (MD)-n

11b   Outer perianth segment tips not distinctly curved

12a   Umbel with 10–20 flowers; perianth red purple, sometimes white, each segment about 3 times as long as wide, widest below the middle; pedicel 10–20 mm long (sometimes on serpentine) . . . . *Allium bolanderi*
Bolander Onion; SCl

12b   Umbel with 10–40 flowers; perianth rose purple, each segment about twice as long as wide, widest near the middle; pedicel 10–35 cm long (in sandy soil) . . . . *Allium crispum*
Crinkled Onion; CC-s

10b   Inner perianth segments nearly smooth

13a   Perianth pink to rose (leaves 2 or 3; umbel with 10–40 flowers; perianth 8–11 mm long; pedicel 7–15 mm long; each filament about 3 times as wide at the base as over most of its length; widespread) . . . . *Allium serra* (pl. 58)
Serrated Onion; La-SCl

13b   Perianth red purple

14a   Leaves 2 or 3; umbel with 5–35 flowers; perianth 10–15 mm long; pedicel 10–40 mm long; stigmas often 3 lobed . . . . *Allium peninsulare*
Peninsular Onion; SFBR-s

14b   Leaves 3–6; umbel with 5–30 flowers; perianth 9–12 mm long; pedicel 5–20 mm long; stigmas not lobed (coastal) . . . . *Allium dichlamydeum*
Coastal Onion; Me-Mo

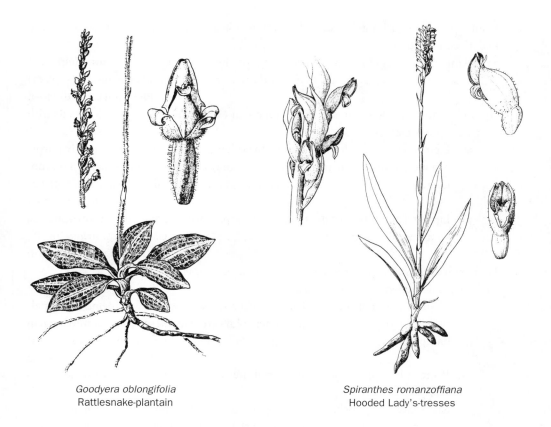

Goodyera oblongifolia
Rattlesnake-plantain

Spiranthes romanzoffiana
Hooded Lady's-tresses

ORCHIDACEAE (ORCHID FAMILY)　　The Orchid Family, consisting entirely of perennials, is one of the largest groups of flowering plants, with over 18,000 species. All of ours, unlike most of those in the tropics, are rooted in soil rather than in bark or in moss growing on trees. The irregular corolla of an orchid is markedly two lipped, the lowest of the three petals (the lip petal) being different from the other two. The three sepals, often petal-like, are usually similar to one another. There is a single anther-bearing stamen in all of our genera, except *Cypripedium,* which has two anther-bearing stamens. The filaments of the stamens are united to the style of the pistil. The ovary is inferior, and the fruiting part of the pistil is partitioned lengthwise into three divisions. Each fruit produces numerous small seeds.

Orchids, like many other plants, have a symbiotic relationship with fungi that penetrate their roots. Some species, in fact, have no chlorophyll and depend entirely on their fungal associates for their nutrition. The complex of orchid and fungus requires very special conditions, and these, at least in the case of terrestrial orchids, are not likely to be met in gardens. Attempts to cultivate terrestrial orchids will almost certainly result in failure, and it is therefore a mistake to disturb these plants in nature, unless they happen to be in the path of a bulldozer. Even then it would be best for experts in botanical gardens to try to transplant them to new situations. However, *Epipactis helleborine* (Helleborine), a European species, is now a benign weed in certain portions of the Pacific Northwest and has become established in some places within the San Francisco Bay Region.

1a Plant saprophytic, lacking chlorophyll, therefore not green; leaves reduced to scales (found in dark coniferous woods)

    2a Plant white, except for a slight yellow tinge on the petals (sepals 12–20 mm long) . . . . *Cephalanthera austiniae [Eburophyton austinae]* (pl. 61)

        Phantom Orchid; Mo-n

    2b Plant mostly brown, but sometimes yellow in *Corallorhiza maculata*, the petals with dark spots or streaks

        3a Lip petal white, with purple spots, the other petals and the sepals mostly purple brown, but in certain areas, some plants may be yellow; sepals usually 8–10 mm long . . . . *Corallorhiza maculata* (pl. 62)

            Spotted Coralroot

        3b Lip petal and other petals, as well as the sepals, with purple or red brown streaks; sepals usually 10–17 mm long . . . . *Corallorhiza striata* (pl. 62)

            Striped Coralroot; SCl-n

1b Plant with green stems and leaves; leaves not reduced to scales

    4a Lip petal inflated, closed to form a sac

        5a Plant with a single basal leaf and almost always a single flower (petals and sepals primarily rose purple, although other colors are evident; sepals 15–22 mm long; in damp woods) . . . . *Calypso bulbosa* (pl. 61)

            Fairy-slipper; SM, Ma-n

        5b Plant with 2 opposite leaves or several alternate leaves and usually with more than 1 flower

            6a Leaves 2, opposite; flowers 1–4; up to 20 cm tall (flower mostly green brown or yellow green, and with brown streaks or margins; upper sepal 1.5–2.5 cm long; in rocky areas) . . . . *Cypripedium fasciculatum*

                Clustered Lady's-slipper; SCr-n; 4

            6b Leaves several, alternate; flowers 1–12; often over 50 cm tall

                7a Flowers 1–3; upper sepal 3–6 cm long; lip petal white, streaked with purple; other 2 petals twisted, up to 6 cm long, mostly dark brown; in moist woods . . . . *Cypripedium montanum*

                    Mountain Lady's-slipper; SCl-n; 4

                7b Flowers 1–12; upper sepal 1.5–2 cm long; lip petal white or pale rose, streaked or spotted with purple; other 2 petals not twisted, less than 2 cm long, mostly green or yellow brown; in wet areas . . . . *Cypripedium californicum* (pl. 62)

                    California Lady's-slipper; Ma-n; 4

    4b Lip petal not inflated, not closed to form a sac, but the basal portion may be deeply concave or drawn out into a long spur

        8a Flower with a distinct pedicel (lip petal with a pouch)

            9a Sepals 12–20 mm long; lip petal 14–20 mm long, with deep grooves, the outside and inside green or yellow, and with red veins; mostly around springs, seepage areas, small streams, and ponds . . . . *Epipactis gigantea* (pl. 62)

                Stream Orchid

9b Sepals 10–13 mm long; lip petal 10–12 mm long, not grooved, the outside white or pink, the inside brown or purple; usually on dry slopes. . . . .
. . . . *Epipactis helleborine* (pl. 62)

Helleborine; eu

8b Flower nearly sessile

10a Lip petal not drawn out into a spur (Coast Ranges)

11a Leaves entirely basal, dark green with white veins, persisting throughout the year; inflorescence usually less than 10 cm long, not obviously twisted; flower 15–20 mm long; in dry areas, mostly in woods (flower white) . . . . *Goodyera oblongifolia* (fig.)

Rattlesnake-plantain; Ma-n

11b Leaves not entirely basal, uniformly light green, usually withering by flowering time; inflorescence up to 14 cm long, obviously twisted; flower 7–12 mm long; in moist or wet areas

12a Flower white or cream colored; lip petal rounded at the tip . . . .
. . . . *Spiranthes romanzoffiana* (fig.)

Hooded Lady's-tresses

12b Flower yellow or cream colored; lip petal usually with a pointed tip
. . . . *Spiranthes porrifolia*

Western Lady's-tresses; Mo-n

10b Lip petal drawn out into a spur (inflorescence 5–40 mm long)

13a Leaves not entirely basal; sepals 4–8 mm long; in wet areas or near streams (flower white or cream colored; spur of lip petal 5–15 mm long, curved)
. . . . *Platanthera leucostachys [Habenaria dilatata* var. *leucostachys]*

Whiteflower Bog Orchid; SLO-n

13b Leaves entirely basal, usually withering before flowering time; sepals 2–6 mm long; mostly in dry areas

14a Flower white or pale green, and with green veins; spur of lip petal 6–14 mm long, twice as long as the lip . . . . *Piperia elegans*
[includes *Habenaria elegans* vars. *elegans* and *maritima]*

Elegant Rein Orchid

14b Flower green; spur of lip petal 1–5 mm long, only slightly longer than the lip (in chaparral, oak woodland, grassland) . . . .
. . . . *Piperia unalascensis [Habenaria unalascensis]* (pl. 62)

Alaska Rein Orchid

POACEAE (GRASS FAMILY) Richard G. Beidleman

With the exception of bamboos, which are woody and sometimes treelike, most grasses are easily recognized as such. In our region, Poaceae (formerly called Gramineae), with about 200 species, is as well represented as Asteraceae (Sunflower Family). Unfortunately, about 40 percent of our grasses were introduced from other parts of North America, Europe (especially the

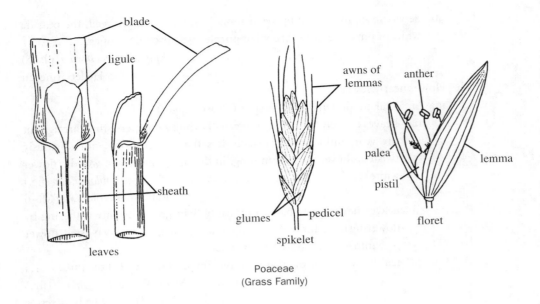

blade

ligule

sheath

leaves

awns of
lemmas

glumes

pedicel

spikelet

anther

palea

pistil

lemma

floret

Poaceae
(Grass Family)

Mediterranean area), Asia, Central and South America, Africa, and Australia. Beginning with the early Spanish colonists, in the eighteenth century, the seeds of exotic grasses arrived with hay, ballast, mattress fillers, and clothing, as well as in the hair of farm animals. Some species, moreover, were introduced intentionally, either as crop plants or for landscaping and erosion control.

The importance of grasses—especially wheat, barley, oats, rice, corn, and rye—in providing food for humans and livestock should be understood by everyone. Many species, furthermore, are cultivated for their ornamental foliage, interesting habit of growth, erosion-controlling attributes, and suitability for lawns. So successfully have introduced grasses taken over wild areas, however, that native species are no longer found in many places. Accounting in part for the decline in the number of native species are overgrazing by livestock, conversion of natural areas into cropland, and urbanization, with its varied impacts. About 20 percent of our native grasses are annual; the rest are perennials. Some of our perennial species are being propagated for use in landscaping. They are not only attractive, but they also survive summer drought and fit in with other native plants.

As common as grasses are, and as simple as they may appear to be, they are often very difficult to identify. The parts that are generally used in identification tend to be small—so small in some cases that even a 10× hand lens is not adequate. Furthermore, certain structures of grasses are unlike those of other plants. Thus they require a special set of terms. In the key presented here, technical terms have been kept to a minimum. Each major unit of the inflorescence is called a spikelet. At the base of the spikelet is a pair of glumes, and within these glumes are one or more florets. A fertile floret typically consists of a pistil, stamens (usually three), or both, enclosed within two bracts. The lower bract, known as the lemma, is almost always larger and better developed than the upper one, called the palea, and at least partially encloses the palea. The pistil becomes a one-seeded dry fruit, and in the following key, this is referred to as the grain. Some florets may be sterile, lacking a pistil. Also, some florets may contain only stamens or a pistil.

The lemmas and glumes are especially important in identification of grasses. It is necessary to examine them carefully to see if they have a hairlike or bristlelike projection, called an awn, at the tip, riblike lengthwise thickenings (veins), or other differentiations.

In the field, when you first encounter a grass, be sure to observe its overall growth form and especially the appearance of parts that may be disturbed upon collection. For instance, the stems and leaves are often grouped together in a dense basal clump; other types are simple individual plants, or form a sod, or spread outward by means of prostrate stems, which may root at the nodes. You will have to look closely at the leaf blades and leaf sheaths. The blades may be flat, folded, long or short, or have some special attributes. At the base of the blade, where this attaches to the sheath, there is usually a collarlike membranous structure, or a circle of long hairs, called the ligule, and there may also be a pair of small lobes, which in some cases are long enough to clasp the stem. Below the blade, the edges of the sheath may overlap or be partly or completely fused. The hairiness of some parts of the plant may be important. The general appearance of the inflorescence is also used extensively in the key. Some inflorescences are very open, with spreading branches and branchlets, whereas others are short branched and compact, at least early in their development.

Many species exhibit considerable variation and may appear in more than one section of the key. Furthermore, certain grass species freely interbreed, and even experts are often unable to agree on identification of a particular specimen. If one section of the key does not seem to fit your specimen, try the couplets in the opposite section. Even if you just come close, you are to be congratulated!

1a Stem at least 1 cm wide, tough and inflexible; leaves often more than 5 cm wide (inflorescence present much of the year, plumelike, mostly over 50 cm long; usually more than 3 m tall)

2a Leaves evenly spaced on the upright stem; inflorescence up to 12 cm wide and 60 cm long (usually in moist places, but sometimes on dry river bottoms) . . . .
. . . . *Arundo donax* (pl. 62)
Giant Reed; eu

2b Leaves mainly basal, forming a dense clump more than 2 m high; inflorescence up to 15 cm wide and more than 100 cm long

3a Leaf sheaths densely hairy; leaf blades dark green, not curled toward the tips; up to 7 m tall (noxious weed). . . . . *Cortaderia jubata*
Hairy Pampas Grass, Jubata Grass; sa

3b Leaf sheaths only sparsely hairy if at all; leaf blades pale green or blue green, curled toward the tips; up to 4 m tall. . . . . *Cortaderia selloana*
Smooth Pampas Grass, Pampas Grass; sa

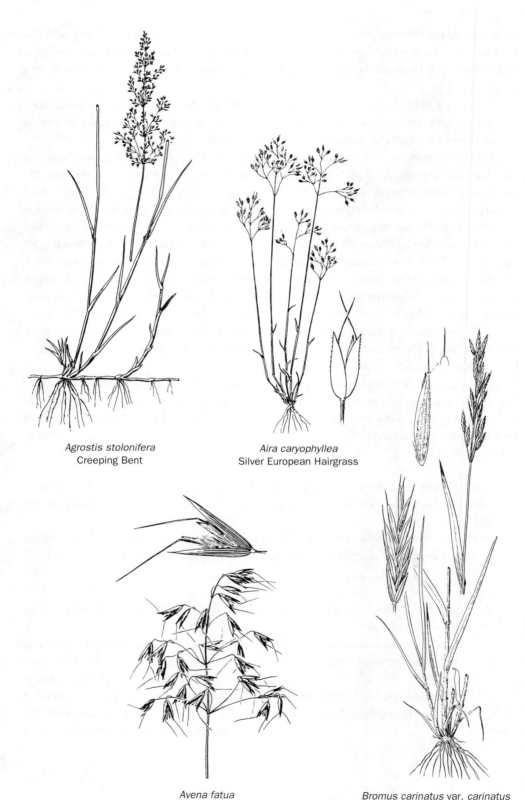

Agrostis stolonifera
Creeping Bent

Aira caryophyllea
Silver European Hairgrass

Avena fatua
Wild Oat

Bromus carinatus var. *carinatus*
California Brome

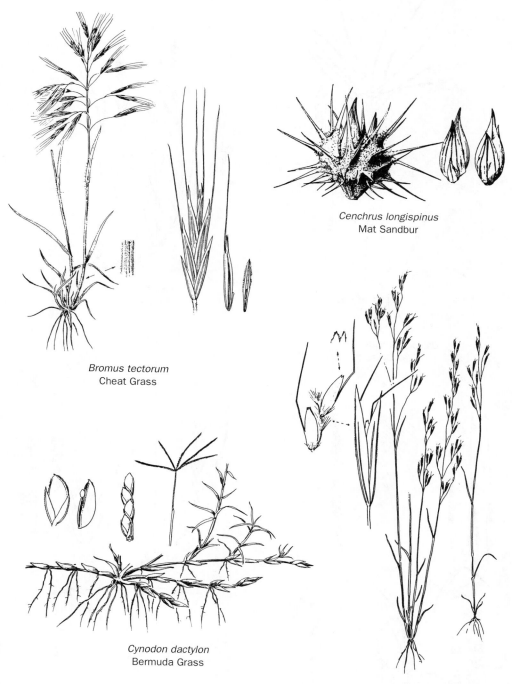

Cenchrus longispinus
Mat Sandbur

Bromus tectorum
Cheat Grass

Cynodon dactylon
Bermuda Grass

Deschampsia danthonioides
Annual Hairgrass

Distichlis spicata
Saltgrass

Digitaria sanguinalis
Crab Grass

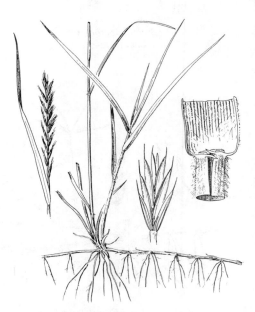

Elymus glaucus
Blue Wildrye

Elytrigia repens
Quackgrass

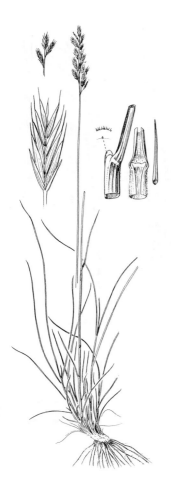

Hordeum marinum ssp. *gussoneanum*
Mediterranean Barley

Festuca rubra
Red Fescue

Koeleria macrantha
Junegrass

Leymus mollis
American Dunegrass

Melica subulata
Alaska Onion Grass

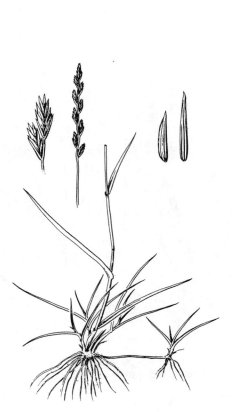

Lolium perenne
Perennial Ryegrass

Poa annua
Annual Bluegrass

Poa pratensis
Kentucky Bluegrass

Vulpia bromoides
Brome Fescue

Vulpia octoflora var. *octoflora*
Sixweeks Fescue

1b Stem usually less than 0.5 cm wide, flexible; leaves not more than 2 cm wide, except in species of *Sorghum* and *Spartina,* in group 4

    4a Spikelets partly concealed in each inflated leaf sheath, the blades falling off early; leaves 3–6 cm long, stiff, spreading outward, usually several on each stem; plant branching extensively at the base, forming mats . . . . *Crypsis vaginiflora [C. niliaca]*

                          Prickle Grass, Sharpleaf Crypsis; eu

    4b Plant not conforming in all respects to the description in choice 4a

        5a Spikelets, florets, or both either enclosed within a spiny bur or situated above stout bristles that arise from the base of the spikelet, the bristles may be modified glumes (see fig. *Hordeum*) (inflorescence sometimes headlike) . . . .

        . . . . POACEAE, GROUP 1

        5b Spikelets, florets, or both neither enclosed within burs nor situated above stout bristles

            6a Each spikelet partly sunken into or pressed against a concave section of the stem, but as it matures, it may bend outward (see fig. *Lolium*) (inflorescence not branched) . . . . POACEAE, GROUP 2

6b Spikelets not sunken into or pressed against the stem

    7a Inflorescence headlike: round, oval, or cylindrical, usually very dense and compact, but sometimes loose and interrupted, either without branches or the spikelets so concentrated that the branches are difficult to distinguish (see *Polypogon,* pl. 64) . . . . POACEAE, GROUP 3

    7b Inflorescence not headlike, usually open and spreading, usually with obvious branches

        8a Inflorescence with obvious branches, these often with branchlets or long pedicels that resemble branchlets (see fig. *Aira*) . . . . .POACEAE, GROUP 4

        8b Inflorescence not branched (lemma 7 veined and with an awn at the tip arising between 2 membranous teeth; prominent teeth present on the upper bract of the floret; glumes papery, without awns; up to 1.5 m tall)

            9a Lemma about 5 mm long, its awn more than 6 mm long; forming clumps (in moist areas) . . . . *Pleuropogon californicus* California Semaphore Grass; Al-n

            9b Lemma about 8 mm long, its awn up to 3 mm long; not forming clumps . . . . *Pleuropogon hooverianus* North Coast Semaphore Grass; Me-Ma; 1b

Poaceae, Group 1: Spikelets, florets, or both either enclosed within a spiny bur or situated above stout bristles that arise from the base of the spikelet, the bristles may be modified glumes; inflorescence sometimes headlike

1a Spikelets enclosed within a spiny bur (annual; stem sometimes falling; up to 50 cm tall; in disturbed areas, especially sandy places)

    2a Bur with a dense ring of needlelike bristles around its base and with larger spines on the inner lobes . . . . *Cenchrus echinatus* Southern Sandbur; na

    2b Bur without a ring of needlelike bristles around its base but with many stout spines throughout

        3a Bur 4–7 mm long, including spines; spines 12, much broadened at their bases . . . . *Cenchrus incertus* Coast Sandbur; na

        3b Bur 6–12 mm long, including spines; spines 14, mostly narrow throughout . . . . *Cenchrus longispinus* (fig.) Mat Sandbur; na

1b Spikelets not enclosed within a bur but situated above or among what appear to be 2 or more bristles (if there are only 2, they may be modified glumes; in some cases, both glumes and florets appear bristlelike)

    4a Spikelets of 2 types in the same inflorescence: fertile spikelet with several florets, 5 mm long, these often falling early, but otherwise hidden among bristly sterile spikelets with empty florets (inflorescence dense, globular or oval, 1 sided; awns up to 2 cm long) . . . . *Cynosurus echinatus* (pl. 63) Hedgehog Dogtail, Dogtail Grass; eu

4b   All spikelets in the inflorescence similar

    5a   Spikelets 2–4, completely hidden within a leaf sheath; stem prostrate and rooting (bristles at base of spikelet long and slender, without plumelike hairs; leaves and stem hairy) . . . . *Pennisetum clandestinum*

                                                             Kikuyu Grass; af

    5b   Spikelets numerous, visible in a loose to compact cylindrical head at stem ends; stem upright

        6a   Bristles numerous, the basal portions covered with delicate, plumelike hairs (bristles 2–4 cm long; inflorescence up to 35 cm long, often lavender; forming dense clumps) . . . . *Pennisetum setaceum*

                                                               Fountain Grass; af

        6b   Bristles 2 or more, the basal portions without plumelike hairs, but sometimes with small, unbranched hairs

            7a   Each node of inflorescence with 1 or 2 spikelets, except *Elymus canadensis* may have 1–4 (spikelets in this choice sometimes difficult to see)

                8a   Spikelet without awns; fertile floret 1, the lower floret sterile; glumes not bristlelike (glumes shorter than the florets; spikelet 2–4 mm long, separating from bristles at maturity)

                    9a   Bristles 3–8 mm long, usually 5–15 below each spikelet (bristles are most easily counted at the base of the inflorescence); leaf blades usually not hairy on the upper surfaces; inflorescence upright, 2–8 cm long; lemma of fertile floret ridged crosswise, the sterile lemma 5 veined . . . . *Setaria pumila [S. glauca]*

                                                               Bristly Foxtail; eu

                    9b   Bristles mostly more than 9 mm long, usually 3 or sometimes 4 or 5 below each spikelet; leaf blades with soft hairs on the upper surfaces; inflorescence drooping, 6–20 cm long; lemma of fertile floret not ridged crosswise, the sterile lemma 7 veined . . . . *Setaria faberi*

                                                              Foxtail; as

            8b   Spikelet with awns, but these may be very short; fertile florets 2 or more; glumes bristlelike

                10a   Awns less than 0.5 cm long; spikelets more or less flattened, with 1 of the broad sides facing the stem

                    11a   Inflorescence usually less than 6 cm long and up to 15 mm wide; spikelet 1 at each node; leaves often extending to or beyond the tip of the inflorescence; up to 60 cm tall (coastal) . . . . *Leymus pacificus [Elymus pacificus]*

                                              Pacific Wildrye, Gould Ryegrass; Me-Mo

                    11b   Inflorescence usually more than 9 cm long and less than 8 mm wide; spikelets usually 2 at each node; leaves usually not extending beyond the inflorescence; up to 130 cm tall (in moist, alkaline areas) . . . . *Leymus triticoides [Elymus triticoides]*

                                              Wet-meadow Wildrye, Beardless Ryegrass

10b Awns up to 10 cm long; spikelets not obviously flattened (spikelets 2 at each node, but may be 1 in *Taeniatherum caput-medusae*)

  12a Each glume and lemma sometimes with more than 1 awn; glumes, including awns, 2.5–8.5 cm long; florets falling with glumes at maturity (glumes veined; lemma awn up to 10 cm long; leaf blades more than 1.5 mm wide; stem 1–3 mm wide, not wirelike)

    13a Glumes divided into 3 or more bristlelike parts; inflorescence dense, usually not hidden in a leaf sheath .... *Elymus multisetus [Sitanion jubatum]* (pl. 64) Big Squirreltail

    13b Glumes undivided, or divided into only 2 parts; inflorescence sparse, often partly hidden in a leaf sheath .... *Elymus elymoides [Sitanion hystrix]* Squirreltail

  12b Each glume and lemma with 1 awn; glumes, including awns, not more than 4 cm long; florets falling from glumes at maturity

    14a Inflorescence nearly as wide as long; bristlelike glumes, including awns, 1.5–4 cm long, less than 1 mm wide at their bases, without obvious veins; lemma awn up to 10 cm long; leaf blades less than 2 mm wide, mostly basal; stem less than 2 mm wide, wirelike .... *Taeniatherum caput-medusae [Elymus caput-medusae]* Medusahead; eu

    14b Inflorescence much longer than wide; bristlelike glumes, including awns, less than 2.5 cm long, about 1 mm wide at their bases, with obvious veins; lemma awn up to 4 cm long; leaf blades up to 15 mm wide, not mostly basal; stem up to 4 mm wide, not wirelike (spikelets sometimes 1–4 at each node) .... *Elymus canadensis* Canadian Wildrye; na

7b Each node of inflorescence with 3 spikelets, these attached together at their bases, the group readily detaching from the stem, but the side spikelets may be reduced to 1–3 bristles (pull off a spikelet group to view characters) (awns usually less than 9 cm long; inflorescence sometimes partly enclosed in a leaf sheath)

  15a Inflorescence including awns almost as wide as long, often nodding; awns 3–9 cm long (bases of leaf blades without overlapping membranous lobes; side florets, if present, much shorter than central floret; bristlelike glumes and awns delicate and eventually spreading) .... *Hordeum jubatum* Foxtail Barley

15b Inflorescence much longer than wide, not nodding; awns usually less than 5 cm long

    16a Some bristlelike glumes fringed with delicate hairs (look carefully); base of each leaf blade with 2 membranous lobes up to 4 mm long clasping the stem; body of side florets not reduced, sometimes longer than body of central floret (awns 1–4 cm long; the following 2 subspecies difficult to separate)

        17a Body of side florets about as long as body of central floret; summer annual; usually less than 40 cm tall . . . .
. . . . *Hordeum murinum* ssp. *glaucum* [*H. glaucum*]
Glaucous Barley; eu

        17b Body of side florets much longer than body of central floret; winter annual; often more than 100 cm tall (widespread) . . . . *Hordeum murinum* ssp. *leporinum*
[*H. leporinum*] (pl. 64)
Hare Barley, Barnyard Foxtail; eu

    16b Bristlelike glumes not fringed with hairs, but there may be small, stiff barbs; base of each leaf blade sometimes with membranous lobes less than 2 mm long, but these not clasping the stem; body of side florets usually reduced, shorter than body of central floret

        18a Bristlelike glumes on central spikelet flat

            19a Side florets greatly reduced, without awns; bases of leaf blades without overlapping membranous lobes (moist, often alkaline areas) . . . . *Hordeum depressum*
Low Barley

            19b Side florets not reduced, with awns up to 8 mm long; bases of leaf blades sometimes with overlapping membranous lobes . . . . *Hordeum marinum*
. . . . ssp. *gussoneanum* [*H. geniculatum*] (fig.)
Mediterranean Barley; eu

        18b Bristlelike glumes on central spikelet round

            20a Awn of central floret up to 18 mm long; bases of leaf blades sometimes with overlapping membranous lobes . . . . *Hordeum marinum* ssp. *gussoneanum*
[*H. geniculatum*] (fig.)
Mediterranean Barley; eu

            20b Awn of central floret less than 8 mm long; bases of leaf blades without overlapping membranous lobes (forming dense clumps)

                21a Awn of central floret up to 8 mm long; leaf blades less than 4 mm wide . . . .
. . . . *Hordeum brachyantherum* ssp. *californicum*
[*H. californicum*]
California Barley

21b Awn of central floret less than 4 mm long; leaf blades up to 8 mm wide (in moist areas) . . . . *Hordeum brachyantherum* ssp. *brachyantherum* Meadow Barley

Poaceae, Group 2: Each spikelet partly sunken into or pressed against a concave section of the stem, but as it matures it may bend outward; inflorescence not branched

1a Spikelet entirely without awns (look at several spikelets) (introduced species)

2a Spikelets sunken into 1 side of the flattened, fibrous stem, but bending out as they mature (lower glume papery, 1 mm long; upper glume about 4 mm long, almost as long as rest of the spikelet; leaf blades up to 7 mm wide, often rounded at the tips; plant creeping; annual; common in lawns) . . . . *Stenotaphrum secundatum* Saint Augustine Grass; sa?

2b Spikelets sunken alternately into both sides of the nearly cylindrical stem (in coastal areas)

3a Spikelet with 1 or 2 florets; annual

4a Inflorescence 6–10 cm long, markedly sickle shaped; basal leaves very congested and curled; mature spikelet so sunken into the stem that it is barely visible (glumes sharp pointed) . . . . *Parapholis incurva* Sickle Grass; eu

4b Inflorescence 10–20 cm long, straight to slightly curved; basal leaves only slightly congested and curled; mature spikelet projecting out, spinelike (inflorescence less than 2 mm wide) . . . . *Hainardia cylindrica [Monerma cylindrica]* Thintail; eu

3b Spikelet with 4–20 florets; perennial

5a Spikelet with the broad surface facing the stem; florets 4–8; both glumes present; inflorescence 4–6 cm long (florets and glumes pointed) . . . . *Elytrigia juncea* ssp. *boreali-atlantica [Agropyron junceum]* Wheat Grass; eu

5b Spikelet with the narrow edge facing the stem and often tilting away from it; florets 4–20; upper glume usually absent; inflorescence 10–30 cm long (widespread) . . . . *Lolium perenne* (pl. 64; fig.) Perennial Ryegrass; eu

1b Spikelet with awns at least 1 mm long (look at several spikelets)

6a Awns mostly more than 1 cm long, sometimes less in *Brachypodium distachyon*

7a Awns up to about 8 cm long, on glumes and sometimes on lemmas (inflorescence with fewer than 10 spikelets, these inflated at maturity, only 1 at each node; each glume spiny, with 2 or 3 awns) . . . . *Aegilops triuncialis* Barbed Goat Grass; me

7b Awns mostly 1–2.5 cm long, on lemmas, but glumes sometimes sharp pointed or awn tipped

8a Spikelet with the narrow edge facing the stem; upper glume often missing, the lower almost as long as or longer than rest of the spikelet, excluding awns (spikelet usually with 5–7 florets; up to 90 cm tall) . . . . . . *Lolium temulentum*
Darnel; me

8b Spikelet with the broad surface facing the stem; glumes 2, somewhat shorter than the whole spikelet

9a Inflorescence 5–20 cm long, usually with more than 30 spikelets, 2 at each node, all along the stem; spikelet, excluding awns, less than 1.5 cm long, with 2–6 florets; forming dense clumps; up to 120 cm tall (widespread) . . . . *Elymus glaucus* (fig.)
Blue Wildrye, Western Ryegrass

9b Inflorescence not more than 7 cm long, with 1–5 spikelets, 1 at each node at stem tip; spikelet, excluding awns, 2–3.5 cm long, with more than 8 florets; not forming dense clumps; up to 45 cm tall . . . .
. . . . *Brachypodium distachyon*
Purple Falsebrome; me

6b Awns mostly less than 1 cm long

10a Spikelet scarcely visible, sunken into the stem, with 1 floret; inflorescence 4–11 cm long; up to 30 cm tall (stem about 1 mm wide; awns 2–5 mm long; annual) . . . . . .
. . . . *Scribneria bolanderi*
Scribner Grass; SLO-n

10b Spikelet readily visible, with more than 1 floret; inflorescence 7–30 cm long; often more than 80 cm tall

11a Spikelet with the broad surface facing the stem, always with 2 glumes (glumes spine tipped and keeled; lemma awn, if present, less than 2 mm long, often spine tipped; perennial) . . . . *Elytrigia repens [Agropyron repens]* (fig.)
Quackgrass; eua

11b Spikelet with the narrow edge facing the stem, usually with only 1 glume

12a Glume(s) almost as long as or even longer than rest of spikelet, excluding awns (florets usually 5–7; annual) . . . . *Lolium temulentum*
Darnel; me

12b Glume(s) much shorter than rest of spikelet, excluding awns (widespread)

13a Most lemmas with awns up to 10 mm long; stem cylindrical; annual or biennial . . . . *Lolium multiflorum* (pl. 64)
Italian Ryegrass; eu

13b Lemmas without awns; stem slightly flattened; perennial . . . .
. . . . *Lolium perenne* (pl. 64; fig.)
Perennial Ryegrass; eu

Poaceae, Group 3:   Inflorescence headlike: round, oval, or cylindrical, usually very dense and compact, but sometimes loose and interrupted, either without branches or the spikelets so concentrated that the branches are difficult to distinguish

1a   Inflorescence a very dense and compact head, usually not interrupted to the extent that stem is visible; spikelets and branches, if present, difficult to see

2a   Awns, if present, not more than 1 mm long

3a   Spikelet usually green striped, sometimes purple, but sometimes difficult to see and sometimes absent in *Phleum pratense* and *Agrostis densiflora* (up to 1 m tall)

4a   Inflorescence widest above the middle, club shaped; each glume notched at the tip, the notch separating 2 unequal teeth; spikelets detaching from the mature inflorescence in clusters of 6 or more, leaving naked sections of stem (inflorescence up to 9 cm long, 1–2 cm wide, often partly or wholly enclosed by a leaf sheath; forming dense clumps) . . . . *Phalaris paradoxa*
Hood Canary Grass, Paradox Canary Grass; me

4b   Inflorescence not widest above the middle; glumes neither notched nor toothed; spikelets not detaching in clusters of 6 or more

5a   Inflorescence narrowly cylindrical; each glume with a stout awn about 1 mm long; glumes, both with awns, forming a U (spikelet flattened, sometimes purple, edged with green; inflorescence less than 1 cm wide, up to 20 cm long). . . . . *Phleum pratense*
Cultivated Timothy; eua

5b   Inflorescence not narrowly cylindrical; glumes without awns

6a   Spikelet not conspicuously flattened, less than 1 mm wide when not yet open; glumes not keeled, 2–4 mm long, not hiding florets, with 1 faint green stripe; inflorescence not very dense, more than 3 times as long as wide, the stem often visible through it (forming dense clumps; coastal) . . . . . *Agrostis densiflora* [includes *A. clivicola* vars.
clivicola and *punta-reyesensis*]
California Bent Grass; SCr-n

6b   Spikelet very flattened, 1.5–4 mm wide when not yet open; glumes keeled, 4–8 mm long, hiding florets, usually with distinct green striping; inflorescence dense, not more than 3 times as long as wide, the stem not visible through it

7a   Inflorescence up to 2 times as long as wide, 6–15 cm long; glumes with a narrow keel (inflorescence 8–15 mm wide; leaves often more than 1 cm wide; annual; in wet areas) . . . .
. . . . *Phalaris angusta*
Narrow Canary Grass; Sn, Sl-s

7b   Inflorescence up to 3 times as long as wide, usually less than 8 cm long; glumes with a broad keel (sterile florets 1 or 2, 1 mm long, hairy, sickle shaped, at the base of the fertile floret)

8a   Glumes 4–6 mm long, often with teeth; sterile floret 1 (inflorescence 10–15 mm wide; annual) . . . *Phalaris minor*
Littleseed Canary Grass; me

8b   Glumes 6–8 mm long, without teeth; sterile florets 2 (in moist to wet areas)

9a   Inflorescence 10–30 mm wide, distinctly squared off across the base; spikelet when not yet open 1–2 mm wide, usually purple tinged; sterile florets white; perennial; forming dense clumps ....

.... *Phalaris californica*

California Canary Grass; SLO-n

9b   Inflorescence 8–18 mm wide, rounded or spindle shaped at the base; spikelet when not yet open 3–4 mm wide, not purple tinged; sterile florets brown; annual; not forming dense clumps ....

.... *Phalaris brachystachys*

Shortspike Canary Grass; me

3b   Spikelet usually not green striped, except occasionally in *Koeleria macrantha* and *Phalaris aquatica*

10a   Spikelet with 1 fertile floret (inconspicuous sterile florets may be present; inflorescence more than 4.5 times as long as wide, often cylindrical)

11a   Spikelet with awns 4 mm long; inflorescence up to 7 cm long and less than 6 mm wide (leaf blades less than 4 mm wide, up to 10 cm long; in wet places, at elevations above 3,000 ft) ....

.... *Alopecurus aequalis* [includes *A. aequalis* var. *sonomensis*]

Shortawn Foxtail

11b   Spikelets without awns; inflorescence up to 30 cm long and at least 10 mm wide

12a   Spikelets 2 mm long, arranged in whorled clusters; inflorescence not cylindrical, loosely and irregularly branched; not more than 75 cm tall (inflorescence 1–4 cm wide; florets falling with glumes at maturity; in moist areas) .... *Agrostis viridis* [*A. semiverticillata*]

Whorled Bent Grass, Water Bent Grass; eu

12b   Spikelets 3–15 mm long, not in whorled clusters; inflorescence cylindrical; up to 200 cm tall

13a   Spikelet 2–4 mm long; glumes much less than 1 mm wide, with small spiny hairs, especially on veins (inflorescence dense, but sometimes interrupted; forming dense clumps; up to 85 cm tall; coastal) .... *Agrostis densiflora* [includes *A. clivicola* vars. *clivicola* and *punta-reyesensis*]

California Bent Grass; SCr-n

13b   Spikelet 4–15 mm long; glumes 1–2 mm wide, not hairy

14a   Inflorescence up to 15 cm long and 1.5 cm wide, widest at the base and tapering toward the tip; spikelet 4–6 mm long; leaf blades flat, up to 15 mm wide; sterile floret 1, 1 mm long at the base of the fertile floret (fertile floret densely hairy; glumes curved at tips, resembling lobster claws; in moist areas) .... *Phalaris aquatica*

Harding Grass; me

14b Inflorescence up to 30 cm long and 3 cm wide, widest in the middle and tapering at both ends; spikelet 11–15 mm long; leaf blades 10 mm wide when flat; sterile floret none (leaf blade margins becoming inrolled as they age; forming clumps; on coastal sand dunes) .... *Ammophila arenaria* (pl. 62) European Beachgrass; eu

10b Spikelet with more than 1 fertile floret

    15a Spikelets, florets, or both of two distinctly different types in the same inflorescence (pry open the spikelet)

        16a Spikelets all similar; inflorescence up to 15 cm long and more than 1.5 cm wide; plant forming dense clumps (florets of 2 types in each spikelet: the lower one with a shiny grain, the upper one purple gray, with a hooked awn) .... *Holcus lanatus* Common Velvet Grass; eu

        16b Spikelets of 2 distinct types, the 2 spikelet types mixed and overlapping all along the stem: one 4–5 mm long, flat, fan shaped, and toothed, without fertile florets, the other 3 mm long, with keeled glumes and 1 fertile floret with a shiny grain; inflorescence 4–8 cm long and up to 1 cm wide; not forming dense clumps .... *Cynosurus cristatus* (pl. 63) Crested Dogtail; eu

    15b Spikelets, florets, or both essentially all alike (florets falling from glumes at maturity; plant forming clumps)

        17a Spikelet flattened; florets 5–10, each stiff and curved; lemma with a spinelike tip and more than 10 prominent veins; leaf blades 2–4 cm long, not wirelike; often prostrate, with stem tips rising; usually less than 8 cm tall, rarely up to 12 cm (inflorescence club shaped; in vernal pools) .... *Tuctoria mucronata [Orcuttia mucronata]* Crampton Tuctoria; 1b

        17b Spikelet only slightly, if at all, flattened; florets 2–6, not as described in choice 17a; some leaf blades more than 5 cm long; upright; at least 10 cm tall

            18a Inflorescence usually less than 5 cm long, oblong to pineapple shaped, less than 4 times as long as wide, not interrupted; spikelet 6–10 mm long; each floret usually with delicate webbed hairs at the base; tips of leaf blades shaped like the bow of a boat; on coastal dunes .... *Poa douglasii* Sand-dune Bluegrass, Dune Bluegrass; Mo-n

            18b Inflorescence up to 15 cm long, somewhat cylindrical, more than 4 times as long as wide, sometimes interrupted; spikelet 3–5 mm long; florets without webbed hairs; tips of leaf blades not shaped like the bow of a boat; in dry areas (leaf blades up to 20 cm long) .... *Koeleria macrantha* (fig.) Junegrass, Koeler Grass

2b Awns present, more than 1 mm long

19a Inflorescence either more or less globular or a short cylinder, usually less than 4 times as long as wide

20a Inflorescence bushy, soft to the touch (annual)

21a Glumes without awns, with dense, gray, feathery hairs hiding florets; lemma with 3 awns, the longest dark, up to 20 mm, attached to middle of lemma, projecting out of inflorescence; inflorescence up to 3 cm long and about as wide; florets falling from glumes at maturity . . . . *Lagurus ovatus*

Hare's-tail; me

21b Glumes with awns up to 10 mm long, without feathery hairs; lemma with 1 awn less than 4 mm long, attached near the tip; inflorescence often more than 3 cm long, 3–4 times as long as wide; florets falling with glumes at maturity (glumes, both with awns, forming a U or V) . . . . *Polypogon monspeliensis* (pl. 64)

Annual Beard Grass, Rabbitfoot Grass; me

20b Inflorescence not bushy, not soft to the touch (awns less than 5 mm long; florets falling from glumes at maturity)

22a Inflorescence compact, not more than 2–3 times as long as wide; glumes, both with awns, forming a U or V, the awns stiff, 2–3 mm long; lemma without an awn; leaf blades up to 8 mm wide, not wirelike; perennial (at elevations of at least 3,000 ft; in wet areas) . . . . *Phleum alpinum*

Mountain Timothy; SF, La-n

22b Inflorescence not compact, up to 4 times as long as wide; glumes without awns; lemma awn 4 mm long, the tip projecting from spikelet; leaf blades less than 2 mm wide, wirelike; annual (spikelet with 2 florets; plant forming clumps) . . . . *Aira praecox*

Early Hair Grass; me

19b Inflorescence cylindrical, usually at least 6 times as long as wide, sometimes less in *Alopecurus saccatus*

23a Awns not more than 2.5 mm long (measure from tip to base of awn) (inflorescence up to 10 cm long and 2.5 cm wide; forming dense clumps)

24a Glumes of fertile spikelets with short, winglike side projections; inflorescence often partly enclosed in a leaf sheath; spikelets detaching in clusters of 6 or more, leaving naked sections of stem (usually only 1 spikelet in each cluster fertile) . . . . *Phalaris paradoxa*

Hood Canary Grass, Paradox Canary Grass; me

24b Glumes without winglike side projections; inflorescence not enclosed in a leaf sheath; spikelets not detaching in clusters (in moist areas) . . . . *Phleum alpinum*

Mountain Timothy; SF, La-n

23b Awns 2.5–10 mm long

    25a Spikelet, excluding awns, 5–12 mm long (lemma awn 3–5 mm long; florets falling from glumes at maturity)

        26a Spikelet 6–7 mm long, with 1 floret; inflorescence not interrupted, with only a very faint sweet smell, if any (inflorescence up to 9 cm long and 1 cm wide; spikelet yellow green above, shiny and cream colored at the swollen base) . . . . *Gastridium ventricosum*
Nit Grass; eu

        26b Spikelet 5–12 mm long, with 2 awned, sterile florets and 1 awnless, fertile floret; inflorescence often interrupted, usually sweet smelling

            27a Inflorescence up to 8 cm long and 1.5 cm wide; spikelet more than 7 mm long; leaf blades more than 3 mm wide; perennial; forming dense clumps . . . .
. . . . *Anthoxanthum odoratum* (pl. 63)
Sweet Vernal Grass; eu

            27b Inflorescence up to 3 cm long and 1 cm wide; spikelet 5–7 mm long; leaf blades not more than 2 mm wide; annual; not forming dense clumps . . . . *Anthoxanthum aristatum*
Vernal Grass; eu

    25b Spikelet less than 4 mm long, sometimes 5 mm in *Alopecurus saccatus*

        28a Awns bent (lemma minutely toothed at tip, its awn attached at the middle or lower; annual; plant forming small clumps)

            29a Spikelet with 1 floret; lemma awn 4–6 mm long; florets falling with glumes at maturity; inflorescence up to 6 cm long and 1 cm wide; glumes very hairy (in moist areas) . . . .
. . . . *Alopecurus saccatus* [includes *A. howellii*]
Pacific Foxtail

            29b Spikelet with 2 florets; lemma awn 2–4 mm long; florets falling from glumes at maturity; inflorescence up to 3 cm long and less than 0.5 cm wide; glumes not hairy, except slightly at the bases (in sandy soils) . . . . *Aira praecox*
Early Hair Grass; me

        28b Awns not bent

            30a Florets falling from glumes at maturity; glumes without awns, but with sharp tips and small spiny hairs, especially on veins; inflorescence up to 10 cm long and 1.5 cm wide, dense but occasionally interrupted (lemma awn up to 3 mm long; forming dense clumps; coastal) . . . . *Agrostis densiflora*
. . . . . . [includes *A. clivicola* vars. *clivicola* and *punta-reyesensis*]
California Bent Grass; SCr-n

            30b Florets falling with glumes at maturity; glumes with awns, sometimes with long hairs on margins; inflorescence up to 15 cm long and 3 cm wide, not interrupted (inflorescence bushy, soft to the touch; flattened glumes, both with awns, forming a U or V)

31a   Glumes without long hairs on margins, the awns originating just below the tips (look carefully); leaf blades 4–7 mm wide; lemma awn up to 4 mm long; inflorescence often more than 15 cm long .... *Polypogon monspeliensis* (pl. 64)

Annual Beard Grass, Rabbitfoot Grass; me

31b   Each glume with long hairs on margins, the awn originating about one-third the distance from the tip to the base; leaf blades usually less than 3 mm wide; lemma without an awn; inflorescence usually less than 9 cm long ....

.... *Polypogon maritimus*

Mediterranean Beard Grass; me

1b   Inflorescence a loose head, sometimes interrupted to the extent that the stem is visible; spikelets and branches, if present, often visible

32a   Spikelet with 1 obvious floret, this fertile, but there may be some much smaller sterile florets

33a   Spikelet without awns

34a   Spikelet more than 7 mm long; inflorescence up to 30 cm long; upright; often more than 120 cm tall (inflorescence up to 3 cm wide; stem up to 1 cm wide at the base; forming dense clumps; on coastal sand dunes) ....

.... *Ammophila arenaria* (pl. 62)

European Beachgrass; eu

34b   Spikelet less than 3 mm long; inflorescence up to 15 cm long; stem sometimes falling down and rooting; up to 75 cm tall

35a   Leaf blades flat, up to 8 mm wide; inflorescence with whorled, lobelike clusters of spikelets; lemma 1 mm long; florets usually falling with glumes at maturity; often prostrate and rooting at the nodes (in moist areas) .... *Agrostis viridis* [*A. semiverticillata*]

Whorled Bent Grass, Water Bent Grass; eu

35b   Leaf blades often wirelike, but if flat then usually less than 5 mm wide; inflorescence without whorled clusters of spikelets; lemma 2–3 mm long; florets falling from glumes at maturity; upright (inflorescence purple and tan) .... *Agrostis pallens* [includes *A. diegoensis*]

Dune Bent Grass, Leafy Bent Grass

33b   Spikelet with awns, these sometimes only on the lemmas

36a   Each floret with a tuft of white hairs at its base (pry out a floret and examine it with a hand lens) (spikelet more than 5 mm long, except in *Muhlenbergia andina;* sometimes forming clumps)

37a   Awns easily visible beyond the tip or side of each spikelet

38a   Awns not bent; spikelet 3–4 mm long, excluding awns; basal hairs of each floret sometimes as long as the floret; glumes sometimes with short awns (inflorescence often purple; lemma awns 2–7 mm long) .... *Muhlenbergia andina*

Foxtail Muhly; SCl-n

38b Awns mostly bent; spikelet 5–10 mm long, excluding awns; basal hairs of each floret not as long as the floret; glumes without awns (lemma tip with 4 minute teeth)

39a Spikelet up to 10 mm long; awn extending 8–10 mm beyond the tip of spikelet (leaf blades 1–3 mm wide, the margins inrolled; coastal) . . . .*Calamagrostis foliosa*
Leafy Reed Grass; Sn-n; 4

39b Spikelet 5–6 mm long; awn extending 2 mm beyond the side of spikelet (look at several) (inflorescence straw colored or purple tinged; on serpentine) . . . . *Calamagrostis ophitidis*
Serpentine Reed Grass; La, Ma, Sn; 4

37b Awns hidden within each spikelet or barely protruding (awn bent, attached below middle of lemma; glumes without awns; inflorescence sometimes purple)

40a Inflorescence 15–30 cm long and up to 3 cm wide; leaf blades 6–12 mm wide; up to 1.5 m tall (in moist areas) . . . . . *Calamagrostis nutkaensis*
Pacific Reed Grass; SLO-n

40b Inflorescence 7–15 cm long and not more than 1.5 cm wide; leaf blades 2–6 mm wide; up to 1 m tall

41a White hairs present where each leaf blade joins the sheath; leaf blades 2–4 mm wide; spikelet about 4 mm long; inflorescence usually less than 1 cm wide . . . .*Calamagrostis rubescens*
Pine Grass; SLO-n

41b White hairs absent where each leaf blade joins the sheath; leaf blades up to 6 mm wide; spikelet about 6 mm long; inflorescence usually 1–1.5 cm wide . . . . *Calamagrostis koelerioides*
Tufted Pine Grass; SLO-n

36b Florets without tufts of white hairs at their bases

42a Glumes with awns 2–8 mm long (inflorescence often interrupted or lobed, usually 7–20 cm long and up to 3.5 cm wide, yellow green to purple; leaf blades 3–10 mm wide)

43a Glumes 3–5 mm long, the tips tapering gradually into awns 2–3 mm long; lemma awn 1–2 mm long; in salt marshes . . *Polypogon elongatus*
Longspike Beardgrass; sa

43b Glumes less than 2 mm long, the tips ending abruptly, the awns 5–8 mm long; lemma awn 2–5 mm long; in moist areas (spikelets often in grapelike clusters) . . . . *Polypogon australis*
Chilean Beard Grass, Australian Beard Grass; sa

42b Glumes without awns, but they may have sharp tips

44a Spikelet with 2 awned, sterile florets on either side of the awnless, fertile floret (the 3 florets together resemble a single lemma with 2 awns) (inflorescence sweet smelling, often interrupted)

45a Inflorescence up to 8 cm long and 1.5 cm wide; spikelet longer than 7 mm; some leaf blades more than 3 mm wide; perennial; forming dense clumps . . . . . . *Anthoxanthum odoratum* (pl. 63)
Sweet Vernal Grass; eu

45b Inflorescence up to 3 cm long and 1 cm wide; spikelet 5–7 mm long; leaf blades less than 2 mm wide; annual; not forming dense clumps . . . . *Anthoxanthum aristatum*
Vernal Grass; eu

44b Spikelet without sterile florets, the fertile floret with an awn (plant forming clumps)

46a Awn originating near the base of the lemma (lemma with short hairs at the base, its awn 6 mm long, bent, protruding about 2 mm beyond the side of the spikelet; inflorescence 7–13 cm long and up to 1.5 cm wide) . . . . *Calamagrostis ophitidis*
Serpentine Reed Grass; La, Ma, Sn; 4

46b Awn originating near the middle or the tip of the lemma

47a Awn 4–5 mm long; lemma with very small teeth; lower glume longer than upper glume; inflorescence up to 8 cm long and 1 cm wide, not interrupted; leaf blades 1–4 cm long (spikelet green above, shiny and cream colored at the swollen base) . . . . *Gastridium ventricosum*
Nit Grass; eu

47b Awn usually less than 3 mm long; lemma without teeth; glumes equal; inflorescence up to 10 cm long and 1.5 cm wide, occasionally interrupted; leaf blades 2–12 cm long (coastal) . . . . *Agrostis densiflora* [includes *A. clivicola* vars. *clivicola* and *punta-reyesensis*]
California Bent Grass; SCr-n

32b Spikelet with more than 1 obvious floret, at least 1 of the florets fertile, but sometimes the sterile florets are as large as the fertile one

48a Awns more than 10 mm long (lemma awn 10–35 mm long)

49a Glumes absent; leaf blades 10–20 mm wide; plant not forming clumps (spikelets 3 or 4 at each node; in shaded areas) . . . .
. . . . *Elymus californicus [Hystrix californica]*
California Bottlebrush Grass; Sn-SCr; 4

49b Glumes present; leaf blades 4–15 mm wide; plant forming dense clumps

50a Inflorescence up to 16 cm long, upright; spikelets 2 at each node; glumes 7–19 mm long, with awns about 1 mm long (widespread) . . . .
. . . . *Elymus glaucus* (fig.)
Blue Wildrye, Western Ryegrass

50b Inflorescence up to 25 cm long, nodding; spikelets 2–4 at each node; glumes 10–25 mm long, with awns about 10 mm long . . . . *Elymus canadensis*
Canadian Wildrye; na

48b Awns, if present, less than 3 mm long

51a Spikelets with awns less than 3 mm long

52a Florets of 2 different types in each spikelet: the lower one with a shiny grain, the upper one with a hooked awn (spikelet 3–6 mm long; florets falling with glumes at maturity; inflorescence up to 15 cm long and more than 1.5 cm wide; often forming dense clumps) . . . . *Holcus lanatus*
Common Velvet Grass; eu

52b Florets similar within each spikelet, but there may be some much smaller sterile florets

    53a Inflorescence 1–3 cm long; annual (inflorescence less than 5 mm wide; spikelets 3–4 mm long, excluding awns; lemma awn originating below the middle, bent; floret 2–4 mm long) ..... *Aira praecox*
        Early Hair Grass; me

    53b Inflorescence more than 5 cm long; perennial (coastal)

        54a Spikelet 15–20 mm long; inflorescence 20–30 cm long; leaf blades up to 12 mm wide; not forming dense clumps, but spreading extensively by rhizomes ....
        .... *Leymus × vancouverensis [Elymus × vancouverensis]*
        Vancouver Wildrye; Ma-n

        54b Spikelet 6–8 mm long; inflorescence up to 20 cm long; leaf blades 1–2 mm wide; forming dense clumps (inflorescence sometimes purple; spikelet usually with 2 florets, these with a pedestal, covered with white hairs, between them; lemma tip with 2–4 tiny teeth) .... *Deschampsia cespitosa*
        .... ssp. *holciformis [D. caespitosa* ssp. *holciformis]*
        Pacific Hairgrass; SLO-n

51b Spikelets without awns

    55a Tips of the leaf blades, especially when folded, curved upward, resembling the bow of a boat; florets sometimes with cottony hairs at their bases (in sandy coastal habitats)

        56a Spikelet 3–5 mm long, with 3 or 4 florets; inflorescence somewhat open, 1–4 cm long, up to 1 cm wide (florets sometimes with sparse cottony hairs at their bases) .... *Poa confinis*
        Beach Bluegrass; Ma-n

        56b Spikelet 6–10 mm long, with 3–8 florets; inflorescence dense, 2–9 cm long, 1–2 cm wide

            57a Inflorescence 2–5 cm long; florets 5–7.5 mm long, often with sparse cottony hairs at their bases; leaf blades 1–2.5 mm wide ..
        .... *Poa douglasii*
        Sand-dune Bluegrass, Dune Bluegrass; Mo-n

            57b Inflorescence 3–9 cm long; florets 3–4.5 mm long, without cottony hairs; leaf blades 1–5 mm wide .... *Poa unilateralis*
        Ocean-bluff Bluegrass, San Francisco Bluegrass; Mo-n

    55b Tips of leaf blades not curved upward; florets without cottony hairs

        58a Spikelet 3–5 mm long, sometimes green striped and purple (inflorescence up to 18 cm long, sometimes interrupted; leaves wire-like; forming dense clumps; up to 60 cm tall) ....
        .... *Koeleria macrantha* (fig.)
        Junegrass, Koeler Grass

58b Spikelet 6–30 mm long, not green striped

    59a Leaves mostly less than 10 cm long, tapering to sharp tips, the blades 1–4 mm wide, the sheaths sometimes overlapping; inflorescence up to 7 cm long; spikelet about 10 mm long (in salt marshes and moist alkaline habitats) .... *Distichlis spicata* [includes *D. spicata* vars. *nana* and *stolonifera*] (fig.)

        Saltgrass

    59b Leaves much more than 10 cm long, not tapering to sharp tips, the blades 5–15 mm wide, the sheaths not overlapping; inflorescence 8–30 cm long; spikelet 15–30 mm long (stem up to 1 cm wide; on backshores of sandy beaches) .... *Leymus mollis [Elymus mollis]* (pl. 63; fig.)

        American Dunegrass; SLO-n

Poaceae, Group 4: Inflorescence with obvious branches, these often with branchlets or long pedicels that resemble branchlets

1a Branches of the inflorescence not divided into branchlets or long pedicels that resemble branchlets, the pedicels, if present, very short

    2a Branches of the inflorescence in 1–4 conspicuous whorls, at least 3 branches in each whorl (spikelets pressed against the branches, sessile or on very short pedicels, in 2 or 3 rows on one side of each branch)

        3a Awns 2–6 mm long; upright (branches 4–14 cm long, in 2–4 whorls; ligules composed of white hairs; spikelets in 2 rows; forming dense clumps) .... *Chloris verticillata*

            Windmill Grass, Finger Grass; na

        3b Awns lacking; often prostrate with stem rooting at nodes

            4a Branches 3–5 cm long, in 1 whorl; ligules composed of white hairs; spikelets in 2 rows, tightly overlapping (common) .... *Cynodon dactylon* (fig.)

                Bermuda Grass; af

            4b Branches 3–15 cm long, in 1–4 whorls; ligules membranous; spikelets in 2 or 3 rows, not tightly overlapping (branches sometimes flattened) .... *Digitaria sanguinalis* (fig.)

                Crab Grass; eu

    2b Branches of the inflorescence not in conspicuous whorls

        5a Spikelet solitary at branch ends, these delicate, much thinner than the main stem (awns up to 4 cm long)

            6a Tip of mature lemma with teeth less than 2 mm long, these soft, membranous, and white .... *Avena fatua* (fig.)

                Wild Oat; eu

            6b Tip of mature lemma with teeth about 4 mm long, these stiff, awnlike, and white, but often turning darker .... *Avena barbata*

                Slender Wild Oat; me

5b   Spikelets numerous on the branches, these not delicate, about as thick as the main stem

    7a   Spikelets 1 or 2 at each node (spikelet sessile)

       8a   Spikelets alternating on both sides of each branch, oriented with narrow edges facing the branch, each with numerous florets; awns up to 10 mm long (widespread)...................... *Lolium multiflorum* (pl. 64)
Italian Ryegrass; eu

       8b   Spikelets in dense groups on 1 side of each branch, each with 1 floret; awns lacking or less than 1 mm long (spikelets flattened and overlapping; in salt marshes)

         9a   Inflorescence dense, the branches not comblike but pressed against the stem and tightly overlapping

           10a   Stem 3–6 mm wide at the base; leaf blade 4–8 mm wide at the base, often inrolled (established in San Francisco Bay).. *Spartina densiflora*
Denseflower Cord Grass; sa

           10b   Stem 7–12 mm wide at base; leaf blade 5–17 mm wide at the base, mostly flat .... *Spartina foliosa*
California Cord Grass

         9b   Inflorescence somewhat open, the branches comblike, not pressed against the stem and not tightly overlapping

           11a   Stem 1–4 mm wide at the base; leaf blade less than 3 mm wide, wirelike; inflorescence branches spreading widely .... *Spartina patens*
Salt-meadow Cord Grass; na

           11b   Stem 5–14 mm wide at the base; leaf blade 4–25 mm wide at the base, not wirelike; inflorescence branches somewhat spreading (spikelets loosely overlapping; established in San Francisco Bay) .... *Spartina alterniflora*
Salt-water Cord Grass; na

    7b   Spikelets 1 to several at each node

       12a   Inflorescence with 2 branches at the stem tip, occasionally another branch below these (branches 3–6 cm long; spikelets without awns, flattened, in 2 rows; leaf blades 2–7 mm wide; plant sometimes creeping and rooting; in coastal, moist habitats)............... *Paspalum distichum*
Knot Grass

       12b   Inflorescence usually with 4 or more branches, these alternating along the stem, sometimes several branches at each node

         13a   Awns 5–25 mm long (spikelets, in 3 or 4 rows on each branch, each with 1 fertile floret, the enclosed grain hard, shiny; glumes and lemma with stiff, bristlelike hairs on the margins; leaf blades 6–20 mm wide) .... *Echinochloa crus-galli [E. crusgalli]* (pl. 64)
Barnyard Grass; eua

13b Awns, if present, up to 3 mm long

    14a Spikelet with up to 12 florets (florets 2–3 mm long; lemma awn 1 mm long; base of inflorescence usually enclosed within a leaf sheath; leaf blade often extending to or beyond the tip of the inflorescence; spikelets often overlapping; in moist, alkaline areas) . . . . *Leptochloa fascicularis*
Bearded Sprangletop; Mo, Sn, SF

    14b Spikelet with only 1 obvious floret (inflorescence with delicate hairs)

        15a Ligules present, membranous; branches 4–10 cm long; lower glume absent, but larger lemmas resemble glumes; grains not shiny; forming clumps . . . .
. . . . *Paspalum dilatatum*
Dallis Grass; sa

        15b Ligules absent; branches less than 3 cm long; both glumes present, the lower one small; grains shiny; not forming clumps (in moist areas) . . . .
. . . . *Echinochloa colona [E. colonum]*
Jungle Rice; eua

1b Branches of the inflorescence usually divided into branchlets or long pedicels that resemble branchlets (bending the branches out may make the branchlets easier to see)

16a Spikelets 2 or 3 at each node, 1 fertile and sessile, the other 1 or 2 sterile and on pedicels, but in *Andropogon*, only a pedicel remains along with the fertile spikelet

    17a Inflorescence delicate, up to about 20 cm long, with dense clusters of spikelets, each partly hidden in an elongated sheath and with silky hairs sometimes more than 10 mm long; spikelets 2, the sterile spikelet reduced to a hairy pedicel; leaf blades up to 5 mm wide (awns up to 20 mm long; perennial; in moist areas) . . . . .
. . . . *Andropogon virginicus*
Broomsedge Bluestem; na

    17b Inflorescence coarse, up to 40 cm long, if dense, not hidden in a sheath and without long silky hairs; spikelets 3, neither the fertile nor the sterile spikelets obviously reduced; leaf blades at least 5 mm wide

        18a Leaf blades 0.5–2 cm wide; inflorescence open and spreading; spikelets elliptical, less than 2 mm wide, not tightly clustered; perennial (spikelets usually colorful). . . . . *Sorghum halepense*
Johnson Grass; me

        18b Leaf blades at least 3 cm wide; inflorescence dense; spikelets globular, 2–4 mm wide, tightly clustered; annual . . . . *Sorghum bicolor*
Sorghum, Milo; af

16b Spikelets, if more than 1 at each node, either all sessile or all on pedicels

    19a Spikelets with 1 obvious floret, this fertile, but there may be smaller sterile florets (in *Poa bulbosa* the apparent floret is actually a bulblet). . . . .
. . . . POACEAE, GROUP 4, SUBKEY 1

    19b Spikelets usually with more than 1 obvious floret, at least one of these fertile, but sometimes sterile florets are as large (pry open a spikelet to see)

20a Lemma awn at least 1 mm long, sometimes missing in *Bromus secalinus* (sub-key 2)

  21a Awn attached at or very near the lemma tip, and in some species it origi-nates between 2 small teeth (glumes without awns)

    22a Glumes of unequal length and usually shorter than rest of the spikelet, excluding awns; florets without long white hairs at their bases; lemma awn either not attached between 2 teeth, or the teeth not awnlike

      23a Spikelet, excluding awns, usually more than 1.5 cm long; edges of each leaf sheath, at least the lower portion, fused together; leaves often more than 10 mm wide, the margins usually flat, sometimes inrolled, but never wirelike; ligules prominent, sometimes torn and jagged ..... POACEAE, GROUP 4, SUBKEY 2
Bromus

      23b Spikelet, excluding awns, usually less than 1.5 cm long; edges of each leaf sheath not fused together, but overlapping; leaves usually less than 2 mm wide, the margins usually inrolled, wire-like; ligules absent or inconspicuous ....
.... POACEAE, GROUP 4, SUBKEY 3
Festuca (in part), *Vulpia*

    22b Glumes of about equal length, and often longer than rest of the spikelet, excluding awns; florets with long white hairs at their bases; lemma awn attached between 2 awnlike teeth (awn bent; glumes pa-pery, with 3 or more prominent veins; ligules consisting of long hairs; forming dense clumps)

      24a Lemma with stiff white hairs on the back and elsewhere, the teeth up to 7 mm long; spikelet pedicels ascending; spikelet 4 mm wide; glumes about 10 mm long .... *Danthonia pilosa*
Hairy Oatgrass; au

      24b Lemma not hairy on the back, the teeth less than 4 mm long; spikelet pedicels spreading outward; spikelet more than 10 mm wide; glumes 15–20 mm long .... *Danthonia californica*
California Oatgrass; Mo-n

  21b Awn attached below the uppermost quarter of the lemma (florets 2 or 3; lemma tip usually with 2–5 very small teeth; awns may bend with age)

    25a Awn less than 4 mm long

      26a Spikelet, excluding awns, less than 4 mm long; florets without a small hairy bristle among them; lemma tip with 2 teeth, but these difficult to see even with a hand lens (lower floret sometimes lack-ing an awn; awns bent, barely projecting beyond the spikelet)

        27a Inflorescence less than 6 mm wide, the lower branches not readily visible (inflorescence oblong in shape, 7–35 mm long) .... *Aira praecox*
Early Hairgrass; me

27b Inflorescence often more than 50 mm wide, the lower branches readily visible

    28a Spikelet 1–1.5 mm long, excluding awns; awn of lower floret much shorter than awn of upper floret, if present, and only the longer awn extending beyond the spikelet . . . . *Aira elegantissima [A. elegans]* Elegant Hairgrass; me

    28b Spikelet 2–3 mm long, excluding awns; awns equal, both extending beyond the spikelet (in sandy coastal areas) . . . . *Aira caryophyllea* (fig.) Silver European Hairgrass; eu

26b Spikelet, excluding awns, at least 4 mm long; florets often with a small hairy bristle among them; lemma tip usually with 4 teeth, sometimes 2, 3, or 5 (florets with long, upright hairs at their bases; awn attached between the middle and the base of the lemma; forming dense clumps; in wet areas)

    29a Inflorescence 1.5–9 cm wide and up to 20 cm long; awn originating near base of lemma, straight to slightly bent, barely extending beyond the spikelet; blades of basal leaves mostly wirelike, but some up to 3 mm wide (coastal) . . . . *Deschampsia cespitosa* ssp. *holciformis* [*D. caespitosa* ssp. *holciformis*] Pacific Hairgrass; SLO-n

    29b Inflorescence about 0.5 cm wide and up to 30 cm long; awn originating near middle of lemma, straight, extending about 2 mm beyond the spikelet; blades of basal leaves threadlike (branches pressed against the stem) . . . . *Deschampsia elongata* Slender Hairgrass

25b Awn at least 4 mm long

    30a Awn up to 4 cm long; spikelet, excluding awns, up to 25 mm long; florets without either hairs at their bases or a small hairy bristle; lemma tip with 2 teeth (glumes longer than the lowest floret and often longer than rest of the spikelet; up to 75 cm tall; widespread)

    31a Tip of mature lemma with teeth less than 2 mm long, these soft, membranous, and white . . . .*Avena fatua* (fig.) Wild Oat; eu

    31b Tip of mature lemma with teeth about 4 mm long, stiff, awnlike, and white, but often turning darker . . . . *Avena barbata* Slender Wild Oat; me

30b Awn not more than 1 cm long; spikelet, excluding awns, not more than 12 mm long; florets with either hairs or a small hairy bristle at their bases, or both; lemma tip with 2–5 teeth

32a Awn 7–10 mm long, curved or bent, attached in the uppermost quarter of the lemma; spikelet 6–10 mm long; glumes much shorter than rest of the spikelet; lemma tip with 2 teeth (teeth needlelike; in moist, shaded areas) ....
.... *Trisetum canescens* [*T. cernuum* var. *canescens*]
Tall Trisetum; SCr-n

32b Awn 4–7 mm long, usually bent, attached near middle of the lemma; spikelet 4–8 mm long; glumes almost as long as or sometimes longer than rest of spikelet; lemma tip usually with 4 teeth, sometimes 2, 3, or 5 (inflorescence up to 9 cm wide; branches spreading; glumes 4–6 mm long; florets with long, upright hairs at their bases)

33a Awns extending beyond the spikelet for about 3–4 mm; leaves few, 1 mm wide, less than 15 cm long; inflorescence with few spikelets; up to 60 cm tall (spikelets at branchlet tips; in moist areas) ....
.... *Deschampsia danthonioides* (fig.)
Annual Hairgrass

33b Awns barely extending beyond the spikelet; leaves numerous, 1.5–3 mm wide, more than 15 cm long; inflorescence with numerous spikelets; up to 120 cm tall (forming dense clumps; in wet, coastal areas) ....
.... *Deschampsia cespitosa* ssp. *cespitosa*
[includes *D. caespitosa* sspp. *beringensis* and *caespitosa*]
Tufted Hairgrass; Ma-n

20b Lemma awn, if present, less than 1 mm long, or hook shaped

34a Spikelet with 2 or 3 florets, each of these different, one often hidden in *Ehrharta erecta* (pull spikelet apart to see florets; inflorescence up to 15 cm long, spreading at maturity; spikelet usually 3 mm long)

35a Spikelet with 3 florets: a large sterile floret with conspicuous cross grooves on its lemma, a sterile floret with a smooth lemma, and a hidden fertile floret with a smooth lemma; florets falling from glumes at maturity; plant not clumped .... *Ehrharta erecta*
Panic Veldt Grass; af

35b Spikelet with 2 florets: the lower one producing a smooth, hard, shiny grain, the upper one staminate, its lemma awn hook shaped; florets falling with glumes at maturity; plant often forming dense clumps (inflorescence gray or purple) .... *Holcus lanatus*
Common Velvet Grass; eu

34b Spikelet with 2 or more florets, all similar

　36a Florets side-by-side in spikelet (florets 3, tan to purple, falling as a group from the glumes; remaining glumes papery, very open; plant sweet smelling; leaf blades up to 17 mm wide; in shaded areas) . . . . .

. . . . *Hierochloë occidentalis*

Sweetgrass, Vanilla Grass; Mo-n

　36b Florets forming an alternating series in each spikelet

　　37a Spikelet inflated and papery, resembling a rattlesnake rattle (spikelet sometimes almost as wide as long)

　　　38a Inflorescence upright, with numerous spikelets; mature spikelet 3 mm long, 4–5 mm wide (widespread) . . . .

. . . . *Briza minor*

Little Quaking Grass; me

　　　38b Inflorescence drooping, with relatively few spikelets; mature spikelet 12–25 mm long, 10–13 mm wide . . . .

. . . . *Briza maxima* (pl. 63)

Big Quaking Grass, Rattlesnake Grass; me

　　37b Spikelet not inflated and papery and not resembling a rattlesnake rattle

　　　39a With a tuft of white hairs where the leaf blade joins the sheath; spikelet with 5–35 florets (florets often crowded; lemma 3 veined, look at several)

　　　　40a Plant low, densely matted and creeping, rooting at the nodes; spikelet with 10–35 florets, these tightly overlapping; leaf blades ascending, usually less than 3 cm long (spikelets less than 3 mm wide; leaf blades with sharp tips; on sand bars) . . . . *Eragrostis hypnoides*

Creeping Lovegrass; Sn-n

　　　　40b Plant sometimes low, but not matted and creeping; spikelet with 6–20 florets, these not tightly overlapping; leaf blades usually spreading outward, usually more than 3 cm long

　　　　　41a Leaf blades usually with minute, brown, glandular bumps along the edges, the glands barely visible with a hand lens; spikelet with 8–20 florets, usually 9–12 (spikelet less than 3 mm wide; inflorescence dark, usually less than 15 cm long) . . .

. . . . *Eragrostis minor* [*E. poaeoides*]

Little Lovegrass; eu

41b Leaf blades without glands; spikelet usually with fewer than 11 florets

42a With a tuft of white hairs where some branches attach to the stem; leaf blades often less than 1 mm wide, the margins inrolled (inflorescence up to 30 cm long, spreading, gray green; branches up to 12 cm long; forming dense clumps) . . . . *Eragrostis curvula* Weeping Lovegrass; af

42b Without white hairs where the branches attach to the stem; leaf blades 1–7 mm wide, the margins flat

43a Spikelet pedicels 1–2 mm long, pressed against the branchlets; leaf blades 1–3 mm wide (inflorescence usually less than 20 cm long, sometimes greenish purple) . . . . *Eragrostis pectinacea [E. diffusa]* Spreading Lovegrass; Ma, Sn, SF

43b Spikelet pedicels 3–15 mm long, not pressed against the branchlets; leaf blades 2–7 mm wide

44a Spikelet oblong, about 1.5 mm long *Eragrostis mexicana* ssp. *mexicana* Mexican Lovegrass

44b Spikelet spindle shaped, less than 1 mm long (inflorescence brown green; usually in moist areas) . . . . *Eragrostis mexicana* ssp. *virescens [E. orcuttiana]* Nonsticky Mexican Lovegrass

39b Without a tuft of white hairs where the leaf blade joins the sheath; spikelet usually with fewer than 12 florets

45a Spikelet about 1 mm long, with not more than 2 florets; inflorescence delicate, often as wide as long; glumes and lemma without prominent veins (inflorescence 5–15 cm wide, delicate, with many threadlike branchlets bearing numerous, very small spikelets; leaves crowded, with overlapping sheaths; in moist, often alkaline areas) . . . . *Muhlenbergia asperifolia* Scratchgrass

45b Spikelet at least 3 mm long, often longer, usually with more than 2 florets; inflorescence usually not as wide as long; glumes and usually lemma, with prominent veins (look at several)

46a Branches and spikelets not concentrated on 1 side of the stem (glumes without awns, somewhat to very unequal, at least the lower one usually shorter than rest of the spikelet) . . . .
. . . . .POACEAE, GROUP 4, SUBKEY 4

46b Branches and spikelets all on 1 side of the stem, most of the stem is visible on the opposite side

  47a Inflorescence 2–18 cm wide; stem flattened; spikelet with 2–4 florets, bunched at branch tips, keeled, and awn tipped; glumes 4–7 mm long, hairy on the margins; leaf sheaths keeled, the edges fused; sometimes forming dense clumps (leaf blades somewhat keeled) . . . . *Dactylis glomerata* (pl. 63)
Orchard Grass; eua

  47b Inflorescence up to 1.5 cm wide; stem cylindrical; spikelet with 4–10 florets, not bunched, but alternating on the branches, not keeled, and without awns; glumes about 2 mm long, not hairy; leaf sheaths not keeled, the edges not fused; not forming dense clumps . . . .
. . . . *Desmazeria rigida [Scleropoa rigida]*
Stiff Grass; me

Poaceae, Group 4, Subkey 1: Inflorescence with obvious branches, these divided into branchlets; spikelet usually with 1 obvious fertile floret, but there may be smaller sterile florets, and the spikelet of *Sporobolus airoides* may seem to consist of 2 florets when the fertile floret splits open

1a Apparent floret a dark purple bulblet enclosed in 2 papery sheaths, these with awnlike projections up to 2 cm long (there may also be bulbs at the bases of the stem) . . *Poa bulbosa* (pl. 63)
Bulbous Bluegrass; eu

1b Single fertile floret normal in appearance

  2a Awns more than 1 mm long, sometimes on the lemma as well as the glumes, but in *Piptatherum miliaceum* they readily fall off

    3a Floret falling with glumes at maturity; mature florets shiny, hard, and nearly round

      4a Inflorescence not especially dense, the main stem and its branches visible; glumes sharp tipped, less than half as long as the fertile floret; sterile floret 1; lemma awn stiff, up to 25 mm long (spikelet 2–3 mm long, crowded into 3 or 4 rows on 1 side of the many branches, these 2–4 cm long; glumes, branches, and branchlets with bristly hairs). . . . .
. . . . *Echinochloa crus-galli [E. crusgalli]* (pl. 64)
Barnyard Grass; eua

4b   Inflorescence dense, the spikelets so crowded that the stem and branches usually not visible except where the inflorescence is interrupted; glumes with awns 2–8 mm long and much longer than the floret; sterile floret absent; lemma awn delicate, 1–5 mm long (inflorescence yellow green, becoming straw colored or occasionally purple; glumes flattened, both with awns, forming a U or V)

   5a   Glumes 3–4 mm long, their tips gradually tapering into awns 2–3 mm long; lemma awn 1–2 mm long (in salt marshes) .... *Polypogon elongatus*
Longspike Beard Grass; sa

   5b   Glumes up to 2 mm long, their tips ending abruptly, but with awns 5–8 mm long; lemma awn 2–5 mm long (spikelet often in clusters; in moist areas) .... *Polypogon australis*
Chilean Beard Grass, Australian Beard Grass; sa

3b   Floret falling from glumes at maturity; mature florets not shiny, hard, and nearly round, except in *Piptatherum miliaceum*

   6a   Lemma awn up to 10 cm long, attached at the tapered tip of the lemma; awns typically bent once or twice when mature (lemma more than 5 mm long; glumes without awns; upper stem and spikelet often purple; forming clumps)

      7a   Floret at least 1 mm wide, spindle shaped; awn stiff, the portion near the tip straight (florets widest at the middle; awn up to 8 cm long; lower glume up to 20 mm long; inflorescence nodding at maturity; the California state grass.)...................... *Nassella pulchra [Stipa pulchra]*
Purple Needlegrass, Nodding Stipa

      7b   Floret less than 1 mm wide, cylindrical; awn flexible, often wavy toward the tip

         8a   Lower glume usually less than 11 mm long; awn less than 5 cm long
.... *Nassella lepida [Stipa lepida]*
Foothill Needlegrass

         8b   Lower glume up to 20 mm long; awn 5–10 cm long, usually more than 6 cm .... *Nassella cernua* [includes *Stipa cernua* and *S. lepida* var. *andersonii*]
Nodding Needlegrass

   6b   Lemma awn less than 2 cm long, attached at the middle of the lemma; awns either straight, wavy, or only slightly bent

      9a   Floret with a tuft of white hairs at its base, but *Agrostis,* in choice 9b, may have inconspicuous hairs (pry out the floret and use a hand lens); glumes sometimes sharp tipped, not awned

         10a Inflorescence open, the spikelets most densely crowded near the branchlet tips; spikelet 3–4 mm long; not forming dense clumps (leaf blades flat, up to 8 mm wide; often in wet areas) ....
.... *Calamagrostis bolanderi*
Bolander Reed Grass; Sn-n; 1b

10b Inflorescence rather compact and somewhat cylindrical, densely crowded with spikelets throughout; spikelet more than 4 mm long; forming dense clumps

    11a Awn protruding at least 2 mm beyond the spikelet (lemma tip with 4 minute teeth)

        12a Spikelet 10 mm long, with awns protruding 8–10 mm (leaf blades l–3 mm wide, the margins inrolled; coastal) . . . .
. . . . *Calamagrostis foliosa*
Leafy Reed Grass; Sn-n; 4

        12b Spikelet 5–6 mm long, with awns protruding about 2 mm (look at several) (inflorescence straw colored or purple tinged; on serpentine soils) . . . . . . *Calamagrostis ophitidis*
Serpentine Reed Grass; La, Ma, Sn; 4

    11b Awn hidden within the spikelet or protruding not more than 1 mm (inflorescence sometimes purple)

        13a Awn about 2 mm long, straight, seldom protruding above the tip of the spikelet; hairs at the base of floret about one-third as long as the floret (inflorescence usually 4–15 cm long; spikelet 3–4 mm long; in wet areas) . . . .
. . . . . *Calamagrostis stricta* ssp. *inexpansa* [*C. crassiglumis*]
Dense Reed Grass; Ma, Me-n; 2

        13b Awn 4 mm long, bent (pry out a floret to see), usually protruding from the side of the spikelet but sometimes hidden; hairs at the base of each floret much less than one-third as long as the floret

            14a Inflorescence loose, up to 3 cm wide and 15–30 cm long; leaf blades 6–12 mm wide (spikelet about 6 mm long; white hairs absent where leaf blade joins the sheath; up to 1.5 m tall; in moist areas) . . . .
. . . . *Calamagrostis nutkaensis*
Pacific Reed Grass; SLO-n

            14b Inflorescence often compact, 1–1.5 cm wide and 5–15 cm long; leaf blades 2–6 mm wide

                15a Inflorescence usually less than 1 cm wide; spikelet 4 mm long; leaf blade less than 4 mm wide, the margins usually inrolled, with delicate white hairs where the blade joins the sheath . . . .
. . . . *Calamagrostis rubescens*
Pine Grass; SLO-n

                15b Inflorescence usually 1–1.5 cm wide; spikelet 6 mm long; leaf blade up to 6 mm wide, the margins flat, without white hairs where the blade joins the sheath . . . . *Calamagrostis koelerioides*
Tufted Pine Grass; SLO-n

9b Floret without a tuft of white hairs at their bases; glumes sometimes with awns

    16a Awn attached to lemma tip (floret shiny and hard, the awn readily falling off; inflorescence spreading; lemma awn 3–4 mm long; glumes with tapering, pointed, awnlike tips; forming dense clumps) . . . . *Piptatherum miliaceum [Oryzopsis miliacea]*
Smilo Grass; eua

    16b Awn attached to near the middle of the lemma, but sometimes near the tip

        17a Awn 4–8 mm long (lemma tip with 2–4 very small teeth, use hand lens; up to 45 cm tall; not forming clumps)

            18a Inflorescence dense, ascending, up to 14 cm long; glumes 2.5–6 mm long, sometimes with short awns; leaf blades up to 3 mm wide, not wirelike; lemma tip with 2–4 teeth (usually in moist areas) . . . . *Agrostis microphylla* [includes *A. microphylla* var. *intermedia* and *A. aristiglumis*]
Small-leaf Bent Grass

            18b Inflorescence open, spreading, up to 20 cm long; glumes up to 2 mm long, without awns; leaf blades about 1 mm wide, wirelike; lemma tip with 2 teeth (branches threadlike) . . . . *Agrostis elliottiana [A. exigua]*
Annual Tickle Grass; Na-n

        17b Awn up to 3 mm long (glumes at least 2 mm long)

            19a Branches often more than 10 cm long, ascending or spreading outward, with branchlets originating near their tips; spikelets few, near branchlet ends (inflorescence delicate, often purple; awn less than 2 mm long, straight; leaf blades 1–3 mm wide; stem wirelike; forming dense clumps; Coast Ranges) . . . . *Agrostis scabra*
Tickle Grass

            19b Branches not more than 5 cm long, usually ascending, but sometimes spreading outward at maturity, with branchlets all along their length; spikelets numerous, present on at least half the length of each branchlet

                20a Inflorescence dense and cylindrical; plant not forming clumps (inflorescence up to 10 cm long, sometimes partly enclosed by a leaf sheath; awns up to 3 mm long; leaves 3–11 mm wide; ligules up to 2 mm long; coastal) . . . . *Agrostis densiflora* [includes *A. clivicola* vars. *clivicola* and *punta-reyesensis*]
California Bent Grass; SCr-n

20b Inflorescence usually open; forming dense clumps

    21a Awn less than 1 mm long, straight; inflorescence up to 20 cm long, with spikelets usually along the entire length of each branchlet; ligules 2–3 mm long (often in sandy areas) ....

        .... *Agrostis pallens* [includes *A. diegoensis*]

        Dune Bent Grass

    21b Awn 2.5–3 mm long, bent or straight; inflorescence up to 30 cm long, with spikelets along only half the length of each branchlet; ligules 3–6 mm long (in moist areas) ... *Agrostis exarata* [includes *A. ampla* and *A. longiligula* var. *australis*]

        Spike Bent Grass; SCr-n

2b Awns lacking or not more than 1 mm long, except in *Agrostis scabra*, in which they may be nearly 2 mm long

  22a Spikelet conspicuously flattened (in wet habitats)

    23a Inflorescence compact, 1–3 cm wide, the spikelets crowded and overlapping on 1 side of each branchlet and arranged in 2 main rows, sometimes with additional spikelets between the rows; glumes present; spikelet roundish, but with a sharp tip, the margins without stiff spines, but there may be fine hairs on the surfaces .... *Beckmannia syzigachne*

        Slough Grass; SFBR-n

    23b Inflorescence open, 5–10 cm wide, the spikelets not crowded or concentrated in 2 rows; glumes absent; spikelet elliptical, without a sharp tip but the margins with small, stiff spines .... *Leersia oryzoides*

        Rice Cutgrass; Na, Me-n

  22b Spikelet not conspicuously flattened

    24a Inflorescence, at maturity, loosely spreading, usually at least 5 cm wide, often more than 10 cm wide

      25a Inflorescence often as long as wide, ball shaped; spikelet about 1 mm long; ligules not more than 1 mm long (inflorescence delicate, with many threadlike branchlets bearing numerous small spikelets; leaves crowded, with overlapping sheaths; in moist areas) ..... *Muhlenbergia asperifolia*

        Scratchgrass

      25b Inflorescence usually longer than wide; spikelet 1–3 mm long; ligules 1–6 mm long

        26a One glume much shorter than the other glume, at least 1 shorter than rest of the spikelet; with long hairs where leaf blade joins the sheath; mature inflorescence often partly hidden in a leaf sheath (inflorescence up to 20 cm wide; fertile floret hard; spikelet usually with green or purple markings)

27a Sterile florets absent; leaf blades up to 4 mm wide, the margins often inrolled, scarcely hairy, except where the blade joins the sheath; floret falling from glumes at maturity; lemma rough, not shiny; forming dense clumps; perennial (fertile floret sometimes splitting at maturity; in moist, often alkaline areas) .... *Sporobolus airoides*
Alkali Dropseed, Alkali Sacaton

27b Sterile florets below fertile floret, prominently veined, resembling the larger glume; leaf blades 5–20 mm wide, flat, hairy; florets falling with glumes at maturity; fertile lemma shiny; not forming dense clumps; annual (mature inflorescence may separate from the stem and roll like a tumbleweed)

    28a Inflorescence up to 25 cm long; fertile floret with a crescent shaped thickening at its base; floret often brown, 2 mm long; leaf blades up to 12 mm wide .... *Panicum hillmanii*
Hillman Witchgrass; na

    28b Inflorescence up to 40 cm long; fertile floret smooth, without a basal thickening; floret ivory white, less than 2 mm long; leaf blades up to 20 mm wide .....*Panicum capillare*
Witchgrass; SF

26b Glumes equal, both longer than rest of the spikelet; leaves without long hairs; mature inflorescence completely exposed

    29a Branchlets at branch tips; spikelets few, near branchlet ends; leaf blades 1–3 mm wide; stems wirelike and usually crowded together; inflorescence often purplish (inflorescence more than 20 cm long; branches upright to spreading, up to 12 cm long; lemma awn, if present, less than 2 mm long; leaves mostly basal; forming dense clumps; up to 75 cm tall; moist areas in Coast Ranges) .... *Agrostis scabra*
Tickle Grass

    29b Branchlets all along branches; spikelets many, often crowded along the branchlets; leaf blades 2–8 mm wide; stems not wirelike and not crowded together; inflorescence often red purple to bronze (in wet areas; the following 2 species difficult to separate)

        30a Inflorescence usually not more than 5 cm wide or 15 cm long, dense; branches ascending; most leaf blades less than 4 mm wide; some stems prostrate and often rooting at the nodes, forming a mat; up to 60 cm tall ....
.... *Agrostis stolonifera* [*A. alba* vars. *alba* (in part) and *palustris*] (fig.)
Creeping Bentgrass; eu

        30b Inflorescence usually up to 10 cm wide and 25 cm long; branches spreading; leaf blades 4–10 mm wide; upright; up to 100 cm tall .... *Agrostis gigantea* [*A. alba* var. *alba* (in part)] Giant Bentgrass, Redtop; eu

24b Inflorescence at maturity not loosely spreading, less than 5 cm wide, but sometimes up to 6 cm wide in *Melica imperfecta*

    31a Branches ascending, usually completely covered with lobelike clusters of spikelets (inflorescence up to 15 cm long; glumes about 2 mm long, longer than the floret; florets falling with glumes at maturity; stem sometimes prostrate and rooting at the nodes; in moist areas) .... *Agrostis viridis* [*A. semiverticillata*]
        Whorled Bentgrass, Water Bentgrass; eu

    31b Branches not covered with lobelike clusters of spikelets

        32a Spikelet up to 6 mm long, with a pedestal of fused sterile florets arising beside the fertile floret (pry open the spikelet); edges of leaf sheaths almost completely fused (glumes somewhat transparent, with conspicuous purple markings; florets falling from glumes at maturity; in dry habitats)

            33a Florets without fine hairs; glumes not prominently veined; pedestal of spikelet slightly wider below the middle; inflorescence up to 6 cm wide .... *Melica imperfecta*
                Coast Range Melic, California Melic; SCl-s

            33b Florets with fine hairs on edges (use hand lens); glumes prominently veined; pedestal of spikelet slightly wider above the middle; inflorescence usually less than 2 cm wide .... *Melica torreyana*
                Torrey Melic; SLO-n

        32b Spikelet usually less than 3 mm long, sometimes up to 5 mm long, without a pedestal; edges of leaf sheath not fused

            34a Glumes unequal, the shorter one much smaller than rest of spikelet; sterile floret present, veined, resembling a glume; spikelet plump and hard; florets falling with glumes at maturity (spikelet about 2 mm long; ligules consisting of upright white hairs; plant varies seasonally, forming a basal cluster, clump, or mat; in moist, sometimes alkaline areas) .... *Panicum acuminatum* [includes *P. occidentale*, *P. pacificum*, and *P. thermale*]
                Marsh Panicum

            34b Glumes nearly equal, longer than rest of the spikelet; sterile floret absent; spikelet delicate; floret falling from glumes at maturity

                35a Floret with an obvious basal tuft of white hairs one-third as long as the floret (inflorescence ascending to spreading, with several branches at each node; spikelet 3–4 mm long; ligules 4–6 mm long; up to 1 m tall; shaded areas in the Coast Ranges) .... *Agrostis hallii*
                    Hall Bent Grass

35b Floret without an obvious basal tuft of white hairs
   36a Leaves usually less than 1 mm wide, wirelike; ligules less than 1.5 mm long (inflorescence dense, less than 8 cm long; up to 30 cm tall, but usually less than 15 cm; often forming dense clumps; on coastal dunes) . . . . . .
   . . . . *Agrostis blasdalei*
   Blasdale Bent Grass; Me-Ma; 1b
   36b Leaves up to 11 mm wide, some of them flat; ligules 1.5–6 mm long (the following species of *Agrostis* difficult to separate)
      37a Some stems prostrate and often rooting at the nodes, forming a mat; inflorescence usually 2–5 cm wide (inflorescence less than 15 cm long; ligules 2–5 mm long; leaf blades less than 5 mm wide; in wet habitats) . . . .
      . . . . *Agrostis stolonifera* [includes *A. alba* vars. *alba* (in part) and *palustris*] (fig.)
      Creeping Bent Grass; eu
      37b All stems upright; inflorescence less than 2 cm wide
         38a Inflorescence up to 30 cm long; ligules 3–6 mm long (leaf blades 2–7 mm wide; forming clumps; in moist areas) . . . .
         . . . . *Agrostis exarata* [includes *A. ampla* and *A. longiligula* var. *australis*]
         Spike Bent Grass, Western Bent Grass
         38b Inflorescence not more than 20 cm long; ligules 1.5–3 mm long
            39a Inflorescence up to 20 cm long, narrow but open, its branches not obscured; ligules 2–3 mm long; stem leaves with blades up to 6 mm wide; not forming clumps (coastal dunes) . . . .
            . . . . *Agrostis pallens* [includes *A. diegoensis*]
            Dune Bent Grass, Leafy Bent Grass
            39b Inflorescence not more than 10 cm long, dense, its branches may be obscured; ligules up to 2 mm long; stem leaves with blades up to 11 mm wide; forming clumps (coastal) . . . .
            . . . . *Agrostis densiflora* [includes *A. clivicola* vars. *clivicola* and *punta-reyesensis*]
            California Bent Grass; SCr-n

Poaceae, Group 4, Subkey 2:   Inflorescence with obvious branches, these divided into branchlets; spikelet, excluding awns, usually more than 1.5 cm long, with more than 1 obvious floret; lemma awn at least 1 mm long, attached near the tip of the lemma; glumes without awns, unequal and usually shorter than rest of the spikelet; leaves often more than 1 cm wide, the margins usually flat; ligules prominent, sometimes torn and jagged; edges of each leaf sheath, at least the lower portion, fused together . . . . *Bromus*

1a   Awn sharply bent and twisted below (look at several) (glumes less than 1 mm wide, pointed, the upper one almost as long as the lowest floret; lemma densely hairy, the tip with bristlelike teeth usually 2–3 mm long; awn 10–17 mm long) . . . . *Bromus trinii*
Chilean Chess, Chilean Brome; sa?

1b   Awn straight or curved

    2a   Awn 3.5–7 cm long (spikelets few, up to 35 cm long, excluding awns; lower branches usually spreading and drooping; ligules white, with jagged teeth; tip of lemma with membranous teeth 3–6 mm long; widespread) . . . . *Bromus diandrus*
Ripgut Brome; eu

    2b   Awn less than 3 cm long

        3a   Spikelet markedly flattened; glumes and lemma with definite keels, thus V shaped in cross section (spikelet less than 2 mm wide; teeth, if present at tip of lemma, very small)

            4a   Awn, if present, less than 3 mm long; lemma 9–12 mm long; lower glume with 5–7 prominent veins; spikelet without soft hairs . . . . *Bromus catharticus*
Rescue Grass; sa

            4b   Awns 4–13 mm long; lemma 12–17 mm long; lower glume usually with 3 prominent veins, sometimes more; spikelet sometimes with soft hairs (inflorescence branches mainly ascending; the following subspecies difficult to separate)

                5a   On coastal dunes and meadows; inflorescence dense toward tip, the upper branches often not visible, usually shorter than the spikelets; spikelets overlapping . . . *Bromus carinatus* var. *maritimus [B. maritimus]*
Marine Brome, Seaside Brome

                5b   In dry habitats; inflorescence somewhat open, all branches visible, usually longer than the spikelets; spikelets not overlapping (widespread) . . . . *Bromus carinatus* var. *carinatus*
[includes *B. breviaristatus* and *B. marginatus*] (fig.)
California Brome

        3b   Spikelet not markedly flattened; glumes and lemma without keels, except near the tips, thus mostly U shaped in cross section

            6a   Tip of lemma with 2 transparent teeth 2–4 mm long (inflorescence often purple; annual; up to 60 cm tall)

                7a   Inflorescence open, up to 8 cm wide, its branches spreading to drooping; spikelets not crowded; awn mostly less than 16 mm long (spikelets often hairy, on delicate branchlets; plant very easily pulled up) . . . . *Bromus tectorum* (fig.)
Cheat Grass, Downy Brome; eu

7b Inflorescence compact, less than 5 cm wide, its branches ascending; spikelets crowded; awn up to 25 mm long

    8a Stem and sheaths with very fine hairs; inflorescence very compact, resembling the head of a broom, the upper branches obscured .... *Bromus madritensis* ssp. *rubens [B. rubens]* (pl. 63)

        Foxtail Chess; me

    8b Stem and sheaths not hairy; inflorescence not so compact that the branches are obscured .... *Bromus madritensis* ssp. *madritensis*

        Spanish Brome; eu

6b Tip of lemma with 2 transparent teeth or lobes, if any, these not more than 2 mm long

    9a Lower glume with 1 prominent vein, sometimes 2 or 3 in *Bromus orcuttianus* (often more than 1 m tall)

        10a Awn 18–30 mm long; leaves 2–5 mm wide; teeth at tip of lemma 2 mm long; annual (upper glume 10–18 mm long; widespread) .... *Bromus sterilis*

            Poverty Brome, Sterile Brome; eua

        10b Awn 4–13 mm long; leaves up to 13 mm wide; teeth at tip of lemma less than 1 mm long; perennial

            11a Awn less than 8 mm long; upper glume up to 10 mm long; ligules less than 2 mm long; many branches of inflorescence less than 2 cm long, ascending (at elevations of at least 3,000 ft) .... *Bromus orcuttianus*

                Orcutt Brome

            11b Awn up to 13 mm long; upper glume usually 10–15 mm long; ligules at least 3 mm long; many branches of inflorescence more than 3 cm long, spreading (in shaded areas in Coast Ranges) .... *Bromus vulgaris*

                Narrowflower Brome; Mo-n

    9b Lower glume with 3–5 prominent veins

        12a Inflorescence often dense, almost headlike, usually less than 10 cm long; branches less than 1 cm long (branches ascending; annual; widespread) .... *Bromus hordeaceus*

            Soft Cheat Grass, Soft Cheat; eu

        12b Inflorescence usually open, up to 20 cm long; branches 2–27 cm long

            13a Upper glume with 3 distinct veins, sometimes 4 or 5 (leaf blades up to 12 mm wide; tip of lemma with small lobes; branches spreading to drooping; perennial) .... *Bromus grandis*

                Grand Brome, Tall Brome

            13b Upper glume with 5–7 distinct veins

                14a Awn up to 16 mm long (glumes and lemma hairy; branches often S shaped; awn often dark; annual) .... *Bromus arenarius*

                Australian Brome, Australian Cheat; au

14b Awn usually less than 10 mm long

15a Lemma 9–15 mm long; lower glume with 3 veins; upper glume up to 11 mm long; leaf blades up to 17 mm wide; perennial (tip of lemma with small lobes; mostly in shaded areas) . . . . *Bromus laevipes* [includes *B. pseudolaevipes*]
Chinook Brome, Woodland Brome

15b Lemma 6–10 mm long; lower glume with 3–5 veins; upper glume up to 8 mm long; leaf blades less than 7 mm wide; annual

16a Many mature lemmas with inrolled margins in the lower half; lemma 6–8 mm long; spikelet rigid, sometimes with fine hairs . . . . *Bromus secalinus*
Chess; eua

16b Mature lemmas without inrolled margins; lemma 7–10 mm long; spikelet not especially rigid, hairy . . . . *Bromus japonicus* [*B. commutatus*]
Japanese Brome, Hairy Chess; eua

Poaceae, Group 4, Subkey 3: Inflorescence with obvious branches, these divided into branchlets; spikelet, excluding awns, usually less than 1.5 cm long, with more than 1 obvious floret; lemma awn at least 1 mm long, attached near the tip; glumes without awns, of unequal length and usually shorter than rest of the spikelet; leaves usually less than 2 mm wide, often wirelike; ligules absent or inconspicuous; edges of each leaf sheath not fused . . . . *Festuca* (in part), *Vulpia*

1a Glumes, lemma, or both, with very small hairs (use lens)

2a Most spikelets with hairy glumes and lemmas (inflorescence 1–6 cm wide; florets 2–4; lemma awn 3–16 mm long; up to 40 cm tall; annual) . . . . *Vulpia microstachys* var. *ciliata* [includes *Festuca eastwoodae* and *F. grayi*]
Hairy Fescue, Gray Fescue

2b Spikelets with either hairy glumes, or hairy lemmas

3a Glumes hairy; lemma not hairy; florets 2 or 3, sometimes 4 (awn 4–14 mm long; up to 40 cm tall; annual) . . . . *Vulpia microstachys* var. *confusa* [*Festuca confusa* and *F. tracyi*]
Hairyleaf Fescue

3b Glumes not hairy; lemma with very small, stiff hairs (look at several spikelets); florets 3–12

4a Spikelet with 5–12 florets, these densely crowded; awn 1–5 mm long (spikelets crowded at stem tips; less than 20 cm tall; annual) . . . . *Vulpia octoflora* var. *hirtella* [*Festuca octoflora* ssp. *hirtella*]
Slender Fescue

4b Spikelet with 3–6 florets; awn 2–15 mm long

5a   Awn usually 3–7 mm long, attached between 2 very small teeth at lemma tip (hard to see with a lens); inflorescence more than 2 cm wide, the branches spreading or drooping; leaf blades up to 4 mm wide, flat; perennial; up to 100 cm tall (florets 3 or 4; in shaded areas) . . . . *Festuca elmeri* [includes *F. elmeri* ssp. *luxurians*]
Elmer Fescue; Mo-n

5b   Awn at least 8 mm long, attached at lemma tip, but not between 2 teeth; inflorescence less than 2 cm wide, the branches upright; leaf blades less than 2 mm wide, wirelike; annual; up to 45 cm tall (spikelet nearly sessile) . . . . *Vulpia myuros* var. *hirsuta* [*Festuca megalura*]
Foxtail Fescue; eu

1b   Glumes and lemma not hairy

     6a   Awn attached at lemma tip between 2 very small teeth (inflorescence often more than 6 cm long, sometimes drooping; leaf blades up to 4 mm wide, flat; florets 3 or 4; awn usually 3–5 mm long; lemma with 5 prominent veins; up to 1 m tall, perennial; in shaded areas) . . . . *Festuca elmeri* [includes *F. elmeri* ssp. *luxurians*]
Elmer Fescue; Mo-n

     6b   Awn attached at lemma tip, but not between 2 teeth

         7a   Awn not more than 4 mm long, except up to 5 mm long in *Festuca idahoensis* (plant forming dense clumps)

             8a   Base of leaf blade with lobes clasping the stem (leaf blades 4–10 mm wide, the margins usually flat; lemma 5–9 mm long, sometimes short awned; inflorescence up to 30 cm long; not forming clumps) . . . . *Festuca arundinacea*
Tall Fescue, Reed Fescue; eu

             8b   Base of leaf blade without lobes clasping the stem

                 9a   Leaf usually hairy where the blade joins the sheath (lower branches of inflorescence spreading, usually paired and becoming 4 cm long before giving rise to the first branchlets; leaf blades up to 4 mm wide; spikelets numerous, each with at least 3 florets; perennial; up to 120 cm tall; in shaded areas) . . . . *Festuca californica*
California Fescue; Mo-n

                 9b   Leaf not hairy where the blade joins the sheath

                     10a   Lemma, excluding awns, 3–5 mm long; spikelet with 6–13 florets; annual; up to 60 cm tall (inflorescence concentrated at stem tip) . . . . *Vulpia octoflora* var. *octoflora* [*Festuca octoflora* ssp. *octoflora*] (fig.)
Sixweeks Fescue

                     10b   Lemma, excluding awns, 5–10 mm long; spikelet with not more than 7 florets; perennial; often more than 80 cm tall

                         11a   Leaf sheath closed, disintegrating into fibers as the plant ages; leaf sheaths usually red purple; stem at base bending outward before turning upward; awns less than 4 mm long; forming loose clumps (branchlets appearing on the lower branches within 2 cm of the stem; in moist areas) . . . *Festuca rubra* (fig.)
Red Fescue; SLO-n

11b Leaf sheath open in the upper half, not disintegrating; leaf sheaths not red purple; stem not bending outward; some awns up to 6 mm long; forming dense clumps (Coast Ranges) . . . . . .
. . . . *Festuca idahoensis*
Idaho Fescue, Blue Bunchgrass; SM-n

7b Some awns at least 5 mm long

12a Inflorescence up to 12 cm wide, the branches spreading or drooping; leaves mostly basal; perennial; up to 100 cm tall (awn 5–12 mm long; florets 3–5; in shaded areas) . . . . *Festuca occidentalis*
Western Fescue; Mo-n

12b Inflorescence less than 3 cm wide, the branches erect or spreading; leaves usually not mostly basal; annual; not more than 60 cm tall

13a Awn less than 6 mm long; spikelet with 5–13 florets (upper glume usually less than 6 mm long; branches upright) . . . .
. . . . . *Vulpia octoflora* var. *octoflora* [*Festuca octoflora* ssp. *octoflora*] (fig.)
Sixweeks Fescue

13b Awn often more than 10 mm long; spikelet usually with fewer than 8 florets

14a Lower glume 1–2 mm long, less than one-half as long as the upper glume; awn up to 17 mm long (inflorescence less than 2 cm wide) . . . . *Vulpia myuros* var. *myuros* [*Festuca myuros*]
Rattail Fescue; eu

14b Lower glume 2–8 mm long, from one-half to as long as the upper glume; awn less than 13 mm long

15a Spikelet usually with 2–4 florets; branches spreading outward or drooping; inflorescence up to 5 cm wide . . . .
. . . . *Vulpia microstachys* var. *pauciflora* [*Festuca pacifica* and *F. reflexa*]
Common Hairyleaf Fescue

15b Spikelet with 4–7 florets; branches upright; inflorescence less than 2 cm wide, sometimes very dense . . . .
. . . . *Vulpia bromoides* [*Festuca dertonensis*] (fig.)
Brome Fescue, Sixweeks Fescue; eu

Poaceae, Group 4, Subkey 4: Inflorescence with obvious branches, these usually divided into branchlets; spikelet with 2 or more similar florets; awns, if present, less than 1 mm long; glumes without awns, somewhat to very unequal, at least the lower one usually shorter than rest of spikelet

1a Leaf blades curved on one side at the tip, forming when folded what looks like the bow of a boat, less than 6 mm wide; glumes and lemma somewhat keeled (lemma usually less than 5 mm long; lower glume usually shorter than the lemma of the lowest floret; spikelet without true awns; florets often with cottony webbing at their bases and sometimes with hairs elsewhere; leaf sheaths open above but often fused below; spikelet often not open at maturity)

2a Spikelet with purple black, teardrop-shaped bulblets, these 2–3 mm long, with papery, awnlike projections 1–2 cm long (there may also be bulblets at the base of the stem, and sometimes true florets are present in the spikelets) . . . . *Poa bulbosa* (pl. 63)
Bulbous Bluegrass; eu

2b   Spikelet without bulblets

   3a   Lemma without cottony webbing at their bases, but there may be other hairs, especially on the veins of the florets, but these not intertwined to form a web

      4a   Spikelet 3–6 mm long; branches spreading, sometimes at right angles; lemma up to 3 mm long, with many very small hairs and 5 prominent veins; annual; usually less than 20 cm tall (leaves 1–3 mm wide, flat; upper glume usually widest at or just above the middle; inflorescence pyramid shaped; often forming mats) . . . . *Poa annua* (fig.)
          Annual Bluegrass; eu

      4b   Spikelet 5–10 mm long; branches upright; lemma up to 4 mm long, only slightly hairy, if at all, and the veins scarcely visible; perennial; up to 100 cm tall

         5a   Spikelets mostly 7–10 mm long, not bunched; each glume and lemma rounded on the back, without an evident keel, thus U shaped in cross section (forming dense clumps) . . . . *Poa secunda [P. scabrella]*
           One-sided Bluegrass, Pine Bluegrass

         5b   Spikelets 5–6 mm long, often in bunches toward branch ends; each glume and lemma usually with an obvious keel, thus V shaped in cross section (in alkaline soils near hot springs) . . . . *Poa napensis*
           Napa Bluegrass; Na; 1b

   3b   Lemma with cottony webbing at their bases

      6a   Glumes distinctly crescent shaped; ligules 3–8 mm long (lemma keel hairy near base; perennial; in moist areas) . . . . *Poa trivialis*
          Rough Bluegrass; eu

      6b   Glumes not crescent shaped; ligules not more than 5 mm long

         7a   Lemma not hairy (lemma 5 veined; inflorescence very open; perennial; in moist, shaded areas) . . . . *Poa kelloggii*
           Kellogg Bluegrass

         7b   Lemma with some hairs

           8a   Ligules less than 1 mm long (lemma with hairs on keel and veins; plant forming dense clumps; perennial; in moist, shaded areas) . . .
              . . . . *Poa nemoralis*
             Wood Bluegrass; eu

           8b   Ligules 1–5 mm long

              9a   Lemma with hairs on the keel and veins and dense cottony webbing at the base; perennial . . . . *Poa pratensis* (fig.)
                Kentucky Bluegrass; eu

              9b   Lemma with short hairs throughout and very sparse cottony webbing at the base; annual (Coast Ranges) . . . .
                . . . . *Poa howellii [P. bolanderi ssp. howellii]*
                Howell Bluegrass

1b Leaf blades tapering gradually to a needlelike tip, up to 15 mm wide, but mostly narrower, and the margins often inrolled; glumes and lemma not keeled, except sometimes in *Festuca viridula*

   10a Spikelet with a clublike pedestal of fused sterile florets among the fertile florets; lemma often with purple splotches; base of stem often either swollen or with a bulb (glumes papery, the lower one with at least 3 veins, upper one with 1–3; lemma rounded on the back; edges of leaf sheaths almost completely fused)

   11a Fertile lemma tapered evenly to a needlelike tip (base of stem with a bulb; glumes unequal; lemma 8–15 mm long; inflorescence branches upright to slightly spreading; ligules closed around stem; pedestal of spikelet only slightly thicker at upper end; in moist areas) .... *Melica subulata* (fig.)
   Alaska Onion Grass; Ma-n

   11b Fertile lemma not evenly tapered, the tip blunt or only short pointed

   12a Base of stem swollen or with a bulb

   13a Inflorescence open, up to 8 cm wide; pedestal of spikelet spindle shaped; glumes very unequal; base of stem with a bulb (tip of pedestal often protruding from spikelet) ... *Melica geyeri* [includes *M. geyeri* var. *aristulata*]
   Geyer Onion Grass; Mo-n

   13b Inflorescence usually less than 2 cm wide; pedestal of spikelet hammer shaped, its knob much shorter than its stalk; glumes almost equal; base of stem usually swollen (Coast Ranges) .... *Melica californica*
   California Melic, Western Melic

   12b Base of stem neither swollen nor with a bulb (inflorescence less than 4 cm wide; glumes equal)

   14a Pedestal of spikelet with a cylindrical knob twice as wide as the stalk; lemma 6–16 mm long, sometimes with a short awn; inflorescence usually less than 4 cm wide (upper edges of ligules with 2 teeth).........
   .... *Melica harfordii*
   Harford Onion Grass, Harford Melic; Mo-n

   14b Pedestal of spikelet with a cylindrical knob scarcely wider than the stalk; lemma 4–6 mm long, without an awn; inflorescence usually less than 2 cm wide...................................... *Melica torreyana*
   Torrey Melic; SLO-n

   10b Spikelet without a clublike pedestal; lemma without purple splotches, but there may be some small purple markings; base of stem neither swollen nor with a bulb

   15a Lemma without prominent parallel veins; lemma narrow, spindle shaped, the tip tapered to a sharp point; lower glume 3.5–6 mm long, upper glume 5–8 mm long (lower glume with 1 vein, upper one with 3; glumes and lemma sometimes keeled; edges of leaf sheaths not fused; usually forming clumps; Coast Ranges) ....
   .... *Festuca viridula*
   Green Fescue, Mountain Bunchgrass; Sn-n

   15b Lemma with 5–9 parallel veins, these sometimes difficult to see; lemma rounded on the back, oval to almond shaped, the tip not sharply pointed; lower glume up to 4 mm long; upper glume up to 5 mm long (in wet areas)

16a Usually associated with freshwater habitats; both glumes with 1 vein; lemma with 7 prominent veins; edges of leaf sheaths fused nearly to the top (glumes papery; leaf blades 4–12 mm wide; up to 1.5 m tall; not forming clumps)

    17a Spikelets 3–5 mm long, oval, not pressed against the stem, with 4–7 florets; inflorescence open, spreading, up to 20 cm or more wide; upper glume less than 2 mm long; lemmas less than 2 mm long ... *Glyceria elata*
        Fowl Mannagrass; Ma, Sn

    17b Spikelets 10–20 mm long, cylindrical, pressed against the stem, with 6–14 florets; inflorescence usually narrow, not more than 5 cm wide; upper glume up to 5 mm long; lemmas 3–5 mm long

        18a Tip of lemma rounded; lower glume less than 2 mm long .... *Glyceria leptostachya*
            Water Mannagrass; Ma, Sn

        18b Tip of lemma jagged; lower glume up to 4 mm long .... *Glyceria occidentalis*
            Western Mannagrass; SM-n

16b Usually in alkaline or saline habitats; lower glume usually with 1 vein, the upper one with 3; lemma with 5 faint veins in *Puccinellia*, and 7–9 prominent veins in *Torreyochloa*; edges of leaf sheaths overlapping or partly fused (glumes shorter than the lowest floret; lower glume less than 2 mm long; upper glume 3 mm long)

    19a Inflorescence less than 0.5 cm wide, the branches upright, each usually with 1 spikelet; annual; less than 20 cm tall (leaf blade less than 3 mm wide; upper glume with 3 prominent veins; lemmas with sharp tips, the longest 3–4 mm long; forming clumps; inland) .... *Puccinellia simplex*
        California Alkali Grass

    19b Inflorescence up to 12 cm wide, the branches spreading, each with numerous spikelets; perennial; more than 60 cm tall

        20a Leaf blades 4–16 mm wide; lemma about 2 mm long, rounded at the tip, with prominent veins (inflorescence spreading, up to 12 cm wide; lower glume less than 1 mm long; upper glume 1–2 mm long; not forming clumps) .... *Torreyochloa pallida* var. *pauciflora* [*Puccinellia pauciflora*]
            Weak Mannagrass; SM-n

        20b Leaf blades less than 4 mm wide; lemma 2–4 mm long, abruptly narrowed toward the tip, with indistinct veins (old leaves persisting; perennial; in salt marshes or alkaline areas)

            21a Inflorescence up to 6 cm wide, the branches upright; spikelet 8–15 mm long; florets 5–12; lemma sparsely hairy at the base, 4 mm long; forming dense clumps .... *Puccinellia nutkaensis* [includes *P. grandis*]
                Alaska Alkali Grass; SM-n

21b Inflorescence less than 10 cm wide, the branches spreading; spikelet 4–7 mm long; florets 3–6; lemma not hairy at the base, less than 3 mm long; forming clumps, but these not dense (lower glume less than 1 mm long; upper glume 1–2 mm long; longest lemma less than 4 mm) . . . . *Puccinellia nuttalliana [P. airoides]*
Nuttall Alkali Grass

PONTEDERIACEAE (PICKEREL-WEED FAMILY) Our only native representative of the Pontederiaceae is *Heteranthera dubia* (fig.) (Water Stargrass), which occurs from Solano and Mendocino counties northward. It is a weak-stemmed aquatic that grows in ponds, ditches, and sluggish streams. The alternate leaves, with prominent sheaths at their bases, are 7–15 cm long but not more than 0.5 cm wide. They are nearly translucent and do not have a distinct midrib. The flowers open at the water surface. Each one originates within a rolled-up bract. The perianth, with a long, slender tube and six lobes about 0.5 cm long, is pale yellow. There are three stamens. The ovary is superior, and the single pistil becomes a many-seeded fruit.

Eichhornia crassipes (Water Hyacinth) is native to South America. This floating plant has inflated leaf petioles and an elongated inflorescence of white to pale blue flowers. It is widely cultivated in garden pools and occasionally becomes established in ponds and lakes in frost-free areas of California. It is a nuisance in Florida and in some other places to which it has been introduced.

Heteranthera dubia
Water Stargrass

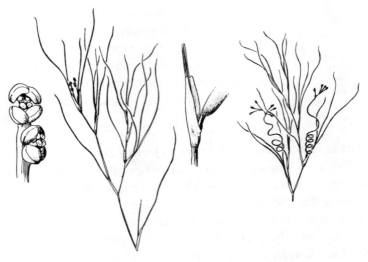

Ruppia cirrhosa
Coiled Ditchgrass

POTAMOGETONACEAE (PONDWEED FAMILY)   The Pondweed Family, consisting of aquatic plants, is represented in our region by two genera. In *Potamogeton*, the leaves are mostly alternate except for those near the tips of the stems. The leaves have a conspicuous scale that resembles a stipule and that often forms a sheath around the stem. In certain species, there are broad floating leaves that are very different from the narrow submerged leaves. The small flowers, usually in rather dense inflorescences at stem ends, typically have four pistils and four stamens. Each stamen arises from the base of a structure that resembles a sepal, but which may in fact be part of the stamen. The pistils ripen into one-seeded fruits.

The genus *Ruppia* (ditchgrasses) is sometimes put into a separate family, Ruppiaceae. The leaves are always slender, although they have rather broad, sheathing bases. The inflorescence originates within the sheath of the uppermost leaves. Each flower has four pistils, as in *Potamogeton*, but there are only two stamens. These have extremely short stalks and do not have sepal-like structures at their bases.

1a   Inflorescence loose, with only a few flowers; leaves all narrow; in brackish water of streams entering salt marshes and bays

 2a   Peduncle of inflorescence not more than 3 cm long, not spiraling; leaves pointed at the tips . . . . *Ruppia maritima*
Straight Ditchgrass

 2b   Peduncle up to 30 cm long, often spiraling or at least showing a tendency to spiral; leaves blunt at the tips . . . . *Ruppia cirrhosa [R. spiralis]* (fig.)
Coiled Ditchgrass

1b   Inflorescence dense, cylindrical; leaves either all narrow or of 2 types, the floating ones much broader than the submerged ones; in freshwater

3a   All leaves submerged and similar (leaves not more than 1 mm wide)

4a   Leaves pointed at the tips; stigma on a style; fruit 2–5 mm long, with a beak......
. . . . *Potamogeton pectinatus*
Fennel-leaf Pondweed

4b   Leaves sometimes pointed at the tips; stigma without a style; fruit 2–3 mm long, without a beak. . . . . *Potamogeton filiformis*
Slenderleaf Pondweed

3b   Leaves of 2 types: firm, floating leaves at least 5 mm wide and submerged leaves 1–5 mm wide

5a   Floating leaves 4–11 cm long and up to 6 cm wide; submerged leaves 10–30 cm long
. . . . *Potamogeton natans* (pl. 64)
Floatingleaf Pondweed; SM-n

5b   Floating leaves mostly 1–3 cm long and up to about 1 cm wide; submerged leaves up to 6 cm long . . . . *Potamogeton diversifolius*
Diverseleaf Pondweed; Sn

TYPHACEAE (CATTAIL FAMILY)   There are two genera in this small family. In *Typha* (cattails), the leaves are somewhat similar to those of grasses, and the pithy, upright stems arise from creeping rhizomes rooted in mud in swamps and at the edges of lakes, ponds, and ditches. The flowers are small and packed tightly into a cylindrical inflorescence 1–2 cm wide. The staminate flowers, intermixed with slender hairs, have two to five stamens and are above

Sparganium eurycarpum
Giant Bur-reed

the pistillate flowers, each of which produces a small, one-seeded dry fruit. There are neither petals nor sepals.

Sparganium (bur-reeds) is often in its own family, Sparganiaceae. The stems and leaves may be mostly submerged or floating, but even when this is the case, the inflorescence is raised above the surface of the water. Globular clusters of pistillate flowers, each of which produces a one-seeded dry fruit, are below the clusters of staminate flowers, which have three to five stamens.

1a   Inflorescence with several separate globular clusters of staminate flowers in its upper portion and clusters of pistillate flowers in its lower portion; fruit 2–20 mm long

     2a   Inflorescence branched; pistil with 2 stigmas; leaves, if straightened, not reaching beyond the top of the inflorescence; up to 2.5 m tall. . . . . . *Sparganium eurycarpum* (fig.)
Giant Bur-reed, Broadfruit Bur-reed

     2b   Inflorescence not branched; pistil with 1 stigma; leaves, if straightened, often reaching beyond the top of the inflorescence; up to 1 m tall . . . .
. . . . *Sparganium emersum* [*S. multipedunculatum*]
Bur-reed; Sn-n

1b   Inflorescence cylindrical, a continuous column of staminate flowers above a column of pistillate flowers; fruit less than 1 mm long

     3a   Staminate and pistillate portions of the inflorescence usually not separated by a gap; pistillate flowers green when fresh; leaves 10–25 mm wide . . . . . *Typha latifolia* (pl. 64)
Broadleaf Cattail

     3b   Staminate and pistillate portions of the inflorescence separated by a gap of at least 5 mm; pistillate flowers yellow or brown when fresh; leaves 4–18 mm wide

         4a   Inflorescence usually shorter than the leaves; pistillate flowers usually brown when fresh; leaves 4–12 mm wide . . . . *Typha angustifolia*
Narrowleaf Cattail

         4b   Inflorescence usually at least as long as the leaves; pistillate flowers yellow to orange brown when fresh; leaves 6–18 mm wide . . . . *Typha domingensis*
Southern Cattail, Narrowleaf Cattail

ZANNICHELLIACEAE (HORNED-PONDWEED FAMILY)   The only member of Zannichelliaceae in our region is *Zannichellia palustris* (fig.) (Horned-pondweed), found in ponds and sluggish streams. It resembles some pondweeds (Potamogetonaceae) because it has extremely slender stems, comparably narrow leaves, and tiny flowers. The leaves, however, are opposite, and the flowers are produced in the leaf axils rather than in inflorescences at stem ends. In each inflorescence, there is usually a staminate flower that consists of a single stamen and a pistillate flower with three or more pistils, these being located above a broad, nearly cup-shaped bract.

ZOSTERACEAE (EELGRASS FAMILY)   Eelgrasses are mostly submerged marine plants, exposed only at low tide. They have creeping rhizomes and narrow alternate leaves that arise in two rows from the stems. The flowers, without petals or sepals, are concentrated on one side of the flattened inflorescence that is at first enclosed within a leaf sheath. Staminate flowers have a single anther; pistillate flowers produce a one-seeded fruit.

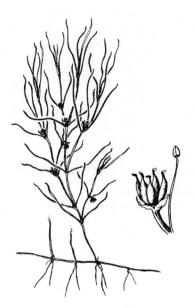

Zannichellia palustris
Horned-pondweed

Phyllospadix scouleri
Scouler Surfgrass

1a Leaves usually more than 4 mm wide, sometimes more than 10 mm wide; each inflorescence with both pistillate and staminate flowers; mostly in bays, rooted in mud or sand . . .
. . . . *Zostera marina* [includes *Z. marina* var. *latifolia*]
Eelgrass

1b Leaves not more than 4 mm wide; pistillate and staminate flowers on separate plants; on rocky shores where there is considerable wave action
    2a Leaves 1–2 mm wide, often at least half as thick as wide; flowering stem 30–100 cm long
. . . . *Phyllospadix torreyi*
Torrey Surfgrass

    2b Leaves 2–4 mm wide and not half as thick as wide; flowering stem up to 20 cm long . .
. . . . *Phyllospadix scouleri* (fig.)
Scouler Surfgrass

Adiantum aleuticum
Fivefinger Fern

Adiantum jordanii
California Maidenhair

Athyrium filix-femina var. *cyclosorum*
Western Lady Fern

Dryopteris arguta
Coastal Wood Fern

Cheilanthes covillei
Coville Lace Fern

Polypodium calirhiza
Common Polypody

Polystichum californicum
California Shield Fern

PLATE 1   FERNS

Pentagramma triangularis
Goldback Fern (Fern)

Pteridium aquilinum var. *pubescens*
Western Bracken (Fern)

Pellaea andromedifolia
Coffee Fern (Fern)

Woodwardia fimbriata
Giant Chain Fern (Fern)

Pellaea mucronata
Bird's-foot Fern (Fern)

Equisetum telmateia ssp. *braunii*
Giant Horsetail (Fern Ally)

PLATE 2   FERNS AND FERN ALLIES

Cupressus macrocarpa
Monterey Cypress

Juniperus californica
California Juniper

Pinus muricata
Bishop Pine

Pinus ponderosa
Pacific Ponderosa Pine

Sequoia sempervirens
Coast Redwood

Pseudotsuga menziesii
Douglas-fir

Tsuga heterophylla
Western Hemlock

PLATE 3   GYMNOSPERMS

Taxus brevifolia
Pacific Yew (Gymnosperm)

Torreya californica
California Nutmeg (Gymnosperm)

Carpobrotus chilensis
Iceplant (Aizoaceae)

Carpobrotus edulis
Hottentot-fig (Aizoaceae)

Tetragonia tetragonioides
New Zealand Spinach (Aizoaceae)

Amaranthus deflexus
Low Amaranth (Amaranthaceae)

Rhus trilobata
Skunkbrush (Anacardiaceae)

Toxicodendron diversilobum
Western Poison-oak (Anacardiaceae)

PLATE 4   GYMNOSPERMS AND DICOTYLEDONS: AIZOACEAE–ANACARDIACEAE

Heracleum lanatum
Cow Parsnip

Lomatium dasycarpum
Hog Fennel

Lomatium utriculatum
Spring-gold

Osmorhiza chilensis
Wood Sweet-cicely

Sanicula arctopoides
Footsteps-of-spring

Sanicula bipinnatifida
Purple Sanicle

Sanicula crassicaulis
Snakeroot

Scandix pecten-veneris
Shepherd's-needle

PLATE 5   DICOTYLEDONS: APIACEAE

Aralia californica
Elk-clover (Araliaceae)

Aristolochia californica
Dutchman's-pipe (Aristolochiaceae)

Asarum caudatum
Wild-ginger (Aristolochiaceae)

Asclepias californica
California Milkweed (Asclepiadaceae)

Asclepias fascicularis
Narrowleaf Milkweed (Asclepiadaceae)

Achillea millefolium
Common Yarrow (Asteraceae)

Achyrachaena mollis
Blow-wives (Asteraceae)

Agoseris apargioides var. *eastwoodiae*
Coast Dandelion (Asteraceae)

PLATE 6   DICOTYLEDONS: ARALIACEAE–ASTERACEAE

Agoseris grandiflora
California Dandelion

Anaphalis margaritacea
Pearly Everlasting

Ambrosia chamissonis
Silvery Beachweed

Arctium minus
Common Burdock

Artemisia douglasiana
Mugwort

Arnica discoidea
Coast Arnica

PLATE 7   DICOTYLEDONS: ASTERACEAE

Artemisia californica
California Sagebrush

Artemisia pycnocephala
Coastal Sagewort

Aster chilensis
Common California Aster

Baccharis pilularis
Coyotebrush

Baccharis salicifolia
Mulefat

Bidens frondosa
Sticktight

PLATE 8   DICOTYLEDONS: ASTERACEAE

Blennosperma nanum var. *nanum*
Common Blennosperma

Calycadenia multiglandulosa
Sticky Rosinweed

Carduus pycnocephalus
Italian Thistle

Centaurea solstitialis
Yellow Star-thistle

Cichorium intybus
Chicory

Cirsium occidentale var. *venustum*
Venus Thistle

Cirsium quercetorum
Brownie Thistle

Coreopsis calliopsidea
Leafy-stemmed Coreopsis

PLATE 9   DICOTYLEDONS: ASTERACEAE

Cotula coronopifolia
Brassbuttons

Crepis capillaris
Smooth Hawk's-beard

Ericameria linearifolia
Interior Goldenbush

Cynara cardunculus
Cardoon

Erigeron petrophilus var. *petrophilus*
Rock Daisy

Erigeron glaucus
Seaside Daisy

Eriophyllum confertiflorum
Golden Yarrow

PLATE 10   DICOTYLEDONS: ASTERACEAE

Eriophyllum lanatum var. *achillaeoides*
Common Woolly Sunflower

Gnaphalium californicum
California Everlasting

Grindelia camporum
Great Valley Gumplant

Helenium puberulum
Rosilla

Filago californica
California Fluffweed

Gnaphalium purpureum
Purple Cudweed

Helianthus annuus
Common Sunflower

PLATE 11   DICOTYLEDONS: ASTERACEAE

Helianthus gracilentus
Slender Sunflower

Hemizonia congesta ssp. *congesta*
Yellow Hayfield Tarweed

Hemizonia congesta ssp. *luzulifolia*
Hayfield Tarweed

Heterotheca grandiflora
Telegraphweed

Heterotheca sessiliflora ssp. *echioides*
Bristly Golden Aster

Holocarpha heermannii
Heermann Tarplant

Hypochaeris radicata
Rough Cat's-ear

Jaumea carnosa
Fleshy Jaumea

PLATE 12   DICOTYLEDONS: ASTERACEAE

Lasthenia californica
California Goldfields

Lactuca serriola
Prickly Lettuce

Madia elegans ssp. *elegans*
Elegant Madia

Layia platyglossa
Tidytips

Picris echioides
Bristly Ox-tongue

Madia gracilis
Slender Madia

Senecio elegans
Purple Ragwort

PLATE 13   DICOTYLEDONS: ASTERACEAE

Silybum marianum
Milk Thistle

Solidago californica
California Goldenrod

Solidago canadensis ssp. *elongata*
Canada Goldenrod

Solidago spathulata
Coast Goldenrod

Stephanomeria virgata
Tall Stephanomeria

Tanacetum camphoratum
Dune Tansy

PLATE 14   DICOTYLEDONS: ASTERACEAE

Tragopogon porrifolius
Salsify (Asteraceae)

Uropappus lindleyi
Silverpuffs (Asteraceae)

Wyethia angustifolia
Narrowleaf Mule-ears (Asteraceae)

Wyethia glabra
Smooth Mule-ears (Asteraceae)

Achlys triphylla
Vanilla-leaf (Berberidaceae)

Berberis pinnata
Shinyleaf Oregon-grape (Berberidaceae)

Vancouveria planipetala
Inside-out-flower (Berberidaceae)

PLATE 15   DICOTYLEDONS: ASTERACEAE–BERBERIDACEAE

Amsinckia tessellata var. *tessellata*
Devil's-lettuce

Amsinckia menziesii var. *intermedia*
Common Fiddleneck

Cryptantha muricata
Prickly Cryptantha

Cryptantha micromeres
Minuteflower Cryptantha

Heliotropium curassavicum
Seaside Heliotrope

Cynoglossum grande
Hound's-tongue

Plagiobothrys nothofulvus
Common Popcornflower

PLATE 16   DICOTYLEDONS: BORAGINACEAE

Arabis blepharophylla
Coast Rock Cress (Brassicaceae)

Cakile maritima
Horned Searocket (Brassicaceae)

Cardamine californica var. *californica*
Common Milkmaids (Brassicaceae)

Erysimum capitatum ssp. *capitatum*
Western Wallflower (Brassicaceae)

Hirschfeldia incana
Summer Mustard (Brassicaceae)

Thysanocarpus curvipes
Hairy Fringepod (Brassicaceae)

Calycanthus occidentalis
Spicebush (Calycanthaceae)

PLATE 17   DICOTYLEDONS: BRASSICACEAE–CALYCANTHACEAE

Lonicera hispidula var. *vacillans*
Hairy Honeysuckle

Lonicera involucrata var. *ledebourii*
Black Twinberry

Sambucus mexicana
Blue Elderberry

Sambucus racemosa
Red Elderberry

Symphoricarpos albus var. *laevigatus*
Common Snowberry

Symphoricarpos mollis
Creeping Snowberry

PLATE 18   DICOTYLEDONS: CAPRIFOLIACEAE

Cerastium arvense
Field Chickweed (Caryophyllaceae)

Silene californica
Indian Pink (Caryophyllaceae)

Silene gallica
Windmill Pink (Caryophyllaceae)

Silene scouleri ssp. *grandis*
Scouler Catchfly (Caryophyllaceae)

Spergularia rubra
Ruby Sand-spurrey (Caryophyllaceae)

Spergularia macrotheca var. *macrotheca*
Perennial Sand-spurrey (Caryophyllaceae)

Euonymus occidentalis
Western Burningbush (Celastraceae)

Paxistima myrsinites
Oregon Boxwood (Celastraceae)

PLATE 19   DICOTYLEDONS: CARYOPHYLLACEAE–CELASTRACEAE

Atriplex leucophylla
Beach Saltbush (Chenopodiaceae)

Atriplex semibaccata
Australian Saltbush (Chenopodiaceae)

Salsola tragus
Russian-thistle (Chenopodiaceae)

Salicornia virginica
Virginia Pickleweed (Chenopodiaceae)

Calystegia malacophylla ssp. *pedicellata*
Hairy Morning-glory (Convolvulaceae)

Calystegia soldanella
Beach Morning-glory (Convolvulaceae)

Calystegia purpurata ssp. *purpurata*
Climbing Morning-glory (Convolvulaceae)

PLATE 20   DICOTYLEDONS: CHENOPODIACEAE–CONVOLVULACEAE

Cornus nuttallii
Mountain Dogwood (Cornaceae)

Cornus sericea ssp. *occidentalis*
Western Creek Dogwood (Cornaceae)

Crassula connata
Sand Pygmyweed (Crassulaceae)

Dudleya cymosa
Common Dudleya (Crassulaceae)

Cuscuta salina var. *major*
Salt Marsh Dodder (Cuscutaceae)

Sedum spathulifolium
Broadleaf Stonecrop (Crassulaceae)

PLATE 21   DICOTYLEDONS: CORNACEAE–CUSCUTACEAE

Dipsacus fullonum
Wild Teasel (Dipsacaceae)

Scabiosa atropurpurea
Pincushion (Dipsacaceae)

Arbutus menziesii
Pacific Madrone (Ericaceae)

Arctostaphylos auriculata
Mount Diablo Manzanita (Ericaceae)

Arctostaphylos glandulosa
Eastwood Manzanita (Ericaceae)

Arctostaphylos pallida
Pallid Manzanita (Ericaceae)

PLATE 22   DICOTYLEDONS: DIPSACACEAE–ERICACEAE

Arctostaphylos tomentosa ssp. *crustacea*
Brittleleaf Manzanita

Arctostaphylos stanfordiana ssp. *stanfordiana*
Stanford Manzanita

Arctostaphylos uva-ursi
Bearberry

Gaultheria shallon
Salal

Rhododendron macrophyllum
Rosebay

Rhododendron occidentale
Western Azalea

Vaccinium ovatum
Evergreen Huckleberry

Vaccinium parvifolium
Red Huckleberry

PLATE 23   DICOTYLEDONS: ERICACEAE

Chamaesyce maculata
Spotted Spurge (Euphorbiaceae)

Croton californicus
(Euphorbiaceae)

Eremocarpus setigerus
Turkey-mullein (Euphorbiaceae)

Euphorbia peplus
Petty Spurge (Euphorbiaceae)

Ricinus communis
Castor-bean (Euphorbiaceae)

Cercis occidentalis
Western Redbud (Fabaceae)

Cytisus scoparius
Scotch Broom (Fabaceae)

Genista monspessulana
French Broom (Fabaceae)

PLATE 24   DICOTYLEDONS: EUPHORBIACEAE–FABACEAE

Lathyrus littoralis
Silky Beach Pea

Lathyrus vestitus
Woodland Pea

Lotus crassifolius
Broadleaf Lotus

Lotus humistratus
Colchita

Lotus purshianus
Spanish Lotus

Lotus scoparius
Deerweed

Lotus wrangelianus
California Lotus

PLATE 25   DICOTYLEDONS: FABACEAE

Lupinus albifrons var. *albifrons*
Silver Lupine

Lupinus arboreus
Yellow Bush Lupine

Lupinus microcarpus var. *densiflorus*
Gully Lupine

Lupinus succulentus
Arroyo Lupine

Pickeringia montana
Chaparral Pea

Melilotus indica
Sourclover

PLATE 26   DICOTYLEDONS: FABACEAE

Pediomelum californicum
Indian Breadroot

Rupertia physodes
California-tea

Spartium junceum
Spanish Broom

Thermopsis macrophylla
False Lupine

Trifolium fucatum
Bull Clover

Trifolium pratense
Red Clover

Trifolium hybridum
Alsike Clover

PLATE 27   DICOTYLEDONS: FABACEAE

Trifolium microcephalum
Small-headed Clover (Fabaceae)

Trifolium willdenovii
Tomcat Clover (Fabaceae)

Vicia sativa ssp. *sativa*
Spring Vetch (Fabaceae)

Vicia gigantea
Giant Vetch (Fabaceae)

Quercus agrifolia
Coast Live Oak (Fagaceae)

Chrysolepis chrysophylla var. *minor*
Golden Chinquapin (Fagaceae)

Frankenia salina
Alkali-heath (Frankeniaceae)

PLATE 28   DICOTYLEDONS: FABACEAE–FRANKENIACEAE

Centaurium muehlenbergii
Monterey Centaury (Gentianaceae)

Gentiana sceptrum
King's Gentian (Gentianaceae)

Erodium botrys
Broadleaf Filaree (Geraniaceae)

Erodium cicutarium
Redstem Filaree (Geraniaceae)

Geranium bicknellii
Bicknell Geranium (Geraniaceae)

Geranium dissectum
Cutleaf Geranium (Geraniaceae)

Geranium robertianum
Herb-Robert (Geraniaceae)

PLATE 29   DICOTYLEDONS: GENTIANACEAE–GERANIACEAE

Ribes aureum var. *gracillimum*
Golden Currant (Grossulariaceae)

Ribes menziesii
Canyon Gooseberry (Grossulariaceae)

Ribes californicum
Hillside Gooseberry (Grossulariaceae)

Ribes sanguineum var. *glutinosum*
Pinkflower Currant (Grossulariaceae)

Ribes speciosum
Fuchsia-flower Gooseberry (Grossulariaceae)

Aesculus californica
California Buckeye (Hippocastanaceae)

PLATE 30   DICOTYLEDONS: GROSSULARIACEAE–HIPPOCASTANACEAE

Emmenanthe penduliflora var. *penduliflora*
Whispering-bells

Eriodictyon californicum
Yerba-santa

Hydrophyllum occidentale
Heliotrope

Phacelia californica
California Phacelia

Nemophila menziesii var. *menziesii*
Baby-blue-eyes

Phacelia ramosissima var. *ramosissima*
Branched Phacelia

Phacelia ciliata
Field Phacelia

PLATE 31   DICOTYLEDONS: HYDROPHYLLACEAE

Hypericum anagalloides
Tinker's-penny (Hypericaceae)

Hypericum concinnum
Goldwire (Hypericaceae)

Lamium amplexicaule
Clasping Henbit (Lamiaceae)

Lamium purpureum
Red Henbit (Lamiaceae)

Lepechinia calycina
Pitcher Sage (Lamiaceae)

Mentha pulegium
Pennyroyal (Lamiaceae)

Monardella villosa ssp. *villosa*
Common Coyotemint (Lamiaceae)

PLATE 32   DICOTYLEDONS: HYPERICACEAE–LAMIACEAE

Pogogyne serpylloides
Thymelike Pogogyne

Prunella vulgaris var. *vulgaris*
European Selfheal

Salvia columbariae
Chia

Salvia mellifera
Black Sage

Scutellaria tuberosa
Blue Skullcap

Salvia spathacea
Hummingbird Sage

PLATE 33   DICOTYLEDONS: LAMIACEAE

Umbellularia californica
California Bay (Lauraceae)

Stachys bullata
California Hedgenettle (Lamiaceae)

Linum bienne
Narrowleaf Flax (Linaceae)

Limnanthes douglasii ssp. *douglasii*
Douglas Meadowfoam (Limnanthaceae)

Mentzelia lindleyi
Lindley Blazing-star (Loasaceae)

Linum lewisii
Western Blue Flax (Linaceae)

Mentzelia micrantha
Golden Blazing-star (Loasaceae)

PLATE 34   DICOTYLEDONS: LAMIACEAE–LOASACEAE

Lavatera arborea
Tree Mallow (Malvaceae)

Lythrum californicum
California Loosestrife (Lythraceae)

Malva neglecta
Common Mallow (Malvaceae)

Malacothamnus fremontii
Fremont Mallow (Malvaceae)

Malva nicaeensis
Bull Mallow (Malvaceae)

Sidalcea malviflora ssp. *malviflora*
Common Checkerbloom (Malvaceae)

Sidalcea calycosa ssp. *calycosa*
Annual Checkerbloom (Malvaceae)

PLATE 35   DICOTYLEDONS: LYTHRACEAE–MALVACEAE

Abronia latifolia
Yellow Sand-verbena (Nyctaginaceae)

Abronia umbellata ssp. *umbellata*
Coast Sand-verbena (Nyctaginaceae)

Camissonia cheiranthifolia
Beach Primrose (Onagraceae)

Camissonia ovata
Suncup (Onagraceae)

Clarkia amoena ssp. *huntiana*
Hunt Clarkia (Onagraceae)

Clarkia biloba
Bilobe Clarkia (Onagraceae)

PLATE 36   DICOTYLEDONS: NYCTAGINACEAE–ONAGRACEAE

Clarkia concinna
Redribbons

Clarkia davyi
Davy Clarkia

Clarkia epilobioides
Willowherb Clarkia

Clarkia franciscana
Presidio Clarkia

Clarkia gracilis ssp. *sonomensis*
Slender Clarkia

Clarkia purpurea ssp. *quadrivulnera*
Winecup Clarkia

Clarkia purpurea ssp. *viminea*
Large Clarkia

Clarkia rhomboidea
Tongue Clarkia

PLATE 37   DICOTYLEDONS: ONAGRACEAE

Clarkia rubicunda
Godetia

Clarkia unguiculata
Elegant Clarkia

Epilobium angustifolium ssp.
circumvagum
Fireweed

Epilobium brachycarpum
Panicled Willowherb

Epilobium canum
California Fuchsia

Oenothera deltoides ssp. *howellii*
Antioch Dunes Evening-primrose

Oenothera glazioviana
Biennial Evening-primrose

PLATE 38   DICOTYLEDONS: ONAGRACEAE

Boschniakia strobilacea
California Groundcone (Orobanchaceae)

Orobanche fasciculata
Clustered Broomrape (Orobanchaceae)

Orobanche uniflora
Naked Broomrape (Orobanchaceae)

Oxalis corniculata
Creeping Oxalis (Oxalidaceae)

Oxalis oregana
Redwood Sorrel (Oxalidaceae)

Oxalis rubra
Windowbox Oxalis (Oxalidaceae)

Dendromecon rigida
Bush Poppy (Papaveraceae)

Dicentra chrysantha
Golden Eardrops (Papaveraceae)

PLATE 39   DICOTYLEDONS: OROBANCHACEAE–PAPAVERACEAE

Dicentra formosa
Bleeding-heart

Eschscholzia caespitosa
Tufted Poppy

Eschscholzia californica
California Poppy

Fumaria officinalis
Fumitory

Platystemon californicus
Creamcups

Stylomecon heterophylla
Wind Poppy

PLATE 40   DICOTYLEDONS: PAPAVERACEAE

Plantago maritima
Sea Plantain (Plantaginaceae)

Plantago erecta
California Plantain (Plantaginaceae)

Armeria maritima ssp. *californica*
California Thrift (Plumbaginaceae)

Limonium californicum
California Sea-lavender
(Plumbaginaceae)

Collomia grandiflora
Largeflower Collomia (Polemoniaceae)

Collomia heterophylla
Variedleaf Collomia (Polemoniaceae)

PLATE 41   DICOTYLEDONS: PLANTAGINACEAE–POLEMONIACEAE

Gilia achilleifolia ssp. *achilleifolia*
California Gilia

Gilia capitata ssp. *capitata*
Globe Gilia

Gilia tricolor
Bird's-eyes

Linanthus bicolor
Bi-colored Linanthus

Linanthus grandiflorus
Largeflower Linanthus

Linanthus parviflorus
Common Linanthus

Navarretia pubescens
Downy Navarretia

Phlox gracilis
Slender Phlox

PLATE 42   DICOTYLEDONS: POLEMONIACEAE

Polygala californica
California Milkwort (Polygalaceae)

Chorizanthe membranacea
Pink Spineflower (Polygonaceae)

Eriogonum fasciculatum
California Buckwheat (Polygonaceae)

Eriogonum latifolium
Coast Buckwheat (Polygonaceae)

Eriogonum luteolum var. *caninum*
Tiburon Buckwheat (Polygonaceae)

Eriogonum nudum var. *nudum*
Nakedstem Buckwheat (Polygonaceae)

Eriogonum umbellatum var. *bahiiforme*
Sulfurflower Buckwheat (Polygonaceae)

Eriogonum vimineum
Wicker Buckwheat (Polygonaceae)

PLATE 43   DICOTYLEDONS: POLYGALACEAE–POLYGONACEAE

Polygonum amphibium var. *emersum*
Swamp Knotweed

Polygonum paronychia
Beach Knotweed

Polygonum persicaria
Lady's-thumb

Rumex acetosella
Sheep Sorrel

Rumex crispus
Curly Dock

Rumex salicifolius
Willow Dock

PLATE 44   DICOTYLEDONS: POLYGONACEAE

Calandrinia ciliata
Redmaids (Portulacaceae)

Claytonia gypsophiloides
Santa Lucia Claytonia (Portulacaceae)

Claytonia sibirica
Candyflower (Portulacaceae)

Lewisia rediviva
Bitterroot (Portulacaceae)

Anagallis arvensis
Scarlet Pimpernel (Primulaceae)

Dodecatheon hendersonii
Mosquito-bills (Primulaceae)

Trientalis latifolia
Starflower (Primulaceae)

PLATE 45   DICOTYLEDONS: PORTULACACEAE–PRIMULACEAE

Aquilegia formosa
Crimson Columbine

Clematis lasiantha
Pipestems

Delphinium nudicaule
Red Larkspur

Delphinium variegatum
Royal Larkspur

Ranunculus californicus
California Buttercup

Ranunculus muricatus
Prickleseed Buttercup

PLATE 46   DICOTYLEDONS: RANUNCULACEAE

Ranunculus repens
Creeping Buttercup (Ranunculaceae)

Ceanothus cuneatus
Buckbrush (Rhamnaceae)

Ceanothus gloriosus var. *gloriosus*
Point Reyes Ceanothus (Rhamnaceae)

Ceanothus griseus
Carmel Ceanothus (Rhamnaceae)

Ceanothus sonomensis
Sonoma Ceanothus (Rhamnaceae)

Ceanothus thyrsiflorus
Blue-blossom (Rhamnaceae)

Rhamnus californica
California Coffeeberry (Rhamnaceae)

Rhamnus crocea
Spiny Redberry (Rhamnaceae)

PLATE 47   DICOTYLEDONS: RANUNCULACEAE–RHAMNACEAE

Adenostoma fasciculatum
Chamise

Amelanchier alnifolia
Serviceberry

Aphanes occidentalis
Western Dewcup

Cercocarpus betuloides
Birchleaf Mountain-mahogany

Fragaria chiloensis
Beach Strawberry

Heteromeles arbutifolia
Toyon

Holodiscus discolor
Creambush

PLATE 48   DICOTYLEDONS: ROSACEAE

Horkelia californica ssp. *californica*
California Horkelia

Physocarpus capitatus
Pacific Ninebark

Potentilla anserina ssp. *pacifica*
Pacific Cinquefoil

Potentilla glandulosa
Sticky Cinquefoil

Potentilla recta
Pale Cinquefoil

Prunus emarginata
Bitter Cherry

Prunus ilicifolia
Hollyleaf Cherry

Rubus spectabilis
Salmonberry

PLATE 49   DICOTYLEDONS: ROSACEAE

Rosa californica
California Wild Rose (Rosaceae)

Rosa gymnocarpa
Wood Rose (Rosaceae)

Galium andrewsii
Phloxleaf Bedstraw (Rubiaceae)

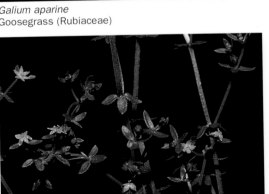

Galium aparine
Goosegrass (Rubiaceae)

Galium californicum
California Bedstraw (Rubiaceae)

Galium porrigens var. *porrigens*
Climbing Bedstraw (Rubiaceae)

Sherardia arvensis
Field Madder (Rubiaceae)

PLATE 50   DICOTYLEDONS: ROSACEAE–RUBIACEAE

Ptelea crenulata
Hoptree (Rutaceae)

Anemopsis californica
Yerba-mansa (Saururaceae)

Lithophragma affine
Woodland-star (Saxifragaceae)

Lithophragma parviflorum
Prairie Starflower (Saxifragaceae)

Tellima grandiflora
Fringecups (Saxifragaceae)

Tolmiea menziesii
Piggy-back-plant (Saxifragaceae)

PLATE 51   DICOTYLEDONS: RUTACEAE–SAXIFRAGACEAE

Antirrhinum kelloggii
Lax Snapdragon

Bellardia trixago

Antirrhinum vexillo-calyculatum ssp. *breweri*
Brewer Snapdragon

Castilleja attenuata
Valley-tassels

Castilleja affinis ssp. *affinis*
Common Indian Paintbrush

Castilleja exserta ssp. *exserta*
Purple Owl's-clover

Castilleja foliolosa
Woolly Indian Paintbrush

PLATE 52   DICOTYLEDONS: SCROPHULARIACEAE

Castilleja wightii
Seaside Paintbrush

Collinsia heterophylla
Chinesehouses

Collinsia sparsiflora var. *sparsiflora*
Fewflower Blue-eyed-Mary

Mimulus aurantiacus
Bush Monkeyflower

Mimulus cardinalis
Scarlet Monkeyflower

Mimulus guttatus
Common Monkeyflower

PLATE 53   DICOTYLEDONS: SCROPHULARIACEAE

Penstemon heterophyllus var. *heterophyllus*
Foothill Penstemon (Scrophulariaceae)

Pedicularis densiflora
Indian-warrior (Scrophulariaceae)

Scrophularia californica
Beeplant (Scrophulariaceae)

Triphysaria eriantha ssp. *eriantha*
Yellow Johnnytuck (Scrophulariaceae)

Nicotiana glauca
Tree-tobacco (Solanaceae)

Solanum umbelliferum
Blue Nightshade (Solanaceae)

Solanum physalifolium
Hairy Nightshade (Solanaceae)

PLATE 54   DICOTYLEDONS: SCROPHULARIACEAE–SOLANACEAE

Fremontodendron californicum
Flannelbush (Sterculiaceae)

Dirca occidentalis
Western Leatherwood (Thymelaeaceae)

Urtica dioica ssp. *holosericea*
Hoary Nettle (Urticaceae)

Plectritis congesta
Seablush (Valerianaceae)

Plectritis macrocera
Long-horned Plectritis (Valerianaceae)

Verbena lasiostachys var. *scabrida*
Robust Verbena (Verbenaceae)

PLATE 55   DICOTYLEDONS: STERCULIACEAE–VERBENACEAE

Viola adunca
Western Dog Violet (Violaceae)

Viola glabella
Stream Violet (Violaceae)

Viola ocellata
Western Heart's-ease (Violaceae)

Viola pedunculata
Johnny-jump-up (Violaceae)

Arceuthobium campylopodum
Western Dwarf Mistletoe (Viscaceae)

Phoradendron villosum
Oak Mistletoe (Viscaceae)

Vitis californica
California Wild Grape (Vitaceae)

Tribulus terrestris
Puncture-vine (Zygophyllaceae)

PLATE 56   DICOTYLEDONS: VIOLACEAE–ZYGOPHYLLACEAE

Cyperus eragrostis
Tall Cyperus (Cyperaceae)

Lysichiton americanum
Yellow Skunk-cabbage (Araceae)

Iris fernaldii
Fernald Iris (Iridaceae)

Iris douglasiana
Douglas Iris (Iridaceae)

Sisyrinchium bellum
Blue-eyed-grass (Iridaceae)

Sisyrinchium californicum
Golden-eyed-grass (Iridaceae)

Luzula comosa
Common Wood Rush (Juncaceae)

PLATE 57   MONOCOTYLEDONS: ARACEAE–JUNCACEAE

Allium acuminatum
Hooker Onion

Allium falcifolium
Sickleleaf Onion

Allium lacunosum
Wild Onion

Allium serra
Serrated Onion

Allium unifolium
Clay Onion

Brodiaea elegans
Elegant Brodiaea

Brodiaea terrestris
Dwarf Brodiaea

Calochortus luteus
Yellow Mariposa Lily

PLATE 58   MONOCOTYLEDONS: LILIACEAE

Calochortus pulchellus
Mount Diablo Fairy-lantern

Calochortus tiburonensis
Tiburon Mariposa Lily

Calochortus umbellatus
Oakland Startulip

Calochortus venustus
Butterfly Mariposa Lily

Camassia quamash
Common Camas

Chlorogalum pomeridianum var. *divaricatum*
Soap-plant

Dichelostemma congestum
Ookow

PLATE 59   MONOCOTYLEDONS: LILIACEAE

Clintonia andrewsiana
Red Bead Lily

Erythronium californicum
California Fawn Lily

Fritillaria recurva
Scarlet Fritillary

Fritillaria affinis
Checker Lily

Muilla maritima
Common Muilla

Lilium pardalinum ssp. *pardalinum*
Leopard Lily

Smilacina racemosa
Fat Solomon

PLATE 60   MONOCOTYLEDONS: LILIACEAE

Trillium albidum
Sweet Trillium (Liliaceae)

Triteleia laxa
Ithuriel's-spear (Liliaceae)

Triteleia lugens
Uncommon Triteleia (Liliaceae)

Zigadenus fremontii
Common Star Lily (Liliaceae)

Zigadenus venenosus
Death Camas (Liliaceae)

Calypso bulbosa
Fairy-slipper (Orchidaceae)

Cephalanthera austiniae
Phantom Orchid (Orchidaceae)

PLATE 61   MONOCOTYLEDONS: LILIACEAE–ORCHIDACEAE

Corallorhiza maculata
Spotted Coralrooot (Orchidaceae)

Corallorhiza striata
Striped Coralroot (Orchidaceae)

Cypripedium californicum
California Lady's-slipper (Orchidaceae)

Epipactis gigantea
Stream Orchid (Orchidaceae)

Epipactis helleborine
Helleborine (Orchidaceae)

Piperia unalascensis
Alaska Rein Orchid (Orchidaceae)

Ammophila arenaria
European Beachgrass (Poaceae)

Arundo donax
Giant Reed (Poaceae)

PLATE 62   MONOCOTYLEDONS: ORCHIDACEAE–POACEAE

Anthoxanthum odoratum
Sweet Vernal Grass

Briza maxima
Big Quaking Grass

Bromus madritensis ssp. *rubens*
Foxtail Chess

Cynosurus cristatus
Crested Dogtail

Cynosurus echinatus
Hedgehog Dogtail

Dactylis glomerata
Orchard Grass

Leymus mollis
American Dunegrass

Poa bulbosa
Bulbous Bluegrass

PLATE 63   MONOCOTYLEDONS: POACEAE

Echinochloa crus-galli
Barnyard Grass (Poaceae)

Elymus multisetus
Big Squirreltail (Poaceae)

Hordeum murinum ssp. *leporinum*
Hare Barley (Poaceae)

Lolium multiflorum
Italian Ryegrass (Poaceae)

Lolium perenne
Perennial Ryegrass (Poaceae)

Polypogon monspeliensis
Annual Beard Grass (Poaceae)

Potamogeton natans
Floatingleaf Pondweed (Potamogetonaceae)

Typha latifolia
Broadleaf Cattail (Typhaceae)

PLATE 64   MONOCOTYLEDONS: POACEAE–TYPHACEAE

ILLUSTRATIONS OF PLANT STRUCTURES

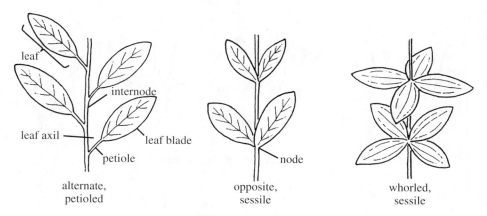

alternate,
petioled

opposite,
sessile

whorled,
sessile

Illustration 1. Leaf arrangements

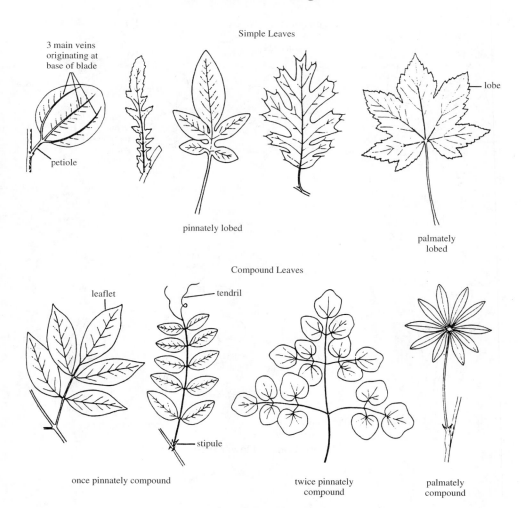

Simple Leaves

pinnately lobed

palmately
lobed

Compound Leaves

once pinnately compound

twice pinnately
compound

palmately
compound

Illustration 2. Types of leaves

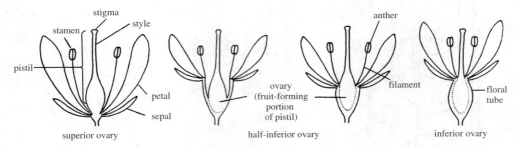

Illustration 3. Structure of a flower

stigma
stamen
style
pistil
anther
filament
superior ovary
petal
sepal
ovary (fruit-forming portion of pistil)
floral tube
half-inferior ovary
inferior ovary

Regular Corolla or Perianth

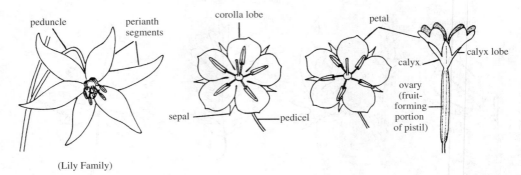

peduncle
perianth segments

corolla lobe

petal

calyx lobe

calyx

sepal
pedicel

ovary (fruit-forming portion of pistil)

(Lily Family)

Irregular Corolla

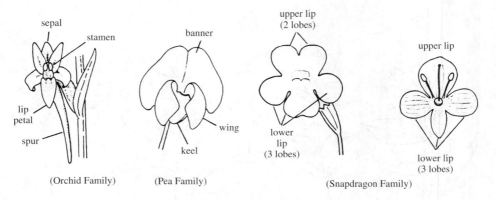

sepal
stamen

banner

upper lip (2 lobes)

upper lip

lip petal

spur

wing

keel

lower lip (3 lobes)

lower lip (3 lobes)

(Orchid Family)

(Pea Family)

(Snapdragon Family)

Illustration 4. Flower types

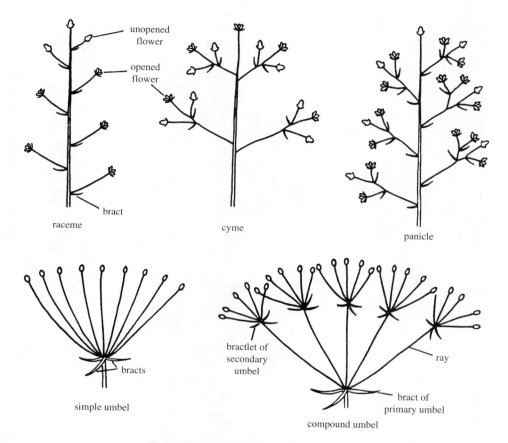

Illustration 5. Types of inflorescences

ABBREVIATIONS AND GLOSSARY

Abbreviations

af	Africa	Mo	Monterey County
Al	Alameda County	MT	Mount Tamalpais
as	Asia	mx	Mexico
au	Australia/New Zealand	n	north
CC	Contra Costa County	na	North America (excluding Mexico)
cm	centimeter (about 2/5 inch)	Na	Napa County
e	east	PR	Point Reyes
eu	Europe	s	south
eua	Eurasia	sa	South and Central America
F	Fahrenheit	SB	San Benito County
ft	feet	SCl	Santa Clara County
in	inch	SCr	Santa Cruz County
La	Lake County	SF	San Francisco County
m	meter (about 39 inches)	SFBR	San Francisco Bay Region
M	monocotyledonous plant	Sl	Solano County
Ma	Marin County	SLO	San Luis Obispo County
MD	Mount Diablo	SM	San Mateo County
me	Mediterranean region	Sn	Sonoma County
Me	Mendocino County	×	a plant considered to be a hybrid but to which a Latin species name has been applied
MH	Mount Hamilton		
mm	millimeter (0.1 centimeter)		
MM	Montara Mountain	?	origin uncertain

Occurrence Status

Abbreviations below, used to designate rare or endangered status, are from Tibor, D. P., ed., 2001. *Inventory of rare and endangered plants of California.* 6th ed. Special publication no. 1. Sacramento: California Native Plant Society.

1a	presumed extinct in California (last recorded sighting shown in parentheses)
1b	plant rare, threatened, or endangered in California and elsewhere
2	plant rare, threatened, or endangered in California, but more common elsewhere
4	plant of limited distribution—a watch list

Glossary

The illustrations referred to in the glossary can be found in "Illustrations of Plant Structures," which precedes the abbreviations.

Achene   A dry and usually hard, one-seeded fruit that does not split open (fig. in Asteraceae).

Acorn   A hard, one-seeded nut whose base is enclosed by a scaly cup (fig. in Fagaceae).

Alternate (leaves or branches)   Leaves or branches that originate singly at the nodes rather than in pairs or whorls (illus. 1).

Annual   A plant that produces flowers and fruit in its first year and then dies.

Anther   The sac or sacs in which a stamen produces pollen (illus. 3).

Awn   A bristle on glumes and lemmas of grass florets and on scales of sedge florets (fig. in Poaceae).

Banner   The uppermost petal in pea flowers (illus. 4).

Basal leaves   Leaves that originate at the very base of a plant. They often form a compact cluster and are sometimes the only substantial leaves on the plant (fig. in Polygonaceae, *Eriogonum nudum* var. *nudum*).

Biennial   A plant that does not produce flowers and fruit until its second year and then dies.

Bract   A modified leaf just below a flower or an inflorescence. It may very much resemble the other leaves, or it may be quite different (illus. 5; figs. in Cyperaceae and Juncaceae).

Bractlet   A bract originating at the base of a secondary umbel (illus. 5), or a bract below an individual flower.

Branchlet   A small branch; in some grasses, it is equivalent to a pedicel.

Bur   A dry fruit covered with spines, scales, or hooks (fig. in Poaceae, *Cenchrus longispinus*).

Calyx (calyces)   The collective term for the sepals of a flower, whether these are separate or united. The calyx is the outermost whorl of an individual flower and is generally green (illus. 4).

Calyx tube   The tube formed by union of the sepals.

Calyx lobes   The free portions of sepals that are united for part of their length (illus. 4).

Capsule   A type of fruit that consists of more than one chamber and that cracks open when dry.

Catkin   A condensed inflorescence of petal-less flowers, each accompanied by a bract; typical of willows, birches, alders, and some other trees and shrubs (fig. in Salicaeae).

Chlorophyll   The green pigment that enables plants to absorb the light energy needed for synthesis of organic compounds.

Compound leaf   A leaf with two or more completely separate leaflets, as in a clover, rose, or pea (illus. 2). In this book, leaves that are divided twice are referred to as "twice compound," and leaves that are divided three or four times are referred to as "3 or 4 times compound."

Compound umbel   *See* umbel.

Corolla   The collective term for the petals of a flower, whether these are separate or united. The corolla is the innermost whorl of an individual flower and is usually not green (illus. 4).

Corolla lobes   The free portions of petals that are united for part of their length (illus. 4).

Corymb   A more or less flat-topped inflorescence in which the pedicels of the outer flowers, which usually open first, are longer than those of flowers closer to the center.

Crown   The persistent base of a perennial herbaceous plant; also applied to the top of a tree.

Cyme   A flat-topped or convex inflorescence in which the central or uppermost flowers open first (illus. 5).

Deciduous   A tree or shrub whose leaves fall at nearly the same time or after their function has been performed; also applied to other plant structures that drop off.

Disk flowers   In Asteraceae, flowers that have tubular corollas, as opposed to ray flowers, in which the corollas are flattened out. Some members of Asteraceae have a central mass of disk flowers surrounded by ray flowers; in others, all the flowers are of the disk-flower or ray-flower type (fig. in Asteraceae).

Evergreen (plant)   A plant that retains most of its leaves from one year to the next. *See* deciduous.

Filament   The slender stalk of a stamen, or any threadlike structure (illus. 3).

Floral tube   A tubelike structure consisting of fused sepals, petals, and stamens. In flowers that have an inferior ovary, the floral tube is fused to the ovary (illus. 3).

Floret   A small flower, especially of a grass or sedge (fig. in Poaceae).

Flower head   In Asteraceae, an aggregation of several to many sessile flowers attached to a disk-shaped, conical, or concave receptacle. The receptacle is usually surrounded by bracts called phyllaries (fig. in Asteraceae).

Fruit   The ripened, seed-containing structure into which a pistil of a flower develops. In some plants, such as apples and pears, the fruit consists partly of a pistil and partly of a fleshy receptacle that envelops the pistil. Some so-called fruit, such as a blackberry, is actually an aggregate of fruit.

Glandular   A hair, bump, or pit that secretes a sticky substance.

Glumes   In grasses, the two bracts below the florets of each spikelet (fig. in Poaceae).

Grain   In Poaceae, the one-seeded fruit developing in the pistil of a grass floret.

Head   An especially compact inflorescence. In this book, primarily used to describe inflorescences of grasses. *See* flower head.

Herb   A plant that does not have woody stems, at least not above the base of the plant.

Indusium (indusia)   In ferns, a fold or shieldlike structure that covers a sorus, at least when the sporangia are young (fig. in Ferns).

Inferior ovary   Fruit-forming portion of a pistil located below the level at which the calyx lobes and corolla lobes originate. A half-inferior ovary is located halfway below the level at which the calyx lobes and corolla lobes originate (illus. 3). *See* ovary, superior ovary.

Inflorescence   A cluster of flowers, or several clusters, on a plant.

Internode   The portion of a stem between two nodes (illus. 1). *See* node.

Involucre   One or more circles of phyllaries (bracts) that partly enclose the flower head (fig. in Asteraceae); in other cases, a circle of bracts below a cluster of flowers or fruit (fig. in Polygonaceae, *Chorizanthe membranacea*).

Irregular (corolla or calyx)   A corolla or calyx in which the petals or sepals (or corolla lobes and calyx lobes) are of unequal size and shape, as the corolla is in a snapdragon or sweet pea (illus. 4).

Keel   A structure formed by the partial or complete union of the two lower petals of a pea flower (illus. 4); in other cases, a ridge, such as is formed when a bract or the glume of a grass is folded or when the midrib of a bract is pronounced.

Leaf   Generally the main food-producing structure of plants, commonly composed of a stalk (petiole) and expanded surface (the blade) (illus. 1).

Leaf axil   The upper side of the junction between a stem and a leaf, where a branch, flower, or inflorescence may originate (illus. 1).

Leaf blade   The flat, expanded portion of a leaf, excluding the petiole (illus. 1).

Leaf sheath   The base of a leaf blade or a broad petiole that wraps around the stem. The character of leaf sheaths is especially important in identification of grasses and sedges (fig. in Poaceae).

Leaflet   Each division of a compound leaf (illus. 2). *See* primary leaflet.

Lemma   In grasses, the lower and usually the larger of two bracts enclosing the stamens and pistil (fig. in Poaceae)

Ligule   In Poaceae, Juncaginaceae, and a few members of Cyperaceae, a collarlike outgrowth that originates at the base of the blade and partly encircles the stem (fig. in Poaceae).

Lip   The upper or lower portion of an irregular corolla or calyx, as in a mint, orchid, or snapdragon (illus. 4).

Lobe   One of the deeply separated divisions of a leaf, such as that of a maple or sycamore (illus. 2). "Twice lobed" refers to leaves that have smaller lobes on the margins of the larger lobes. *See* calyx lobes, corolla lobes.

Needle   A narrow, stiff leaf of a pine, fir, or other cone-bearing tree.

Node   The "joint" of a stem, where one or more leaves are attached and where a branch, flower, or inflorescence may develop (illus. 1); in grasses, the place where a spikelet is attached to the rachis of the inflorescence.

Nutlet   A small, dry, one-seeded fruit, or comparable structure into which a fruit separates early in its development, as is typical of borages and mints.

Opposite (leaves or branches)   Leaves or branches that originate in pairs or whorls at the nodes (illus. 1).

Ovary   The part of a pistil that forms the fruit. This term, although entrenched in botanical literature, is a misnomer, and is used here with reluctance. The ovary of a flowering plant does not directly produce eggs. Deep within its tissues it produces spores that develop into microscopic female plants of the sexual generation. Each contains a gamete that corresponds to an egg. If this is fertilized by a sperm nucleus that reaches it by way of a tube formed by a pollen grain after this has landed on the stigma of the pistil, it will develop into an embryo of the spore-producing generation. *See* inferior ovary, superior ovary.

Palmate   Used to describe plant structure such as a leaf, leaflet, lobe, or vein that is divided in such a way as to resemble a hand with fingers spread (illus. 2). *See* pinnate.

Palmately compound, palmately lobed   *See* palmate.

Panicle   A corymb, cyme, or raceme type of inflorescence in which the pedicels branch (in other words, a compound corymb, cyme, or raceme) (illus. 5).

Pappus   Modified sepals, appearing as scalelike, bristlelike, hairlike, or plumose structures at the top of the achene (fig. in Asteraceae).

Parasitic (plant)   A plant that draws nourishment from another plant, as does a mistletoe.

Pedicel   The stalk of an individual flower (illus. 4) or the stalk of a spikelet (fig. in Poaceae).

Peduncle   The stalk of a flower that is borne singly, or the stalk of an inflorescence that consists of several to many flowers (illus. 4).

Perennial   A plant that lives indefinitely, as opposed to one that lives only one to two years.

Perianth   The complex formed by the corolla, the calyx, or both (illus. 4). The term is especially useful in connection with flowers in which the corolla and the calyx are very similar, as in some lilies.

Perianth segments   The divisions of a corolla or calyx, whether they are completely separate or not (illus. 4).

Petals   The individual segments of a corolla, which may be completely separate or at least partly united (illus. 4).

Petiole   The stalk of a leaf (illus. 1).

Phyllaries   In Asteraceae, the bracts that surround the receptacle of a flower head (fig. in Asteraceae).

Pinnate   Used to describe plant structures such as leaflets, lobes, or veins that are arranged in two rows on opposite sides of a central axis, as in a pea or rose (illus. 2). *See* palmate.

Pinnately compound, pinnately lobed   *See* pinnate.

Pistil   The portion of a flower in which a seed or seeds are eventually produced (illus. 3). A flower may have more than one pistil.

Pistillate   A flower that has one or more pistils but no stamens; also, a plant that produces only pistillate flowers.

Pod   A usually narrow fruit that splits open when dry, as does a pea pod.

Pollen   Microscopic reproductive structures produced by stamens of flowering plants and by certain short-lived conelike structures of gymnosperms. Each pollen grain is at first a one-celled spore. Before it is released, however, the nucleus of the cell divides, initiating the formation of a microscopic male plant. Development is not completed until the pollen grain reaches the stigma of a flower or the scale of a cone that is destined to produce seeds.

Prickle   A sharp outgrowth derived from the outermost cell layer of a stem. *See* spine, thorn.

Primary leaflet   In some ferns and flowering plants, a leaflet that is further divided into secondary leaflets.

Prostrate (stem or plant)   A stem or plant that lies on the soil; used to describe the way a plant grows.

Raceme   An inflorescence in which the flowers are borne all along the peduncle, on short pedicels of more or less equal length (illus. 5). The flowers at the base of a raceme open first.

Rachis   The main axis of a compound or nearly compound leaf; in Poaceae and Cyperaceae, the main axis of a spikelet.

Ray   The flattened corolla of a ray flower (fig. in Asteraceae); also, a stalk in the primary umbel of a compound umbel (illus. 5).

Ray flowers   In Asteraceae, the marginal flowers, with flattened corolla, that surround the disk flowers in the central portion of the flower head. The distinction between ray flowers and disk flowers is clearly seen in daisies, asters, and sunflowers. Some members of the family (such as thistles) lack ray flowers; in others, all flowers are of this type (fig. in Asteraceae).

Receptacle   The portion of a flower peduncle or pedicel on which the flower parts are mounted (fig. in Asteraceae).

Regular   Applied to a corolla or calyx that has perfect or nearly perfect radial symmetry such that it can be divided into two equal halves three or more different ways. The petals or sepals (or corolla lobes and calyx lobes) are of equal size and shape as in a poppy or wild rose (illus. 4).

Resin pit   A microscopic glandular pit on the scalelike leaves of some conifers of Cupressaceae.

Rhizome   A horizontal underground stem from which leaves or other stems arise.

Saprophyte   A plant that lacks chlorophyll (the leaves therefore not green), subsisting on decaying organic matter, usually with the aid of fungi that penetrate its roots.

Secondary leaflet   *See* primary leaflet.

Secondary umbel   *See* umbel.

Sepals   The individual segments of a calyx, which may be completely separate or at least partly united (illus. 4).

Sessile (leaves or flowers)   Attached directly to a stem or some other structure (i.e., a leaf that lacks a petiole and a flower that lacks a pedicel) (illus. 1).

Shrub   A perennial plant whose stems are woody throughout; usually does not have a distinct trunk and is not often more than 3 m tall. Some large shrubs may have the form of a small tree, so the distinction between the two categories is not absolute.

Smooth-margined (leaf blade, leaflet)   A leaf blade or leaflet without any teeth or lobes on its margins.

Sorus (sori)   In ferns, a cluster of sporangia (fig. in Ferns). *See* sporangium.

Spikelet   In grasses, each unit of an inflorescence, consisting of a pair of glumes and one to several florets; in sedges, a similar group of florets, but without glumes (fig. in Poaceae).

Spine   A sharp projection from a leaf, fruit, bract, or phyllary. *See* prickle, thorn.

Sporangium (sporangia)   In ferns, fern allies, and other organisms not covered in this book (i.e., mosses and fungi), a structure within which spores are formed (fig. in Ferns). Part of a pistil and the anther of a stamen are also technically sporangia.

Spur   A hollow, saclike, or tubular extension of a petal or sepal, as in a larkspur, columbine, or violet (illus. 4).

Stamen The pollen-producing part of a flower, usually consisting of an anther and filament. Most flowers have at least two stamens, and some have many (illus. 3).

Staminate A flower that has stamens but no pistils; also a plant whose flowers are staminate

Sterile (stamen or flower) A modified stamen that does not produce pollen, or a flower that does not function in reproduction.

Stigma The sticky tip of a pistil, which traps pollen (illus. 3).

Stipules Paired appendages at the base of a petiole; these may resemble leaves or be a pair of glands or scales (illus. 2).

Style The usually slender upper portion of a pistil, at the tip of which the stigma is located. In many plants, the pistil has more than one style (illus. 3).

Succulent (leaves and stems) Leaves and stems that are thick and juicy.

Superior ovary Fruit-forming portion of the pistil located above the level at which the calyx lobes and corolla lobes originate (illus. 3). *See* inferior ovary, ovary.

Tendril A slender, twining structure, usually part of a leaf, by which a climbing plant, such as a pea or grape vine, clings to its support (illus. 2).

Thorn A short branch that has become specialized as a stiff, sharp-tipped structure. *See* prickle, spine.

Two-lipped (flower) A flower in which the corolla, consisting of united petals and often also the calyx, has distinctly different upper and lower portions, as in a snapdragon or mint (illus. 4).

Umbel A nearly flat-topped inflorescence in which the pedicels of the flowers originate at the top of the peduncle; in compound umbels, the peduncle branches into rays (the primary umbel) that support the pedicels (the secondary umbel) (illus. 5).

Vegetative (plant structure) Nonreproductive.

Vein In a leaf, petal, or lemma, a structure, usually branched, consisting of tissues that distribute water and nutrients. In the case of petals, the term is often applied to colored lines (illus. 2).

Vernal (ponds) Ponds that become filled with water in winter or spring and then dry out in late spring or summer.

Whorl A type of leaf or branch arrangement in which three or more of these structures originate from a single node (illus. 1).

Wing The two side petals of flowers in Fabaceae (illus. 4); also, a thin expansion of a dry fruit, as in that of a maple.

INDEX

Numbers in boldface indicate color plates. Numbers in italics are page numbers of figures.

ABOUT THE AUTHORS

Linda H. Beidleman has been an instructor at the Jepson Herbarium, University of California, Berkeley; Aspen Center for Environmental Studies; and The Rocky Mountain Nature Association; among other institutions. She is coauthor of *Plants of Rocky Mountain National Park* (2000, with Richard G. Beidleman and Beatrice E. Willard), and *Annotated Bibliography of Colorado Vertebrate Zoology, 1776–1995* (2000, with Richard G. Beidleman and Reba R. Beidleman). She collaborated with Eugene N. Kozloff on *Marine Invertebrates of the Pacific Northwest* (1987).

Eugene N. Kozloff is Professor at the University of Washington and author of *Invertebrates* (1990), *Seashore Life of the Northern Pacific Coast: An Illustrated Guide to Northern California, Oregon, Washington, and British Columbia* (1983), and *Plants and Animals of the Pacific Northwest: An Illustrated Guide to the Natural History of Western Oregon, Washington, and British Columbia* (1976).

Compositor:	Impressions Book and Journal Services, Inc.
Text:	10/13.5 Minion
Display:	Franklin Gothic
Printer:	C&C Offset
Binder:	C&C Offset